D1288089

Structural Dynamics
THEORY AND
COMPUTATION

Structural Dynamics
THEORY AND COMPUTATION

MARIO PAZ

Professor of Civil Engineering
University of Louisville

Third Edition

VNR VAN NOSTRAND REINHOLD
New York

The computer programs used in this book are available in several versions of the BASIC language. To order diskettes for IBM-compatible microcomputer formats, use the order form on the back of the book or write to MICROTEXT, P.O. Box 35101, Louisville, KY 40232. Technical questions, corrections, and request for additional information should be directed to this address.

Extreme care has been taken in preparing the programs used in this book. Extensive testing and checking have been performed to insure the accuracy and effectiveness of computer solutions. However, neither the author nor the publisher shall be held responsible or liable for any damages arising from the use of any of the programs in this book.

Copyright © 1991 by Van Nostrand Reinhold

Library of Congress Catalog Number 90–33615
ISBN 0-442-31894-4

Printed in the United States of America

Van Nostrand Reinhold
115 Fifth Avenue
New York, New York 10003

Chapman & Hall
2-6 Boundary Row
London SE1 8HN, England

Thomas Nelson Australia
102 Dodds Street
South Melbourne, Victoria 3205, Australia

Nelson Canada
1120 Birchmount Road
Scarborough, Ontario M1K 5G4, Canada

16   15   14   13   12   11   10   9   8   7   6   5   4   3   2

Library of Congress Cataloging-in-Publication Data

Paz, Mario.
    Structural dynamics: theory and computation / Mario Paz.—3rd
ed.
       p.   cm.
    ISBN 0-442-31894-4
    1. Structural dynamics. I. Title.
    TA654.P39   1985
    624.1'71—dc20

                                                            90-33615
                                                                CIP

כַּבֵּד אֶת־ אָבִ֙יךָ֙ וְאֶת־אִמֶּ֔ךָ כַּאֲשֶׁ֥ר צִוְּךָ֖ יְהוָ֣ה אֱלֹהֶ֑יךָ לְמַ֣עַן ׀ יַאֲרִכֻ֣ן
יָמֶ֗יךָ וּלְמַ֙עַן֙ יִ֣יטַב לָ֔ךְ עַ֚ל הָֽאֲדָמָ֔ה אֲשֶׁר־יְהוָ֥ה אֱלֹהֶ֖יךָ נֹתֵ֥ן לָֽךְ

Honor your father and your mother, as the Lord your God
has commanded you, that you may long endure and
that you may fare well...

Exodus 20:12

TO THE MEMORY OF MY PARENTS

Benjamin Maman Paz
Salma Misri Paz

CONTENTS

Preface to the Third Edition

The basic structure of the two previous editions is maintained in this third edition, although numerous revisions and additions have been introduced. Three new chapters on earthquake-resistant design of buildings have been incorporated into Part VI of the book. The computer programs formerly written in FORTRAN for execution on mainframe computers have been expanded and rewritten in BASIC for implementation on microcomputers. Two independent packages of computer programs are used throughout the book: A structural dynamics package of 20 interactive computer programs and an earthquake-resistant design package of 10 interactive programs.

The structural dynamics package includes programs to determine the response of a single oscillator in the time domain, or in the frequency domain using the FFT (Fast Fourier Transform). It also includes a program to determine the response of an inelastic system with elastoplastic behavior, and another program for the development of seismic response spectral charts. A set of seven computer programs is included for use in modeling structures as two-dimensional and three-dimensional frames and trusses. Finally, other programs, incorporating modal superposition or a step-by-step time history solution, are provided for calculation of the responses to forces or motions exciting the structure.

The new chapters in earthquake-resistant design of buildings describe the provisions of both the 1985 and 1988 versions of the UBC (Uniform Building Code) for the static lateral force method and for the dynamic lateral force method. Other revisions of the book include the presentation of the Newmark beta method to obtain the time history response of dynamic systems, and the direct integration method in which the response is found assuming that the excitation function is linear for a specified time interval. A modification of the dynamic condensation method, which has been developed recently by the author for the reduction of eigenproblems, is presented in Chapter 13. The proposed modification substantially reduces the numerical operation required in the implementation of the dynamic condensation method.

The subjects in this new edition are organized in six parts. Part I deals with structures modeled as single degree-of-freedom systems. It introduces basic concepts and presents important methods for the solution of such dynamic systems. Part II introduces important concepts and methodology for multi-degree-of-freedom systems through the use of structures modeled as *shear buildings*. Part III describes methods for the dynamic analysis of framed structures modeled as discrete systems with many degrees of freedom. Part IV presents the mathematical solution for some simple structures modeled as systems with distributed properties, thus having an infinite number of degrees of freedom. Part V, which contains one chapter, introduces the reader to the fascinating topic of random vibration. Finally, Part VI presents the important current topic of earthquake engineering with applications to earthquake-resistant design of buildings following the provisions of the Uniform Building Codes of 1985 and 1988.

Several considerations—different sizes of diskettes ($3^1/_2$ inch or $5^1/_4$ inch), different versions of the programs (compiled or not compiled), and the anticipation of future updating—contributed to a final decision not to include in the book the program diskettes. Diskettes with the computer programs, in several versions, are available directly from the author. A convenient form to order the selected version of the programs is provided in the back of the book.

The author believes that, just as knowledge of a combination of figures and turns is needed to open a safe, a combination of knowledge on the subjects of mathematics, theory of structures, and computer programming is needed today for successful professional practice in engineering. To provide the reader with such a combination of knowledge has been the primary objective of this book. The reader may wish to inform the author on the extent to which this objective has been fulfilled.

Many students, colleagues, and practicing professionals have suggested improvements, identified typographical errors, and recommended additional topics for inclusion. All these suggestions were carefully considered and have been included in this third edition whenever possible.

During the preparation of this third edition, I became indebted to many people to whom I wish to express my appreciation. I am grateful to John D. Hooper and to Robert D. Anderson, consulting engineers on the west coast, who most diligently reviewed the three new chapters in earthquake engineering. Their commentary and suggestions were most useful in making the presentation of the material in these chapters closer to present engineering practice. I am thankful to my colleague Dr. Joseph Hagerty for carefully reading and editing the manuscript of the new chapters. I am also grateful to Dean Leo Jenkins of Speed Scientific School and to Dr. Manuel Schwartz, of the Department of Physics, who reviewed the manuscript at the early stages of its preparation.

A special acknowledgment of gratitude is extended to my friend Dr. Edwin A. Tuttle, Professor of Education, who most kindly spent many hours checking my English grammar. My thanks also go to Miss Debbie Gordon for her competent typing of the manuscript.

To those people whom I recognized in the prefaces to the first and second editions for their help, I again express my wholehearted appreciation. Finally, I give due recognition to my wife, Jean, who with infinite patience and dedication helped me in editing the whole manuscript. Any remaining errors are mine.

MARIO PAZ

Preface to
the First Edition

Natural phenomena and human activities impose forces of time-dependent variability on structures as simple as a concrete beam or a steel pile, or as complex as a multistory building or a nuclear power plant constructed from different materials. Analysis and design of such structures subjected to dynamic loads involve consideration of time-dependent inertial forces. The resistance to displacement exhibited by a structure may include forces which are functions of the displacement and the velocity. As a consequence, the governing equations of motion of the dynamic system are generally nonlinear partial differential equations which are extremely difficult to solve in mathematical terms. Nevertheless, recent developments in the field of structural dynamics enable such analysis and design to be accomplished in a practical and efficient manner. This work is facilitated through the use of simplifying assumptions and mathematical models, and of matrix methods and modern computational techniques.

In the process of teaching courses on the subject of structural dynamics, the author came to the realization that there was a definite need for a text which would be suitable for the advanced undergraduate or the beginning graduate engineering student being introduced to this subject. The author is familiar with the existence of several excellent texts of an advanced nature but generally these texts are, in his view, beyond the expected comprehension of the student. Consequently, it was his principal aim in writing this book to incorporate modern methods of analysis and techniques adaptable to computer programming in a manner as clear and easy as the subject permits. He felt that computer programs should be included in the book in order to assist the student in the application of modern methods associated with computer usage. In addition, the author hopes that this text will serve the practicing engineer for purposes of self-study and as a reference source.

In writing this text, the author also had in mind the use of the book as a possible source for research topics in structural dynamics for students working toward an advanced degree in engineering who are required to write a thesis. At Speed Scientific School, University of Louisville, most engineering students complete a fifth year of study with a thesis requirement leading to a Master in Engineering degree. The author's experience as a thesis advisor leads him to believe that this book may well serve the students in their search and selection of topics in subjects currently under investigation in structural dynamics.

Should the text fulfill the expectations of the author in some measure, particularly the elucidation of this subject, he will then feel rewarded for his efforts in the preparation and development of the material in this book.

MARIO PAZ

December, 1979

PART I

Structures Modeled as a Single Degree-of-Freedom System

1

Undamped Single Degree-of-Freedom System

It is not always possible to obtain rigorous mathematical solutions for engineering problems. In fact, analytical solutions can be obtained only for certain simplified situations. For problems involving complex material properties, loading, and boundary conditions, the engineer introduces assumptions and idealizations deemed necessary to make the problem mathematically manageable but still capable of providing sufficiently approximate solutions and satisfactory results from the point of view of safety and economy. The link between the real physical system and the mathematically feasible solution is provided by the *mathematical model* which is the symbolic designation for the substitute idealized system including all the assumptions imposed on the physical problem.

1.1 DEGREES OF FREEDOM

In structural dynamics the number of independent coordinates necessary to specify the configuration or position of a system at any time is referred to as the number of degrees of freedom. In general, a continuous structure has an infinite *number of degrees of freedom*. Nevertheless, the process of idealization or selection of an appropriate mathematical model permits the reduction in the number of degrees of freedom to a discrete number and in some cases to just a single degree of freedom. Figure 1.1 shows some examples of structures which may be represented for dynamic analysis as *one-degree-of-freedom systems*, that is, structures modeled as systems with a single displacement coordinate. These one-degree-of-freedom systems may be described conveniently by the mathematical model shown in Fig. 1.2 which has the following elements: (1) a mass element *m* representing the mass and inertial characteristic of the structure; (2) a spring element *k* representing the elastic restoring force and potential energy capacity of the structure; (3) a damping element *c* representing the frictional characteristics and energy losses of the structure; and (4) an excitation force *F(t)* representing the external forces acting on the structural system. The force *F(t)* is written this way to indicate that it is a function of time. In adopting the mathematical model shown in Fig. 1.2, it is assumed that each element in the system represents a single property; that is, the mass *m* represents only the property of inertia and not elasticity or energy dissipation, whereas the spring *k* represents exclusively elasticity and not inertia or energy dissipation. Finally, the damper *c* only dissipates energy. The reader certainly realizes that such "pure" elements do not exist in our physical world and that mathematical models are only conceptual idealizations of real structures. As such, mathematical models may provide complete and accurate knowledge of the behavior of the model itself, but only limited or approximate information on the behavior of the real physical system. Nevertheless, from a practical point of view, the information acquired from the analysis of the mathematical model may very well be sufficient for an adequate understanding of the dynamic behavior of the physical system, including design and safety requirements.

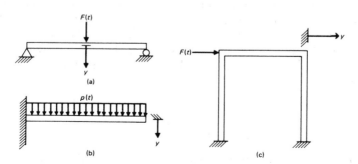

Fig. 1.1 Examples of structures modeled as one-degree-of-freedom systems.

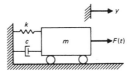

Fig. 1.2 Mathematical model for one-degree-of-freedom systems.

1.2 UNDAMPED SYSTEM

We start our study of structural dynamics with the analysis of a fundamental and simple system, the one-degree-of-freedom system in which we disregard or "neglect" frictional forces or damping. In addition, we consider the system, during its motion or vibration, to be free from external actions or forces. Under these conditions, the system is in motion governed only by the influence of the so-called *initial conditions*, that is, the given displacement and velocity at time $t = 0$ when the study of the system is initiated. This undamped, one-degree-of-freedom system is often referred to as the *simple undamped oscillator*. It is usually represented as shown in Fig. 1.3(a) or Fig. 1.3(b) or any similar arrangements. These two figures represent mathematical models which are dynamically equivalent. It is only a matter of personal preference to adopt one or the other. In these models the mass m is restrained by the spring k and is limited to rectilinear motion along one coordinate axis.

The mechanical characteristic of a spring is described by the relation between the magnitude of the force F_s applied to its free end and the resulting end displacement y, as shown graphically in Fig. 1.4 for three different springs. The curve labeled (a) in Fig. 1.4 represents the behavior of a "hard spring," in which the force required to produce a given displacement becomes increasingly greater as the spring is deformed. The second spring (b) is designated a *linear spring* because the deformation is directly proportional to the force and the graphical representation of its characteristic is a straight line. The con-

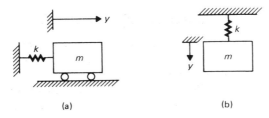

(a)   (b)

Fig. 1.3 Alternate representations of mathematical models for one-degree-of-freedom systems.

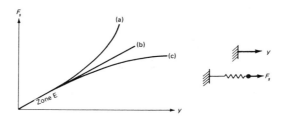

Fig. 1.4 Force displacement relation. (a) Hard spring. (b) Linear spring. (c) Soft spring.

stant of proportionality between the force and displacement [slope of line (b)] of a linear spring is referred to as the *spring constant*, usually designated by the letter k. Consequently, we may write the following relation between force and displacement for a linear spring.

$$F_s = ky \qquad (1.1)$$

A spring with characteristics shown by curve (c) in Fig. 1.4 is known as a "soft spring." For such a spring the incremental force required to produce additional deformation decreases as the spring deformation increases. Undoubtedly, the reader is aware from his previous exposure to mathematical modeling of physical systems that the linear spring is the simplest type to manage analytically. It should not come as a surprise to learn that most of the technical literature on structural dynamics deals with models using linear springs. In other words, either because the elastic characteristics of the structural system are, in fact, essentially linear, or simply because of analytical expediency, it is usually assumed that the force-deformation properties of the system are linear. In support of this practice, it should be noted that in many cases the displacements produced in the structure by the action of external forces or disturbances are small in magnitude (Zone E in Fig. 1.4), thus rendering the linear approximation close to the actual structural behavior.

1.3   SPRINGS IN PARALLEL OR IN SERIES

Sometimes it is necessary to determine the equivalent spring constant for a system in which two or more springs are arranged in parallel as shown in Fig. 1.5(a) or in series as in Fig. 1.5(b).

For two springs in parallel the total force required to produce a relative displacement of their ends of one unit is equal to the sum of their spring constants. This total force is by definition the equivalent spring constant k_e and is given by

$$k_e = k_1 + k_2 \qquad (1.2)$$

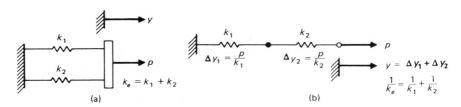

Fig. 1.5 Combination of springs. (a) Springs in parallel. (b) Springs in series.

In general for n springs in parallel

$$k_e = \sum_{i=1}^{n} k_i \tag{1.3}$$

For two springs assembled in series as shown in Fig. 1.5(b), the force P produces the relative displacements in the springs

$$\Delta y_1 = \frac{P}{k_1}$$

and

$$\Delta y_2 = \frac{P}{k_2}$$

Then, the total displacement y of the free end of the spring assembly is equal to $y = \Delta y_1 + \Delta y_2$, or substituting Δy_1 and Δy_2,

$$y = \frac{P}{k_1} + \frac{P}{k_2} \tag{1.4}$$

Consequently, the force necessary to produce one unit displacement (equivalent spring constant) is given by

$$k_e = \frac{P}{y}$$

Substituting y from this last relation into eq. (1.4), we may conveniently express the reciprocal value of the equivalent spring constant as

$$\frac{1}{k_e} = \frac{1}{k_1} + \frac{1}{k_2} \tag{1.5}$$

In general for n springs in series the equivalent spring constant may be obtained from

$$\frac{1}{k_e} = \sum_{i=1}^{n} \frac{1}{k_i} \tag{1.6}$$

1.4   NEWTON'S LAW OF MOTION

We continue now with the study of the simple oscillator depicted in Fig. 1.3. The objective is to describe its motion, that is, to predict the displacement or velocity of the mass m at any time t, for a given set of initial conditions at time $t = 0$. The analytical relation between the displacement, y, and time, t, is given by Newton's Second Law of Motion, which in modern notation may be expressed as

$$F = ma \tag{1.7}$$

where F is the resultant force acting on a particle of mass m and a is its resultant acceleration. The reader should recognize that eq. (1.7) is a vector relation and as such it can be written in equivalent form in terms of its components along the coordinate axes x, y, and z, namely

$$\sum F_x = ma_x \tag{1.8a}$$

$$\sum F_y = ma_y \tag{1.8b}$$

$$\sum F_z = ma_z \tag{1.8c}$$

The acceleration is defined as the second derivative of the position vector with respect to time; it follows that eqs. (1.8) are indeed differential equations. The reader should also be reminded that these equations as stated by Newton are directly applicable only to bodies idealized as particles, that is, bodies which possess mass but no volume. However, as is proved in elementary mechanics, Newton's Law of Motion is also directly applicable to bodies of finite dimensions undergoing translatory motion.

For plane motion of a rigid body which is symmetric with respect to the reference plane of motion (x-y plane), Newton's Law of Motion yields the following equations:

$$\sum F_x = m(a_G)_x \tag{1.9a}$$

$$\sum F_y = m(a_G)_y \tag{1.9b}$$

$$\sum M_G = I_G \alpha \tag{1.9c}$$

In the above equations $(a_G)_x$ and $(a_G)_y$ are the acceleration components, along the x and y axes, of the center of mass G of the body; α is the angular acceleration; I_G is the mass moment of inertia of the body with respect to an axis through G, the center of mass; and $\sum M_G$ is the sum of the moments of all the forces acting on the body with respect to an axis through G, perpendicular to the x-y plane. Equations (1.9) are certainly also applicable to the motion of a rigid body in pure rotation about a fixed axis. For this particular type of plane motion, alternatively, eq. (1.9c) may be replaced by

$$\sum M_0 = I_0 \alpha \qquad (1.9d)$$

in which the mass moment of inertia I_0 and the moment of the forces M_0 are determined with respect to the fixed axis of rotation. The general motion of a rigid body is described by two vector equations, one expressing the relation between the forces and the acceleration of the mass center, and another relating the moments of the forces and the angular motion of the body. This last equation expressed in its scalar components is rather complicated, but seldom needed in structural dynamics.

1.5   FREE BODY DIAGRAM

At this point, it is advisable to follow a method conducive to an organized and systematic analysis in the solution of dynamics problems. The first and probably the most important practice to follow in any dynamic analysis is to draw a free body diagram of the system, prior to writing a mathematical description of the system.

The free body diagram (FBD), as the student may recall, is a sketch of the body isolated from all other bodies, in which all the forces external to the body are shown. For the case at hand, Fig. 1.6(b) depicts the FBD of the mass m of the oscillator, displaced in the positive direction with reference to coordinate y, and acted upon by the spring force $F_s = ky$ (assuming a linear spring). The weight of the body mg and the normal reaction N of the supporting surface are also shown for completeness, though these forces, acting in the vertical direction, do not enter into the equation of motion written for the y direction. The application of Newton's Law of Motion gives

$$-ky = m\ddot{y} \qquad (1.10)$$

where the spring force acting in the negative direction has a minus sign, and where the acceleration has been indicated by $\ddot{y}$. In this notation, double

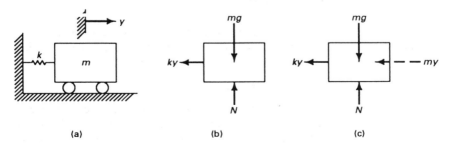

Fig. 1.6 Alternate free body diagrams: (a) Single degree-of-freedom system. (b) Showing only external forces. (c) Showing external and inertial forces.

overdots denote the second derivative with respect to time and obviously a single overdot denotes the first derivative with respect to time, that is, the velocity.

1.6 D'ALEMBERT'S PRINCIPLE

An alternative approach to obtain eq. (1.10) is to make use of *D'Alembert's Principle* which states that a system may be set in a state of *dynamic equilibrium* by adding to the external forces a fictitious force which is commonly known as the *inertial force.*

Figure 1.6(c) shows the FBD with inclusion of the inertial force $m\ddot{y}$. This force is equal to the mass multiplied by the acceleration, and should always be directed negatively with respect to the corresponding coordinate. The application of D'Alembert's Principle allows us to use equations of equilibrium in obtaining the equation of motion. For example, in Fig. 1.6(c), the summation of forces in the y direction gives directly

$$m\ddot{y} + ky = 0 \qquad (1.11)$$

which obviously is equivalent to eq. (1.10).

The use of D'Alembert's Principle in this case appears to be trivial. This will not be the case for a more complex problem, in which the application of D'Alembert's Principle, in conjunction with the *Principle of Virtual Work*, constitutes a powerful tool of analysis. As will be explained later, the Principle of Virtual Work is directly applicable to any system in equilibrium. It follows then that this principle may also be applied to the solution of dynamic problems, provided that D'Alembert's Principle is used to establish the dynamic equilibrium of the system.

Example 1.1. Show that the same differential equation is obtained for a spring-supported body moving vertically as for the same body vibrating along a horizontal axis, as shown in Figs. 1.7(a) and 1.7(b).

Solution: The FBDs for these two representations of the simple oscillator are shown in Figs. 1.7(c) and 1.7(e), where the inertial forces are included. Equating to zero the sum of the forces in Fig. 1.7(c), we obtain

$$m\ddot{y} + ky = 0 \qquad (a)$$

When the body in Fig. 1.7(d) is in the static equilibrium position, the spring is stretched y_0 units and exerts a force $ky_0 = W$ upward on the body, where W is the weight of the body. When the body is displaced a distance y downward from this position of equilibrium the magnitude of the spring force is given by $F_s = k\,(y_0 + y)$ or $F_s = W + ky$, since $ky_0 = W$. Using this result

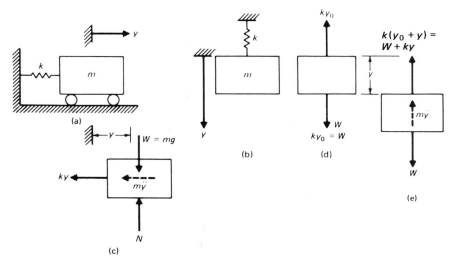

Fig. 1.7 Two representations of the simple oscillator and corresponding free body diagrams.

and applying it to the body in Fig. 1.7(e), we obtain from Newton's Second Law of Motion

$$-(W + ky) + W = m\ddot{y}$$

or   (b)

$$m\ddot{y} + ky = 0$$

which is identical to eq. (a).

1.7   SOLUTION OF THE DIFFERENTIAL EQUATION OF MOTION

The next step toward our objective is to find the solution of the differential equation (1.11). We should again adopt a systematic approach and proceed to first classify this differential equation. Since the dependent variable y and its second derivative $\ddot{y}$ appear in the first degree in eq. (1.11), this equation is classified as linear and of second order. The fact that the coefficients of y and $\ddot{y}$ (k and m, respectively) are constants and the second member (right-hand side) of the equation is zero further classifies the equation as homogeneous with constant coefficients. We should recall, probably with a certain degree of satisfaction, that a general procedure exists for the solution of linear differential equations (homogeneous or nonhomogeneous) of any order. For this simple, second-order differential equation we may proceed directly by assuming a trial solution given by

$$y = A \cos \omega t \qquad (1.12)$$

or

$$y = B \sin \omega t \qquad (1.13)$$

where A and B are constants depending on the initiation of the motion while ω is a quantity denoting a physical characteristic of the system as it will be shown next. The substitution of eq. (1.12) into eq. (1.11) gives

$$(-m\omega^2 + k) A \cos \omega t = 0 \qquad (1.14)$$

If this equation is to be satisfied at any time, the factor in parentheses must be equal to zero or

$$\omega^2 = \frac{k}{m} \qquad (1.15)$$

The reader should verify that eq. (1.13) is also a solution of the differential equation (1.11), with ω also satisfying eq. (1.15).

The positive root of eq. (1.15),

$$\omega = \sqrt{k/m} \qquad (1.16)$$

is known as the *natural frequency* of the system for reasons that will soon be apparent.

Since either eq. (1.12) or eq. (1.13) is a solution of eq. (1.11), and since this differential equation is linear, the superposition of these two solutions, indicated by eq. (1.17) below, is also a solution. Furthermore, eq. (1.17), having two constants of integration, A and B, is, in fact, the general solution for this second-order differential equation,

$$y = A \cos \omega t + B \sin \omega t. \qquad (1.17)$$

The expression for velocity, $\dot{y}$, is found simply by differentiating eq. (1.17) with respect to time; that is,

$$\dot{y} = -A \omega \sin \omega t + B \omega \cos \omega t. \qquad (1.18)$$

Next, we should determine the constants of integration A and B. These constants are determined from known values for the motion of the system which almost invariably are the displacement y_0 and the velocity v_0 at the initiation of the motion, that is, at time $t = 0$. These two conditions are referred to as *initial conditions*, and the problem of solving the differential equation for the initial conditions is called an *initial value problem*.

After substituting, for $t = 0$, $y = y_0$, and $\dot{y} = v_0$ into eqs. (1.17) and (1.18) we find that

$$y_0 = A \tag{1.19a}$$
$$v_0 = B\omega \tag{1.19b}$$

Finally, the substitution of A and B from eqs. (1.19) into eq. (1.17) gives

$$y = y_0 \cos \omega t + \frac{v_0}{\omega} \sin \omega t \tag{1.20}$$

which is the expression of the displacement y of the simple oscillator as a function of the time variable t; thus we have accomplished our objective of describing the motion of the simple undamped oscillator modeling structures with a single degree of freedom.

1.8   FREQUENCY AND PERIOD

An examination of eq. (1.20) shows that the motion described by this equation is *harmonic* and, therefore, periodic; that is, it can be expressed by a sine or cosine function of the same frequency ω. The period may easily be found since the functions sine and cosine both have a period of 2π. The *period T* of the motion is determined from

$$\omega T = 2\pi$$

or $\qquad\qquad\qquad\qquad\qquad\qquad\qquad\qquad\qquad\qquad\qquad$ (1.21)

$$T = \frac{2\pi}{\omega}$$

The period is usually expressed in seconds per cycle or simply in seconds with the tacit understanding that it is "per cycle." The value reciprocal to the period is the *natural frequency f*. From eq. (1.21)

$$f = \frac{1}{T} = \frac{\omega}{2\pi} \tag{1.22}$$

The natural frequency f is usually expressed in hertz or cycles per second (cps). Because the quantity ω differs from the natural frequency f only by the constant factor, 2π, ω also is sometimes referred to as the natural frequency. To distinguish between these two expressions for natural frequency, ω may be called the *circular* or *angular* natural frequency. Most often, the distinction is understood from the context or from the units. The natural frequency f is measured in cps as indicated, while the circular frequency ω should be given in radians per second (rad/sec).

Example 1.2.   Determine the natural frequency of the system shown in Fig. 1.8 consisting of a weight of $W = 50.7$ lb attached to a horizontal cantilever beam through the coil spring k_2. The cantilever beam has a thickness $t =$

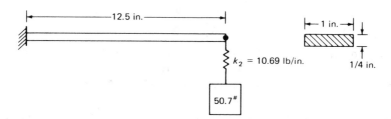

Fig. 1.8 System for Example 1.2.

$\frac{1}{4}$ in, a width $b = 1$ in modulus of elasticity $E = 30 \times 10^6$ psi, and a length $L = 12.5$ in. The coil spring has a stiffness, $k_2 = 10.69$ (lb/in).

Solution: The deflection Δ at the free end of a cantilever beam acted upon by a static force P at the free end is given by

$$\Delta = \frac{PL^3}{3EI}$$

The corresponding spring constant k_1 is then

$F = kx$

$P = k\Delta$

$$k_1 = \frac{P}{\Delta} = \frac{3EI}{L^3}$$

where $I = \frac{1}{12}bt^3$ (for rectangular section). Now, the cantilever and the coil spring of this system are connected as springs in series. Consequently, the equivalent spring constant as given from eq. (1.5) is

$$\frac{1}{k_e} = \frac{1}{k_1} + \frac{1}{k_2}$$

Substituting corresponding numerical values, we obtain

$$I = \frac{1}{12} \times 1 \times \left(\frac{1}{4}\right)^3 = \frac{1}{768} \ (\text{in})^4$$

$$k_1 = \frac{3 \times 30 \times 10^6}{(12.5)^3 \times 768} = 60 \ \text{lb/in}$$

and

$$\frac{1}{k_e} = \frac{1}{60} + \frac{1}{10.69}$$

$$k_e = 9.07 \ \text{lb/in}$$

The natural frequency for this system is then given by eq. (1.16) as

$$\omega = \sqrt{k_e/m} \quad (m = W/g \text{ and } g = 386 \text{ in/sec}^2)$$

$$\omega = \sqrt{9.07 \times 386/50.7}$$

$$\omega = 8.31 \text{ rad/sec}$$

or using eq. (1.22)

$$f = 1.32 \text{ cps} \tag{Ans.}$$

1.9   AMPLITUDE OF MOTION

Let us now examine in more detail eq. (1.20), the solution describing the free vibratory motion of the undamped oscillator. A simple trigonometric transformation may show us that we can rewrite this equation in the equivalent forms, namely

$$y = C \sin (\omega t + \alpha) \tag{1.23}$$

or

$$y = C \cos (\omega t - \beta) \tag{1.24}$$

where

$$C = \sqrt{y_0^2 + (v_0/\omega)^2} \tag{1.25}$$

$$\tan \alpha = \frac{y_0}{v_0/\omega} \tag{1.26}$$

and

$$\tan \beta = \frac{v_0/\omega}{y} \tag{1.27}$$

The simplest way to obtain eq. (1.23) or eq. (1.24) is to multiply and divide eq. (1.20) by the factor C defined in eq. (1.25) and to define α (or β) by eq. (1.26) [or eq. (1.27)]. Thus

$$y = C\left(\frac{y_0}{C} \cos \omega t + \frac{v_0/\omega}{C} \sin \omega t\right) \tag{1.28}$$

With the assistance of Fig. 1.9, we recognize that

$$\sin \alpha = \frac{y_0}{C} \tag{1.29}$$

and

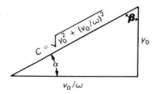

Fig. 1.9 Definition of angle α.

$$\cos \alpha = \frac{v_0/\omega}{C} \tag{1.30}$$

The substitution of eqs. (1.29) and (1.30) into eq. (1.28) gives

$$y = C(\sin \alpha \cos \omega t + \cos \alpha \sin \omega t) \tag{1.31}$$

The expression within the parentheses of eq. (1.31) is identical to $\sin(\omega t + \alpha)$, which yields eq. (1.23). Similarly, the reader should verify, without difficulty, the form of solution given by eq. (1.24).

The value of C in eq. (1.23) [or eq. (1.24)] is referred to as the amplitude of motion and the angle α (or β) as the phase angle. The solution for the motion of the simple oscillator is shown graphically in Fig. 1.10.

Example 1.3.   Consider the frame shown in Fig. 1.11(a). This is a rigid steel frame to which a horizontal dynamic force is applied at the upper level. As part of the overall structural design it is required to determine the natural frequency of the frame. Two assumptions are made: (1) the masses of the columns and walls are negligible; and (2) the horizontal members are sufficiently rigid to prevent rotation at the tops of the columns. These assumptions are not mandatory for the solution of the problem, but they serve to simplify the analysis. Under these conditions, the frame may be modeled by the spring-mass system shown in Fig. 1.11(b).

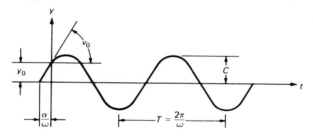

Fig. 1.10 Undamped free-vibration response.

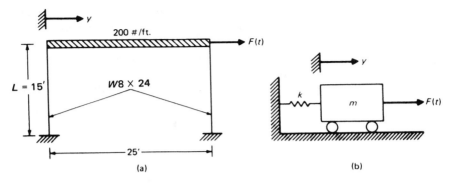

Fig. 1.11 One-degree-of-freedom frame and corresponding mathematical model for Example 1.3.

Solution: The parameters of this model may be computed as follows:

$$W = 200 \times 25 = 5000 \text{ lb}$$

$$I = 82.5 \text{ in}^4$$

$$E = 30 \times 10^6 \text{ psi}$$

$$k = \frac{12E(2I)}{L^3} = \frac{12 \times 30 \times 10^6 \times 165}{(15 \times 12)^3}$$

$$k = 10,185 \text{ lb/in}$$

Note: A unit displacement of the top of a fixed column requires a force equal to $12EI/L^3$. Therefore, the natural frequency from eqs. (1.16) and (1.22) is

$$f = \frac{1}{2\pi} \sqrt{\frac{kg}{W}} = \frac{1}{2\pi} \sqrt{\frac{10,185 \times 386}{5000}}$$

$$f = 4.46 \text{ cps} \qquad \qquad \text{(Ans.)}$$

1.10   SUMMARY

Several basic concepts were introduced in this chapter.

(1)   The mathematical model of a structure is an idealized representation for its analysis.
(2)   The number of degrees of freedom of a system is equal to the number of independent coordinates necessary to describe its position.
(3)   The free body diagram (FBD) for dynamic equilibrium (to allow appli-

cation of D'Alembert's Principle) is a diagram of the system isolated from all other bodies, showing all the external forces on the system, including the inertial force.

(4) The stiffness or spring constant of a linear system is the force necessary to produce a unit displacement.

(5) The differential equation of the undamped simple oscillator in free motion is

$$m\ddot{y} + ky = 0$$

and its general solution is

$$y = A \cos \omega t + B \sin \omega t$$

where A and B are constants of integration determined from initial conditions:

$$A = y_0$$

$$B = v_0/\omega$$

$$\omega = \sqrt{k/m} \quad \text{is the natural frequency in rad/sec}$$

$$f = \frac{\omega}{2\pi} \quad \text{is the natural frequency in cps}$$

$$T = \frac{1}{f} \quad \text{is the natural period in seconds}$$

(6) The equation of motion may be written in the alternate forms:

$$y = C \sin (\omega t + \alpha)$$

or

$$y = C \cos (\omega t - \beta)$$

where

$$C = \sqrt{y_0^2 + (v_0/\omega)^2}$$

and

$$\tan \alpha = \frac{y_0}{v_0/\omega}$$

$$\tan \beta = \frac{v_0/\omega}{y_0}$$

PROBLEMS

1.1 Determine the natural period for the system in Fig. 1.12. Assume that the beam and springs supporting the weight W are massless.

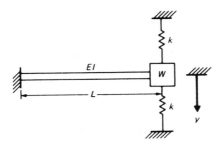

Fig. 1.12

1.2 The following numerical values are given in Problem 1.1: $L = 100$ in, $EI = 10^8$ (lb·in²), $W = 3000$ lb, and $k = 2000$ lb/in. If the weight W has an initial displacement of $y_0 = 1.0$ in and an initial velocity $v_0 = 20$ in/sec, determine the displacement and the velocity 1 sec later.

1.3 Determine the natural frequency for horizontal motion of the steel frame in Fig. 1.13. Assume the horizontal girder to be infinitely rigid and neglect the mass of the columns.

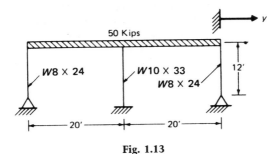

Fig. 1.13

1.4 Calculate the natural frequency in the horizontal mode of the steel frame in Fig. 1.14 for the following cases: (a) the horizontal member is assumed to be infinitely rigid; (b) the horizontal member is flexible and made of steel—W10 × 33.

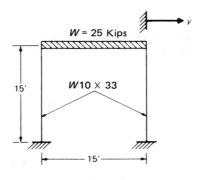

Fig. 1.14

1.5   Determine the natural frequency of the fixed beam in Fig. 1.15 carrying a concentrated weight W at its center. Neglect the mass of the beam.

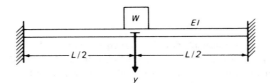

Fig. 1.15

1.6   The numerical values for Problem 1.5 are given as: $L = 120$ in, $EI = 10^9$ (lb · in²), and $W = 5000$ lb. If the initial displacement and the initial velocity of the weight are, respectively, $y_0 = 0.5$ in and $v_0 = 15$ in/sec, determine the displacement, velocity, and acceleration of W when time $t = 2$ sec.

1.7   Consider the simple pendulum of weight W illustrated in Fig. 1.16. A simple pendulum is a particle or concentrated weight that oscillates in a vertical arc and is supported by a weightless cord. The only forces acting are those of gravity and cord tension (i.e., frictional resistance is neglected). If the cord length is L,

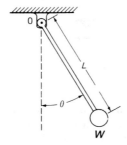

Fig. 1.16

determine the motion if the maximum oscillation angle θ is small and the initial displacement and velocity are θ_0 and $\dot{\theta}_0$, respectively.

1.8   Write the differential equation of motion for the inverted pendulum shown in Fig. 1.17 and determine its natural frequency. Assume small oscillations, and neglect the mass of the rod.

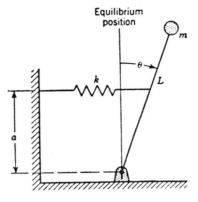

Fig. 1.17

1.9   A vertical pole of length L and flexual rigidity EI carries a mass m at its top, as shown in Fig. 1.18. Neglecting the weight of the pole, derive the differential equation for small horizontal vibrations of the mass, and find the natural frequency. Assume that the effect of gravity is small and nonlinear effects may be neglected.

Fig. 1.18

1.10 Determine an expression for the natural frequency of the weight W in each of the cases shown in Fig. 1.19. The beams are uniform of cross-sectional moment of inertia I and modulus of elasticity E. Neglect the mass of the beams.

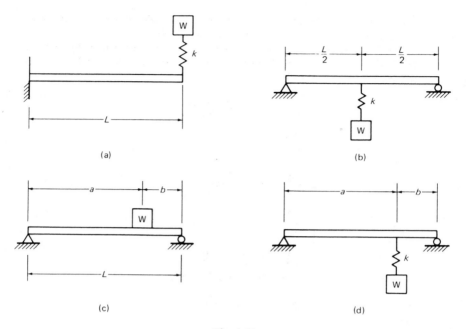

Fig. 1.19

1.11 A system (see Fig. 1.20) is modeled by two freely vibrating masses m_1 and m_2 interconnected by a spring having a constant k. Determine for this system the differential equation of motion for the relative displacement $u = y_2 - y_1$ between the two masses. Also determine the corresponding natural frequency of the system.

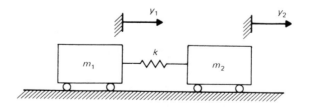

Fig. 1.20

2

Damped Single Degree-of-Freedom System

We have seen in the preceding chapter that the simple oscillator under idealized conditions of no damping, once excited, will oscillate indefinitely with a constant amplitude at its natural frequency. Experience indicates, however, that it is not possible to have a device which vibrates under these ideal conditions. Forces designated as frictional or damping forces are always present in any physical system undergoing motion. These forces dissipate energy; more precisely, the unavoidable presence of these frictional forces constitutes a mechanism through which the mechanical energy of the system, kinetic or potential energy, is transformed to other forms of energy such as heat. The mechanism of this energy transformation or dissipation is quite complex and is not completely understood at this time. In order to account for these dissipative forces in the analysis of dynamic systems, it is necessary to make some assumptions about these forces, on the basis of experience.

2.1 VISCOUS DAMPING

In considering damping forces in the dynamic analysis of structures, it is usually assumed that these forces are proportional to the magnitude of the velocity, and opposite to the direction of motion. This type of damping is known as *viscous damping*; it is the type of damping force that could be developed in a body restrained in its motion by a surrounding viscous fluid.

There are situations in which the assumption of viscous damping is realistic and in which the dissipative mechanism is approximately viscous. Nevertheless, the assumption of viscous damping is often made regardless of the actual dissipative characteristics of the system. The primary reason for such wide use of this method is that it leads to a relatively simple mathematical analysis.

2.2 EQUATION OF MOTION

Let us assume that we have modeled a structural system as a simple oscillator with viscous damping, as shown in Fig. 2.1(a). In this figure, m and k are, respectively, the mass and spring constant of the oscillator and c is the viscous damping coefficient. We proceed, as in the case of the undamped oscillator, to draw the free body diagram (FBD) and apply Newton's Law to obtain the differential equation of motion. Figure 2.1(b) shows the FBD of the damped oscillator in which the inertial force $m\ddot{y}$ is also shown, so that we can use D'Alembert's Principle. The summation of forces in the y direction gives the differential equation of motion,

$$m\ddot{y} + c\dot{y} + ky = 0 \qquad (2.1)$$

The reader may verify that a trial solution $y = A \sin \omega t$ or $y = B \cos \omega t$ will not satisfy eq. (2.1). However, the exponential function $y = Ce^{pt}$ does satisfy this equation. Substitution of this function into eq. (2.1) results in the equation

$$mCp^2 \, e^{pt} + cCp \, e^{pt} + kC \, e^{pt} = 0$$

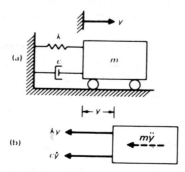

Fig. 2.1 (a) Viscous damped oscillator. (b) Free body diagram.

which, after cancellation of the common factors, reduces to an equation called *the characteristic equation* for the system, namely

$$mp^2 + cp + k = 0 \tag{2.2}$$

The roots of this quadratic equation are

$$\begin{matrix} p_1 \\ p_2 \end{matrix} = -\frac{c}{2m} \pm \sqrt{\left(\frac{c}{2m}\right)^2 - \frac{k}{m}} \tag{2.3}$$

Thus the general solution of eq. (2.1) is given by the superposition of the two possible solutions, namely

$$y(t) = C_1 e^{p_1 t} + C_2 e^{p_2 t} \tag{2.4}$$

where C_1 and C_2 are constants of integration to be determined from the initial conditions.

The final form of eq. (2.4) depends on the sign of the expression under the radical in eq. (2.3). Three distinct cases may occur: the quantity under the radical may either be zero, positive, or negative. The limiting case in which the quantity under the radical is zero is treated first. The damping present in this case is called *critical damping.*

2.3   CRITICALLY DAMPED SYSTEM

For a system oscillating with critical damping, as defined above, the expression under the radical in eq. (2.3) is equal to zero; that is,

$$\left(\frac{c_{cr}}{2m}\right)^2 - \frac{k}{m} = 0 \tag{2.5}$$

or

$$c_{cr} = 2\sqrt{km} \tag{2.6}$$

where c_{cr} designates the critical damping value. Since the natural frequency of the undamped system is designated by $\omega = \sqrt{k/m}$, the critical damping coefficient given by eq. (2.6) may also be expressed in alternative notation as

$$c_{cr} = 2m\omega = \frac{2k}{\omega} \tag{2.7}$$

In a critically damped system the roots of the characteristic equation are equal, and, from eq. (2.3), they are

$$p_1 = p_2 = -\frac{c_{cr}}{2m} \tag{2.8}$$

Since the two roots are equal, the general solution given by eq. (2.4) would provide only one independent constant of integration, hence, one independent solution, namely

$$y_1(t) = C_1 \, e^{-(c_{cr}/2m)t} \tag{2.9}$$

Another independent solution may be found by using the function

$$y_2(t) = C_2 t y_1(t) = C_2 \, t e^{-(c_{cr}/2m)t} \tag{2.10}$$

This equation, as the reader may verify, also satisfies the differential equation (2.1). The general solution for a critically damped system is then given by the superposition of these two solutions,

$$y(t) = (C_1 + C_2 t) \, e^{-(c_{cr}/2m)t} \tag{2.11}$$

2.4   OVERDAMPED SYSTEM

In an overdamped system, the damping coefficient is greater than the value for critical damping, namely

$$c > c_{cr} \tag{2.12}$$

Therefore, the expression under the radical of eq. (2.3) is positive, thus the two roots of the characteristic equation are real and distinct, and consequently the solution is given directly by eq. (2.4). It should be noted that, for the overdamped or the critically damped system, the resulting motion is not oscillatory; the magnitude of the oscillations decays exponentially with time to zero. Figure 2.2 depicts graphically the response for the simple oscillator with critical damping. The response of the overdamped system is similar to the motion of the critically damped system of Fig. 2.2, but the return toward the neutral position requires more time as the damping is increased.

2.5   UNDERDAMPED SYSTEM

When the value of the damping coefficient is less than the critical value $(c < c_{cr})$, which occurs when the expression under the radical is negative, the roots of the characteristic eq. (2.3) are complex conjugates, so that

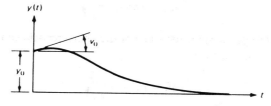

Fig. 2.2 Free-vibration response with critical damping.

$$\begin{matrix} p_1 \\ p_2 \end{matrix} = -\frac{c}{2m} \pm i \sqrt{\frac{k}{m} - \left(\frac{c}{2m}\right)^2} \tag{2.13}$$

where $i = \sqrt{-1}$ is the imaginary unit.

For this case, it is convenient to make use of Euler's equations which relate exponential and trigonometric functions, namely

$$e^{ix} = \cos x + i \sin x$$
$$e^{-ix} = \cos x - i \sin x \tag{2.14}$$

The substitution of the roots p_1 and p_2 from eq. (2.13) into eq. (2.4) together with the use of eq. (2.14) gives the following convenient form for the general solution of the underdamped system:

$$y(t) = e^{-(c/2m)t} \left(A \cos \omega_D t + B \sin \omega_D t\right) \tag{2.15}$$

where A and B are redefined constants of integration and ω_D, the damped frequency of the system, is given by

$$\omega_D = \sqrt{\frac{k}{m} - \left(\frac{c}{2m}\right)^2} \tag{2.16}$$

or

$$\omega_D = \omega\sqrt{1 - \xi^2} \tag{2.17}$$

This last result is obtained after substituting, in eq. (2.16), the expression for the undamped natural frequency

$$\omega = \sqrt{\frac{k}{m}} \tag{2.18}$$

and defining the *damping ratio* of the system as

$$\xi = \frac{c}{c_{cr}} \tag{2.19}$$

where the critical damping coefficient c_{cr} is given by eq. (2.6).

Finally, when the initial conditions of displacement and velocity, y_0 and v_0, are introduced, the constants of integration can be evaluated and substituted into eq. (2.15), giving

$$y(t) = e^{-\xi\omega t}\left(y_0 \cos \omega_D t + \frac{v_0 + y_0 \xi\omega}{\omega_D} \sin \omega_D t\right) \tag{2.20}$$

Alternatively, this expression can be written as

$$y(t) = Ce^{-\xi\omega t} \cos (\omega_D t - \alpha) \tag{2.21}$$

where

$$C = \sqrt{y_0^2 + \frac{(v_0 + y_0\,\xi\omega)^2}{\omega_D^2}} \qquad (2.22)$$

and

$$\tan \alpha = \frac{(v_0 + y_0\,\xi\omega)}{\omega_D y_0} \qquad (2.23)$$

A graphical record of the response of an underdamped system with initial displacement y_0 but starting with zero velocity $(v_0 = 0)$ is shown in Fig. 2.3. It may be seen in this figure that the motion is oscillatory, but not periodic. The amplitude of vibration is not constant during the motion but decreases for successive cycles; nevertheless, the oscillations occur at equal intervals of time. This time interval is designated as the damped period of vibration and is given from eq. (2.17) by

$$T_D = \frac{2\pi}{\omega_D} = \frac{2\pi}{\omega\sqrt{1 - \xi^2}} \qquad (2.24)$$

The value of the damping coefficient for real structures is much less than the critical damping coefficient and usually ranges between 2 to 10% of the critical damping value. Substituting for the extreme value $\xi = 0.10$ into eq. (2.17),

$$\omega_D = 0.995\omega \qquad (2.25)$$

It can be seen that the frequency of vibration for a system with as much as a 10% damping ratio is essentially equal to the undamped natural frequency. Thus, in practice, the natural frequency for a damped system may be taken to be equal to the undamped natural frequency.

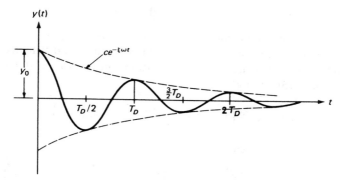

Fig. 2.3 Free vibration response for underdamped system.

2.6   LOGARITHMIC DECREMENT

A practical method for determining experimentally the damping coefficient of a system is to initiate free vibration, obtain a record of the oscillatory motion, such as the one shown in Fig. 2.4, and measure the rate of decay of the amplitude of motion. The decay may be conveniently expressed by the *logarithmic decrement* δ which is defined as the natural logarithm of the ratio of any two successive peak amplitudes, y_1 and y_2, in free vibration, that is,

$$\delta = \ln \frac{y_1}{y_2} \tag{2.26}$$

The evaluation of damping from the logarithmic decrement follows. Consider the damped vibration motion represented graphically in Fig. 2.4 and given analytically by eq. (2.21) as

$$y(t) = C\, e^{-\xi \omega t} \cos (\omega_D t - \alpha)$$

We note from this equation that, when the cosine factor is unity, the displacement is on points of the exponential curve $y(t) = C\, e^{-\xi \omega t}$ as shown in Fig. 2.4. However, these points are near but not equal to the positions of maximum displacement. The points on the exponential curve appear slightly to the right of the points of maximum amplitude. For most practical problems, the discrepancy is negligible and the displacement curve may be assumed to coincide at the peak amplitude, with the curve $y(t) = C\, e^{-\xi \omega t}$ so that we may write, for two consecutive peaks, y_1 at time t_1 and y_2 at T_D seconds later,

$$y_1 = C\, e^{-\xi \omega t_1}$$

and

$$y_2 = C\, e^{-\xi \omega (t_1 + T_D)}$$

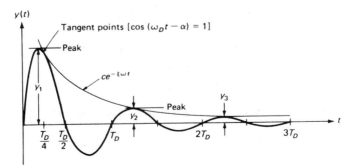

Fig. 2.4 Curve showing peak displacements and displacements at the points of tangency.

Dividing these two peak amplitudes and taking the natural logarithm, we obtain

$$\delta = \ln \frac{y_1}{y_2} = \xi\omega T_D \tag{2.27}$$

or by substituting T_D, the damped period, from eq. (2.24),

$$\delta = 2\pi\xi/\sqrt{1 - \xi^2} \tag{2.28}$$

As we can see, the damping ratio ξ can be calculated from eq. (2.28) after determining experimentally the amplitudes of two successive peaks of the system in free vibration. For small values of the damping ratio, eq. (2.28) can be approximated by

$$\delta \approx 2\pi\xi \tag{2.29}$$

Example 2.1. A vibrating system consisting of a weight of $W = 10$ lb and a spring with stiffness $k = 20$ lb/in is viscously damped so that the ratio of two consecutive amplitudes is 1.00 to 0.85. Determine: (a) the natural frequency of the undamped system, (b) the logarithmic decrement, (c) the damping ratio, (d) the damping coefficient, and (e) the damped natural frequency.

Solution: (a) The undamped natural frequency of the system in radians per second is

$$\omega = \sqrt{k/m} = \sqrt{(20 \text{ lb/in} \times 386 \text{ in/sec}^2)/10 \text{ lb}} = 27.78 \text{ rad/sec}$$

or in cycles per second

$$f = \frac{\omega}{2\pi} = 4.42 \text{ cps}$$

(b) The logarithmic decrement is given by eq. (2.26) as

$$\delta = \ln \frac{y_1}{y_2} = \ln \frac{1.00}{0.85} = 0.163$$

(c) The damping ratio from eq. (2.29) is approximately equal to

$$\xi \approx \frac{\delta}{2\pi} = \frac{0.163}{2\pi} = 0.026$$

(d) The damping coefficient is obtained from eqs. (2.6) and (2.19) as

$$c = \xi c_{cr} = 2 \times 0.026\sqrt{(10 \times 20)/386} = 0.037 \frac{\text{lb} \cdot \text{sec}}{\text{in}}$$

(e) The natural frequency of the damped system is given by eq. (2.17), so that

$$\omega_D = \omega\sqrt{1 - \xi^2}$$

$$\omega_D = 27.78\sqrt{1 - (0.026)^2} = 27.77 \text{ rad/sec}$$

Example 2.2. A platform of weight W = 4000 lb is being supported by four equal columns which are clamped to the foundation as well as to the platform. Experimentally it has been determined that a static force of F = 1000 lb applied horizontally to the platform produces a displacement of Δ = 0.10 in. It is estimated that damping in the structures is of the order of 5% of the critical damping.

Determine for this structure the following: (a) undamped natural frequency, (b) absolute damping coefficient, (c) logarithmic decrement, and (d) the number of cycles and the time required for the amplitude of motion to be reduced from an initial value of 0.1 in to 0.01 in.

Solution: (a) The stiffness coefficient (force per unit displacement) is computed as

$$k = \frac{F}{\Delta} = \frac{1000}{0.1} = 10,000 \text{ lb/in}$$

and the undamped natural frequency

$$\omega = \sqrt{\frac{k}{W/g}} = \sqrt{\frac{10,000 \times 386}{4000}} = 31.06 \text{ rad/sec}$$

(b) The critical damping is

$$c_{cr} = 2\sqrt{km} = 2\sqrt{10,000 \times 4000/386} = 643.8 \frac{\text{lb} \cdot \text{sec}}{\text{in}}$$

and the absolute damping

$$c = \xi c_{cr} = 0.05 \times 643.8 = 32.19 \frac{\text{lb} \cdot \text{sec}}{\text{in}}$$

(c) Approximately, the logarithmic decrement is

$$\delta = \ln\left(\frac{y_0}{y_1}\right) \simeq 2\pi\xi = 2\pi(0.05) = 0.314$$

and the ratio of two consecutive amplitudes

$$\frac{y_0}{y_1} = 1.37$$

(d) The ratio between the first amplitude y_0 and the amplitude y_k after k cycles may be expressed as

$$\frac{y_0}{y_k} = \frac{y_0}{y_1} \cdot \frac{y_1}{y_2} \cdots \frac{y_{k-1}}{y_k}$$

Then taking the natural logarithm, we obtain

$$\ln \frac{y_0}{y_k} = \delta + \delta + \ldots + \delta = k\delta$$

$$\ln \frac{0.1}{0.01} = 0.314k$$

$$k = \frac{\ln 10}{0.314} = 7.33 \rightarrow 8 \text{ cycles}$$

The damped frequency ω_D is given by

$$\omega_D = \omega\sqrt{1 - \xi^2} = 31.06\sqrt{1 - (0.05)^2} = 31.02 \text{ rad/sec}$$

and the damped period T_D by

$$T_D = \frac{2\pi}{\omega_D} = \frac{2\pi}{31.02} = 0.2025 \text{ sec}$$

Then the time for eight cycles is

$$t(8 \text{ cycles}) = 8T_D = 1.62 \text{ sec}$$

2.7 SUMMARY

Real structures dissipate energy while undergoing vibratory motion. The most common and practical method for considering this dissipation of energy is to assume that it is due to viscous damping forces. These forces are assumed to be proportional to the magnitude of the velocity but acting in the direction opposite to the motion. The factor of proportionality is called the *viscous damping coefficient*. It is expedient to express this coefficient as a fraction of the *critical damping* in the system (the damping ratio $\xi = c/c_{cr}$). The critical damping may be defined as the least value of the damping coefficient for which the system will not oscillate when disturbed initially, but it simply will return to the equilibrium position.

The differential equation of motion for the free vibration of a damped single degree-of-freedom system is given by

$$m\ddot{y} + c\dot{y} + ky = 0$$

The analytical expression for the solution of this equation depends on the magnitude of the damping ratio. Three cases are possible: (1) critically damped system ($\xi = 1$), (2) underdamped system ($\xi < 1$), and (3) overdamped system

$(\xi > 1)$. For the underdamped system $(\xi < 1)$ the solution of the differential equation of motion may be written as

$$y(t) = e^{-\xi\omega t}\left[y_0 \cos \omega_D t + \frac{v_0 + y_0\, \xi\omega}{\omega_D} \sin \omega_D t \right]$$

in which

$$\omega = \sqrt{k/m} \text{ is the undamped frequency}$$

$$\omega_D = \omega\sqrt{1 - \xi^2} \text{ is the damped frequency}$$

$$\xi = c/c_{cr} \text{ is the damping ratio}$$

$$c_{cr} = 2\sqrt{km} \text{ is the critical damping}$$

and y_0 and v_0 are, respectively, the initial displacement and velocity.

A common method of determining the damping present in a system is to evaluate experimentally the logarithmic decrement, which is defined as the natural logarithm of the ratio of two consecutive peaks in free vibration, that is,

$$\delta = \ln \frac{y_1}{y_2}$$

The damping ratio in structural systems is usually less than 10% of the critical damping $(\xi < 0.1)$. For such systems, the damped frequency is approximately equal to the undamped frequency.

PROBLEMS

2.1 Repeat Problem 1.2 assuming that the system has 15% of critical damping.

2.2 Repeat Problem 1.6 assuming that the system has 10% of critical damping.

2.3 The amplitude of vibration of the system shown in Fig. 2.5 is observed to decrease 5% on each consecutive cycle of motion. Determine the damping coefficient c of the system. $k = 200$ lb/in and $m = 10$ lb·sec²/in.

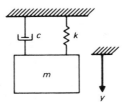

Fig. 2.5

2.4 It is observed experimentally that the amplitude of free vibration of a certain structure, modeled as a single degree-of-freedom system, decreases from 1 to 0.4 in 10 cycles. What is the percentage of critical damping?

2.5 Show that the displacement for critical and overcritical damped systems with initial displacement y_0 and velocity v_0 may be written as

$$y = e^{-\omega t} [y_0 (1 + \omega t) + v_0 t] \quad \text{for} \quad \xi = 1$$

$$y = e^{-\xi\omega t} \left[y_0 \cosh \omega_D' t + \frac{v_0 + y_0 \xi\omega}{\omega_D'} \sinh \omega_D' t \right] \quad \text{for} \quad \xi > 1$$

where $\omega_D' = \omega\sqrt{\xi^2 - 1}$

2.6 A structure is modeled as a damped oscillator with spring constant $k = 30$ Kips/in and undamped natural frequency $\omega = 25$ rad/sec. Experimentally it was found that a force 1 kip produced a relative velocity of 1.0 in/sec in the damping element. Find: (a) the damping ratio ξ, (b) the damped period T_D, (c) the logarithmic decrement δ, and (d) the ratio between two consecutive amplitudes.

2.7 In Fig. 2.4 it is indicated that the tangent points to the displacement curve correspond to $\cos(\omega_D t - \alpha) = 1$. Therefore the difference in $\omega_D t$ between any two consecutive tangent points is 2π. Show that the difference in $\omega_D t$ between any two consecutive peaks of the curve is also 2π.

2.8 Show that for an underdamped system in free vibration the logarithmic decrement may be written as

$$\delta = \frac{1}{k} \ln \frac{y_i}{y_{i+k}}$$

where k is the number of cycles separating two measured peak amplitudes y_i and y_{i+k}.

2.9 A single degree-of-freedom system consists of a mass with a weight of 386 lb and a spring of stiffness $k = 3000$ lb/in. By testing the system it was found that a force of 100 lb produces a relative velocity 12 in/sec. Find (a) the damping ratio ξ, (b) the damped frequency of vibration f_D, (c) logarithmic decrement δ, and (d) the ratio of two consecutive amplitudes.

2.10 Solve Problem 2.9 when the damping coefficient is $c = 2$ lb·sec/in.

2.11 A system is modeled by two freely vibrating masses m_1 and m_2 interconnected by a spring and a damper element as shown in Fig. 2.6. Determine for this system

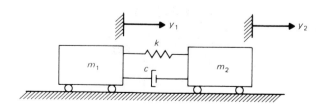

Fig. 2.6

the differential equation of motion in terms of the relative motion of the masses $u = y_2 - y_1$.

2.12 Determine the relative motion $u = y_2 - y_1$ for the system shown in Fig. 2.6 in terms of the natural frequency ω, damped frequency ω_D, and relative damping. Hint: Define equivalent mass as $M = m_1 m_2 / (m_1 + m_2)$.

3

Response of One-Degree-of-Freedom System to Harmonic Loading

In this chapter, we will study the motion of structures idealized as single degree-of-freedom systems excited harmonically, that is, structures subjected to forces or displacements whose magnitudes may be represented by a sine or cosine function of time. This type of excitation results in one of the most important motions in the study of mechanical vibrations as well as in applications to structural dynamics. Structures are very often subjected to the dynamic action of rotating machinery which produces harmonic excitations due to the unavoidable presence of mass eccentricities in the rotating parts of such machinery. Furthermore, even in those cases when the excitation is not a harmonic function, the response of the structure may be obtained using the *Fourier Method*, as the superposition of individual responses to the harmonic components of the external excitation. This approach will be dealt with in Chapter 5.

3.1 UNDAMPED SYSTEM : HARMONIC EXCITATION

The impressed force $F(t)$ acting on the simple oscillator in Fig. 3.1 is assumed to be harmonic and equal to $F_0 \sin \bar{\omega} t$, where F_0 is the peak amplitude and

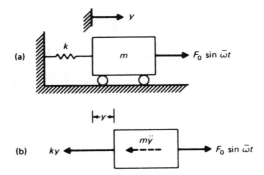

Fig. 3.1 (a) Undamped oscillator harmonically excited. (b) Free body diagram.

$\bar{\omega}$ is the frequency of the force in radians per second. The differential equation obtained by summing all the forces in the free body diagram of Fig. 3.1(b) is

$$m\ddot{y} + ky = F_0 \sin \bar{\omega}t \qquad (3.1)$$

The solution of eq. (3.1) can be expressed as

$$y(t) = y_c(t) + y_p(t) \qquad (3.2)$$

where $y_c(t)$ is the complementary solution satisfying the homogeneous equation, that is, eq. (3.1) with the left-hand side set equal to zero; and $y_p(t)$ is the particular solution based on the solution satisfying the nonhomogeneous differential equation (3.1). The complementary solution, $y_c(t)$, is given by eq. (1.17) as

$$y_c(t) = A \cos \omega t + B \sin \omega t \qquad (3.3)$$

where $\omega = \sqrt{k/m}$.

The nature of the forcing function in eq. (3.1) suggests that the particular solution be taken as

$$y_p(t) = Y \sin \bar{\omega}t \qquad (3.4)$$

where Y is the peak value of the particular solution. The substitution of eq. (3.4) into eq. (3.1) followed by cancellation of common factors gives

$$-m\bar{\omega}^2 Y + kY = F_0$$

or

$$Y = \frac{F_0}{k - m\bar{\omega}^2} = \frac{F_0/k}{1 - r^2} \qquad (3.5)$$

in which r represents the ratio (frequency ratio) of the applied forced frequency to the natural frequency of vibration of the system, that is,

$$r = \frac{\bar{\omega}}{\omega} \tag{3.6}$$

Combining eqs. (3.3) through (3.5) with eq. (3.2) yields

$$y(t) = A \cos \omega t + B \sin \omega t + \frac{F_0/k}{1 - r^2} \sin \bar{\omega} t \tag{3.7}$$

If the initial conditions at time $t = 0$ are taken as zero ($y_0 = 0$, $v_0 = 0$), the constants of integration determined from eq. (3.7) are

$$A = 0, \quad B = -\frac{rF_0/k}{1 - r^2}$$

which, upon substitution in eq. (3.7), gives

$$y(t) = \frac{F_0/k}{1 - r^2} (\sin \bar{\omega} t - r \sin \omega t) \tag{3.8}$$

As we can see from eq. (3.8), the response is given by the superposition of two harmonic terms of different frequencies. The resulting motion is not harmonic; however, in the practical case, damping forces will always be present in the system and will cause the last term, i.e., the free frequency term in eq. (3.8), to vanish eventually. For this reason, this term is said to represent the *transient response*. The forcing frequency term in eq. (3.8), namely

$$y(t) = \frac{F_0/k}{1 - r^2} \sin \bar{\omega} t \tag{3.9}$$

is referred to as the *steady-state response*. It is clear from eq. (3.8) that in the case of no damping in the system, the transient will not vanish and the response is then given by eq. (3.8). It can also be seen from eq. (3.8) or eq. (3.9) that when the forcing frequency is equal to the natural frequency ($r = 1.0$), the amplitude of the motion becomes infinitely large. A system acted upon by an external excitation of frequency coinciding with the natural frequency is said to be at *resonance*. In this circumstance, the amplitude will increase gradually to infinite. However, materials that are commonly used in practice are subjected to strength limitations and in actual structures failures occur long before extremely large amplitudes can be attained.

3.2 DAMPED SYSTEM: HARMONIC EXCITATION

Now consider the case of the one-degree-of-freedom system in Fig. 3.2 vibrating under the influence of viscous damping. The differential equation of motion is obtained by equating to zero the sum of the forces in the free body diagram of Fig. 3.2(b). Hence

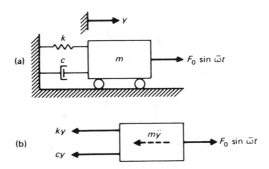

Fig. 3.2 (a) Damped oscillator harmonically excited. (b) Free body diagram.

$$m\ddot{y} + c\dot{y} + ky = F_0 \sin \bar{\omega}t \qquad (3.10)$$

The complete solution of this equation again consists of the complementary solution $y_c(t)$ and the particular solution $y_p(t)$. The complementary solution is given for the underdamped case $(c < c_{cr})$ by eqs. (2.15) and (2.19) as

$$y_c(t) = e^{-\xi \omega t}(A \cos \omega_D t + B \sin \omega_D t) \qquad (3.11)$$

The particular solution may be found by substituting y_p, in this case assumed to be of the form

$$y_p(t) = C_1 \sin \bar{\omega}t + C_2 \cos \bar{\omega}t \qquad (3.12)$$

into eq. (3.10) and equating the coefficients of the sine and cosine functions. Here we follow a more elegant approach using Euler's relation, namely

$$e^{i\bar{\omega}t} = \cos \bar{\omega}t + i \sin \bar{\omega}t$$

For this purpose, the reader should realize that we can write eq. (3.10) as

$$m\ddot{y} + c\dot{y} + ky = F_0 e^{i\bar{\omega}t} \qquad (3.13)$$

with the understanding that only the imaginary component of $F_0 e^{i\bar{\omega}t}$, i.e., the force component of $F_0 \sin \bar{\omega}t$, is acting and, consequently, the response will then consist only of the imaginary part of the total solution of eq. (3.13). In other words, we obtain the solution of eq. (3.13) which has real and imaginary components, and disregard the real component.

It is reasonable to expect that the particular solution of eq. (3.13) will be of the form

$$y_p = C e^{i\bar{\omega}t} \qquad (3.14)$$

Substitution of eq. (3.14) into eq. (3.13) and cancellation of the factor $e^{i\bar{\omega}t}$ gives

$$-m\bar{\omega}^2 C + ic\bar{\omega}C + kC = F_0$$

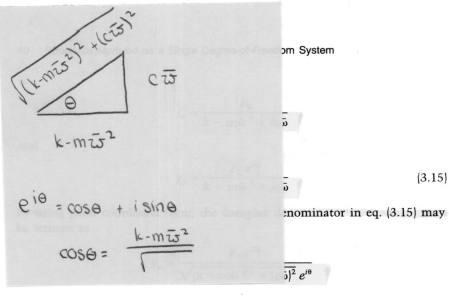

$$\sqrt{(k-m\bar{\omega}^2)^2+(c\bar{\omega})^2}$$

$$c\bar{\omega}$$

$$\theta$$

$$k-m\bar{\omega}^2$$

(3.15)

$$e^{i\theta}=\cos\theta+i\sin\theta$$

enominator in eq. (3.15) may

$$\cos\theta=\frac{k-m\bar{\omega}^2}{\sqrt{}}$$

$$)^2 e^{i\theta}$$

or

$$y_p=\frac{F_0\,e^{i(\bar{\omega}t-\theta)}}{\sqrt{(k-m\bar{\omega}^2)^2+(c\bar{\omega})^2}} \tag{3.16}$$

where

$$\tan\theta=\frac{c\bar{\omega}}{k-m\bar{\omega}^2} \tag{3.17}$$

The response to the force $F_0\sin\bar{\omega}t$ (the imaginary component of $F_0\,e^{i\bar{\omega}t}$) is then the imaginary component of eq. (3.16), namely,

$$y_p=\frac{F_0\sin(\bar{\omega}t-\theta)}{\sqrt{(k-m\bar{\omega}^2)^2+(c\bar{\omega})^2}} \tag{3.18}$$

or

$$y_p=Y\sin(\bar{\omega}t-\theta) \tag{3.19}$$

where

$$Y=\frac{F_0}{\sqrt{(k-m\bar{\omega}^2)^2+(c\bar{\omega})^2}}$$

is the amplitude of the steady-state motion. Equations (3.18) and (3.17) may conveniently be written in terms of dimensionless ratios as

$$y_p=\frac{y_{st}\sin(\bar{\omega}t-\theta)}{\sqrt{(1-r^2)^2+(2\xi r)^2}} \tag{3.20}$$

and

$$\tan \theta = \frac{2\xi r}{1 - r^2} \tag{3.21}$$

where $y_{st} = F_0/k$ is seen to be the static deflection of the spring acted upon by the force F_0; $\xi = c/c_{cr}$, the damping ratio; and $r = \bar{\omega}/\omega$, the frequency ratio. The total response is then obtained by summing the complementary solution (transient response) from eq. (3.11) and the particular solution (steady-state response) from eq. (3.20), that is,

$$y(t) = e^{-\xi\omega t}(A \cos \omega_D t + B \sin \omega_D t) + \frac{y_{st} \sin(\bar{\omega}t - \theta)}{\sqrt{(1 - r^2)^2 + (2r\xi)^2}} \tag{3.22}$$

The reader should be warned that the constants of integration A and B must be evaluated from initial conditions using the total response given by eq. (3.22) and not from just the transient component of the response given in eq. (3.11). By examining the transient component of response, it may be seen that the presence of the exponential factor $e^{-\xi\omega t}$ will cause this component to vanish, leaving only the steady-state motion which is given by eq. (3.20).

The ratio of the steady-state amplitude of $y_p(t)$ to the static deflection y_{st} defined above is known as the *dynamic magnification factor D*, and is given from eqs. (3.19) and (3.20) by

$$D = \frac{Y}{y_{st}} = \frac{1}{\sqrt{(1 - r^2)^2 + (2r\xi)^2}} \tag{3.23}$$

It may be seen from eq. (3.23) that the *dynamic magnification factor* varies with the frequency ratio r and the damping ratio ξ. Parametric plots of the dynamic magnification factor are shown in Fig. 3.3. The phase angle θ, given in eq. (3.21), also varies with the same quantities as it is shown in the plots of Fig. 3.4. We note in Fig. 3.3 that for a lightly damped system, the peak amplitude occurs at a frequency ratio very close to 1; that is, the dynamic magnification factor has its maximum value virtually at resonance ($r = 1$). It can also be seen from eq. (3.23) that at resonance the dynamic magnification factor is inversely proportional to the damping ratio, that is,

$$D(r = 1) = \frac{1}{2\xi} \tag{3.24}$$

Although the dynamic magnification factor evaluated at resonance is close to its maximum value, it is not exactly the maximum response for a damped system. However, for moderate amounts of damping, the difference between the approximate value of eq. (3.24) and the exact maximum is negligible.

Example 3.1. A simple beam supports at its center a machine having a weight $W = 16,000$ lb. The beam is made of two standard S8 × 23 sections

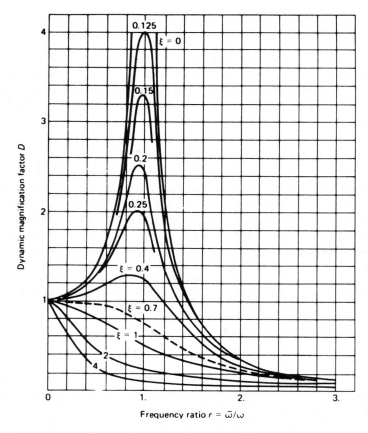

Fig. 3.3 Dynamic magnification factor as a function of the frequency ratio for various amounts of damping.

with a clear span $L = 12$ ft and total cross-sectional moment of inertia $I = 2 \times 64.2 = 128.4$ in^4. The motor runs at 300 rpm, and its rotor is out of balance to the extent of $W' = 40$ lb at a radius of $e_0 = 10$ in. What will be the amplitude of the steady-state response if the equivalent viscous damping for the system is assumed 10% of the critical?

Solution: This dynamic system may be modeled by the damped oscillator. The distributed mass of the beam will be neglected in comparison with the large mass of the machine. Figures 3.5 and 3.6 show, respectively, the schematic diagram of a beam-machine system and the adapted model. The force at the center of a simply supported beam necessary to deflect this point one unit (i.e., the stiffness coefficient) is given by the formula

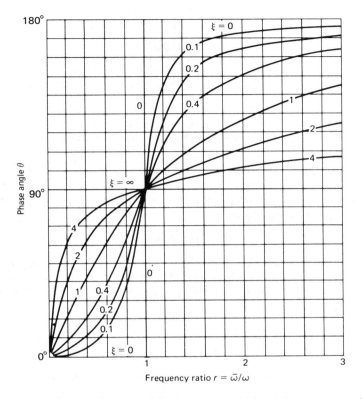

Fig. 3.4 Phase angle θ as a function of the frequency ratio for various amounts of damping.

$$k = \frac{48EI}{L^3} = \frac{48 \times 30 \times 10^6 \times 128.4}{(144)^3} = 61{,}920 \text{ lb/in}$$

The natural frequency of the system (neglecting the mass of the beam) is

$$\omega = \sqrt{\frac{k}{m}} = \sqrt{\frac{61{,}920}{16{,}000/386}} = 38.65 \text{ rad/sec}$$

the force frequency

$$\bar{\omega} = \frac{300 \times 2\pi}{60} = 31.41 \text{ rad/sec}$$

and the frequency ratio

$$r = \frac{\bar{\omega}}{\omega} = \frac{31.41}{38.65} = 0.813$$

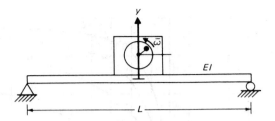

Fig. 3.5 Diagram for beam-machine system of Example 3.1.

Referring to Fig. 3.6, let m be the total mass of the motor and m' the unbalanced rotating mass. Then, if y is the vertical displacement of the non-rotating mass $(m - m')$ from the equilibrium position, the displacement y_1 of m' as shown in Fig. 3.6 is

$$y_1 = y + e_0 \sin \bar{\omega}t \qquad (a)$$

The equation of motion is then obtained by summing forces along the vertical direction in the free body diagram of Fig. 3.6(b), where the inertial forces of both the nonrotating mass and the unbalanced mass are also shown. This summation yields

$$(m - m')\ddot{y} + m'\ddot{y}_1 + c\dot{y} + ky = 0 \qquad (b)$$

Substitution of y_1 from eq. (a) gives

$$(m - m')\ddot{y} + m'(\ddot{y} - e_0\bar{\omega}^2 \sin \bar{\omega}t) + c\dot{y} + ky = 0$$

and with a rearrangement of terms

$$m\ddot{y} + c\dot{y} + ky = m'e_0\bar{\omega}^2 \sin \bar{\omega}t \qquad (c)$$

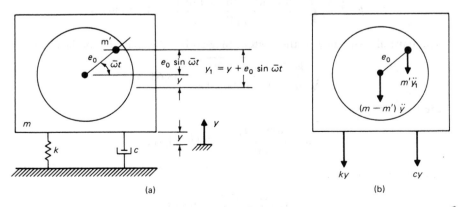

Fig. 3.6 (a) Mathematical model for Example 3.1. (b) Free body diagram.

This last equation is of the same form as the equation of motion (3.10) for the damped oscillator excited harmonically by a force of amplitude

$$F_0 = m'e_0\bar{\omega}^2 \tag{d}$$

Substituting in this equation the numerical values for this example, we obtain

$$F_0 = (40)(10)(31.41)^2/386 = 1022 \text{ lb}$$

The amplitude of the steady-state resulting motion from eq. (3.19) is then

$$Y = \frac{1022/61,920}{\sqrt{(1 - 0.813^2)^2 + (2 \times 0.813 \times 0.1)^2}}$$

$$Y = 0.044 \text{ in} \tag{Ans.}$$

Example 3.2.   The steel frame shown in Fig. 3.7 supports a rotating machine which exerts a horizontal force at the girder level, $F(t) = 200 \sin 5.3t$ lb. Assuming 5% of critical damping, determine: (a) the steady-state amplitude of vibration and (b) the maximum dynamic stress in the columns. Assume that the girder is rigid. ✱ homework

Solution: This structure may be modeled for dynamic analysis as the damped oscillator shown in Fig. 3.7(b). The parameters in this model are computed as follows:

$$\text{(a)} \quad k^* = \frac{3E(2I)}{L^3} = \frac{3 \times 30 \times 10^6 \times 2 \times 69.2}{(12 \times 15)^3} = 2136 \text{ lb/in}$$

$$\xi = 0.05$$

$$y_{st} = \frac{F_0}{k} = \frac{200}{2136} = 0.0936 \text{ in}$$

$$\omega = \sqrt{\frac{k}{m}} = \sqrt{\frac{2136 \times 386}{15,000}} = 7.41 \text{ rad/sec}$$

$$r = \frac{\bar{\omega}}{\omega} = \frac{5.3}{7.41} = 0.715$$

The steady-state amplitude from eqs. (3.19) and (3.20) is

$$Y = \frac{y_{st}}{\sqrt{(1 - r^2)^2 + (2r\xi)^2}} = 0.189 \text{ in} \tag{Ans.}$$

---

*A unit displacement at the top of a pinned supported column requires a force equal to $3EI/L^3$.

$$\bar{F} = kx \quad \rightarrow \quad k = \frac{F}{x} \quad \text{where} \quad x = 1$$

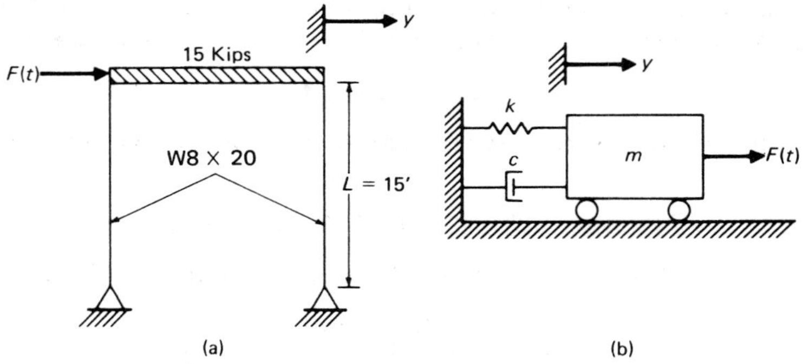

Fig. 3.7 (a) Diagram of frame for Example 3.2. (b) Mathematical model.

(b) Then, the maximum shear force in the columns is

$$V_{max} = \frac{3EIY}{L^3} = 201.8 \text{ lb}$$

the maximum bending moment

$$M_{max} = V_{max}L = 36,324 \text{ lb} \cdot \text{in}$$

and the maximum stress

$$\sigma_{max} = \frac{M_{max}}{I/c} = \frac{36,324}{17} = 2136 \text{ psi} \qquad \text{(Ans.)}$$

in which I/c is the section modulus.

3.3 EVALUATION OF DAMPING AT RESONANCE

We have seen in Chapter 2 that the free-vibration decay curve permits the evaluation of damping of a single degree-of-freedom system by simply calculating the logarithm decrement as shown in eq. (2.28). Another technique for determining damping is based on observations of steady-state harmonic response, which requires harmonic excitations of the structure in a range of frequencies in the neighborhood of resonance. With the application of a harmonic force $F_0 \sin \bar{\omega}t$ at closely spaced values of frequencies, the response curve for the structure can be plotted, resulting in displacement amplitudes as a function of the applied frequencies. A typical response curve for such a moderately damped structure is shown in Fig. 3.8. It is seen from eq. (3.24) that, at resonance, the damping ratio is given by

$$\xi = \frac{1}{2D(r = 1)} \qquad (3.25)$$

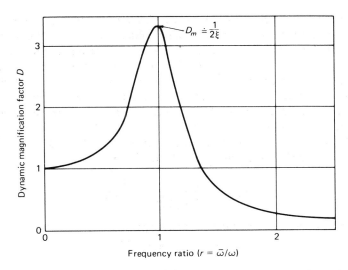

Fig. 3.8 Frequency response curve for moderately damped system.

where $D(r = 1)$ is the dynamic magnification factor evaluated at resonance. In practice, the damping ratio ξ is determined from the dynamic magnification factor evaluated at the maximum amplitude, namely,

$$\xi = \frac{1}{2D_m} \qquad (3.26)$$

where

$$D_m = \frac{Y_m}{y_{st}}$$

and Y_m is the maximum amplitude. The error involved in evaluating the damping ratio ξ using the approximate eq. (3.26) is not significant in ordinary structures. This method of determining the damping ratio requires only some simple equipment to vibrate the structure in a range of frequencies that span the resonance frequency and a transducer for measuring amplitudes; nevertheless, the evaluation of the static displacement $y_{st} = F_0/k$ may present a problem since, frequently, it is difficult to apply a static lateral load to the structure.

3.4 BANDWIDTH METHOD (HALF-POWER) TO EVALUATE DAMPING

An examination of the response curves in Fig. 3.3 shows that the shape of these curves is controlled by the amount of damping present in the system;

in particular, the *bandwidth*, that is, the difference between two frequencies corresponding to the same response amplitude, is related to the damping in the system. A typical frequency amplitude curve obtained experimentally for a moderately damped structure is shown in Fig. 3.9. In the evaluation of damping, it is convenient to measure the bandwidth at $1/\sqrt{2}$ of the peak amplitude as shown in this figure. The frequencies corresponding in this bandwidth f_1 and f_2 are also referred to as *half-power* points and are shown in Fig. 3.9. The values of the frequencies for this bandwidth can be determined by setting the response amplitude in eq. (3.20) equal to $1/\sqrt{2}$ times the resonant amplitude given by eq. (3.24), that is,

$$\frac{y_{st}}{\sqrt{(1-r^2)^2 + (2r\xi)^2}} = \frac{1}{\sqrt{2}}\frac{y_{st}}{2\xi}$$

Squaring both sides and solving for the frequency ratio results in

$$r^2 = 1 - 2\xi^2 \pm 2\xi\sqrt{1 + \xi^2}$$

or by neglecting ξ^2 in the square root terms

$$r_1^2 \simeq 1 - 2\xi^2 - 2\xi$$

$$r_2^2 \simeq 1 - 2\xi^2 + 2\xi$$

$$r_1 \simeq 1 - \xi - \xi^2$$

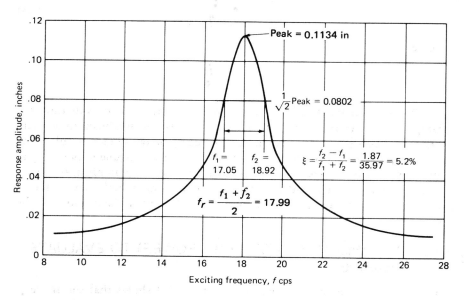

Fig. 3.9 Experimental frequency response curve of Example 3.3.

$$r_2 \approx 1 + \xi - \xi^2$$

Finally, the damping ratio is given approximately by half the difference between these half-power frequency ratios, namely

$$\xi = \frac{1}{2}(r_2 - r_1)$$

or

$$\xi = \frac{1}{2}\frac{\bar{\omega}_2 - \bar{\omega}_1}{\omega} = \frac{f_2 - f_1}{f_2 + f_1} \qquad (3.27)$$

since

$$r = \frac{\bar{\omega}}{\omega} = \frac{\bar{f}}{f} \quad \text{and} \quad \bar{f} \approx \frac{f_1 + f_2}{2}$$

Example 3.3. Experimental data for the frequency response of a single degree-of-freedom system are plotted in Fig. 3.9. Determine the damping ratio of this system.

Solution: From Fig. 3.9 the peak amplitude is 0.1134 in; hence the amplitude at half-power is equal to

$$0.1134/\sqrt{2} = 0.0802 \text{ in}$$

The frequencies at this amplitude obtained from Fig. 3.9 are

$$f_1 = 17.05$$

$$f_2 = 18.92$$

The damping ratio is then calculated from eq. (3.27) as

$$\xi \approx \frac{f_2 - f_1}{f_2 + f_1}$$

$$\xi \approx \frac{18.92 - 17.05}{18.92 + 17.05} = 5.2\% \qquad \text{(Ans.)}$$

3.5 RESPONSE TO SUPPORT MOTION

There are many actual cases where the foundation or support of a structure is subjected to time varying motion. Structures subjected to ground motion by earthquakes or other excitations such as explosions or dynamic action of machinery are examples in which support motions may have to be considered in the analysis of dynamic response. Let us consider in Fig. 3.10 the case

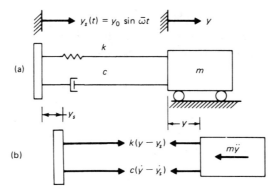

Fig. 3.10 (a) Damped simple oscillator harmonically excited through its support. (b) Free body diagram including inertial force.

where the support of the simple oscillator modeling the structure is subjected to a harmonic motion given by the expression

$$y_s(t) = y_0 \sin \bar{\omega} t \tag{3.28}$$

where y_0 is the maximum amplitude and $\bar{\omega}$ is the frequency of the support motion. The differential equation of motion is obtained by setting equal to zero the sum of the forces (including the inertial force) in the corresponding free body diagram shown in Fig. 3.10(b). The summation of the forces in the horizontal direction gives

$$m\ddot{y} + c(\dot{y} - \dot{y}_s) + k(y - y_s) = 0 \tag{3.29}$$

The substitution of eq. (3.28) into eq. (3.29) and the rearrangement of terms result in

$$m\ddot{y} + c\dot{y} + ky = ky_0 \sin \bar{\omega} t + c\bar{\omega} y_0 \cos \bar{\omega} t \tag{3.30}$$

The two harmonic terms of frequency $\bar{\omega}$ in the right-hand side of this equation may be combined and eq. (3.30) rewritten as [similarly to eqs. (1.20) and (1.23)]

$$m\ddot{y} + c\dot{y} + ky = F_0 \sin (\bar{\omega} t + \beta) \tag{3.31}$$

where

$$F_0 = y_0 \sqrt{k^2 + (c\bar{\omega})^2} = y_0 k \sqrt{1 + (2r\xi)^2} \tag{3.32}$$

and

$$\tan \beta = c\bar{\omega}/k = 2r\xi \tag{3.33}$$

It is apparent that eq. (3.31) is the differential equation for the oscillator excited by the harmonic force $F_0 \sin(\bar{\omega}t + \beta)$ and is of the same form as eq. (3.10). Consequently, the steady-state solution of eq. (3.31) is given as before by eq. (3.20), except for the addition of the angle β in the argument of the sine function, that is,

$$y(t) = \frac{F_0/k \sin(\bar{\omega}t + \beta - \theta)}{\sqrt{(1 - r^2)^2 + (2r\xi)^2}}$$ (3.34)

or substituting F_0 from eq. (3.32)

$$\frac{y(t)}{y_0} = \frac{\sqrt{1 + (2r\xi)^2}}{\sqrt{(1 - r^2)^2 + (2r\xi)^2}} \sin(\bar{\omega}t + \beta - \theta)$$ (3.35)

Equation (3.35) is the expression for the relative transmission of the support motion to the oscillator. This is an important problem in vibration isolation in which equipment must be protected from harmful vibrations of the supporting structure. The degree of relative isolation is known as *transmissibility* and is defined as the ratio of the amplitude of motion Y of the oscillator to the amplitude y_0, the motion of the support. From eq. (3.35), transmissibility T_r is then given by

$$T_r = \frac{Y}{y_0} = \frac{\sqrt{1 + (2r\xi)^2}}{\sqrt{(1 - r^2)^2 + (2r\xi)^2}}$$ (3.36)

A plot of transmissibility as a function of the frequency ratio and damping ratio is shown in Fig. 3.11. The curves in this figure are similar to curves in Fig. 3.3, representing the frequency response of the damped oscillator, the major difference being that all of the curves in Fig. 3.11 pass through the same point at a frequency ratio $r = \sqrt{2}$. It can be seen in Fig. 3.11 that damping tends to reduce the effectiveness of vibration isolation for frequencies greater than this ratio, that is, for r greater than $\sqrt{2}$.

Equation (3.34) provides the absolute response of the damped oscillator to a harmonic motion of its base. Alternatively, we can solve the differential equation (3.29) in terms of the relative motion between the mass m and the support given by

$$u = y - y_s$$ (3.37)

which substituted into eq. (3.29) gives

$$m\ddot{u} + c\dot{u} + ku = F_{eff}(t)$$ (3.38)

where $F_{eff}(t) = -m\ddot{y}_s$ may be interpreted as the effective force acting on the mass of the oscillator, and its displacement is indicated by coordinate u. Using eq. (3.28) to obtain $\ddot{y}_s$ and substituting in eq. (3.38) results in

$$m\ddot{u} + c\dot{u} + ku = my_0\bar{\omega}^2 \sin \bar{\omega}t$$ (3.39)

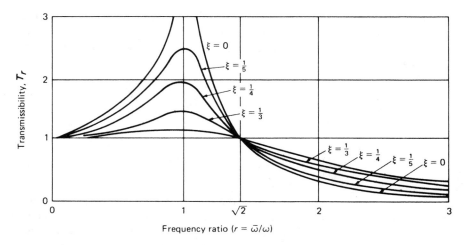

Fig. 3.11 Transmissibility versus frequency ratio for vibration isolation.

Again, eq. (3.39) is of the same form as eq. (3.10) with $F_0 = my_0\bar{\omega}^2$. Then, from eq. (3.20), the steady-state response in terms of relative motion is given by

$$u(t) = \frac{my_0\bar{\omega}^2/k \, \sin{(\bar{\omega}t - \theta)}}{\sqrt{(1 - r^2)^2 + (2r\xi)^2}} \qquad (3.40)$$

or substituting

$$\frac{\bar{\omega}^2}{k/m} = \frac{\bar{\omega}^2}{\omega^2} = r^2$$

we obtain

$$\frac{u(t)}{y_0} = \frac{r^2 \, \sin{(\bar{\omega}t - \theta)}}{\sqrt{(1 - r^2)^2 + (2r\xi)^2}} \qquad (3.41)$$

where θ is given in eq. (3.21).

Example 3.4. If the frame of Example 3.2 (Fig. 3.7) is subjected to a sinusoidal ground motion $y_s(t) = 0.2 \sin 5.3t$, determine: (a) the transmissibility of motion to the girder, (b) the maximum shearing force in the supporting columns, and (c) maximum stresses in the columns.

Solution: (a) The parameters for this system are calculated in Example 3.2 as

$$k = 2136 \text{ lb/in}$$

$$\xi = 0.05$$

$$y_0 = 0.2 \text{ in}$$

$$y_{st} = 0.0936 \text{ in}$$

$$\omega = 7.41 \text{ rad/sec}$$

$$\bar{\omega} = 5.3 \text{ rad/sec}$$

$$r = 0.715$$

The transmissibility from eq. (3.36) is

$$T_r = \sqrt{\frac{1 + (2r\xi)^2}{(1 - r^2)^2 + (2r\xi)^2}} = 2.1 \qquad \text{(Ans.)}$$

(b) The maximum relative displacement U from eq. (3.41) is

$$U = \frac{y_0 r^2}{\sqrt{(1 - r^2)^2 + (2r\xi)^2}} = 0.206 \text{ in}$$

Then the maximum shear force in each column is

$$V_{max} = \frac{kU}{2} \; 219.8 \text{ lb} \qquad \text{(Ans.)}$$

(c) The maximum bending moment

$$M_{max} = V_{max}L = 39{,}567 \text{ lb} \cdot \text{in}$$

and the corresponding stress

$$\sigma_{max} = \frac{M_{max}}{I/c} = \frac{39{,}567}{17} = 2327 \text{ psi} \qquad \text{(Ans.)}$$

in which I/c is the section modulus.

3.6 FORCE TRANSMITTED TO THE FOUNDATION

In the preceding section, we determined the response of the structure to a harmonic motion of its foundation. In this section we shall consider a similar problem of vibration isolation; the problem now, however, is to find the force transmitted to the foundation. Consider again the damped oscillator with a harmonic force $F(t) = F_0 \sin \bar{\omega} t$ acting on its mass as shown in Fig. 3.2. The differential equation of motion is

$$m\ddot{y} + c\dot{y} + ky = F_0 \sin \bar{\omega} t$$

with the steady-state solution, eq. (3.20),

$$y = Y \sin(\bar{\omega}t - \theta)$$

where

$$Y = \frac{F_0/k}{\sqrt{(1 - r^2)^2 + (2r\xi)^2}} \qquad (3.42)$$

and

$$\tan \theta = \frac{2\xi r}{1 - r^2}$$

The force transmitted to the support through the spring is ky and through the damping element is $c\dot{y}$. Hence the total force transmitted F_T is

$$F_T = ky + c\dot{y} \qquad (3.43)$$

Differentiating eq. (3.19) and substituting in eq. (3.43) yield

$$F_T = Y[k \sin(\bar{\omega}t - \theta) + c\bar{\omega} \cos(\bar{\omega}t - \theta)]$$

or

$$F_T = Y\sqrt{k^2 + c^2\bar{\omega}^2} \sin(\bar{\omega}t - \theta + \beta) \qquad (3.44)$$

$$F_T = Y\sqrt{k^2 + c^2\bar{\omega}^2} \sin(\bar{\omega}t - \phi) \qquad (3.45)$$

in which

$$\tan \beta = \frac{c\bar{\omega}}{k} = 2\xi r \qquad (3.46)$$

and

$$\phi = \theta - \beta \qquad (3.47)$$

Then from eqs. (3.42) and (3.45), the maximum force A_T transmitted to the foundation is

$$A_T = F_0 \sqrt{\frac{1 + (2\xi r)^2}{(1 - r^2)^2 + (2\xi r)^2}} \qquad (3.48)$$

In this case, the transmissibility T_r is defined as the ratio between the amplitude of the force transmitted to the foundation and the amplitude of the applied force. Hence from eq. (3.48)

$$T_r = \frac{A_T}{F_0} = \sqrt{\frac{1 + (2\xi r)^2}{(1 - r^2)^2 + (2r\xi)^2}} \qquad (3.49)$$

It is interesting to note that both the transmissibility of motion from the foundation to the structure, eq. (3.36), and the transmissibility of force from the structure to the foundation, eq. (3.49), are given by exactly the same function. Hence the curves of transmissibility in Fig. 3.11 represent either type of transmissibility. An expression for the total phase angle ϕ in eq. (3.45) may be determined by taking the tangent function to both members of eq. (3.47), so that

$$\tan \phi = \frac{\tan \theta - \tan \beta}{1 + \tan \theta \tan \beta}$$

Then substituting $\tan \theta$ and $\tan \beta$, respectively, from eqs. (3.21) and (3.46), we obtain

$$\tan \phi = \frac{2\xi r^3}{1 - r^2 + 4\xi^2 r^2} \tag{3.50}$$

Example 3.5. A machine of weight $W = 3860$ lb is mounted on a simple supported steel beam as shown in Fig. 3.12. A piston that moves up and down in the machine produces a harmonic force of magnitude $F_0 = 7000$ lb and frequency $\bar{\omega} = 60$ rad/sec. Neglecting the weight of the beam and assuming 10% of the critical damping, determine: (a) the amplitude of the motion of the machine, (b) the force transmitted to the beam supports, and (c) the corresponding phase angle.

Solution: The damped oscillator in Fig. 3.12(b) is used to model the system. The following parameters are calculated:

$$k = \frac{48EI}{L^3} = 10^5 \text{ lb/in}$$

$$\omega = \sqrt{\frac{k}{m}} = 100 \text{ rad/sec}$$

$$\xi = 0.1$$

$$r = \frac{\bar{\omega}}{\omega} = 0.6$$

$$y_{st} = \frac{F_0}{k} = 0.07 \text{ in}$$

(a) From eq. (3.19), the amplitude of motion is

$$Y = \frac{y_{st}}{\sqrt{(1 - r^2)^2 + (2r\xi)^2}} = 0.1075 \text{ in} \qquad \text{(Ans.)}$$

with a phase angle from eq. (3.21)

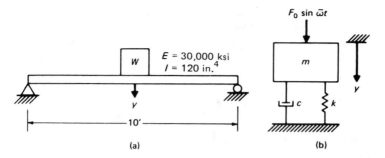

Fig. 3.12 (a) Beam-machine system for Example 3.5. (b) Mathematical model.

$$\theta = \tan^{-1} \frac{2r\xi}{1 - r^2} = 10.6°$$

(b) From eq. (3.49), the transmissibility is

$$T_r = \frac{A_T}{F_0} = \sqrt{\frac{1 + (2r\xi)^2}{(1 - r^2)^2 + (2r\xi)^2}} = 1.547$$

Hence the amplitude of the force transmitted to the foundation is

$$A_T = F_0 T_r = 10,827 \text{ lb} \tag{Ans.}$$

(c) The corresponding phase angle from eq. (3.50) is

$$\phi = \tan^{-1} \frac{2\xi r^3}{1 - r^2 + (2r\xi)^2} = 3.78° \tag{Ans.}$$

3.7   SEISMIC INSTRUMENTS

When a system of the type shown in Fig. 3.13 is used for the purpose of vibration measurement, the relative displacement between the mass and the base is ordinarily recorded. Such an instrument is called a *seismograph* and it can be designed to measure either the displacement or the acceleration of the base. The peak relative response U/y_0 of the seismograph depicted in Fig. 3.13, for harmonic motion of the base, is given from eq. (3.41) by

$$\frac{U}{y_0} = \frac{r^2}{\sqrt{(1 - r^2)^2 + (2r\xi)^2}} \tag{3.51}$$

A plot of this equation as a function of the frequency ratio and damping ratio is shown in Fig. 3.14. It may be seen from this figure that the response is essentially constant for frequency ratios $r > 1$ and damping ratio $\xi = 0.5$. Consequently, the response of a properly damped instrument of this type is

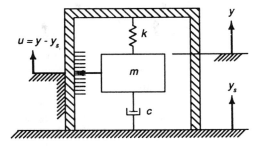

Fig. 3.13 Model of seismograph.

essentially proportional to the base-displacement amplitude for high frequencies of motion of the base. The instrument will thus serve as a displacement meter for measuring such motions. The range of applicability of the instrument is increased by reducing the natural frequency, i.e., by reducing the spring stiffness or increasing the mass.

Now consider the response of the same instrument to a harmonic acceleration of the base $\ddot{y}_s = y_0 \sin \bar{\omega}t$. The equation of motion of this system is obtained from eq. (3.38) as

$$m\ddot{u} + c\dot{u} + ku = -m\ddot{y}_0 \sin \bar{\omega}t \qquad (3.52)$$

The steady-state response of this system expressed as the dynamic magnification factor is then given from eq. (3.23) by

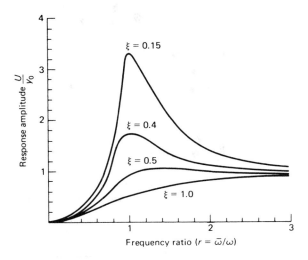

Fig. 3.14 Response of seismograph to harmonic motion of the base.

$$D = \frac{U}{m\ddot{y}_0/k} = \frac{1}{\sqrt{(1-r^2)^2 + (2r\xi)^2}} \tag{3.53}$$

This equation is represented graphically in Fig. 3.3. In this case, it can be seen from this figure that for a damping ratio $\xi = 0.7$, the value of the response is nearly constant in the frequency range $0 < r < 0.6$. Thus it is clear from eq. (3.53) that the response indicated by this instrument will be directly proportional to the base-acceleration amplitude for frequencies up to about six-tenths of the natural frequency. Its range of applicability will be increased by increasing the natural frequency, that is, by increasing the stiffness of the spring or by decreasing the mass of the oscillator. Such instrument is an accelerometer.

3.8 SUMMARY

In this chapter, we have determined the response of a single degree-of-freedom system subjected to harmonic loading. This type of loading is expressed as a sine, cosine, or exponential function and can be handled mathematically with minimum difficulty for the undamped or damped structure. The differential equation of motion for a linear single degree-of-freedom system is the second-order differential equation

$$m\ddot{y} + c\dot{y} + ky = F_0 \sin \bar{\omega}t \tag{3.10}$$

or

$$\ddot{y} + 2\xi\omega\dot{y} + \omega^2 y = \frac{F_0}{m} \sin \bar{\omega}t$$

in which $\bar{\omega}$ is the forced frequency,

$$\xi = \frac{c}{c_{cr}} \quad \text{is the damping ratio}$$

and

$$\omega = \sqrt{\frac{k}{m}} \quad \text{is the natural frequency}$$

The general solution of eq. (3.10) is obtained as the summation of the complementary (transient) and the particular (steady-state) solutions, namely

$$y = \underbrace{e^{-\xi\omega t}(A \cos \omega_D t + B \sin \omega_D t)}_{\text{transient solution}} + \underbrace{\frac{F_0/k \sin (\bar{\omega}t - \theta)}{\sqrt{(1-r^2)^2 + (2r\xi)^2}}}_{\text{steady-state solution}}$$

in which A and B are constants of integration,

$$r = \frac{\bar{\omega}}{\omega} \quad \text{is the frequency ratio}$$

$$\omega_D = \omega\sqrt{1 - r^2} \quad \text{is the damped natural frequency}$$

and

$$\theta = \tan^{-1}\left(\frac{2r\xi}{1 - r^2}\right) \text{ is the phase angle}$$

The transient part of the solution vanishes rapidly to zero because of the negative exponential factor leaving only the steady-state solution. Of particular significance is the condition of resonance $(r = \bar{\omega}/\omega = 1)$ for which the amplitudes of motion become very large for the damped system and tend to become infinity for the undamped system.

The response of the structure to support or foundation motion can be obtained in terms of the absolute motion of the mass or of its relative motion with respect to the support. In this latter case, the equation assumes a much simpler and more convenient form, namely

$$m\ddot{u} + c\dot{u} + ku = F_{\text{eff}}(t) \tag{3.38}$$

in which

$$F_{\text{eff}}(t) = -m\ddot{y}_s(t) \quad \text{is the effective force}$$

and

$$u = y - y_s \quad \text{is the relative displacement}$$

For harmonic excitation of the foundation, the solution of eq. (3.38) in terms of the relative motion is of the same form as the solution of eq. (3.10) in which the force is acting on the mass.

In this chapter, we have also shown that the damping in the system may be evaluated experimentally either from the peak amplitude or from the bandwidth obtained from a plot of the amplitude-frequency curve when the system is forced to harmonic vibration. Two related problems of vibrating isolation were discussed in this chapter: (1) the motion transmissibility, that is, the relative motion transmitted from the foundation to the structure; and (2) the force transmissibility which is the relative magnitude of the force transmitted from the structure to the foundation. For both of these problems, the transmissibility is given by

$$T_r = \sqrt{\frac{1 + (2r\xi)^2}{(1 - r^2)^2 + (2r\xi)^2}}$$

PROBLEMS

3.1 An electric motor of total weight W = 1000 lb is mounted at the center of a simply supported beam as shown in Fig. 3.15. The unbalance in the rotor is $W'e$ = 1 lb·in. Determine the steady-state amplitude of vertical motion of the motor for a speed of 900 rpm. Assume that the damping in the system is 10% of the critical damping. Neglect the mass of the supporting beam.

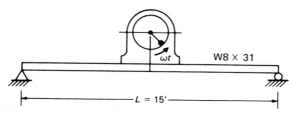

Fig. 3.15

3.2 Determine the maximum force transmitted to the supports of the beam in Problem 3.1.

3.3 Determine the steady-state amplitude for the horizontal motion of the steel frame in Fig. 3.16. Assume the horizontal girder to be infinitely rigid and neglect both the mass of the columns and damping.

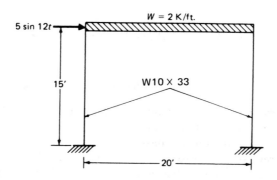

Fig. 3.16

3.4 Solve for Problem 3.3 assuming that the damping in the system is 8% of the critical damping.

3.5 For Problem 3.4 determine: (a) the maximum force transmitted to the foundation and (b) the transmissibility.

3.6 A delicate instrument is to be spring mounted to the floor of a test laboratory where it has been determined that the floor vibrates vertically with harmonic motion of amplitude 0.1 in at 10 cps. If the instrument weighs 100 lb, determine the stiffness of the isolation springs required to reduce the vertical motion amplitude of the instrument to 0.01 in. Neglect damping.

3.7   Consider the water tower shown in Fig. 3.17 which is subjected to ground motion
produced by a passing train in the vicinity of the tower. The ground motion is
idealized as a harmonic acceleration of the foundation of the tower with an amplitude of 0.1 g at a frequency of 10 cps. Determine the motion of the tower
relative to the motion of its foundation. Assume an effective damping coefficient
of 10% of the critical damping in the system.

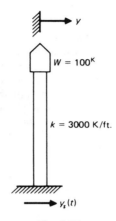

$W = 100^K$

$k = 3000 \ K/ft.$

$y_s(t)$

Fig. 3.17

3.8   Determine the transmissibility in Problem 3.7.

3.9   An electric motor of total weight $W = 3330$ lb is mounted on a simple supported
beam with overhang as shown in Fig. 3.18. The unbalance of the rotor is $W'e =
50$ lb · in. (a) Find the amplitudes of forced vertical vibration of the motor for
speeds 800, 1000, and 1200 rpm. (b) Draw a rough plot of the amplitude versus
rpm. Assume damping equal to 10% of the critical damping.

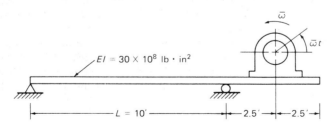

$EI = 30 \times 10^8 \ lb \cdot in^2$

$\bar{\omega}$

$\bar{\omega} t$

$L = 10'$ $2.5'$ $2.5'$

Fig. 3.18

3.10   A machine of mass m rests on an elastic floor as shown in Fig. 3.19. In order to
find the natural frequency of the vertical motion, a mechanical shaker of mass
m_s is bolted to the machine and run at various speeds until the resonant frequency f_r is found. Determine the natural frequency f_n of the floor-machine system in terms of f_r and the given data.

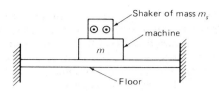

Fig. 3.19

3.11 Determine the frequency at which the peak amplitude of a damped oscillator will occur. Also, determine the peak amplitude and corresponding phase angle.

3.12 A structure modeled as a damped spring-mass system (Fig. 3.20) with $mg = 2520$ lb, $k = 89,000$ lb/in, and $c = 112$ lb·sec/in is subjected to a harmonic exciting force. Determine: (a) the natural frequency, (b) the damping ratio, (c) the amplitude of the exciting force when the peak amplitude of the vibrating mass is measured to be 0.37 in, and (d) the amplitude of the exciting force when the amplitude measured is at the peak frequency assumed to be the resonant frequency.

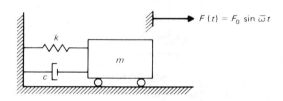

Fig. 3.20

3.13 A structural system modeled as a damped oscillator is subjected to the harmonic excitation produced by an eccentric rotor. The spring constant k and the mass m are known but not the damping and the amount of unbalance in the rotor. From measured amplitudes Y_r at resonance and Y_1 at a frequency ratio $r_1 \neq 1$, determine expressions to calculate the damping ratio ξ and the amplitude of the exciting force F_r at resonance.

3.14 A system is modeled by two vibrating masses m_1 and m_2 interconnected by a spring k and damper element c (Fig. 3.21). For harmonic force $F = F_0 \sin \bar{\omega}t$ acting on mass m_2 determine: (a) equation of motion in terms of the relative motion of the two masses, $u = y_2 - y_1$; (b) the steady-state solution of the relative motion.

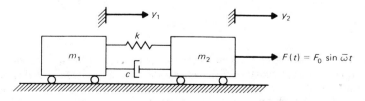

Fig. 3.21

4

Response to General Dynamic Loading

In the preceding chapter we studied the response of a single degree-of-freedom system with harmonic loading. Though this type of loading is important, real structures are often subjected to loads which are not harmonic. In the present chapter we shall study the response of the single degree-of-freedom system to a general type of force. We shall see that the response can be obtained in terms of an integral which for some simple load functions can be evaluated analytically. For the general case, however, it will be necessary to resort to a numerical integration procedure.

4.1 IMPULSIVE LOADING AND DUHAMEL'S INTEGRAL

An impulsive loading is a load which is applied during a short duration of time. The corresponding impulse of this type of load is defined as the product of the force and the time of its duration. For example, the impulse of the force $F(\tau)$ depicted in Fig. 4.1 at time τ during the interval $d\tau$ is represented by the shaded area and it is equal to $F(\tau)\, d\tau$. This impulse acting on a body of mass m produces a change in velocity which can be determined from Newton's Law of Motion, namely

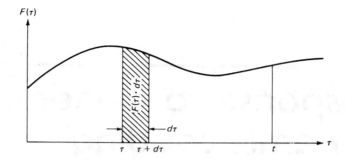

Fig. 4.1 General load function as impulsive loading.

$$m \frac{dv}{d\tau} = F(\tau)$$

Rearrangement yields

$$dv = \frac{F(\tau) \, d\tau}{m} \tag{4.1}$$

where $F(\tau) \, d\tau$ is the impulse and dv is the incremental velocity. This incremental velocity may be considered to be an initial velocity of the mass at time τ. Now let us consider this impulse $F(\tau) \, d\tau$ acting on the structure represented by the undamped oscillator. At the time τ the oscillator will experience a change of velocity given by eq. (4.1). This change in velocity is then introduced in eq. (1.20) as the initial velocity v_0 together with the initial displacement $y_0 = 0$ at time τ producing a displacement at a later time t given by

$$dy(t) = \frac{F(\tau) \, d\tau}{m\omega} \sin \omega(t - \tau) \tag{4.2}$$

The loading function may then be regarded as a series of short impulses at successive incremental times $d\tau$, each producing its own differential response at time t of the form given by eq. (4.2). Therefore, we conclude that the total displacement at time t due to the continuous action of the force $F(\tau)$ is given by the summation or integral of the differential displacements $dy(t)$ from time $t = 0$ to time t, that is,

$$y(t) = \frac{1}{m\omega} \int_0^t F(\tau) \sin \omega(t - \tau) \, d\tau \tag{4.3}$$

The integral in this equation is known as *Duhamel's integral*. Equation (4.3) represents the total displacement produced by the exciting force $F(\tau)$ acting on the undamped oscillator; it includes both the steady-state and the transient components of the motion corresponding to zero initial conditions, y_0 = 0 and v_0 = 0. If the function $F(\tau)$ cannot be expressed analytically, the integral of eq. (4.3) can always be evaluated approximately by suitable numerical methods. To include the effect of initial displacement y_0 and initial velocity v_0 at time t = 0, it is only necessary to add to eq. (4.3) the solution given by eq. (1.20) for the effects due to the initial conditions. Thus the total displacement of an undamped single degree-of-freedom system with an arbitrary load is given by

$$y(t) = y_0 \cos \omega t + \frac{v_0}{\omega} \sin \omega t + \frac{1}{m\omega} \int_0^t F(\tau) \sin \omega(t - \tau) \, d\tau \qquad (4.4)$$

Applications of eq. (4.4) for some simple forcing functions for which it is possible to obtain the explicit integration of eq. (4.4) are presented below.

4.1.1   Constant Force

Consider the case of a constant force of magnitude F_0 applied suddenly to the undamped oscillator at time t = 0 as shown in Fig. 4.2. For both initial displacement and initial velocity equal to zero, the application of eq. (4.4) to this case gives

$$y(t) = \frac{1}{m\omega} \int_0^t F_0 \sin \omega(t - \tau) \, d\tau$$

and integration yields

$$y(t) = \frac{F_0}{m\omega^2} |\cos \omega(t - \tau)|_0^t$$

$$y(t) = \frac{F_0}{k} (1 - \cos \omega t) = y_{st}(1 - \cos \omega t) \qquad (4.5)$$

where $y_{st} = F_0/k$. The response for such a suddenly applied constant load is shown in Fig. 4.3. It will be observed that this solution is very similar to the solution for the free vibration of the undamped oscillator. The major differ-

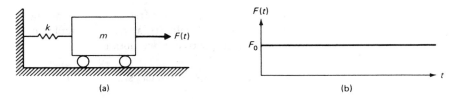

Fig. 4.2 Undamped oscillator acted upon by a constant force.

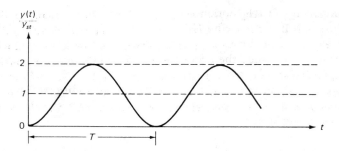

Fig. 4.3 Response of an undamped single degree-of-freedom system to a suddenly applied constant force.

ence is that the coordinate axis t has been shifted by an amount equal to y_{st} $= F_0/k$. Also, it should be noted that the maximum displacement $2y_{st}$ is exactly twice the displacement which the force F_0 would produce if it were applied statically. We have found an elementary but important result: the maximum displacement of a linear elastic system for a constant force applied suddenly is twice the displacement caused by the same force applied statically (slowly). This result for displacement is also true for the internal forces and stresses in the structure.

4.1.2   Rectangular Load

Let us consider a second case, that of a constant force F_0 suddenly applied but only during a limited time duration t_d as shown in Fig. 4.4. Up to the time t_d, eq. (4.5) applies and at that time the displacement and velocity are

$$y_d = \frac{F_0}{k}(1 - \cos \omega t_d)$$

and

$$v_d = \frac{F_0}{k} \omega \sin \omega t_d$$

For the response after time t_d we apply eq. (1.20) for free vibration, taking as the initial conditions the displacement and velocity at t_d. After replacing t by $t - t_d$, and y_0 and v_0 by y_d and v_d, respectively, we obtain

$$y(t) = \frac{F_0}{k}(1 - \cos \omega t_d) \cos \omega(t - t_d) + \frac{F_0}{k} \sin \omega t_d \sin \omega(t - t_d)$$

which can be reduced to

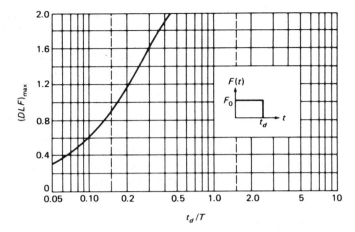

Fig. 4.4 Maximum dynamic load factor for the undamped oscillator acted upon by a rectangular force.

$$y(t) = \frac{F_0}{k} \{\cos \omega(t - t_d) - \cos \omega t\} \tag{4.6}$$

If the dynamic load factor (DLF) is defined as the displacement at any time t divided by the static displacement $y_{st} = F_0/k$, we may write eqs. (4.5) and (4.6) as

$$\text{DLF} = 1 - \cos \omega t, \qquad t \le t_d$$

and

$$\text{DLF} = \cos \omega(t - t_d) - \cos \omega t, \qquad t \ge t_d. \tag{4.7}$$

It is often convenient to express time as a dimensionless parameter by simply using the natural period instead of the natural frequency $(\omega = 2\pi/T)$. Hence eq. (4.7) may be written as

$$\text{DLF} = 1 - \cos 2\pi \frac{t}{T}, \qquad t \le t_d$$

and

$$\text{DLF} = \cos 2\pi \left(\frac{t}{T} - \frac{t_d}{T}\right) - \cos 2\pi \frac{t}{T}, \qquad t \ge t_d \tag{4.8}$$

The use of dimensionless parameters in eq. (4.8) serves to emphasize the fact that the ratio of duration of the time the constant force is applied to the natural period rather than the actual value of either quantity is the important parameter. The maximum dynamic load factor $(\text{DLF})_{max}$, obtained by maximizing eq. (4.8), is plotted in Fig. 4.4. It is observed from this figure that the maximum dynamic load factor for loads of duration $t_d/T \geq 0.5$ is the same as if the load duration had been infinite.

Charts, as shown in Fig. 4.4, which give the maximum response of a single degree-of-freedom system for a given loading function, are called *response spectral* charts. These charts are extremely useful for design purposes, as will be discussed in Chapter 8. Response spectral charts for impulsive loads of short duration are often presented for the undamped system. For short duration of the load, damping does not have a significant effect on the response of the system. The maximum dynamic load factor usually corresponds to the first peak of response and the amount of damping normally found in structures is not sufficient to appreciably decrease this value.

4.1.3   Triangular Load

We consider now a system represented by the undamped oscillator, initially at rest and subjected to a force $F(t)$ which has an initial value F_0 and which decreases linearly to zero at time t_d (Fig. 4.5). The response may be computed by eq. (4.4) in two intervals. For the first interval, $\tau \leq t_d$, the force is given by

$$F(\tau) = F_0\left(1 - \frac{\tau}{t_d}\right)$$

and the initial conditions by

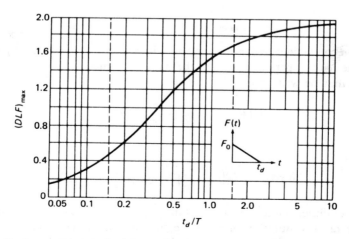

Fig. 4.5 Maximum dynamic load factor for the undamped oscillator acted upon by a triangular force.

$$y_0 = 0, \qquad v_0 = 0$$

The substitution of these values in eq. (4.4) and integration gives

$$y = \frac{F_0}{k}(1 - \cos \omega t) + \frac{F_0}{kt_d}\left(\frac{\sin \omega t}{\omega} - t\right) \qquad (4.9)$$

or in terms of the dynamic load factor and dimensionless parameters

$$DLF = \frac{y}{y_{st}} = 1 - \cos (2\pi t/T) + \frac{\sin (2\pi t/T)}{2\pi t_d/T} - \frac{t}{t_d} \qquad (4.10)$$

which defines the response before time t_d. For the second interval $(t \geq t_d)$, we obtain from eq. (4.9) the displacement and velocity at time t_d as

$$y_d = \frac{F_0}{k}\left(\frac{\sin \omega t_d}{\omega t_d} - \cos \omega t_d\right)$$

and

$$v_d = \frac{F_0}{k}\left(\omega \sin \omega t_d + \frac{\cos \omega t_d}{t_d} - \frac{1}{t_d}\right) \qquad (4.11)$$

These values may be considered as the initial conditions at time $t = t_d$ for this second interval. Replacing in eq. (1.20) t by $t - t_d$ and y_0 and v_0, respectively, by y_d and v_d and noting that $F(\tau) = 0$ in this interval, we obtain the response as

$$y = \frac{F_0}{k\omega t_d}\{\sin \omega t - \sin \omega(t - t_d)\} - \frac{F_0}{k}\cos \omega t$$

and upon dividing by $y_{st} = F_0/k$ gives

$$DLF = \frac{1}{\omega t_d}\{\sin \omega t - \sin \omega(t - t_d)\} - \cos \omega t \qquad (4.12)$$

In terms of the dimensionless time parameter, this last equation may be written as

$$DLF = \frac{1}{2\pi t_d/T}\left\{\sin 2\pi \frac{t}{T} - \sin 2\pi \left(\frac{t}{T} - \frac{t_d}{T}\right)\right\} - \cos 2\pi \frac{t}{T} \qquad (4.13)$$

The plot of the maximum dynamic load factor as a function of the relative time duration t_d/T for the undamped oscillator is given in Fig. 4.5. As would be expected, the maximum value of the dynamic load factor approaches 2 as t_d/T becomes large; that is, the effect of the decay of the force is negligible for the time required for the system to reach the maximum peak.

We have studied the response of the undamped oscillator for two simple impulse loadings: the rectangular pulse and the triangular pulse. Extensive charts have been prepared by the U.S. Army Corps of Engineers[1], and are available for a variety of other loading pulses.

In the next section, we will determine the response for forcing functions which do not permit an analytical solution of Duhamel's integral. In these cases it is necessary to resort to a numerical evaluation of Duhamel's integral in order to obtain the response of the system.

4.2   NUMERICAL EVALUATION OF DUHAMEL'S INTEGRAL—UNDAMPED SYSTEM

In many practical cases the applied loading function is known only from experimental data as in the case of seismic motion and the response must be evaluated by a numerical method. For this purpose we use the trigonometric identity $\sin \omega(t - \tau) = \sin \omega t \cos \omega \tau - \cos \omega t \sin \omega \tau$, in Duhamel's integral. Then, assuming zero initial conditions, we obtain Duhamel's integral, eq. (4.4), in the form

$$y(t) = \sin \omega t \, \frac{1}{m\omega} \int_0^t F(\tau) \cos \omega \tau \, d\tau - \cos \omega t \, \frac{1}{m\omega} \int_0^t F(\tau) \sin \omega \tau \, d\tau$$

or

$$y(t) = \{A(t) \sin \omega t - B(t) \cos \omega t\}/m\omega \qquad (4.14)$$

where

$$A(t) = \int_0^t F(\tau) \cos \omega \tau \, d\tau$$

$$B(t) = \int_0^t F(\tau) \sin \omega \tau \, d\tau \qquad (4.15)$$

The calculation of Duhamel's integral thus requires the evaluation of the integrals $A(t)$ and $B(t)$ numerically.

Several numerical integration techniques have been used for this evaluation. In these techniques the integrals are replaced by a suitable summation of the function under the integral and evaluated for convenience at n equal time increments, $\Delta\tau$. The most popular of these methods are the trapezoidal rule and the Simpson's rule. Consider the integration of a general function $I(\tau)$

$$A(t) = \int_0^t I(\tau) \, d\tau$$

[1] U.S. Army Corps of Engineers, *Design of Structures to Resist the Effects of Atomic Weapons*, Manuals 415, 415, and 416, March 15, 1957; Manuals 417 and 419, January 15, 1958; Manuals 418, 420, 421, January 15, 1960.

The elementary operation required for the trapezoidal rule is

$$A(t) = \Delta\tau \tfrac{1}{2} (I_0 + 2I_1 + 2I_2 + \ldots + 2I_{n-1} + I_n) \qquad (4.16)$$

and for Simpson's rule

$$A(t) = \Delta\tau \tfrac{1}{3} (I_0 + 4I_1 + 2I_2 + \ldots + 4I_{n-1} + I_n) \qquad (4.17)$$

where $n = t/\Delta\tau$ must be an even number for Simpson's rule. The implementation of these rules is straightforward. The response obtained will be approximate since these rules are based on the substitution of the function $I(\tau)$ for a piecewise linear function for the trapezoidal rule, or piecewise parabolic function for Simpson's rule. An alternative approach to the evaluation of Duhamel's integral is based on obtaining the exact analytical solution of the integral for the loading function assumed to be given by a succession of linear segments. This method does not introduce numerical approximations for the integration other than those inherent in the round off error, so in this sense it is an exact method.

In using this method, it is assumed that $F(\tau)$, the forcing function, may be approximated by a segmentally linear function as shown in Fig. 4.6. To provide a complete response history, it is more convenient to express the integrations in eq. (4.15) in incremental form, namely

$$A(t_i) = A(t_{i-1}) + \int_{t_{i-1}}^{t_i} F(\tau) \cos \omega\tau \, d\tau \qquad (4.18)$$

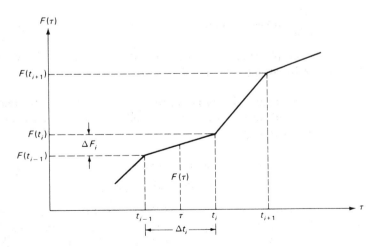

Fig. 4.6 Segmentally linear loading function.

$$B(t_i) = B(t_{i-1}) + \int_{t_{i-1}}^{t_i} F(\tau) \sin \omega\tau \, d\tau \tag{4.19}$$

where $A(t_i)$ and $B(t_i)$ represent the values of the integrals in eq. (4.15) at time t_i. Assuming that the forcing function $F(\tau)$ is approximated by a piecewise linear function as shown in Fig. 4.6, we may write

$$F(\tau) = F(t_{i-1}) + \frac{\Delta F_i}{\Delta t_i}(\tau - t_{i-1}), \qquad t_{i-1} \le \tau \le t_i \tag{4.20}$$

where

$$\Delta F_i = F(t_i) - F(t_{i-1})$$

and

$$\Delta t_i = t_i - t_{i-1}$$

The substitution of eq. (4.20) into eq. (4.18) and integration yield

$$A(t_i) = A(t_{i-1}) + \left(F(t_{i-1}) - t_{i-1}\frac{\Delta F_i}{\Delta t_i} \right)(\sin \omega t_i - \sin \omega t_{i-1})/\omega$$

$$+ \frac{\Delta F_i}{\omega^2 \Delta t_i} \{\cos \omega t_i - \cos \omega t_{i-1} + \omega(t_i \sin \omega t_i - t_{i-1} \sin \omega t_{i-1})\} \tag{4.21}$$

Analogously from eq. (4.19),

$$B(t_i) = B(t_{i-1}) + \left(F(t_{i-1}) - t_{i-1}\frac{\Delta F_i}{\Delta t_i} \right)(\cos \omega t_{i-1} - \cos \omega t_i)/\omega$$

$$+ \frac{\Delta F_i}{\omega^2 \Delta t_i} \{\sin \omega t_i - \sin \omega t_{i-1} - \omega(t_i \cos \omega t_i - t_{i-1} \cos \omega t_{i-1})\} \tag{4.22}$$

Equations (4.21) and (4.22) are recurrent formulas for the evaluation of the integrals in eq. (4.15) at any time $t = t_i$.

Example 4.1.   Determine the dynamic response of a tower subjected to a blast loading. The idealization of the structure and the blast loading are shown in Fig. 4.7. Neglect damping.

Solution: For this system, the natural frequency is

$$\omega = \sqrt{k/m} = \sqrt{100,000/100} = 31.62 \text{ rad/sec}$$

Since the loading is given as a segmented linear function, the response obtained using Duhamel's integral, eq. (4.14), with the coefficients $A(t)$ and $B(t)$ determined from eqs. (4.21) and (4.22), will be exact. The necessary calculations are presented in a convenient tabular format in Table 4.1 for a few time

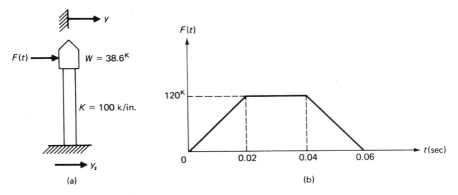

Fig. 4.7 Idealized structure and loading for Example 4.1.

steps. The integrals in eqs. (4.18) and (4.19) are labeled $\Delta A(t)$ and $\Delta B(t)$ in this table, since

$$\Delta A(t_i) = A(t_i) - A(t_{i-1}) = \int_{t_{i-1}}^{t_i} F(\tau) \cos \omega\tau \; d\tau$$

and

$$\Delta B(t_i) = B(t_i) - B(t_{i-1}) = \int_{t_{i-1}}^{t_i} F(\tau) \sin \omega\tau \; d\tau$$

Since the blast terminates at $t = 0.060$ sec, the values of A and B remain constant after this time. Consequently, the free vibration which follows is obtained by substituting these values of A and B evaluated at $t = 0.060$ sec into eq. (4.14), that is,

$$y(t) = (2571 \sin 31.62t - 3585 \cos 31.62t)/3162$$

or

$$y(t) = 0.8130 \sin 31.62t - 1.1338 \cos 31.62t$$

for $t \geq 0.060$ sec.

TABLE 4.1   Numerical Calculation of the Response for Example 4.1

t (sec)	$F(\tau)$	ωt	$\Delta A(t)$	$A(t)$	$\Delta B(t)$	$B(t)$	$y(t)$ (in)
0.000	0	0	0	0	0	0	0
0.020	120,000	0.6324	1082	1082	486	486	0.078
0.040	120,000	1.2649	1376	2458	1918	2404	0.512
0.060	0	1.8974	113	2571	1181	3585	1.134
0.080	0	2.5298	0	2571	0	3585	1.395
0.100	0	3.1623	0	2571	0	3585	1.117

4.3   NUMERICAL EVALUATION OF DUHAMEL'S INTEGRAL—DAMPED SYSTEM

The response of a damped system expressed by the Duhamel's integral is obtained in a manner entirely equivalent to the undamped analysis except that the impulse $F(\tau)\,d\tau$ producing an initial velocity $dv = F(\tau)\,d\tau/m$ is substituted into the corresponding damped free-vibration equation. Setting $y_0 = 0$, $v_0 = F(\tau)\,d\tau/m$, and substituting t for $t - \tau$ in eq. (2.20), we obtain the differential displacement at time t as

$$dy(t) = e^{-\xi\omega(t-\tau)}\,\frac{F(\tau)\,d\tau}{m\omega_D}\,\sin\omega_D(t - \tau) \qquad (4.23)$$

Summing these differential response terms over the entire loading interval results in

$$y(t) = \frac{1}{m\omega_D}\int_0^t F(\tau)\,e^{-\xi\omega(t-\tau)}\,\sin\omega_D(t - \tau)\,d\tau \qquad (4.24)$$

which is the response for a damped system in terms of the Duhamel's integral. For numerical evaluation, we proceed as in the undamped case and obtain from eq. (4.24)

$$y(t) = \{A_D(t)\,\sin\omega_D t - B_D(t)\,\cos\omega_D t\}\,\frac{e^{-\xi\omega t}}{m\omega_D} \qquad (4.25)$$

where

$$A_D(t_i) = A_D(t_{i-1}) + \int_{t_{i-1}}^{t_i} F(\tau)\,e^{\xi\omega\tau}\,\cos\omega_D\tau\,d\tau \qquad (4.26)$$

$$B_D(t_i) = B_D(t_{i-1}) + \int_{t_{i-1}}^{t_i} F(\tau)\,e^{\xi\omega\tau}\,\sin\omega_D\tau\,d\tau \qquad (4.27)$$

For a linear piecewise loading function, $F(\tau)$ given by eq. (4.20) is substituted into eqs. (4.26) and (4.27) which requires the evaluation of the following integrals:

$$I_1 = \int_{t_{i-1}}^{t_i} e^{\xi\omega\tau}\,\cos\omega_D\tau\,d\tau = \frac{e^{\xi\omega\tau}}{(\xi\omega)^2 + \omega_D^2}\,(\xi\omega\,\cos\omega_D\tau + \omega_D\,\sin\omega_D\tau)\,\Big|_{t_{i-1}}^{t_i} \qquad (4.28)$$

$$I_2 = \int_{t_{i-1}}^{t_i} e^{\xi\omega\tau}\,\sin\omega_D\tau\,d\tau = \frac{e^{\xi\omega\tau}}{(\xi\omega)^2 + \omega_D^2}\,(\xi\omega\,\sin\omega_D\tau - \omega_D\,\cos\omega_D\tau)\,\Big|_{t_{i-1}}^{t_i} \qquad (4.29)$$

$$I_3 = \int_{t_{i-1}}^{t_i} \tau e^{\xi\omega\tau}\,\sin\omega_D\tau\,d\tau = \left(\tau - \frac{\xi\omega}{(\xi\omega)^2 + \omega_D^2}\right)I_2' + \frac{\omega_D}{(\xi\omega)^2 + \omega_D^2}\,I_1'\,\Big|_{t_{i-1}}^{t_i} \qquad (4.30)$$

$$I_4 = \int_{t_{i-1}}^{t_i} \tau e^{\xi\omega\tau} \cos \omega_D\tau \; d\tau = \left(\tau - \frac{\xi\omega}{(\xi\omega)^2 + \omega_D^2}\right)I_1' - \frac{\omega_D}{(\xi\omega)^2 + \omega_D^2} I_2' \Bigg|_{t_{i-1}}^{t_i} \qquad (4.31)$$

where I_1' and I_2' are the integrals indicated in eqs. (4.28) and (4.29) before their evaluation at the limits. In terms of these integrals, $A_D(t_i)$ and $B_D(t_i)$ may be evaluated after substituting eq. (4.20) into eqs. (4.26) and (4.27) as

$$A_D(t_i) = A_D(t_{i-1}) + \left(F(t_{i-1}) - t_{i-1}\frac{\Delta F_i}{\Delta t_i}\right)I_1 + \frac{\Delta F_i}{\Delta t_i} I_4 \qquad (4.32)$$

$$B_D(t_i) = B_D(t_{i-1}) + \left(F(t_{i-1}) - t_{i-1}\frac{\Delta F_i}{\Delta t_i}\right)I_2 + \frac{\Delta F_i}{\Delta t_i} I_3 \qquad (4.33)$$

Finally, the substitution of eqs. (4.32) and (4.33) into eq. (4.25) gives the displacement at time t_i as

$$y(t_i) = \frac{e^{-\xi\omega t_i}}{m\omega_D} \{A_D(t_i) \sin \omega_D t_i - B_D(t_i) \cos \omega_D t_i\} \qquad (4.34)$$

4.4 RESPONSE BY DIRECT INTEGRATION

The differential equation of motion for a one-degree-of-freedom system represented by the damped simple oscillator, as shown in Fig. 4.8(a), is obtained by establishing the dynamic equilibrium of the forces in the free body diagram, Fig. 4.8(b):

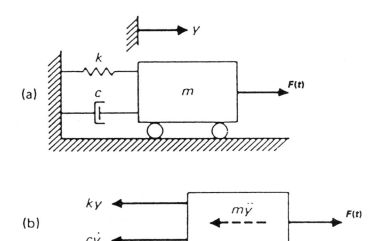

Fig. 4.8 (a) Damped simple oscillator excited by the force $F(t)$. (b) Free body diagram.

$$m\ddot{y} + c\dot{y} + ky = F(t) \tag{4.35}$$

in which the function $F(t)$ represents the force applied to the mass of the oscillator.

When the structure, modeled by the simple oscillator, is excited by a motion at its support, as is shown in Fig. 4.9(a), the equation of motion obtained using the free body diagram in Fig. 4.9(b) is

$$m\ddot{y} + c(\dot{y} - \dot{y}_s) + k(y - y_s) = 0 \tag{4.36}$$

In this case, it is convenient to express the displacement of the mass relative to the displacement y_s of the support, namely

$$u = y - y_s \tag{4.37}$$

The substitution u and its derivatives from eq. (4.37) into eq. (4.36) results in

$$m\ddot{u} + c\dot{u} + ku = -m\ddot{y}_s(t) \tag{4.38}$$

Comparison of eqs. (4.35) and (4.38) reveals that both equations are mathematically equivalent if the right-hand side of eq. (4.38) is interpreted as the effective force

$$F_{\text{eff}}(t) = -m\ddot{y}_s(t) \tag{4.39}$$

Equation (4.38) may then be written as

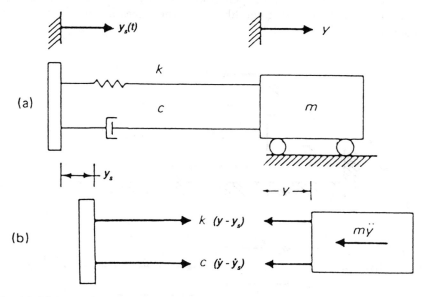

Fig. 4.9 (a) Damped simple oscillator excited by the displacement $y_s(t)$ at its support. (b) Free body diagram.

$$m\ddot{u} + c\dot{u} + ku = F_{\text{eff}}(t) \tag{4.40}$$

Consequently, the solution of the second order differential equation (4.35) or eq. (4.40) gives the response in terms of the absolution motion y, for the case in which the mass is excited by a force, or in terms of the relative motion $u = y - y_s$ for the structure excited at its support.

4.4.1 Solution of the Equation of Motion

The method of solution for the differential equation of motion implemented in the computer program presented in this chapter is exact for an excitation function described by linear segments. The process of solution requires for convenience that the excitation function be calculated at equal time intervals Δt. This result is accomplished by a linear interpolation between points defining the excitation. Thus, the time duration of the excitation, including a suitable extension of time after cessation of the excitation, is divided into N equal time intervals of duration Δt. For each interval Δt, the response is calculated by considering the initial conditions at the beginning of that time interval and the linear excitation during the interval. The initial conditions are, in this case, the displacement and velocity at the end of the preceding time interval. Assuming that the excitation function $F(t)$ is approximated by a piecewise linear function as shown in Fig. 4.6, we may express this function by

$$F(t) = \left(1 - \frac{t - t_i}{\Delta t}\right)F_i + \left(\frac{t - t_i}{\Delta t}\right)F_{i+1}, \qquad t_i \leq t \leq t_{i+1} \tag{4.41}$$

in which $t_i = i \cdot \Delta t$ for equal intervals of duration Δt and $i = 1, 2, 3, \ldots, N$. The differential equation of motion, eq. (4.35), is then given by

$$m\ddot{y} + c\dot{y} + ky = \left(1 - \frac{t - t_i}{\Delta t}\right)F_i + \left(\frac{t - t_i}{\Delta t}\right)F_{i+1}, \qquad t_i \leq t \leq t_{i+1} \tag{4.42}$$

The solution of eq. (4.42) may be expressed as the sum of complementary solution y_c, for which the second member of eq. (4.42) is set equal to zero, and the particular solution y_p, that is,

$$y = y_c + y_p \tag{4.43}$$

The complementary solution is given in general by eq. (2.15), which for the interval $t_i \leq t \leq t_i + \Delta t$ is

$$y_c = e^{-\xi\omega(t - t_i)}[C_i \cos \omega_D (t - t_i) + D_i \sin \omega_D (t - t_i)] \tag{4.44}$$

On the other hand, the particular solution of eq. (4.42) takes the form

$$y_p = B_i + A_i(t - t_i) \tag{4.45}$$

which upon its substitution into eq. (4.42) gives

$$cA_i + k[B_i + A_i(t - t_i)] = \left(1 - \frac{t - t_i}{\Delta t}\right)F_i + \left(\frac{t - t_i}{\Delta t}\right)F_{i+1}$$

where A_i and B_i are constants for the interval $t_i \leq t \leq t_i + \Delta t$ and where we use the notation $F_i = F(t_i)$ and $F_{i+1} = F(t_i + \Delta t)$. Establishing the identity of terms between the left- and right-hand sides, that is, between the constant terms and the terms with a factor t and solving the resulting equations, we obtain

$$A_i = \frac{F_{i+1} - F_i}{k\Delta t}$$

(4.46)

$$B_i = \frac{F_i - cA_i}{k}$$

The substitution into eq. (4.43) of the complementary solution y_c from eq. (4.44) and of the particular solution y_p from eq. (4.45) gives the total solution as

$$y = e^{-\xi\omega(t - t_i)}[C_i \cos \omega_D (t - t_i) + D_i \sin \omega_D (t - t_i)] + B_i + A_i(t - t_i) \quad (4.47)$$

The velocity is then given by the derivative of eq. (4.47) as

$$\dot{y} = e^{-\xi\omega(t - t_i)}[(\omega_D D_i - \xi\omega C_i) \cos \omega_D (t - t_i)$$

$$-(\omega_D C_i + \xi\omega D_i) \sin \omega_D (t - t_i)] + A_i \quad (4.48)$$

The constants of integration C_i and D_i are obtained from eqs. (4.47) and (4.48) introducing the initial conditions for the displacement y_i and for the velocity $\dot{y}_i$ at the beginning of the interval Δt, that is, at time t_i. Thus, introducing into eqs. (4.47) and (4.48) the initial conditions and solving the resulting relations yields

$$C_i = y_i - B_i$$

$$D_i = \frac{\dot{y}_i - A_i - \xi\omega C_i}{\omega_D}$$

(4.49)

The evaluation of eqs. (4.47) and (4.48) at time $t_i + \Delta t$ results in the displacement y_{i+1} and the velocity $\dot{y}_{i+1}$ at time t_{i+1}. Namely,

$$y_{i+1} = e^{-\xi\omega\Delta t}[C_i \cos \omega_D \Delta t + D_i \sin \omega_D \Delta t] + B_i + A_i\Delta t \quad (4.50)$$

and

$$\dot{y}_{i+1} = e^{-\xi\omega\Delta t}[D_i(\omega_D \cos \omega_D \Delta t - \xi\omega \sin \omega_D \Delta t)$$

$$- C_i(\xi\omega \cos \omega_D \Delta t + \omega_D \sin \omega_D \Delta t)] + A_i \quad (4.51)$$

Finally, the acceleration at time $t_{i+1} = t_i + \Delta t$ is obtained directly after substituting y_{i+1} and $\dot{y}_{i+1}$ from eqs. (4.50) and (4.51) into the differential eq. (4.35) and letting $t = t_i + \Delta t$. Specifically,

$$\ddot{y}_{i+1} = \frac{1}{m}(F_{i+1} - c\dot{y}_{i+1} - ky_{i+1}) \tag{4.52}$$

Example 4.2.   Determine the dynamic response of a tower subjected to a blast loading. The idealization of the structure and the blast loading are shown in Fig. 4.10. Assume damping equal to 20% of the critical damping.

Solution: Since the loading is given as a segmental linear function, the response obtained using the direct method will be exact. The necessary calculations are presented in a convenient tabular format in Table 4.2. For this system, the natural frequency is

$$\omega = \sqrt{k/m} = \sqrt{100,000/100} = 31.62 \text{ rad/sec}$$

Hence, the natural period is

$$T = \frac{2\pi}{\omega} = \frac{2\pi}{31.62} = 0.20 \text{ sec}$$

Recommended practice is to select $\Delta t \leq T/10$. Specifically, we select $\Delta t = 0.02$ sec. We calculate the constants quantities in eqs. (4.47) and (4.48) as follows:

$$c = c_{cr}\xi = 2\sqrt{km} \; \xi = 2 \sqrt{100,000 \times 100} \times 0.20 = 1265 \text{ lb} \cdot \text{sec/in}$$

$$\omega_D = \omega\sqrt{1 - \xi^2} = 31.62 \sqrt{1 - 0.2^2} = 30.989 \text{ rad/sec}$$

Also we calculate the following terms:

$$C = e^{-\xi\omega\Delta t} \cos \omega_D \Delta t = 0.71730$$

$$S = e^{-\xi\omega\Delta t} \sin \omega_D \Delta t = 0.51184$$

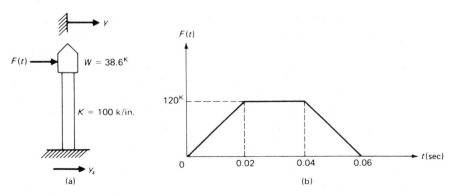

Fig. 4.10 Idealized structure and loading for Example 4.2.

TABLE 4.2   Calculation of the Response for Example 4.2

t_i sec	y_i in	$\dot{y}_i$ in/sec	$\ddot{y}_i$ in/sec^2	F_i lb	A_i	B_i	C_i	D_i
0	0	0	0	0	60	−0.759	0.7590	−1.7816
0.02	0.074	10.692	991.023	120000	0	1.2	−1.1263	0.1151
0.04	0.451	25.155	430.768	120000	−60	1.959	−1.5080	2.4405
0.06	0.926	17.096	−1142.511	0	0	0	0.9262	0.7409
0.08	1.044	−4.821	−982.581	0	0	0	1.0436	0.0574
0.10	0.778	−20.191	−522.555	0	0	0	0.7780	−0.4929

$$U = e^{-\xi\omega\Delta t}(\omega_D \cos \omega_D \Delta t - \xi\omega \sin \omega_D \Delta t) = 18.99140$$

$$V = e^{-\xi\omega\Delta t}(\xi\omega \cos \omega_D \Delta t + \omega_D \sin \omega_D \Delta t) = 20.39785$$

With initial conditions $y_0 = 0$ and $v_0 = 0$, we obtain using eqs. (4.46) through (4.49) the values

$$A_0 = 60, \; B_0 = -0.759, \; C_0 = 0.7590, \; D_0 = -1.7816$$

which substituted into eqs. (4.50) through (4.52) gives the response at time t_1 = 0.02 sec as

$$y_1 = 0.074 \text{ in}, \quad \dot{y}_1 = 10.692 \text{ in/sec}, \quad \text{and} \quad \ddot{y}_1 = 991.023 \text{ in/sec}^2$$

thus completing the first cycle of calculations in the direct method of solution. Considering the calculated values y_1, $\dot{y}_1$, and $\ddot{y}_1$ as the initial conditions for the next time interval and using eqs. (4.46) through (4.52), we obtain the response at time $t_2 = 0.04$. The continuation of this process results in the response of this system as indicated in Table 4.2 up to 0.10 sec.

4.5   PROGRAM 2—RESPONSE BY DIRECT INTEGRATION

The computer program described in this section calculates the response of the simple oscillator excited by a time-dependent external force acting on the mass or by an acceleration applied to the support. The excitation is assumed to be piecewise linear between defining points. The response consists of a table giving at equal increments of time the displacement, the velocity, and the acceleration of the mass for the case of the oscillator excited by a force applied to the mass. For the case of the oscillator excited at its support, the response is given in terms of the displacement and the velocity of the mass relative respectively to the displacement and velocity of the support and of the absolute acceleration of the mass. The program implements the direct integration method to calculate the response. The program has been written in an interactive mode with the user. It starts by requesting information about the data needed in the solution of the problem, stores the data in a file, and then calculates and prints the response at equal increments of time. Maximum values for the response are also calculated and printed. A description of the programs used in this book and a form to order them is provided in Appendix II.

Example 4.3.   Solve Example 4.2 Using Program 2.

Solution: From Fig. 4.10, we have the following data:

Mass: $m = w/g = (38.6 \times 1000)/386 = 100$ (lb·sec²/in)

Spring constant: $k = 100 \times 1000 = 100,000$ (lb·in)

Damping coefficient: $c = 2\xi\sqrt{km} = 1265$ (lb·sec/in)

Natural period: $T = 2\pi\sqrt{m/k} = 0.20$ sec

Select time step for integration: $\Delta t = T/10 = 0.02$ sec

Input Data and Output Results

```
PROGRAM 2: DIRECT INTEGRATION          DATA FILE:D2

INPUT DATA:

NUMBER OF POINTS DEFINING THE EXCITATION              NE = 5
MASS                                                  AM = 100
SPRING CONSTANT                                       AK = 100000
DAMPING COEFFICIENT                                   C = 1265
TIME STEP OF INTEGRATION                              H = .02
INDEX (ACCELERATION OF GRAVITY OR ZERO)               G = 0

  TIME   EXCITATION   TIME   EXCITATION   TIME   EXCITATION   TIME   EXCITATION
  0.000     0.00     0.020  120000.00    0.040  120000.00    0.060     0.00
  0.260     0.00

   PRINT TIME-RESPONSE TABLE Y/N? Y

OUTPUT RESULTS:

        TIME            DISPL.           VELOC.            ACC.

        0.000          0.000            0.000            0.000
        0.020          0.074           10.692          991.023
        0.040          0.451           25.155          430.768
        0.060          0.926           17.096        -1142.511
        0.080          1.044           -4.821         -982.581
        0.100          0.778          -20.191         -522.555
        0.120          0.306          -25.225           13.248
        0.140         -0.165          -20.511          424.754
        0.160         -0.475           -9.840          599.095
        0.180         -0.553            1.809          529.696
        0.200         -0.424           10.235          294.760
        0.220         -0.180           13.280           11.592
        0.240          0.072           11.105         -212.241
        0.260          0.242            5.621         -313.497

MAX.DISPLACEMENT =              1.04
MAX. VELOCITY =                25.22
MAX. ACCELERATION =          1142.51
```

Example 4.4. Consider the tower shown in Fig. 4.10, but now subjected to a constant impulsive acceleration of magnitude $\ddot{y}_s = 0.5$ g during 0.5 sec applied at the foundation of the tower. Determine the response of the tower in terms of the displacement and velocity of the mass relative to the motion of the foundation. Also, determine the maximum acceleration of the mass.

Solution:

The following data is obtained from Fig. 4.10:

Mass: $m = 100$ (lb · sec²/in)

Spring constant: $k = 100,000$ (lb/in)

Damping coefficient: $c = 1265$ (lb · sec/in)

Acceleration of gravity: $g = 386$ (in/sec²)

Selected time step: $\Delta t = 0.02$ sec

Excitation:	Time (sec)	Support Acceleration (g)
	0	0.5
	0.5	0.5

Input Data and Output Results

```
     PROGRAM 2: DIRECT INTEGRATION        DATA FILE:D4.4
     INPUT DATA:

NUMBER OF POINTS DEFINING THE EXCITATION        NE = 2
MASS                                            AM = 100
SPRING CONSTANT                                 AK = 100000
DAMPING COEFFICIENT                             C = 1265
TIME STEP OF INTEGRATION                        H = .02
INDEX (ACCELERATION OF GRAVITY OR ZERO)         G = 386

   TIME           EXCITATION        TIME      EXCITATION
   0.00             0.500           0.50        0.500

     PRINT TIME-RESPONSE TABLE Y/N? N


OUTPUT RESULTS:

MAX.DISPLACEMENT²=        0.29
MAX. VELOCITY² =         4.65
MAX. ACCELERATION=      299.33
```

[2]For excitation at the support: Displacement and velocity are relative to the support, whereas the acceleration is absolute value.

4.6   PROGRAM 3—RESPONSE TO IMPULSIVE EXCITATION

Program 3, the same as Program 2, gives the response of the simple oscillator to an excitation specified as linear segments between points defining the loading function. Program 2 may be used for any excitation function, while Program 3 is restricted to certain specific excitation functions predefined in the program. A total of eight possible excitation functions are implemented in the program. There is also a provision for a ninth function labeled as "new function" which may be implemented by the user. The excitation functions in the program are shown graphically in Fig. 4.11 and given analytically by the following expressions:

$$E_1(T) = P(1) * \sin [P(2) * T - P(3)] \qquad T \geq 0 \qquad (4.53)$$

$$E_2(T) = P(1) \qquad T \geq 0 \qquad (4.54)$$

$$E_3(T) = P(1) * T \qquad 0 \leq T \leq P(2) \qquad (4.55)$$

$$= 0 \qquad T > P(2)$$

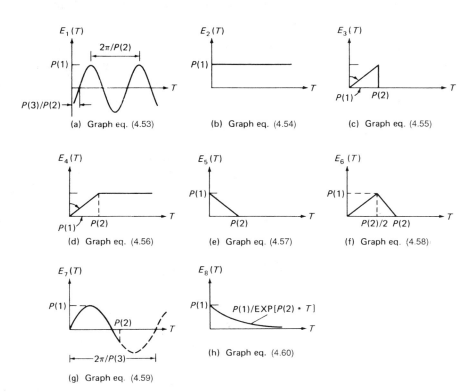

(a) Graph eq. (4.53)   (b) Graph eq. (4.54)   (c) Graph eq. (4.55)

(d) Graph eq. (4.56)   (e) Graph eq. (4.57)   (f) Graph eq. (4.58)

(g) Graph eq. (4.59)   (h) Graph eq. (4.60)

Fig. 4.11 Graphs for excitation functions given by eqs. (4.53) through (4.60).

$$E_4(T) \quad = P(1) * T \qquad\qquad\qquad\qquad 0 \le T \le P(2) \qquad (4.56)$$

$$= P(1) * P(2) \qquad\qquad\qquad\qquad T > P(2)$$

$$E_5(T) \quad = P(1) * [1 - T/P(2)] \qquad\qquad 0 \le T \le P(2) \qquad (4.57)$$

$$= 0 \qquad\qquad\qquad\qquad\qquad T > P(2)$$

$$E_6(T) \quad = P(1) * [2 * T/P(2)] \qquad\qquad 0 \le T \le P(2)/2 \qquad (4.58)$$

$$= 2 * P(1) * [1 - T/P(2)], \qquad P(2)/2 < T \le P(2)$$

$$= 0 \qquad\qquad\qquad\qquad\qquad T > P(2)$$

$$E_7(T) \quad = P(1) * \sin [P(3) * T] \qquad\qquad 0 \le T \le P(2) \qquad (4.59)$$

$$= 0 \qquad\qquad\qquad\qquad\qquad T > P(2)$$

$$E_8(T) \quad = P(1)/\mathrm{EXP}[P(2) * T] \qquad\qquad T \ge 0 \qquad (4.60)$$

$$E_9(T) \quad = \text{NEW FUNCTION}$$

In eqs. (4.53) through (4.60) T is the time variable and $P(1)$, $P(2)$, and $P(3)$ are constants or parameters to be input as data in order to completely define the specific excitation function selected.

Example 4.5. (a) Determine the dynamic response of the tower shown in Fig. 4.12 subjected to the sinusoidal force $F(t) = F_0 \sin \bar{\omega} t$ applied at its top for 0.30 sec. (b) Check results using the exact solution which in this case is available in closed form. Neglect damping.

Solution: (a) The following data is obtained from Fig. 4.12:

Mass: $m = w/g = (38.6 \times 1000)/386 = 100$ (lb · sec²/in)

Spring constant: $k = 100 \times 1000 = 100,000$ (lb/in)

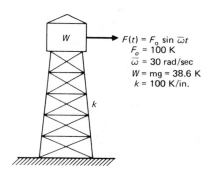

Fig. 4.12 Tower for Example 4.5

Natural frequency: $\omega = \sqrt{k/m} = 31.623$ rad/sec

Natural period: $T = 2\pi/\omega = 0.20$ sec

Select time step: $\Delta T = 0.01$ sec

Input Data and Output Results

```
PROGRAM 3--RESPONSE TO IMPULSIVE EXCITATION--  DATA FILE:D3

      INPUT DATA:

MASS                                    M = 100
SPRING CONSTANT                         K = 100000
DAMPING COEFFICIENT                     C = 0
TIME STEP OF INTEGRATION                H = .01
RESPONSE MAX. TIME                   TMAX = .3
INDEX (GRAVITY OR ZERO)                 G = 0

      PARAMETERS OF THE EXCITATION:
                        E(T) = P(1)*SIN[P(2)*T-P(3)]      T>0
PARAMETER P( 1 ) = 100000
PARAMETER P( 2 ) = 30
PARAMETER P( 3 ) = 0

         PRINT TIME--RESPONSE TABLE Y/N? Y

OUTPUT RESULTS:

      TIME        DISPL.      VELOC.          ACC.

      0.000       0.000       0.000           0.000
      0.010       0.005       1.465         290.619
      0.020       0.038       5.585         526.458
      0.030       0.123      11.570         660.487
      0.040       0.272      18.229         660.204
      0.050       0.485      24.142         512.573

      0.060       0.747      27.868         226.485
      0.070       1.031      28.165        -167.612
      0.080       1.296      24.189        -620.916
      0.090       1.499      15.654       -1071.980
      0.100       1.595       2.916       -1454.280
      0.110       1.547     -13.012       -1704.758
      0.120       1.330     -30.544       -1772.318
      0.130       0.937     -47.675       -1625.244
      0.140       0.385     -62.205       -1256.580
      0.150      -0.291     -72.003        -686.736
      0.160      -1.033     -75.279          37.082
      0.170      -1.771     -70.833         844.732
      0.180      -2.423     -58.254        1649.939
      0.190      -2.910     -38.039        2359.407
```

0.200	-3.163	-11.605	2883.269
0.210	-3.129	18.794	3145.744
0.220	-2.783	50.259	3094.781
0.230	-2.131	79.525	2709.542
0.240	-1.211	103.295	2004.786
0.250	-0.093	118.604	1031.496
0.260	1.125	123.168	-126.486
0.270	2.330	115.674	-1359.642
0.280	3.398	95.996	-2543.164
0.290	4.213	65.276	-3549.554
0.300	4.674	25.889	-4262.176
0.310	4.713	-18.736	-4588.323

```
MAX.DISPLACEMENT=              4.71
MAX. VELOCITY =              123.17
MAX. ACCELERATION=          4588.32
```

(b) The exact solution for the response of a simple oscillator to the sinusoidal force $F_0 = \sin \bar{\omega} t$, with zero initial displacement and velocity, from eq. (3.8) is

$$y(t) = \frac{F_0}{k - m\bar{\omega}^2} \left(\sin \bar{\omega} t - \frac{\bar{\omega}}{\omega} \sin \omega t \right)$$

where ω is the natural frequency in rad/sec, $\bar{\omega}$ the forced frequency also in rad/sec, and F_0 the amplitude of the sinusoidal force.

Substituting corresponding numerical values for this example yields

$$y(t) = \frac{100000}{100000 - 100(30)^2} \left(\sin 30t - \frac{30}{31.623} \sin 31.623t \right)$$

$$= 10(\sin 30t - 0.94868 \sin 31.623t)$$

The velocity and acceleration functions are then given by

$$\dot{y}(t) = 300 \cos 30t - 300 \cos 31.623t$$

and

$$\ddot{y}(t) = -9000 \sin 30t + 948.7 \sin 31.623t$$

The evaluation of the response at specific values of time results in the following table:

t (sec)	$y(t)$ (in)	$\dot{y}(t)$ (in/sec)	$\ddot{y}(t)$ (in/sec^2)
0.1	1.6076	2.9379	-1466.51
0.2	-3.1865	-11.6917	2907.53
0.3	4.7420	26.0822	-4298.04

Results shown in the above table are sufficiently close to corresponding values given by the computer in part (a) of this problem.

Example 4.6.   Solve Example 4.5 selecting the time step Δt equal to 0.02, 0.01, and 0.005. Then compare the displacements at time $t = 0.1, 0.2,$ and 0.3 sec with the response obtained in Example 4.5 using the exact solution of the differential equation.

Solution: In solving the problem using Program 3, it is only necessary to modify existing file by selecting option 2 (Modify Existing Data File) at the computer request for file information. The only modification necessary in the data is to change the time step to the values prescribed by this problem. The following table records the results provided by the computer and their comparison with the exact solution calculated in Example 4.5:

Time (sec)	Exact Displacement (in)	$\Delta t = 0.02$ sec		$\Delta t = 0.01$ sec		$\Delta t = 0.005$ sec	
		Displ. (in)	% error	Displ. (in)	% error	Displ. (in)	% error
0.1	1.6076	1.560	2.96	1.595	0.78	1.604	0.22
0.2	-3.1865	3.092	2.96	-3.163	0.73	-3.181	0.17
0.3	4.7420	4.570	3.62	4.674	1.43	4.701	0.86

Example 4.7.   A structural system modeled in Fig. 4.13(a) by the simple oscillator with 10% ($\xi = 0.10$) of critical damping is subjected to the impulsive load as shown in Fig. 4.13(b). Determine the response.

Solution: The following data is obtained from Fig. 4.13:

Mass: $m = 10$ lb$\cdot$sec^2/in

Spring constant: $k = 10000$ lb/in

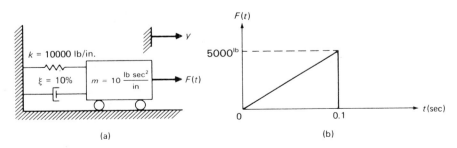

(a)   (b)

Fig. 4.13 Mathematical model and load function for Example 4.7.

Damping coefficient: $c = \xi c_{cr} = 2\xi\sqrt{km} = 63.25$ lb·sec/in

Natural period: $T = 2\pi\sqrt{m/k} = 0.20$ sec

Select time step: $\Delta T = 0.02$ sec

Excitation function: $F(t) = 50000t$ lb, $0 \le t \le 0.1$ sec

$$= 0 \qquad\qquad t > 0.1 \text{ sec}$$

Maximum time: $t_{max} = 0.2$ sec

Input Data and Output Results

```
PROGRAM 3--RESPONSE TO IMPULSIVE EXCITATION--  DATA FILE:D4.7

        INPUT DATA:

MASS                                M= 10
SPRING CONSTANT                     K= 10000
DAMPING COEFFICIENT                 C= 63.25
TIME STEP OF INTEGRATION            H= .02
RESPONSE MAX. TIME               TMAX= .2
INDEX (GRAVITY OR ZERO)             G= 0

     PARAMETERS OF THE EXCITATION:

         E(T) = P(1)*T           0<T<P(2)
              = 0                  T>P(2)
PARAMETER P( 1 )= 50000
PARAMETER P( 2 )= .1


      PRINT TIME-RESPONSE TABLE Y/N? Y


OUTPUT RESULTS:
        TIME        DISPL.       VELOC.        ACC.
        0.000       0.000        0.000        0.000
        0.020       0.006        0.928        87.797
        0.040       0.046        3.225       133.258
        0.060       0.138        5.894       124.894
        0.080       0.278        7.922        72.140
        0.100       0.446        8.646        -0.560
        0.120       0.576        2.393      -591.199
        0.140       0.511       -8.433      -457.841
        0.160       0.268      -14.907      -173.960
        0.180      -0.043      -15.195       139.402
        0.200      -0.302       -9.928       364.872
        0.220      -0.420       -1.679       430.965

MAX.DISPLACEMENT=           0.58
MAX. VELOCITY =            15.20
MAX. ACCELERATION=        591.20
```

4.7   SUMMARY

In this chapter, we have shown that the differential equation of motion for a linear system can be solved for any forcing function in terms of Duhamel's integral. The numerical evaluation of this integral can be accomplished by any standard method such as the trapezoidal or the Simpson's rule. We have preferred the use of a numerical integration by simply assuming that the forcing function is a linear function between defining points, and, on this basis, we have obtained the exact response for each time increment. The computer programs described in this chapter use the direct integration method. In this method, the differential equation of motion is solved for each time increment for the conditions existent at the end of the preceding interval (initial conditions for the new interval) and for the action of the excitation applied during the interval, which is assumed to be linear.

Two programs were presented in this chapter: (1) Program 2 to calculate the response of a single degree-of-freedom system excited by a force applied to its mass (or by an acceleration at the support). (2) Program 3 to determine the response of a single degree-of-freedom system excited by one of the impulsive functions specified in the program. These programs allow us to obtain the response in terms of displacement, velocity, and acceleration as a function of time for any single degree-of-freedom system of linear-elastic behavior when subjected to a general force function of time applied to the mass or to an acceleration applied to the support.

PROBLEMS

4.1   The steel frame shown in Fig. 4.14 is subjected to a horizontal force applied at the girder level. The force decreases linearly from 5 kip at time $t = 0$ to zero at

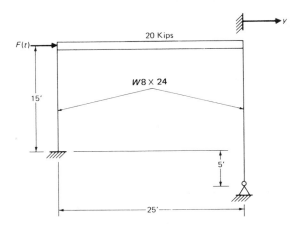

Fig. 4.14

$t = 0.6$ sec. Determine: (a) the horizontal deflection at $t = 0.5$ sec and (b) the maximum horizontal deflection. Assume the columns massless and the girder rigid. Neglect damping.

4.2   Repeat Problem 4.1 for 10% of critical damping.

4.3   For the load-time function in Fig. 4.15, derive the expression for the dynamic load factor for the undamped simple oscillator as a function of t, ω, and t_d.

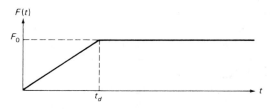

Fig. 4.15

4.4   The frame shown in Fig. 4.14 is subjected to a sudden acceleration of 0.5 g applied to its foundation. Determine the maximum shear force in the columns. Neglect damping.

4.5   Repeat Problem 4.4 for 10% of critical damping.

4.6   For the dynamic system shown in Fig. 4.16, determine and plot the displacement as a function of time for the interval $0 \le t \le 0.5$ sec. Neglect damping.

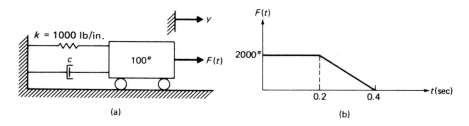

Fig. 4.16

4.7   Repeat Problem 4.6 for 10% of critical damping.

4.8   The tower of Fig. 4.17(a) is subjected to horizontal ground acceleration $a(t)$ shown in Fig. 4.17(b). Determine the relative displacement at the top of the tower at time $t = 1.0$ sec. Neglect damping.

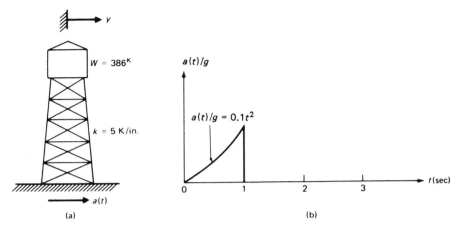

Fig. 4.17

4.9 Repeat Problem 4.8 for 20% of critical damping.

4.10 Determine for the tower of Problem 4.9, the maximum displacement at the top of the tower relative to the ground displacement.

4.11 The frame of Fig. 4.18(a) is subjected to horizontal support motion shown in Fig. 4.18(b). Determine the maximum absolute deflection at the top of the frame. Assume no damping.

4.12 Repeat Problem 4.11 for 10% of critical damping.

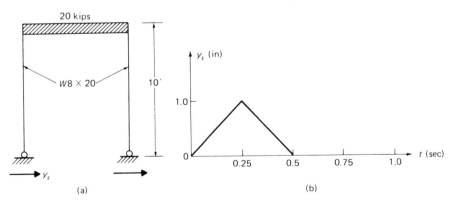

Fig. 4.18

4.13 A structural system modeled by the simple oscillator with 10% ($\xi = 0.10$) of critical damping is subjected to the impulsive load as shown in Fig. 4.19. Determine the response.

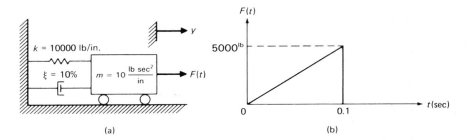

Fig. 4.19

4.14 A water tower modeled as shown in Fig. 4.20(a) is subjected to ground shock given by the function depicted in Fig. 4.20(b). Determine: (a) the maximum displacement at the top of the tower, and (b) the maximum shear force at the base of the tower. Neglect damping. Use time step for integration $\Delta t = 0.005$ sec.

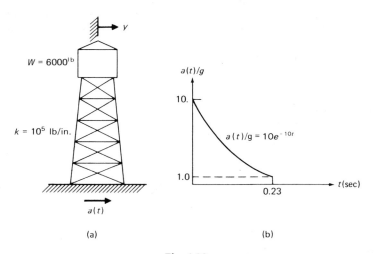

Fig. 4.20

4.15 Repeat Problem 4.14 for 20% of the critical damping.
4.16 Determine the maximum response of the tower of Problem 4.14 when subjected to the impulsive ground acceleration depicted in Fig. 4.21.

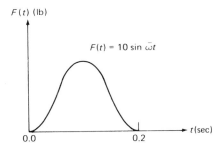

Fig. 4.21

4.17 The steel frame in Fig. 4.22 is subjected to the ground motion produced by a passing train in its vicinity. The ground motion is idealized as a harmonic acceleration of the foundation of the frame with amplitude 0.1 g at frequency 10 cps. Determine the maximum response in terms of the relative displacement of the girder of the frame and the motion of the foundation. Assume 10% of the critical damping.

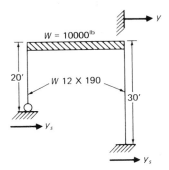

Fig. 4.22

4.18 Determine the maximum stresses in the columns of the frame in Problem 4.17 using the maximum response in terms of relative motion. Also check that the same results may be obtained using the response in terms of the maximum absolute acceleration.

4.19 A machine having a weight $W = 3000$ lb is mounted through coil springs to a steel beam of rectangular cross section as shown in Fig. 4.23(a). Due to malfunctioning, the machine produces a shock force represented in Fig. 4.23(b). Neglecting the mass of the beam and damping in the system, determine the maximum displacement of the machine.

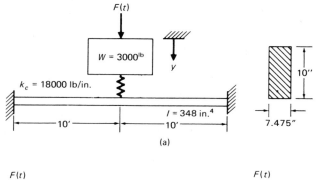

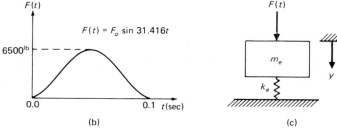

Fig. 4.23

4.20 For Problem 4.19 determine (a) the maximum tensile and compressive stresses in the beam and (b) the maximum force experienced by the coil springs during the shock.

5

Fourier Analysis and Response in the Frequency Domain

This chapter presents the application of Fourier series to determine: (1) the response of a system to periodic forces, and (2) the response of a system to nonperiodic forces in the frequency domain as an alternate approach to the usual analysis in the time domain. In either case, the calculations require the evaluation of integrals which, except for some relatively simple loading functions, employ numerical methods for their computation. Thus, in general, to make practical use of the Fourier method, it is necessary to replace the integrations with finite sums.

5.1 FOURIER ANALYSIS

The subject of Fourier series and Fourier analysis has extensive ramifications in its application to many fields of science and mathematics. We begin by considering a single degree-of-freedom system under the action of a periodic loading, that is, a forcing function that repeats itself at equal intervals of time T (the period of the function). Fourier has shown that a periodic function may be expressed as the summation of an infinite number of sine and cosine terms. Such a sum is known as a Fourier series.

For a periodic function, such as the one shown in Fig. 5.1, the Fourier series may be written as

$$F(t) = a_0 + a_1 \cos \bar{\omega}t + a_2 \cos 2\bar{\omega}t + a_3 \cos 3\bar{\omega}t + \ldots a_n \cos n\bar{\omega}t + \ldots$$
$$+ b_1 \sin \bar{\omega}t + b_2 \sin 2\bar{\omega}t + b_3 \sin 3\bar{\omega}t + \ldots b_n \sin n\bar{\omega}t + \ldots \qquad (5.1)$$

or

$$F(t) = a_0 + \sum_{n=1}^{\infty} \{a_n \cos n\bar{\omega}t + b_n \sin n\bar{\omega}t\} \qquad (5.2)$$

where $\bar{\omega} = 2\pi/T$ is the frequency and T the period of the function. The evaluation of the coefficients a_0, a_n, and b_n for a given function $F(t)$ is determined from the following expressions:

$$a_0 = \frac{1}{T} \int_{t_1}^{t_1+T} F(t)\, dt$$

$$a_n = \frac{2}{T} \int_{t_1}^{t_1+T} F(t) \cos n\bar{\omega}t\, dt$$

$$b_n = \frac{2}{T} \int_{t_1}^{t_1+T} F(t) \sin n\bar{\omega}t\, dt \qquad (5.3)$$

where t_1 in the limits of the integrals may be any value of time, but is usually equal to either $-T/2$ or zero. The constant a_0 represents the average of the periodic function $F(t)$.

5.2   RESPONSE TO A LOADING REPRESENTED BY FOURIER SERIES

The response of a single degree-of-freedom system to a periodic force represented by its Fourier series is found as the superposition of the response to each component of the series. When the transient is omitted, the response of an undamped system to any sine term of the series is given by eq. (3.9) as

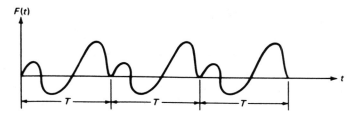

Fig. 5.1 Arbitrary periodic function.

$$y_n(t) = \frac{b_n/k}{1 - r_n^2} \sin n\bar{\omega}t \tag{5.4}$$

where $r_n = n\bar{\omega}/\omega$ and $\omega = \sqrt{k/m}$. Similarly, the response to any cosine term is

$$y_n(t) = \frac{a_n/k}{1 - r_n^2} \cos n\bar{\omega}t \tag{5.5}$$

The total response of an undamped, single degree-of-freedom system may then be expressed as the superposition of the responses to all the force terms of the series, including the response a_0/k (steady-state response) to the constant force a_0. Hence we have

$$y(t) = \frac{a_0}{k} + \sum_{n=1}^{\infty} \frac{1}{1 - r_n^2} \left(\frac{a_n}{k} \cos n\bar{\omega}t + \frac{b_n}{k} \sin n\bar{\omega}t \right) \tag{5.6}$$

When the damping in the system is considered, the steady-state response for the general sine term of the series is given from eq. (3.20) as

$$y_n(t) = \frac{b_n/k \sin (n\bar{\omega}t - \theta)}{\sqrt{(1 - r_n^2)^2 + (2r_n\xi)^2}} \tag{5.7}$$

or

$$y_n(t) = \frac{b_n}{k} \cdot \frac{\sin n\bar{\omega}t \cos \theta - \cos n\bar{\omega}t \sin \theta}{\sqrt{(1 - r_n^2)^2 + (2r_n\xi)^2}}$$

The substitution of $\sin \theta$ and $\cos \theta$ from eq. (3.21) gives

$$y_n(t) = \frac{b_n}{k} \cdot \frac{(1 - r_n^2) \sin n\bar{\omega}t - 2r_n\xi \cos n\bar{\omega}t}{(1 - r_n^2)^2 + (2r_n\xi)^2} \tag{5.8}$$

Similarly, for a cosine term of the series, we obtain

$$y_n(t) = \frac{a_n}{k} \cdot \frac{(1 - r_n^2) \cos n\bar{\omega}t + 2r_n\xi \sin n\bar{\omega}t}{(1 - r_n^2)^2 + (2r_n\xi)^2} \tag{5.9}$$

Finally, the total response is then given by the superposition of the terms expressed by eqs. (5.8) and (5.9) in addition to the response to the constant term of the series. Therefore, the total response of a damped single degree-of-freedom system may be expressed as

$$y(t) = \frac{a_0}{k} + \frac{1}{k} \sum_{n=1}^{\infty} \left\{ \frac{a_n 2r_n\xi + b_n(1 - r_n^2)}{(1 - r_n^2)^2 + (2r_n\xi)^2} \sin n\bar{\omega}t \right.$$

$$\left. + \frac{a_n(1 - r_n^2) - b_n 2r_n\xi}{(1 - r_n^2)^2 + (2r_n\xi)^2} \cos n\bar{\omega}t \right\} \tag{5.10}$$

Example 5.1. As an application of the use of Fourier series in determining the response of a system to a periodic loading, consider the undamped simple oscillator in Fig. 5.2(a) which is acted upon by the periodic force shown in Fig. 5.2(b).

Solution: The first step is to determine the Fourier series expansion of $F(t)$. The corresponding coefficients are determined from eqs. (5.3) as follows:

$$a_0 = \frac{1}{T} \int_0^T \frac{F_0}{T} t \, dt = \frac{F_0}{2}$$

$$a_n = \frac{2}{T} \int_0^T \frac{F_0}{T} t \cos n\bar{\omega}t \, dt = 0$$

$$b_n = \frac{2}{T} \int_0^T \frac{F_0}{T} t \sin n\bar{\omega}t \, dt = -\frac{F_0}{n\pi}$$

The response of the undamped system is then given from eq. (5.6) as

$$y(t) = \frac{F_0}{2k} - \sum_{n=1}^{\infty} \frac{F_0 \sin n\bar{\omega}t}{n\pi k(1 - r_n^2)}$$

or in expanded form as

$$y(t) = \frac{F_0}{2k} - \frac{F_0 \sin \bar{\omega}t}{\pi k(1 - r_1^2)} - \frac{F_0 \sin 2\bar{\omega}t}{2\pi k(1 - 4r_1^2)} - \frac{F_0 \sin 3\bar{\omega}t}{3\pi k(1 - 9r_1^2)} - \cdots$$

where

$$r_1 = \bar{\omega}/\omega, \quad \omega = \sqrt{k/m}, \text{ and } \bar{\omega} = 2\pi/T.$$

5.3 FOURIER COEFFICIENTS FOR PIECEWISE LINEAR FUNCTIONS

Proceeding as before in the evaluation of Duhamel's integral, we can represent the forcing function by piecewise linear function as shown in Fig. 5.3.

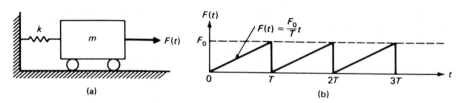

(a)   (b)

Fig. 5.2 Undamped oscillator acted upon by a periodic force.

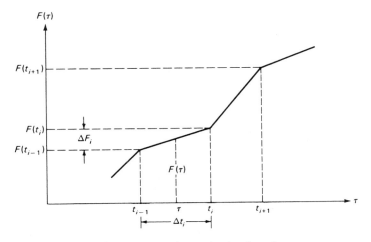

Fig. 5.3 Piecewise linear forcing function.

The calculation of Fourier coefficients, eq. (5.3), is then obtained as a summation of the integrals evaluated for each linear segment of the forcing function, that is, as

$$a_0 = \frac{1}{T} \sum_{i=1}^{N} \int_{t_{i-1}}^{t_i} F(t) \, dt \qquad (5.11)$$

$$a_n = \frac{2}{T} \sum_{i=1}^{N} \int_{t_{i-1}}^{t_i} F(t) \cos n\bar{\omega}t \, dt \qquad (5.12)$$

$$b_n = \frac{2}{T} \sum_{i=1}^{N} \int_{t_{i-1}}^{t_i} F(t) \sin n\bar{\omega}t \, dt \qquad (5.13)$$

where N is the number of segments of the piecewise forcing function. The forcing function in any interval $t_{i-1} \le t \le t_i$ is expressed by eq. (4.20) as

$$F(t) = F(t_{i-1}) + \frac{\Delta F_i}{\Delta t_i} (t - t_{i-1}) \qquad (5.14)$$

in which $\Delta F_i = F(t_i) - F(t_{i-1})$ and $\Delta t_i = t_i - t_{i-1}$. The integrals required in the expressions of a_n and b_n have been evaluated in eqs. (4.21) and (4.22) and designated as $A(t_i)$ and $B(t_i)$ in the recurrent expressions (4.18) and (4.19). The use of eqs. (4.18) through (4.22) to evaluate the coefficients a_n and b_n yields

$$a_n = \frac{2}{T} \sum_{i=1}^{N} \left\{ \frac{1}{n\bar{\omega}} \left(F(t_{i-1}) - t_{i-1} \frac{\Delta F_i}{\Delta t_i} \right) (\sin n\bar{\omega}t_i - \sin n\bar{\omega}t_{i-1}) \right.$$

$$+ \frac{\Delta F_i}{n^2\bar{\omega}^2\Delta t_i} ((\cos n\bar{\omega}t_i - \cos n\bar{\omega}t_{i-1})$$

$$\left. + n\bar{\omega}(t_i \sin n\bar{\omega}t_i - t_{i-1} \sin n\bar{\omega}t_{i-1})) \right\} \tag{5.15}$$

$$b_n = \frac{2}{T} \sum_{i=1}^{N} \left\{ \frac{1}{n\bar{\omega}} \left(F(t_{i-1}) - t_{i-1} \frac{\Delta F_i}{\Delta t_i} \right) (\cos n\bar{\omega}t_{i-1} - \cos n\bar{\omega}t_i \right.$$

$$+ \frac{\Delta F_i}{n^2\bar{\omega}^2\Delta t_i} ((\sin n\bar{\omega}t_i - \sin n\bar{\omega}t_{i-1})$$

$$\left. - n\bar{\omega}(t_i \cos n\bar{\omega}t_i - t_{i-1} \cos n\bar{\omega}t_{i-1})) \right\} \tag{5.16}$$

The integral appearing in the coefficient a_0 of eq. (5.3) is readily evaluated after substituting $F(t)$ from eq. (5.14) into eq. (5.11). This evaluation yields

$$a_0 = \frac{1}{T} \sum_{i=1}^{N} \{ \Delta t_i (F_i + F_{i-1})/2 \} \tag{5.17}$$

5.4   EXPONENTIAL FORM OF FOURIER SERIES

The Fourier series expression (5.2) may also be written in exponential form by substituting the trigonometric functions using Euler's relationships:

$$\sin n\bar{\omega} = \frac{e^{in\bar{\omega}} - e^{-in\bar{\omega}}}{2i}$$

$$\cos n\bar{\omega} = \frac{e^{in\bar{\omega}} + e^{-in\bar{\omega}}}{2} \tag{5.18}$$

The result of this substitution may be written as

$$F(t) = \sum_{n=-\infty}^{\infty} C_n e^{in\bar{\omega}t} \tag{5.19}$$

where

$$C_n = \frac{1}{T} \int_0^T F(t) e^{-in\bar{\omega}t} \, dt \tag{5.20}$$

The interval of integration in eq. (5.20) has been selected from zero to T for the periodic function. It should be noted that the exponential form for the Fourier series in eq. (5.19) has the advantage of simplicity when compared to the equivalent trigonometric series, eq. (5.2). The exponential form of the Fourier series can be used as before to determine the dynamic response of structural systems. However, a more effective method is available for the determination of the coefficients C_n as well as for the calculation of the response for a single degree of freedom excited by the force expanded as in eq. (5.19). This method, which is based on Fourier analysis for the discrete case, is presented in the next sections.

5.5   DISCRETE FOURIER ANALYSIS

When the periodic function $F(t)$ is supplied only at N equally spaced time intervals $(\Delta t = T/N)$ t_0, t_1, t_2, ..., t_{N-1}, where $t_j = j\Delta t$, the integrals in eqs. (5.3) may be replaced approximately by the summations

$$a_n = \frac{1}{T} \sum_{j=0}^{N-1} F(t_j) \cos n\bar{\omega} t_j \Delta t$$

$$b_n = \frac{1}{T} \sum_{j=0}^{N-1} F(t_j) \sin n\bar{\omega} t_j \, \Delta t, \qquad n = 0, 1, 2, \ldots \tag{5.21}$$

where $\bar{\omega} = 2\pi/T$. The above definitions for the Fourier coefficients have been slightly altered by omitting the factor 2 in the expressions for a_n and b_n. In this case eq. (5.2) is then written as

$$F(t) = 2 \sum_{n=0}^{\infty} \{a_n \cos n\bar{\omega} t + b_n \sin n\bar{\omega} t\} \tag{5.22}$$

If we use complex notation, eqs. (5.21) can be combined into a single form by defining

$$C_n = a_n - ib_n \tag{5.23}$$

and using Euler's relationship

$$e^{-in\bar{\omega} t_j} = \cos n\bar{\omega} t_j - i \sin n\bar{\omega} t_j \tag{5.24}$$

to obtain after substituting eq. (5.21) into eq. (5.23)

$$C_n = \frac{1}{T} \sum_{j=0}^{N-1} F(t_j) \, e^{-in\bar{\omega} t_j} \Delta t. \tag{5.25}$$

Substituting $t_j = j\Delta t$, $T = N\Delta t$, and $\bar{\omega} = 2\pi/T$ into eq. (5.25), we obtain

$$C_n = \frac{1}{N} \sum_{j=0}^{N-1} F(t_j) \, e^{-2\pi i(nj/N)}, \qquad n = 0, 1, 2, \ldots \tag{5.26}$$

Equation (5.26) may be considered as an approximate formula for calculating the complex Fourier coefficients in eq. (5.20). The discrete coefficients given by eq. (5.26) do not provide sufficient information to obtain a continuous function for $F(t)$; however, it is a most important fact that it does allow us to obtain all the discrete values of the series $\{F(t_j)\}$ exactly [Newland, D.E., 1984]. This fact leads to the formal definition of the *discrete Fourier transform* of the series $\{F(t_j)\}$, $j = 0, 1, 2, ..., N - 1$, given by

$$C_n = \frac{1}{N} \sum_{j=0}^{N-1} F(t_j)\, e^{-2\pi i(nj/N)}, \quad n = 0, 1, 2, ..., (N-1) \tag{5.27}$$

and its *inverse discrete Fourier transform* by

$$F(t_j) = \sum_{n=0}^{N-1} C_n\, e^{2\pi i(nj/N)}, \quad j = 0, 1, 2, ..., (N-1) \tag{5.28}$$

The range of the coefficients C_n has been limited to 0 to $(N-1)$ in order to maintain the symmetry of transform pair eqs. (5.27) and (5.28). It is important to realize that in the calculation of the summation indicated in eq. (5.28), the frequencies increase with increasing index n up to $n = N/2$. It will be shown very shortly that, for $n > N/2$, the corresponding frequencies are equal to the negative of frequencies of order $N - n$. This fact restricts the harmonic components that may be represented in the series to a maximum of $N/2$. The frequency corresponding to this maximum order $\omega_{N/2} = (N/2)\bar{\omega}$ is known as the *Nyquist frequency* or sometimes as the *folding frequency*. Moreover, if there are harmonic components above $\omega_{N/2}$ in the original function, these higher components will introduce distortions in the lower harmonic components of the series. This phenomenon is called *aliasing* [Newland, D.E., 1984, p. 118]. In view of this fact, it is recommended that the number of intervals or sampled points N should be at least twice the highest harmonic component present in the function.

The Nyquist frequency ω_y is given in radians per second by

$$\omega_y = \frac{2\pi N/2}{T} = \frac{2\pi N/2}{N\Delta t} = \frac{\pi}{\Delta t} \left(\frac{\text{rad}}{\text{sec}}\right) \tag{5.29}$$

and in cycles per second by

$$f_y = \frac{\omega_y}{2\pi} = \frac{1}{2\Delta t} \text{ (cps)} \tag{5.30}$$

As a matter of interest, Example 5.4 is presented later in this chapter to illustrate the importance of choosing the number of sampling points N for the excitation function sufficiently large to avoid spurious results due to aliasing.

Having represented an arbitrary discrete function by a finite sum, we may then also obtain as a discrete function the response of a simple oscillator excited by the harmonic components of the loading function. Again, only the steady-state response will be considered. The introduction of the unit exponential forcing function $E_n = e^{i\omega_n t}$ into the equation of motion, eq. (3.13), leads to

$$m\ddot{y} + c\dot{y} + ky = e^{i\omega_n t} \tag{5.31}$$

which has a steady-state solution of the form

$$y(t) = H(\omega_n)e^{i\omega_n t} \tag{5.32}$$

When eq. (5.32) is introduced into eq. (5.31), it is found that the function $H(\omega_n)$, which will be designated as the *complex frequency response* function, takes the form

$$H(\omega_n) = \frac{1}{k - m\omega_n^2 + ic\omega_n} \tag{5.33}$$

Upon introducing the frequency ratio

$$r_n = \frac{\omega_n}{\omega}$$

and the damping ratio

$$\xi = \frac{c}{c_{cr}} = \frac{c}{2\sqrt{km}}$$

eq. (5.33) becomes

$$H(\omega_n) = \frac{1}{k(1 + r_n^2 + 2ir_n\xi)}$$

Therefore, the response $y_n(t_j)$ at time $t_j = j\Delta t$ to a harmonic force component of amplitude C_n indicated in eq. (5.28) is given by

$$y_n(t_j) = \frac{C_n e^{2\pi i(nj/N)}}{k(1 - r_n^2 + 2ir_n\xi)} \tag{5.34}$$

and the total response due to the N harmonic force components by

$$y(t_j) = \sum_{n=0}^{N-1} \frac{C_n \, e^{2\pi i(nj/N)}}{k(1 - r_n^2 + 2ir_n\xi)} \tag{5.35}$$

where C_n is expressed in discrete form by eq. (5.27). In the determination of the response $y(t_j)$ using eq. (5.35), it is necessary to bear in mind that in eq. (5.28) the force component of the frequency of order n is equal to the negative of the component of the frequency of order $N - n$. This fact may be verified by substituting $-(N - n)$ for n in the exponential factor of eq. (5.28). In this case we obtain

$$e^{-2\pi i[(N-n)j/N]} = e^{-2\pi ij}e^{2\pi i(nj/N)} = e^{2\pi i(nj/N)} \tag{5.36}$$

since $e^{-2\pi i j} = \cos 2\pi j - i \sin 2\pi j = 1$ for all integer values of j. Equation (5.36) together with eq. (5.28) shows that harmonic components of the force corresponding to frequencies of orders n and $-(N - n)$ have the same value. As a consequence of this fact, $r_n = \omega_n/\omega$, where $\omega = \sqrt{K/m}$ should be evaluated (selecting N as an even number) as

$$\omega_n = n\bar{\omega} \qquad \text{for} \qquad n \leqslant N/2$$

and

$$\omega_n = -(N - n)\bar{\omega} \qquad \text{for} \qquad n > N/2$$

where the frequency corresponding to $n = N/2$, as already mentioned, is the highest frequency that can be considered in the discrete Fourier series.

The evaluation of the sums necessary to determine the response using the discrete Fourier transform is greatly simplified by the fact that the exponential functions involved are harmonic and extend over a range of N^2, as demonstrated in the next section.

5.6   FAST FOURIER TRANSFORM

A numerical technique is available which is efficient for computer determination of the response in the frequency domain. This method is known as the *fast Fourier transform* (FFT) [Cooley, P. M., et. al. (1965)]. The corresponding computer program is reproduced as a subroutine of computer Program 4. The response in frequency domain of a single degree-of-freedom system to a general force is given by eq. (5.35) and the coefficients required are computed from eq. (5.27). It can be seen that either eq. (5.35) or eq. (5.27) may be represented, except for sign, by the exponential function as

$$A(j) = \sum_{n=0}^{N-1} A^{(0)}(n) W_N^{jn} \tag{5.37}$$

where

$$W_N = e^{2\pi i/N} \tag{5.38}$$

The evaluation of the sum will be most efficient if the number of time increments N into which the period T is divided is a power of 2, that is,

$$N = 2^M \tag{5.39}$$

where M is an integer. In this case, the integers j and n can be expressed in binary form. For the purpose of illustration, we will consider a very simple case where the load period is divided into only eight time increments, that is, $N = 8$, $M = 3$. In this case, the indices in eqs. (5.27) and (5.35) will have the binary representation

$$j = j_0 + 2j_1 + 4j_2$$

$$n = n_0 + 2n_1 + 4n_2 \tag{5.40}$$

and eq. (5.37) may be written as

$$A(j) = \sum_{n_2=0}^{1} \sum_{n_1=0}^{1} \sum_{n_0=0}^{1} A^{(0)}(n) \, W_8^{(j_0+2j_1+4j_2)(n_0+2n_1+4n_2)} \tag{5.41}$$

The exponential factor can be written as

$$W_8^{jn} = W_8^{8(j_1n_2+2j_2n_2+j_2n_1)} \, W_8^{4n_2j_0} \, W_8^{2n_1(2j_1+j_0)} \, W_8^{n_0(4j_2+2j_1+j_0)}$$

We note that the first factor on the right-hand side is unity since from eq. (5.38)

$$W_8^{8I} = e^{2\pi i(8/8)I} = \cos 2\pi I + i \sin 2\pi I = 1$$

where $I = j_1 n_2 + 2j_2 n_2 + j_2 n_1$ is an integer. Therefore, only the remaining three factors need to be considered in the summations. These summations may be performed conveniently in sequence by introducing a new notation to indicate the successive steps in the summation process. Thus the first step can be indicated by

$$A^{(1)}(j_0, n_1, n_0) = \sum_{n_2=0}^{1} A^{(0)}(n_2, n_1, n_0) \, W_8^{4n_2j_0}$$

where $A^{(0)}(n_2, n_1, n_0) = A^{(0)}(n)$ in eq. (5.37). Similarly, the second step is

$$A^{(2)}(j_0, j_1, n_0) = \sum_{n_1=0}^{1} A^{(1)}(j_0, n_1, n_0) \, W_8^{2n_1(2j_1+j_0)}$$

and the third step (final step for $M = 3$) is

$$A^{(3)}(j_0, j_1, j_2) = \sum_{n_0=0}^{1} A^{(2)}(j_0, j_1, n_0) W_8^{n_0(4j_2+2j_1+j_0)}$$

The final result $A^{(3)}(j_0, j_1, j_2)$ is equal to $A(j)$ in eq. (5.37). This process, indicated for $N = 8$, can readily be extended to any integer $N = 2^M$. The method is particularly efficient because the results of one step are immediately used in the next step, thus reducing storage requirements and also because the exponential takes the value of unity in the first factor of the summation. The reduction in computational time which results from this formulation is significant when the time interval is divided into a large number of increments. The comparative times required for computing the Fourier series by a conventional program and by the fast Fourier transform algorithm are illustrated in Fig. 5.4. It is seen here how, for large values of N, one can rapidly consume so much computer time as to make the conventional method unfeasible.

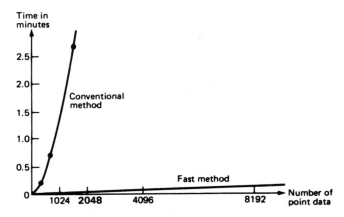

Fig. 5.4 Time required for Fourier transform using conventional and fast method (from Cooley, J. W., Lewis, P. A. W., and Welch, P. D. (1969), *IEEE Trans. Education*, **E-12** (1).

5.7   PROGRAM 4—RESPONSE IN THE FREQUENCY DOMAIN

The computer program presented in this chapter calculates the response in the frequency domain for a damped single-degree-of-freedom system. The excitation is input as a discrete function of time. The computer output consists of two tables: (1) a table giving the first N complex Fourier coefficients of the series expansions of the excitation and of the response and (2) a table giving the displacement history of the steady-state motion of the response. This second table also gives the excitation function as calculated by eq. (5.28), thus providing a check of the computations. The main body of this program performs the tasks of calculating, using the FFT algorithm, the coefficients C_n and the function $F(t_j)$ in eqs. (5.27), (5.28), and the response $y(t_j)$ in eq. (5.35).

Example 5.2.   Determine the response of the tower shown in Fig. 5.5(a) subjected to the impulsive load of duration 0.48 sec as shown in Fig. 5.5(b). Assume damping equal to 10% of the critical damping.

Solution:
Problem Data:

Mass: $m = 38600/386 = 100$ (lb · sec^2/in)

Spring constant: $k = 100,000$ (lb/in)

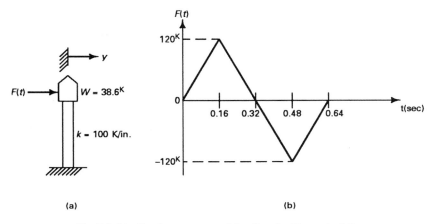

Fig. 5.5 Idealized structure and loading for Example 5.2.

Damping coefficient: $c = 2 \, \xi \sqrt{km} = 632$ (lb · sec/in)

Select M such that $2^M = 8$: $M = 3$

Excitation function:

Time (sec)	Force (lb)
0.00	0
0.16	120000
0.48	−120000
0.64	0

Input Data and Output Results

```
PROGRAM 4:--RESPONSE IN THE FREQUENCY DOMAIN--      FILE:D4

        INPUT DATA:

NUMBER OF POINTS DEFINING THE EXCITATION              NE = 4
MASS                                                  AM = 100
SPRING CONSTANT                                       AK = 100000
DAMPING COEFFICIENT                                    C = 632
EXPONENT OF 2^M FOR FFT                                M = 3
INDEX (GRAVITY OR ZERO)                                G = 0

 TIME  EXCITATION  TIME  EXCITATION  TIME  EXCITATION  TIME  EXCITATION

 0.000     0.00    0.160  120000.00  0.480  -120000.00 0.640     0.00
```

```
OUTPUT RESULTS:
```

N	FORCE FOURIER COEFFICIENTS REAL	IMAG	RESPONSE FOURIER COEFFICIENTS REAL	IMAG
0	0.0015	0.0000	0.0000	0.0000
1	-0.0003	-51213.2100	-0.0387	-0.5641
2	0.0000	0.0015	0.0000	0.0000
3	-0.0006	8786.7950	0.3132	0.2230
4	-0.0015	0.0000	0.0000	0.0000
5	-0.0023	-8786.7970	0.3132	-0.2230
6	0.0000	-0.0015	0.0000	-0.0000
7	0.0033	51213.2100	-0.0387	0.5641

TIME	DISPL. REAL	DISPL. IMAG.	FORCE REAL	FORCE IMAG.
0.000	0.5490	-0.0000	0.0000	0.0000
0.080	-0.0154	0.0000	60000.0100	-0.0020
0.160	1.5743	0.0000	120000.0000	-0.0052
0.240	0.9800	-0.0000	60000.0200	-0.0020
0.320	-0.5490	0.0000	0.0000	0.0000
0.400	0.0154	-0.0000	-60000.0100	0.0020
0.480	-1.5743	-0.0000	-120000.0000	0.0052
0.560	-0.9800	0.0000	-60000.0000	0.0020

Example 5.3.   Determine the response of the simple oscillator shown 5.6(a) when subjected to the forcing function depicted in Fig. 5.6(b). Use $M = 4$ for the exponent in $N = 2^M$. Assume 15% of the critical damping.

Solution:
Problem Data:

Mass: $m = 100/386 = 0.259$ (kip $\cdot$ sec^2/in)

Spring Constant: $k = 200$ kip/in

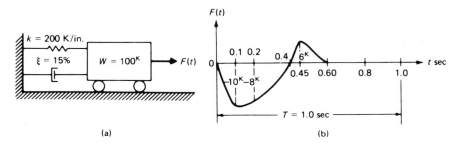

Fig. 5.6 Simple oscillator and loading for Example 5.3.

Damping coefficient: $c = 2\xi\sqrt{km}$

$$c = 2 \times 0.15 \sqrt{200 \times 0.259} = 2.159 \text{ (kip} \cdot \text{sec/in)}$$

Exponent of $N = 2^M$: $M = 4$

Gravitational Index: $G = 0$ (force on the mass)

Excitation function:

Time (sec)	Force (kip)
0.00	0
0.10	−10
0.20	−8
0.40	0
0.45	6
0.60	0
1.00	0

Input Data and Output Results

```
PROGRAM 4: --RESPONSE IN THE FREQUENCY DOMAIN--     FILE: D5.3

        INPUT DATA:

NUMBER OF POINTS DEFINING THE EXCITATION                NE = 7
MASS                                                    AM = .259
SPRING CONSTANT                                         AK = 200
DAMPING COEFFICIENT                                     C = 2.159
EXPONENT OF 2^M FOR FFT                                 M = 4
INDEX (GRAVITY OR ZERO)                                 G = 0
   TIME      EXCITATION      TIME    EXCITATION      TIME     EXCITATION
   0.000        0.00         0.100     -10.00        0.200      -8.00
   0.400        0.00         0.450       6.00        0.600       0.00
   1.000        0.00

        OUTPUT RESULTS:

FORCE FOURIER COEFFICIENTS              RESPONSE FOURIER COEFFICIENTS

N         REAL         IMAG           REAL            IMAG

0       -1.5313       0.0000        -0.0077          0.0000
1       -1.4466       1.5952        -0.0070          0.0089
2        1.1332       1.1500         0.0081          0.0058
3        0.1069      -0.0043         0.0009         -0.0004
4        0.5313       0.2813         0.0081         -0.0044
5        0.1444      -0.1824        -0.0027          0.0000
6        0.1168       0.0875        -0.0004         -0.0007
7        0.1953      -0.0828        -0.0007          0.0001
```

8	−0.0313	0.0000	0.0001	0.0000
9	0.1953	0.0828	−0.0007	−0.0001
10	0.1168	−0.0875	−0.0004	0.0007
11	0.1444	0.1824	−0.0027	−0.0000
12	0.5313	−0.2813	0.0081	0.0044
13	0.1069	0.0043	0.0009	0.0004
14	1.1332	−1.1500	0.0081	−0.0058
15	−1.4466	−1.5952	−0.0070	−0.0089

TIME	DISPL. REAL	DISPL. IMAG.	FORCE REAL	FORCE IMAG.
0.0000	0.0052	0.0000	0.0000	0.0000
0.0625	−0.0096	−0.0000	−6.2500	−0.0000
0.1250	−0.0571	0.0000	−9.5000	−0.0000
0.1875	−0.0638	−0.0000	−8.2500	0.0000
0.2500	−0.0254	0.0000	−6.0000	−0.0000
0.3125	−0.0092	−0.0000	−3.5000	0.0000
0.3750	−0.0144	0.0000	−1.0000	0.0000
0.4375	0.0056	−0.0000	4.5000	0.0000
0.5000	0.0430	0.0000	4.0000	0.0000
0.5625	0.0210	−0.0000	1.5000	−0.0000
0.6250	−0.0167	0.0000	0.0000	0.0000
0.6875	−0.0076	−0.0000	0.0000	0.0000
0.7500	0.0116	0.0000	0.0000	0.0000
0.8125	0.0018	−0.0000	0.0000	0.0000
0.8750	−0.0071	0.0000	0.0000	−0.0000
0.9375	0.0001	−0.0000	−0.0000	−0.0000

Example 5.4. Consider a single degree-of-freedom undamped system in which $k = 200$ lb/in, $m = 100$ lb $\cdot$ sec^2/in subjected to a force expressed as

$$P(t) = \sum_{n=1}^{16} 100 \cos 2\pi nt \qquad \text{(a)}$$

Determine the steady-state response of the system using Program 4 with $M = 3, 4, 5$, and 6 corresponding to $N = 8, 16, 32$, and 64 sampled points. Then discuss the results in relation to the limitations imposed by the Nyquist frequency.

Solution: The fundamental frequency of the excitation function, eq. (a), is $\omega_1 = 2\pi$ and its period $T = \omega_1/2\pi = 1$ sec. Since the highest component in eq. (a) is of order $\omega_{16} = 16\omega_1$, to avoid aliasing, the number of sampled points should be at least twice that order, that is, the minimum number of sampled points should be $N = 32$.

With a simple modification of Program 4, the applied force is calculated in the program, instead of being supplied through a numerical table as normally required by the program. The results given by the computer for this example are conveniently arranged in two tables: Table 5.1, giving the displacement response to the excitation having all 16 harmonic components as prescribed

TABLE 5.1   Displacement Response for Example 5.4 (Excitation having 16 harmonics)

Time (sec.)	Number of Sampling Points for the Excitation			
	$N = 8$	$N = 16$	$N = 32$	$N = 64$
0	0.8531	0.4201	−0.0416	−0.0416
0.125	0.9357	0.4698	−0.0153	−0.0153
0.250	1.022	0.5107	0.0052	0.0052
0.375	1.071	0.5358	0.0178	0.0178
0.500	1.089	0.5443	0.0221	0.0221
0.625	1.071	0.5358	0.0178	0.0178
0.750	1.022	0.5107	0.0052	0.0052
0.875	0.9357	0.4698	−0.0153	−0.0153
1.000	0.8531	0.4201	−0.0416	−0.0416

for this problem; and Table 5.2, showing the displacement response to a reduced number of harmonic terms in the excitation function.

For this example, in which the exciting force is supplied in 16 harmonic components, the response given in Table 5.1 corresponding to $N = 32$ or $N = 64$ may be considered the exact solution. A comparison of the response shown for sample points $N = 8$ or $N = 16$ with the exact solution ($N = 32$) dramatically demonstrates the risk of not choosing N sufficiently large enough so that none of the frequencies of the components in the exciting force exceed the Nyquist frequency. The response obtained for $N = 8$ or $N = 16$ gives spurious numerical results.

TABLE 5.2   Displacement Response for Example 5.4 (Excitation force sampled at $N = 8$ points)

Time (sec.)	Number of Harmonic Components in the Excitation Force		
	$N = 4$	$N = 8$	$N = 16$
0	−0.0375	0.4246	0.8531
0.125	−0.0153	0.4679	0.9357
0.250	0.0048	0.5112	1.022
0.375	0.0184	0.5353	1.071
0.500	0.0215	0.5446	1.089
0.625	0.0184	0.5353	1.071
0.750	0.0048	0.5112	1.022
0.875	−0.0153	0.4679	0.9357
1.000	−0.0375	0.4246	0.8531

Results in Table 5.2, which were obtained using $N = 8$ sampled points, also verify that when the exciting force contains harmonic components higher than the Nyquist frequency which corresponds, in this case, to $N_y = 4$, the results are again spurious.

A final comment is in order. The exampled presented, having equal amplitude for all the components of the exciting force, serves to emphasize the importance of choosing the number of sampling points N sufficiently large to avoid aliasing. In practical situations normally the higher harmonics have a much smaller amplitude than that of the fundamental or lower frequencies. Consequently, the distortion in the response might not be as dramatic as shown in Tables 5.1 and 5.2.

5.8   SUMMARY

In general, any periodic function may be expanded into a Fourier series, eq. (5.1), whose terms are sine and cosine functions of successive multiples of the fundamental frequency. The coefficients of these functions may be calculated by integrating over a period the product of the periodic function multiplied by a sine or cosine function, eq. (5.3). The response of the dynamic system is then obtained as the superposition of the response for each term of the Fourier series expansion of the excitation function. The extension of the Fourier series to nonperiodic functions results in integrals which are known as Fourier transforms. The discrete form of these transforms, eqs. (5.27) and (5.28), permits their use in numerical applications. An extremely efficient algorithm known as the fast Fourier transform (FFT) can save as much as 99% of the computer time otherwise consumed in the evaluation of Fourier complex coefficients for the excitation function and for the response of a dynamic system.

PROBLEMS

5.1   Determine the first three terms of the Fourier series expansion for the time-varying force shown in Fig. 5.7.

5.2   Determine the steady-state response for the damped spring-mass system shown in Fig. 5.8 that is acted upon by the forcing function of Problem 5.1.

5.3   The spring-mass system of Fig. 5.8 is acted upon by the time-varying force shown in Fig. 5.9. Assume that the force is periodic of period $T = 1$ sec and determine the steady-state response of the system by applying Fourier series expansion of $F(t)$.

5.4   The cantilever beam shown in Fig. 5.10(a) carries a concentrated weight at its free end and it is subjected to a periodic acceleration at its support which is the rectified sine function of period $T = 0.4$ sec and amplitude $\ddot{y}_0 = 180$ in/sec^2. Determine: (a) the Fourier series expansion of the forcing function, and (b) the

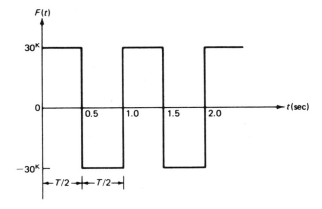

Fig. 5.7

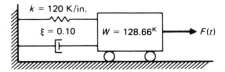

Fig. 5.8

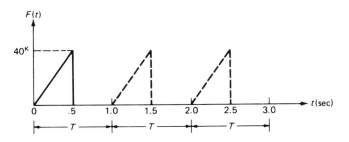

Fig. 5.9

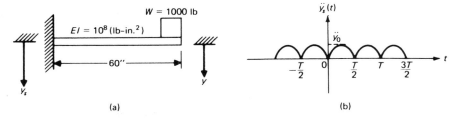

(a)   (b)

Fig. 5.10

steady-state response considering only three terms of the series. Neglect damping in the system and assume the beam massless.

5.5   Solve Problem 5.4 using Program 4. Take 16 Fourier terms. Input the values of the excitation functions at intervals of 0.025 sec.

5.6   Solve Problem 5.4 in the frequency domain using Program 4. Take the exponent of $N = 2^M$, $M = 4$. Input the effective force, $F_{eff} = -m\ddot{y}_s(t)$ calculated every 0.025 sec.

5.7   Repeat Problem 5.6 assuming 20% of critical damping.

5.8   The forcing function shown in Fig. 5.11(a) is assumed to be periodic in the extended interval $T = 1.4$ sec. Use Program 4 to determine the first eight Fourier coefficients and the steady-state response of a structure modeled by the undamped oscillator shown in Fig. 5.11(b).

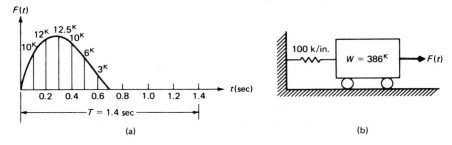

Fig. 5.11

5.9   Solve Problem 5.8 in the frequency domain using Program 4. Take $M = 4$ for the exponent in $N = 2^M$.

5.10   Use Program 4 to determine: (1) the Fourier series expansion of the forcing function shown in Fig. 5.12(a), and (2) the steady-state response calculated in the frequency domain for the spring-mass system in Fig. 5.12(b). Assume 15% of the critical damping. Take $M = 3$ for the exponent in $N = 2^M$.

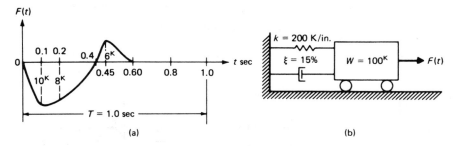

Fig. 5.12

5.11   Solve Problem 5.10 in the frequency domain using Program 4. Take $M = 4$ for the exponent in $N = 2^M$.

5.12   Consider the system shown in Fig. 5.13 and its loading with assumed period $T = 2$ sec. Determine: (a) the first four terms of the Fourier series expansion for the forcing function in terms of P_0; (b) the first four terms of the Fourier series expansion for the response.

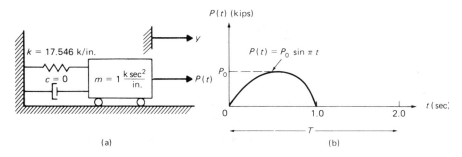

Fig. 5.13

5.13   Repeat Problem 5.12 using Program 4. Take 32 terms of Fourier series; input the force at intervals of 0.10 sec.

5.14   Repeat Problem 5.13 assuming that the system has 20% of the critical damping.

6

Generalized Coordinates and Rayleigh's Method

In the preceding chapters we concentrated our efforts in obtaining the response to dynamic loads of structures modeled by the simple oscillator, that is, structures which may be analyzed as a damped or undamped spring-mass system. Our plan in the present chapter is to discuss the conditions under which a structural system consisting of multiple interconnected rigid bodies or having distributed mass and elasticity can still be modeled as a one-degree-of-freedom system. We begin by presenting an alternative method to the direct application of Newton's Law of Motion, the *principle of virtual work*.

6.1 PRINCIPLE OF VIRTUAL WORK

An alternative approach to the direct method employed thus far for the formulation of the equations of motion is the use of the principle of virtual work. This principle is particularly useful for relatively complex structural systems which contain many interconnected parts. The principle of virtual work was originally stated for a system in equilibrium. Nevertheless, the principle can be readily applied to dynamic systems by the simple recourse

116

to D'Alembert's Principle, which establishes dynamic equilibrium by the inclusion of the inertial forces in the system.

The principle of virtual work may be stated as follows: For a system that is in equilibrium, the work done by all the forces during an assumed displacement (virtual displacement) which is compatible with the system constraints is equal to zero. In general, the equations of motion are obtained by introducing virtual displacements corresponding to each degree of freedom and equating the resulting work done to zero.

To illustrate the application of the principle of virtual work to obtain the equation of motion for a single degree-of-freedom system, let us consider the damped oscillator shown in Fig. 6.1(a) and its corresponding free body diagram in Fig. 6.1(b). Since the inertial force has been included among the external forces, the system is in "equilibrium" (dynamic equilibrium). Consequently, the principle of virtual work is applicable. If a virtual displacement δy is assumed to have taken place, the total work done by the forces shown in Fig. 6.1(b) is equal to zero, that is,

$$m\ddot{y}\delta y + c\dot{y}\delta y + ky\delta y - F(t)\,\delta y = 0$$

or

$$\{m\ddot{y} + c\dot{y} + ky - F(t)\}\,\delta y = 0 \qquad (6.1)$$

Since δy is arbitrarily selected as not equal to zero, the other factor in eq. (6.1) must equal zero. Hence,

$$m\ddot{y} + c\dot{y} + ky - F(t) = 0 \qquad (6.2)$$

Thus we obtained in eq. (6.2) the differential equation for the motion of the damped oscillator.

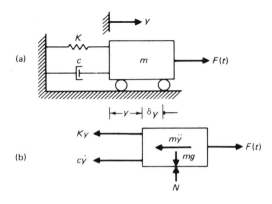

Fig. 6.1 Damped simple oscillator undergoing virtual displacement δy.

6.2 GENERALIZED SINGLE DEGREE-OF-FREEDOM SYSTEM—RIGID BODY

Most frequently the configuration of a dynamic system is specified by coordinates indicating the linear or angular positions of elements of the system. However, coordinates do not necessarily have to correspond directly to displacements; they may in general be any independent quantities which are sufficient in number to specify the position of all parts of the system. These coordinates are usually called *generalized coordinates* and their number is equal to the number of degrees of freedom of the system.

The example of the rigid-body system shown in Fig. 6.2 consists of a rigid bar with distributed mass supporting a circular plate at one end. The bar is supported by springs and dampers in addition to a single frictionless support. Dynamic excitation is provided by a transverse load $F(x,t)$ varying linearly on the portion AB of the bar. Our purpose is to obtain the differential equation of motion and to identify the corresponding expressions for the parameters of the simple oscillator representing this system.

Since the bar is rigid, the system in Fig. 6.2 has only one degree of freedom, and, therefore, its dynamic response can be expressed with one equation of motion. The generalized coordinate could be selected as the vertical displacement of any point such as A, B, or C along the bar, or may be taken as the angular position of the bar. This last coordinate designated $\theta(t)$ is selected as the generalized coordinate of the system. The corresponding free body diagram showing all the forces including the inertial forces and the inertial moments is shown in Fig. 6.3. In evaluating the displacements of the different forces, it is assumed that the displacements of the system are small and, therefore, vertical displacements are simply equal to the product of the distance to support D multiplied by the angular displacement $\theta = \theta(t)$.

The displacements resulting at the points of application of the forces in Fig. 6.3 due to a virtual displacement $\delta\theta$ are indicated in this figure. By the principle of virtual work, the total work done by the forces during this virtual displacement is equal to zero. Hence

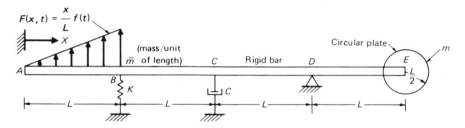

Fig. 6.2 Example of single degree-of-freedom rigid system.

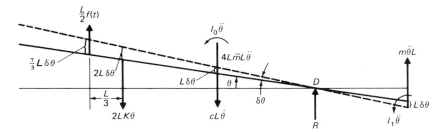

Fig. 6.3 Displacements and resultant forces for system in Fig. 6.2.

$$\delta\theta[I_0\ddot\theta + I_1\ddot\theta + 4L^3\bar m\ddot\theta + mL^2\ddot\theta + cL^2\dot\theta + 4kL^2\theta - \tfrac{7}{6}L^2\,f(t)] = 0$$

or, since $\delta\theta$ is arbitrarily set not equal to zero, it follows that

$$(I_0 + I_1 + 4L^3\bar m + mL^2)\,\ddot\theta + cL^2\dot\theta + 4kL^2\theta - \tfrac{7}{6}L^2f(t) = 0, \qquad (6.3)$$

where

$$I_0 = \tfrac{1}{12}(4\bar mL)(4L)^2 = \text{mass moment of inertia of the rod}$$

$$I_1 = \tfrac{1}{2}m\left(\frac{L}{2}\right)^2 = \text{mass moment of inertia of the circular plate}$$

The differential equation (6.3) governing the motion of this system may conveniently be written as

$$I^*\ddot\theta + C^*\dot\theta + K^*\theta = F^*(t) \qquad (6.4)$$

where I^*, C^*, K^*, and $F^*(t)$ are, respectively, the generalized inertia, generalized damping, generalized stiffness, and generalized load for this system. These quantities are given in eq. (6.3) by the factors corresponding to the acceleration, velocity, displacement, and force terms, namely,

$$I^* = I_0 + I_1 + 4\bar mL^3 + mL^2$$

$$C^* = cL^2$$

$$K^* = 4kL^2$$

$$F^*(t) = \tfrac{7}{6}L^2\,f(t)$$

Example 6.1. For the system shown in Fig. 6.4, determine the generalized physical properties M^*, C^*, K^* and generalized loading $F^*(t)$. Let $Y(t)$ at the point A_2 in Fig. 6.4 be the generalized coordinate of the system.

Solution: The free body diagram for the system is depicted in Fig. 6.5 which shows all the forces on the two bars of the sytem including the inertial force

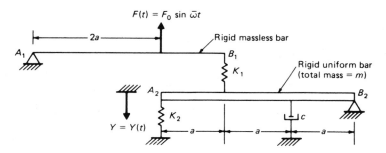

Fig. 6.4 System for Example 6.1.

and the inertial moment. The generalized coordinate is $Y(t)$ and the displacement of any point in the system should be expressed in terms of this coordinate; nevertheless, for convenience, we select also the auxiliary coordinate $Y_1(t)$ as indicated in Fig. 6.5.

The summation of the moments about point A_1 of all the forces acting on bar A_1-B_1, and the summation of moments about B_2 of the forces on bar A_2-B_2, give the following equations:

$$k_1(\tfrac{2}{3}Y - Y_1)3a = 2aF_0 \sin \bar{\omega}t \tag{6.5}$$

$$\frac{I_0}{3a}\ddot{Y} + \frac{3}{4}ma\ddot{Y} + \frac{a}{3}c\dot{Y} + k_1\left(\frac{2}{3}Y - Y_1\right)2a + 3ak_2Y = 0 \tag{6.6}$$

Substituting Y_1 from eq. (6.5) into eq. (6.6), we obtain the differential equation for the motion of the system in terms of the generalized coordinate $Y(t)$, namely

$$M^*\ddot{Y}(t) + C^*\dot{Y}(t) + K^*Y(t) = F^*(t)$$

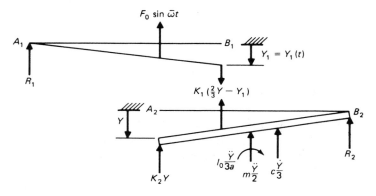

Fig. 6.5 Displacements and resultant forces for Example 6.1.

where the generalized quantities are given by

$$M^\star = \frac{I_0}{3a^2} + \frac{3m}{4}$$

$$C^\star = \frac{c}{3}$$

$$K^\star = 3k_2$$

and

$$F^\star(t) = -\tfrac{4}{3} F_0 \sin \bar{\omega} t$$

6.3 GENERALIZED SINGLE DEGREE-OF-FREEDOM SYSTEM—DISTRIBUTED ELASTICITY

The example presented in the preceding section had only one degree of freedom in spite of the complexity of the various parts of the system because the two bars were interconnected through a spring and one of the bars was massless so that only one coordinate sufficed to completely specify the motion. If the bars were not rigid, but could deform in flexure, the system would have an infinite number of degrees of freedom. However, a single degree-of-freedom analysis could still be made, provided that only a single shape could be developed during motion, that is, provided that the knowledge of the displacement of a single point in the system determines the displacement of the entire system.

As an illustration of this method for approximating the analysis of a system with an infinite number of degrees of freedom with a single degree of freedom, consider the cantilever beam shown in Fig. 6.6. In this illustration, the physical properties of the beam are the flexural stiffness $EI(x)$ and its mass per unit of length $m(x)$. It is assumed that the beam is subjected to an arbitrary distributed forcing function $p(x, t)$ and to an axial compressive force N.

In order to approximate the motion of this system with a single coordinate, it is necessary to assume that the beam deflects during its motion in a prescribed shape. Let $\phi(x)$ be the function describing this shape and, as a generalized coordinate, $Y(t)$ the function describing the displacement of the motion corresponding to the free end of the beam. Therefore, the displacement at any point x along the beam is

$$y(x, t) = \phi(x) Y(t) \tag{6.7}$$

where $\phi(L) = 1$.

The equivalent one-degree-of-freedom system [Fig. 6.6(b)] may be defined simply as the system for which the kinetic energy, potential energy (strain

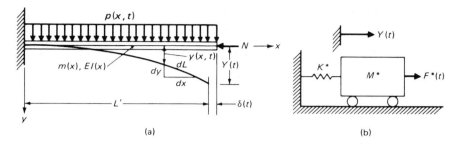

Fig. 6.6 Single degree-of-freedom continuous system.

energy), and work done by the external forces have at all times the same values in the two systems.

The kinetic energy T of the beam in Fig. 6.6 vibrating in the pattern indicated by eq. (6.7) is

$$T = \int_0^L \tfrac{1}{2} m(x)\{\phi(x)\dot{Y}(t)\}^2 \, dx \tag{6.8}$$

Equating this expression for the kinetic energy of the continuous system to the kinetic energy of the equivalent single degree-of-freedom system $\tfrac{1}{2}M^* \dot{Y}(t)^2$ and solving the resulting equation for the generalized mass, we obtain

$$M^* = \int_0^L m(x)\phi^2(x) \, dx \tag{6.9}$$

The flexural strain energy V of a prismatic beam may be determined as the work done by the bending moment $M(x)$ undergoing an angular displacement $d\theta$. This angular displacement is obtained from the well-known formula for the flexural curvature of a beam, namely

$$\frac{d^2y}{dx^2} = \frac{d\theta}{dx} = \frac{M(x)}{EI} \tag{6.10}$$

or

$$d\theta = \frac{M(x)}{EI} \, dx \tag{6.11}$$

since $dy/dx = \theta$, where θ, being assumed small, is taken as the slope of the elastic curve. Consequently, the strain energy is given by

$$V = \int_0^L \tfrac{1}{2} M(x) \, d\theta \tag{6.12}$$

The factor $\frac{1}{2}$ is required for the correct evaluation of the work done by the flexural moment increasing from zero to its final value $M(x)$ [average value $M(x)/2$]. Now, utilizing eqs. (6.10) and (6.11) in eq. (6.12), we obtain

$$V = \int_0^L \frac{1}{2} EI(x) \left(\frac{d^2y}{dx^2} \right)^2 dx \tag{6.13}$$

Finally, equating the potential energy, eq. (6.13), for the continuous system to the potential energy of the equivalent system and using eq. (6.7) results in

$$\frac{1}{2} K^* Y(t)^2 = \int_0^L \frac{1}{2} EI(x)\{\phi''(x)Y(t)\}^2 \, dx$$

or

$$K^* = \int_0^L EI(x)\{\phi''(x)\}^2 \, dx \tag{6.14}$$

where

$$\phi''(x) = \frac{d^2y}{dx^2}$$

The generalized force $F^*(t)$ may be found from the virtual displacement $\delta Y(t)$ of the generalized coordinate $Y(t)$ upon equating the work performed by the external forces in the structure to the work done by the generalized force in the equivalent single degree-of-freedom system. The work of the distributed external force $p(x, t)$ due to this virtual displacement is given by

$$W = \int_0^L p(x, t) \, \delta y \, dx$$

Substituting $\delta y = \phi(x)\delta Y$ from eq. (6.7) gives

$$W = \int_0^L p(x, t)\phi(x) \, \delta Y \, dx \tag{6.15}$$

The work of the generalized force $F^*(t)$ in the equivalent system corresponding to the virtual displacement δY of the generalized coordinate is

$$W^* = F^*(t) \, \delta Y \tag{6.16}$$

Equating eq. (6.15) with eq. (6.16) and canceling the factor δY, which is taken to be different from zero, we obtain the generalized force as

$$F^*(t) = \int_0^L p(x, t)\phi(x) \, dx \tag{6.17}$$

Similarly, to determine the generalized damping coefficient, assume a virtual displacement and equate the work of the damping forces in the physical system with the work of the damping force in the equivalent single degree-of-freedom system. Hence

$$C^* \dot{Y} \delta Y = \int_0^L c(x) \dot{y} \, \delta y \, dx$$

where $c(x)$ is the distributed damping coefficient per unit length along the beam. Substituting $\delta y = \phi(x) \, \delta Y$ and $\dot{y} = \phi(x) \, \dot{Y}$ from eq. (6.7) and canceling the common factors, we obtain

$$C^* = \int_0^L c(x)[\phi(x)]^2 \, dx \tag{6.18}$$

which is the expression for the generalized damping coefficient.

To calculate the potential energy of the axial force N which is unchanged during the vibration of the beam and consequently is a conservative force, it is necessary to evaluate the horizontal component of the motion $\delta(t)$ of the free end of the beam as indicated in Fig. 6.6. For this purpose, we consider a differential element of length dL along the beam as shown in Fig. 6.6(a). The length of this element may be expressed as

$$dL = (dx^2 + dy^2)^{1/2}$$

or

$$dL = (1 + (dy/dx)^2)^{1/2} \, dx \tag{6.19}$$

Now, integrating over the horizontal projection of the length of beam (L') and expanding in series the binomial expression, we obtain

$$L = \int_0^{L'} \left(1 + \left(\frac{dy}{dx} \right)^2 \right)^{1/2} dx$$

$$= \int_0^{L'} \left\{ 1 + \frac{1}{2} \left(\frac{dy}{dx} \right)^2 - \frac{1}{8} \left(\frac{dy}{dx} \right)^4 + \ldots \right\} dx$$

Retaining only the first two terms of the series results in

$$L = L' + \int_0^L \frac{1}{2} \left(\frac{dy}{dx} \right)^2 dx \tag{6.20}$$

or

$$\delta(t) = L - L' = \int_0^L \frac{1}{2} \left(\frac{dy}{dx} \right)^2 dx \tag{6.21}$$

The reader should realize that eqs. (6.20) and (6.21) involve approximations since the series was truncated and the upper limit of the integral in the final expression was conveniently set equal to the initial length of the beam L instead of to its horizontal component L'.

Now we define a new stiffness coefficient to be called the generalized geometric stiffness K_G^* as the stiffness of the equivalent system required to store the same potential energy as the potential energy stored by the normal force N, that is,

$$\tfrac{1}{2} K_G^* Y(t)^2 = N\delta(t)$$

Substituting $\delta(t)$ from eq. (6.21) and the derivative dy/dx from eq. (6.7), we have

$$\frac{1}{2} K_G^* Y(t)^2 = \frac{1}{2} N \int_0^L \left\{ Y(t) \frac{d\phi}{dx} \right\}^2 dx$$

or

$$K_G^* = N \int_0^L \left(\frac{d\phi}{dx} \right)^2 dx \tag{6.22}$$

Equations (6.9), (6.14), (6.17), (6.18), and (6.22) give, respectively, the generalized expression for the mass, stiffness, force, damping, and geometric stiffness for a beam with distributed properties and load, modeling it as a simple oscillator.

For the case of an axial compressive force, the potential energy in the beam decreases with a loss of stiffness in the beam. The opposite is true for a tensile axial force, which results in an increase of the flexural stiffness of the beam. Customarily, the geometric stiffness is determined for a compressive axial force. Consequently, the combined generalized stiffness K_c^* is then given by

$$K_c^* = K^* - K_G^* \tag{6.23}$$

Finally, the differential equation for the equivalent system may be written as

$$M^* \ddot{Y}(t) + C^* \dot{Y}(t) + K_c^* Y(t) = F^*(t) \tag{6.24}$$

The critical buckling load N_{cr} is defined as the axial compressive load that reduces the combined stiffness to zero, that is,

$$K_c^* = K^* - K_G^* = 0$$

The substitution of K^* and K_G^* from eqs. (6.14) and (6.22) gives

$$\int_0^L EI \left(\frac{d^2\phi}{dx^2} \right)^2 dx - N_{cr} \int \left(\frac{d\phi}{dx} \right)^2 dx = 0$$

and solving for the critical buckling load, we obtain

$$N_{cr} = \frac{\displaystyle\int_0^L EI(d^2\phi/dx^2)^2\, dx}{\displaystyle\int_0^L (d\phi/dx)^2\, dx} \qquad (6.25)$$

To provide an example of the determination of the equivalent one degree of freedom for a system with distributed mass and stiffness, consider the water tower in Fig. 6.7 to have uniformly distributed mass $\bar{m}$ and stiffness EI along its length with a concentrated mass $M = \bar{m}L$ at the top. The tower is subjected to an earthquake ground motion excitation of acceleration $a_g(t)$ and to an axial compressive load due to the weight of its distributed mass and concentrated mass at the top. Neglect damping in the system. Assume that during the motion the shape of the tower is given by

$$\phi(x) = 1 - \cos\frac{\pi x}{2L} \qquad (6.26)$$

Selecting the lateral displacement $Y(t)$ at the top of the tower as the generalized coordinate as shown in Fig. 6.7, we obtain for the displacement at any point

$$y(x, t) = Y(t)\phi(x) = Y(t)\left(1 - \cos\frac{\pi x}{2L}\right) \qquad (6.27)$$

The generalized mass and the generalized stiffness of the tower are computed, respectively, from eqs. (6.9) and (6.14) as

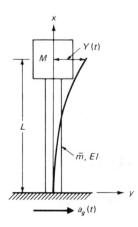

Fig. 6.7 Water tower with distributed properties for Example 6.2.

$$M^\star = \bar{m}L + \bar{m} \int_0^L \left(1 - \cos\frac{\pi x}{2L}\right)^2$$

$$M^\star = \frac{\bar{m}L}{2\pi}(5\pi - 8) \tag{6.28}$$

and

$$K^\star = \int_0^L EI\left(\frac{\pi}{2L}\right)^4 \cos^2\frac{\pi x}{2L}\,dx$$

$$K^\star = \frac{\pi^4 EI}{32L^3} \tag{6.29}$$

The axial force is due to the weight of the tower above a particular section, including the concentrated weight at the top, and may be expressed as

$$N(x) = \bar{m}Lg\left(2 - \frac{x}{L}\right) \tag{6.30}$$

where g is the gravitational acceleration. Since the normal force in this case is a function of x, it is necessary in using eq. (6.22) to include $N(x)$ under the integral sign. The geometric stiffness coefficient $K_G^\star$ is then given by

$$K_G^\star = \int_0^L \bar{m}Lg\left(2 - \frac{x}{L}\right)\left(\frac{\pi}{2L}\right)^2 \sin^2\frac{\pi x}{2L}\,dx$$

which upon integration yields

$$K_G^\star = \frac{\bar{m}g}{16}(3\pi^2 - 4) \tag{6.31}$$

Consequently, the combined stiffness from eqs. (6.29) and (6.31) is

$$K_c^\star = K^\star - K_G^\star = \frac{\pi^4 EI}{32L^3} - \frac{\bar{m}g}{16}(3\pi^2 - 4) \tag{6.32}$$

By setting $K_c^\star = 0$, we obtain

$$\frac{\pi^4 EI}{32L^3} - \frac{\bar{m}g}{16}(3\pi^2 - 4) = 0$$

which gives the critical load

$$(\bar{m}g)_{cr} = \frac{\pi^4 EI}{2(3\pi^2 - 4)L^3} \tag{6.33}$$

The equation of motion in terms of the relative motion $u = y(t) - y_g(t)$ is given by eq. (3.38) for the undamped system as

$$M^* \ddot{u} + K_c^* u = F_{\text{eff}}^*(t) \tag{6.34}$$

where M^* is given by eq. (6.28), K_c^* by eq. (6.32), and the effective force by eq. (6.17) for the effective distributed force and by $-\bar{m}La_g(t)$ for the effective concentrated force at the top of the tower. Hence

$$F_{\text{eff}}^*(t) = \int_0^L p_{\text{eff}}(x, t)\phi(x)\, dx - \bar{m}La_g(t)$$

where $p_{\text{eff}}(x, t) = -\bar{m}a_g(t)$ is the effective distributed force. Then

$$F_{\text{eff}}^* = \int_0^L -\bar{m}a_g(t)\phi(x)\, dx - \bar{m}La_g(t)$$

Substitution of $\phi(x)$ from eq. (6.26) into the last equation yields upon integration

$$F_{\text{eff}}^* = -\frac{2\bar{m}a_g(t)L}{\pi}(\pi - 1) \tag{6.35}$$

Example 6.2   As a numerical example of calculating the response of a system with distributed properties, consider the water tower shown in Fig. 6.7 excited by a sinusoidal ground acceleration $a_g(t) = 20 \sin 6.36t$ (in/sec^2).

Model the structures by assuming the shape given by eq. (6.26) and determine the response.

Solution: The numerical values for this example are:

$$\bar{m} = 0.1 \text{ k} \cdot \text{sec}^2/\text{in per unit of length}$$

$$EI = 1.2\ 10^{13} \text{ k} \cdot \text{in}^2$$

$$L = 100 \text{ ft} = 1200 \text{ in}$$

$$\bar{\omega} = 6.36 \text{ rad/sec}$$

From eq. (6.28) the generalized mass is

$$M^* = \frac{0.1 \times 1200}{2\pi}(5\pi - 8) = 147.21\ \frac{\text{k} \cdot \text{sec}^2}{\text{in}}$$

and from eq. (6.32) the generalized stiffness is

$$K_c^* = \frac{\pi^4 1.2\ 10^{13}}{32 \times (1200)^3} - \frac{0.1 \times 386}{16}(3\pi^2 - 4) = 21{,}077 \text{ k/in}$$

The natural frequency is

$$\omega = \sqrt{K_c^*/M^*} = 11.96 \text{ rad/sec}$$

and the frequency ratio

$$r = \frac{\bar{\omega}}{\omega} = \frac{6.36}{11.96} = 0.532$$

From eq. (6.35) the effective force is

$$F_{\text{eff}}^{\star} = -\frac{2(0.1)(1200)(\pi - 1)}{\pi} \, 20.0 \sin 6.36t$$

or

$$F_{\text{eff}}^{\star} = -3272 \sin 6.36t \;\text{(kip)}$$

Hence the response (neglecting damping) in terms of relative motion is given from eq. (3.9) as

$$u = \frac{F_{\text{eff}}^{\star}/k^{\star}}{1 - r^2} \sin \bar{\omega}t$$

$$= -\frac{3272/21{,}077}{1 - (0.532)^2} \sin 6.36t$$

$$= -0.217 \sin 6.36t \;\text{in} \qquad\qquad \text{(Ans.)}$$

6.4 RAYLEIGH'S METHOD

In the preceding sections of this chapter the differential equation for a vibrating system was obtained by application of the principle of virtual work as an alternative method of considering the dynamic equilibrium of the system. However, the differential equation of motion for an undamped system in free vibration may also be obtained with the application of the *Principle of Conservation of Energy*. This principle may be stated as follows: If no external forces are acting on the system and there is no dissipation of energy due to damping, then the total energy of the system must remain constant during motion and consequently its derivative with respect to time must be equal to zero.

To illustrate the application of the Principle of Conservation of Energy in obtaining the differential equation of motion, consider the spring-mass system shown in Fig. 6.8. The total energy in this case consists of the sum of the kinetic energy of the mass and the potential energy of the spring. In this case the kinetic energy T is given by

$$T = \tfrac{1}{2} m \dot{y}^2 \qquad\qquad (6.36)$$

where $\dot{y}$ is the instantaneous velocity of the mass.

The force in the spring, when displaced y units from the equilibrium po-

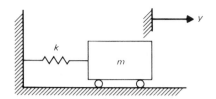

Fig. 6.8 Spring-mass system in free vibration.

sition, is ky and the work done by this force on the mass for an additional displacement dy is $- ky\, dy$. This work is negative because the force ky acting on the mass is opposite to the incremental displacement dy given in the positive direction of coordinate y. However, by definition, the potential energy is the value of this work but with opposite sign. It follows then that the total potential energy V in the spring for a final displacement y will be

$$V = \int_0^y ky\,dy = \tfrac{1}{2} ky^2 \tag{6.37}$$

Adding eq. (6.36) and (6.37), and setting this sum equal to a constant, will give

$$\tfrac{1}{2} m\dot{y}^2 + \tfrac{1}{2}ky^2 = C_0 \tag{6.38}$$

Differentiation with respect to time yields

$$m\dot{y}\ddot{y} + ky\dot{y} = 0$$

Since $\dot{y}$ cannot be zero for all values of t, it follows that

$$m\ddot{y} + ky = 0 \tag{6.39}$$

This equation is identical with eq. (1.11) of Chapter 1 obtained by application of Newton's Law of Motion. Used in this manner, the energy method has no particular advantage over the equilibrium method. However, in many practical problems it is only the natural frequency that is desired. Consider again the simple oscillator of Fig. 6.8, and assume that the motion is harmonic. This assumption leads to the equation of motion of the form

$$y = C \sin (\omega t + \alpha) \tag{6.40}$$

and velocity

$$\dot{y} = \omega C \cos (\omega t + \alpha) \tag{6.41}$$

where C is the maximum displacement and ωC the maximum velocity. Then, at the neutral position $(y = 0)$, there will be no force in the spring and the potential energy is zero. Consequently, the entire energy is then kinetic energy and

$$T_{\text{max}} = \tfrac{1}{2} m(\omega C)^2 \tag{6.42}$$

At the maximum displacement the velocity of the mass is zero and all the energy is then potential energy, thus

$$V_{\text{max}} = \tfrac{1}{2} k C^2 \tag{6.43}$$

The energy in the system changes gradually over one-quarter of the cycle from purely kinetic energy, as given by eq. (6.42), to purely potential energy, as given by eq. (6.43). If no energy has been added or lost during the quarter cycle, the two expressions for this energy must be equal. Thus

$$\tfrac{1}{2} m \omega^2 C^2 = \tfrac{1}{2} k C^2 \tag{6.44}$$

Canceling common factors and solving eq. (6.44) will give

$$\omega = \sqrt{\frac{k}{m}} \tag{6.45}$$

which is the natural frequency for the simple oscillator obtained previously from the differential equation of motion. This method, in which the natural frequency is obtained by equating maximum kinetic energy with maximum potential energy, is known as *Rayleigh's Method*.

Example 6.3. In the previous calculations on the spring-mass system, the mass of the spring was assumed to be so small that its effect on the natural frequency could be neglected. A better approximation to the true value of the natural frequency may be obtained using Rayleigh's Method. The distributed mass of the spring could easily be considered in the calculation by simply assuming that the deflection of the spring along its length is linear. In this case, consider in Fig. 6.9 the spring-mass system for which the spring has a length L and a total mass m_s. Use Rayleigh's Method to determine the fraction of the spring mass that should be added to the vibrating mass.

Solution: The displacement of an arbitrary section of the spring at a distance s from the support will now be assumed to be $u = sy/L$. Assuming that

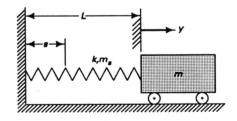

Fig. 6.9 Spring-mass system with heavy spring.

the motion of the mass m is harmonic and given by eq. (6.40), we obtain

$$u = \frac{s}{L} y = \frac{s}{L} C \sin{(\omega t + \alpha)} \qquad (6.46)$$

The potential energy of the uniformly stretched spring is given by eq. (6.37) and its maximum value is

$$V_{max} = \tfrac{1}{2} k C^2 \qquad (6.47)$$

A differential element of the spring of length ds has mass equal to $m_s ds/L$ and maximum velocity $\dot{u}_{max} = \omega u_{max} = \omega s C/L$. Consequently the total kinetic energy in the system at its maximum value is

$$T_{max} = \int_0^L \frac{1}{2} \frac{m_s}{L} ds \left(\omega \frac{s}{L} C \right)^2 + \frac{1}{2} m \omega^2 C^2 \qquad (6.48)$$

After integrating eq. (6.48) and equating it with eq. (6.47), we obtain

$$\frac{1}{2} k C^2 = \frac{1}{2} \omega^2 C^2 \left(m + \frac{m_s}{3} \right) \qquad (6.49)$$

Solving for the natural frequency yields

$$\omega = \sqrt{\frac{k}{m + m_s/3}} \qquad (6.50)$$

or in cycles per second (cps),

$$f = \frac{1}{2\pi} \sqrt{\frac{k}{m + m_s/3}} \qquad (6.51)$$

The application of Rayleigh's Method shows that a better value for the natural frequency may be obtained by adding one-third of the mass of the spring to that of the main vibrating mass.

Rayleigh's Method may also be used to determine the natural frequency of a continuous system provided that the deformed shape of the structure is described as a generalized coordinate. The deformed shape of continuous structures and also of discrete structures of multiple degrees of freedom could in general be assumed arbitrarily. However, in practical applications, the success of the method depends on how close the assumed deformed shape will come to match the actual shape of the structure during vibration. Once the deformed shape has been specified, the maximum kinetic energy and the maximum potential energy may be determined by application of pertinent equations such as eqs. (6.8) and (6.13). However, if the deformed shape has been defined as the shape resulting from statically applied forces, it would be simpler to calculate the work done by the external forces, instead of directly determining the potential energy. Consequently, in this case, the maximum kinetic energy is equated to the work of the forces applied statically. The following examples illustrate the application of Rayleigh's Method to systems with distributed properties.

Example 6.4.   Determine the natural frequency of vibration of a cantilever beam with a concentrated mass at its end when the distributed mass of the beam is taken into account. The beam has a total mass m_b and length L. The flexural rigidity of the beam is EI and the concentrated mass at its end is m, as shown in Fig. 6.10.

Solution: It will be assumed that the shape of deflection curve of the beam is that of the beam acted upon by a concentrated force F applied at the free end as shown in Fig. 6.10(b). For this static load the deflection at a distance x from the support is

$$u = \frac{3y}{L^3}\left(\frac{Lx^2}{2} - \frac{x^3}{6}\right) \tag{6.52}$$

where y = deflection at the free end of the beam. Upon substitution into eq. (6.52) of $y = C \sin(\omega t + \alpha)$, which is the harmonic deflection of the free end, we obtain

$$u = \frac{3x^2L - x^3}{2L^3} C \sin(\omega t + \alpha) \tag{6.53}$$

The potential energy is equated to the work done by the force F as it gradually increases from zero to the final value F. This work is equal to $\frac{1}{2}Fy$, and its maximum value which is equal to the maximum potential energy is then

$$V_{max} = \tfrac{1}{2} FC = \frac{3EI}{2L^3} C^2 \tag{6.54}$$

since the force F is related to the maximum deflection by the formula from elementary strength of materials,

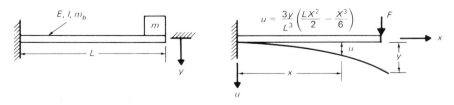

Fig. 6.10 (a) Cantilever beam of uniform mass with a mass concentrated at its tip. (b) Assumed deflection curve.

$$y_{max} = C = \frac{FL^3}{3EI} \tag{6.55}$$

The kinetic energy due to the distributed mass of the beam is given by

$$T = \int_0^L \frac{1}{2}\left(\frac{m_b}{L}\right) \dot{u}^2 \, dx \tag{6.56}$$

and using eq. (6.53) the maximum value for total kinetic energy will then be

$$T_{max} = \frac{m_b}{2L} \int_0^L \left(\frac{3x^2L - x^3}{2L^3}\,\omega C\right)^2 dx + \frac{m}{2}\omega^2 C^2 \tag{6.57}$$

After integrating eq. (6.57) and equating it with eq. (6.54), we obtain

$$\frac{3EI}{2L^3}C^2 = \frac{1}{2}\omega^2 C^2\left(m + \frac{33}{140}m_b\right) \tag{6.58}$$

and the natural frequency becomes

$$f = \frac{\omega}{2\pi} = \frac{1}{2\pi}\sqrt{\frac{3EI}{L^3\left(m + \dfrac{33}{140}m_b\right)}} \tag{6.59}$$

It is seen, then, that by concentrating a mass equal to $(33/140)\,m_b$ at the end of the beam, a more accurate value for the natural frequency of the cantilever beam is obtained compared to the result obtained by simply neglecting its distributed mass. In practice the fraction $33/140$ is rounded to $1/4$, thus approximating the natural frequency of a cantilever beam by

$$f = \frac{1}{2\pi}\sqrt{\frac{3EI}{L^3\left(m + \dfrac{m_b}{4}\right)}} \tag{6.60}$$

The approximation given by either eq. (6.59) or eq. (6.60) is a good one even for the case in which $m = 0$. For this case the error given by these formulas is about 1.5% compared to the exact solution which will be presented in Chapter 20.

Example 6.5.   Consider in Fig. 6.11 the case of a simple beam carrying several concentrated masses. Neglect the mass of the beam and determine an expression for the natural frequency by application of Rayleigh's Method.

Solution: In the application of Rayleigh's Method, it is necessary to choose a suitable curve to represent the deformed shape that the beam will have during vibration. A choice of a shape that gives consistently good results is

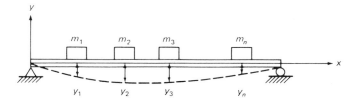

Fig. 6.11 Simple beam carrying concentrated masses.

the curve produced by forces proportional to the magnitude of the masses acting on the structure. For the simple beam, these forces could be assumed to be the weights $W_1 = m_1g$, $W_2 = m_2g$, ..., $W_N = m_Ng$ due to gravitational action on the concentrated masses. The static deflections under these weights may then be designated by y_1, y_2, ..., y_N. The potential energy is then equal to work done during the loading of the beam, thus,

$$V_{max} = \tfrac{1}{2}\,W_1y_1 + \tfrac{1}{2}\,W_2y_2 + \ldots + \tfrac{1}{2}\,W_Ny_N \tag{6.61}$$

For harmonic motion in free vibration, the maximum velocities under the weights would be ωy_1, ωy_2, ..., ωy_N, and therefore the maximum kinetic energy would be

$$T_{max} = \frac{1}{2}\frac{W_1}{g}(\omega y_1)^2 + \frac{1}{2}\frac{W_2}{g}(\omega y_2)^2$$

$$+ \ldots + \frac{1}{2}\frac{W_N}{g}(\omega y_N)^2 \tag{6.62}$$

When the maximum potential energy, eq. (6.61), is equated with the maximum kinetic energy, eq. (6.62), the natural frequency is found to be where

$$\omega = \sqrt{\frac{g(W_1y_1 + W_2y_2 + \ldots + W_Ny_N)}{W_1y_1^2 + W_2y_2^2 + \ldots + W_Ny_N^2}} \tag{6.63}$$

or

$$\omega = \sqrt{\frac{g\displaystyle\sum_{i=1}^{N} W_iy_i}{\displaystyle\sum_{i=1}^{N} W_iy_i^2}}$$

y_i is the deflection at coordinate i and W_i the weight at this coordinate.

This method is directly applicable to any beam, but in applying the method, it must be remembered that these are not gravity forces at all but substituted forces for the inertial forces. For example, in the case of a simple beam with

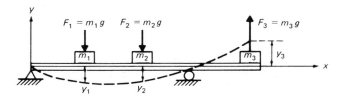

Fig. 6.12 Overhanging massless beam carrying concentrated masses.

overhang (Fig. 6.12) the force at the free end should be proportional to m_3 (F_3 = m_3g) but directed upward in order to obtain the proper shape for the deformed beam.

In the application of Rayleigh's method, the forces producing the deflected shape do not necessarily have to be produced by gravitational forces. The only requirement is that these forces produce the expected deflection shape for the fundamental mode. For example, if the deflected shape for the beam shown in Fig. 6.11 is produced by forces designated by f_1, f_2, ..., f_N instead of the gravitation forces W_1, W_2, ..., W_N, we will obtain, as in eq. (6.61), the maximum potential energy

$$V_{max} = \tfrac{1}{2} f_1 y_1 + \tfrac{1}{2} f_2 y_2 + \dots + \tfrac{1}{2} f_N y_N \tag{6.64}$$

which equated to the maximum kinetic energy, eq. (6.62), will result in the following formula for the fundamental frequency:

$$\omega = \sqrt{\dfrac{g \displaystyle\sum_{i=1}^{N} f_i y_i}{\displaystyle\sum_{i=1}^{N} W_i y_i^2}} \tag{6.65}$$

Consequently, the fundamental period could be calculated as

$$T = \dfrac{2\pi}{\omega} = 2\pi \sqrt{\dfrac{\displaystyle\sum_{i=1}^{N} W_i y_i^2}{g \displaystyle\sum_{i=1}^{N} f_i y_i}} \tag{6.66}$$

6.5   IMPROVED RAYLEIGH'S METHOD

The concept of applying inertial forces as static loads in determining the deformed shape for Rayleigh's Method may be used in developing an improved scheme for the method. In the application of the improved Rayleigh's Method,

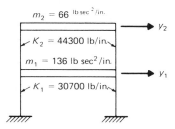

Fig. 6.13 Two-story frame for Example 6.6.

one would start from an assumed deformation curve followed by the calculation of the maximum values for the kinetic energy and for the potential energy of the system. An approximate value for natural frequency is calculated by equating maximum kinetic energy with the maximum potential energy. Then an improved value for the natural frequency may be obtained by loading the structure with the inertial loads associated with the assumed deflection. This load results in a new deformed shape which is used in calculating the maximum potential energy. The method is better explained with the aid of numerical examples.

Example 6.6.   By Rayleigh's Method, determine the natural frequency (lower or fundamental frequency) of the two-story frame shown in Fig. 6.13. Assume that the horizontal members are very rigid compared to the columns of the frame. This assumption reduces the system to only two degrees of freedom, indicated by coordinates y_1 and y_2 in the figure. The mass of the structure, which is lumped at the floor levels, has values $m_1 = 136$ lb·sec²/in and $m_2 = 66$ lb·sec²/in. The total stiffness of the first story is $k_1 = 30{,}700$ lb/in and of the second story $k_2 = 44{,}300$ lb/in, as indicated in Fig. 6.13.

Solution: This structure may be modeled by the two mass systems shown in Fig. 6.14. In applying Rayleigh's Method, let us assume a deformed shape for which $y_1 = 1$ and $y_2 = 2$. The maximum potential energy is then

$$V_{max} = \tfrac{1}{2}K_1 y_1^2 + \tfrac{1}{2}K_2(y_2 - y_1)^2$$

$$= \tfrac{1}{2}(30{,}700)(1)^2 + \tfrac{1}{2}(44{,}300)(1)^2$$

$$= 37{,}500 \text{ lb} \cdot \text{in} \tag{a}$$

and the maximum kinetic energy

$$T_{max} = \tfrac{1}{2}m_1(\omega y_1)^2 + \tfrac{1}{2}m_2(\omega y_2)^2$$

$$= \tfrac{1}{2}(136)\omega^2 + \tfrac{1}{2}(66)(2\omega)^2$$

$$= 200\omega^2 \tag{b}$$

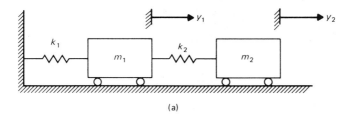

(a)

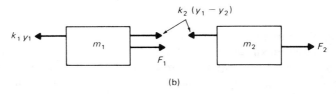

(b)

Fig. 6.14 Mathematical model for structure of Example 6.6.

Equating maximum potential energy with maximum kinetic energy and solving for the natural frequency gives

$$\omega = 13.69 \text{ rad/sec}$$

or

$$f = \frac{\omega}{2\pi} = 2.18 \text{ cps}$$

The natural frequency calculated as $f = 2.18$ cps is only an approximation to the exact value, since the deformed shape was assumed for the purpose of applying Rayleigh's Method. To improve this calculated value of the natural frequency, let us load the mathematical model in Fig. 6.14 with the inertial load calculated as

$$F_1 = m_1\omega^2 y_1 = (136)(13.69)^2(1) = 25,489$$

$$F_2 = m_2\omega^2 y_2 = (66)(13.69)^2(2) = 24,739$$

The equilibrium equations obtained by equating to zero the sum of the forces in the free body diagram shown in Fig. 6.14(b) gives

$$30,700 y_1 - 44,300(y_2 - y_1) = 25,489$$

$$44,300(y_2 - y_1) = 24,739$$

and solving

$$y_1 = 1.64$$
$$y_2 = 2.19$$

or in the ratio

$$y_1 = 1.00$$
$$y_2 = 1.34 \tag{c}$$

Introducing these improved values for the displacements y_1 and y_2 into eqs. (a) and (b) to recalculate the maximum potential energy and maximum kinetic energy results in

$$V_{max} = 25{,}293 \tag{d}$$

$$T_{max} = 160.03\omega^2 \tag{e}$$

and upon equating V_{max} and T_{max}, we obtain

$$\omega = 12.57 \text{ rad/sec}$$

or

$$f = 2.00 \text{ cps}$$

This last calculated value for the natural frequency $f = 2.00$ cps could be further improved by applying a new inertial load in the system based on the last value of the natural frequency and repeating a new cycle of calculations. Table 6.1 shows results obtained for four cycles.

The exact natural frequency and deformed shape, which are calculated for this system in Chapter 10, Example 10.1, as a two-degrees-of-freedom system, checks with the values obtained in the last cycle of the calculations shown in Table 6.1.

6.6   SHEAR WALLS

Horizontal forces in buildings, such as those produced by earthquake motion or wind action, are often resisted by structural elements called *shear walls*. These structural elements are generally designed as reinforced concrete walls fixed at the foundation. A single cantilever shear wall can be expected to

TABLE 6.1   Improved Rayleigh's Method Applied to Example 6.6

Cycle	Deformed Shape	Inertial Load		Natural Frequency
		F_1	F_2	
1	1:2.00			2.18 cps
2	1:1.34	25,489	24,739	2.00
3	1:1.32	21,489	18,725	1.88
4	1:1.27	19,091	12,230	1.88

behave as an ordinary flexural member if its length-to-depth ratio (L/D) is greater than about 2. For short shear walls $(L/D < 2)$, the shear strength assumes preeminence and both flexural and shear deformations should be considered in the analysis.

When the floor system of a multistory building is rigid, the structure's weights or masses at each floor may be treated as concentrated loads, as shown in Fig. 6.15 for a three-story building. The response of the structure is then a function of these masses and of the stiffness of the shear wall. In practice a mathematical model is developed in which the mass as well as the stiffness of the structure are combined at each floor level. The fundamental frequency (lowest natural frequency) for such a structure can then be obtained using Rayleigh's Method, as shown in the following illustrative example.

Example 6.7.   Determine, using Rayleigh's Method, the natural period of the three-story building shown in Fig. 6.15. All the floors have equal weight W. Assume the mass of the wall negligible compared to the floor masses and consider only flexural deformations $(L/D \geq 2)$.

Solution: The natural frequency can be calculated using eq. (6.63), which is repeated here for convenience:

$$\omega = \sqrt{\dfrac{g \displaystyle\sum_{i=1}^{N} W_i y_i}{\displaystyle\sum_{i=1}^{N} W_i y_i^2}} \tag{a}$$

The deformed shape equation is assumed as the deflection curve produced on a cantilever beam supporting three concentrated weights W, as shown in

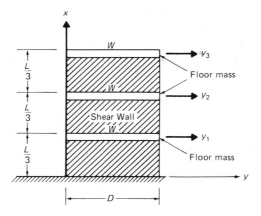

Fig. 6.15 Mathematical model for shear wall and rigid floors.

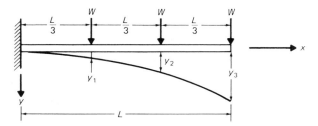

Fig. 6.16 Assumed deflection curve for Example 6.7.

Fig. 6.16. The static deflections y_1, y_2, and y_3 calculated by using basic knowledge of strength of materials are

$$y_1 = \frac{15}{162} \frac{WL^3}{EI} = 0.0926 \frac{WL^3}{EI}$$

$$y_2 = \frac{49}{162} \frac{WL^3}{EI} = 0.3025 \frac{WL^3}{EI}$$

$$y_3 = \frac{92}{162} \frac{WL^3}{EI} = 0.5679 \frac{WL^3}{EI} \qquad (b)$$

The natural frequency by eq. (6.63) is then calculated as

$$\omega = \sqrt{\frac{386(0.0926 + 0.3025 + 0.5679)}{(0.0926^2 + 0.3025^2 + 0.5679^2)} \cdot \frac{EI}{WL^3}}$$

$$= 29.66 \sqrt{\frac{EI}{WL^3}} \text{ rad/sec}$$

or

$$f = \frac{\omega}{2\pi} = 4.72 \sqrt{\frac{EI}{WL^3}} \text{ cps}$$

Example 6.8. For the mathematical model of a three-story building shown in Fig. 6.15, determine the total deflections at the floor levels considering both flexural and shear deformations.

Solution: The lateral deflection Δy_s, considering only shear deformation for a beam segment of length Δx, is given by

$$\Delta y_s = \frac{V \Delta x}{\alpha A G} \qquad (a)$$

where

V = shear force

A = cross-sectional area

α = shape constant (α = 1.2 for rectangular sections)

G = shear modulus of elasticity

At the first story, $V = 3W$. Therefore, by eq. (a) the shear deflection at the first story is

$$y_{s1} = \frac{3W(L/3)}{\alpha A G} = \frac{WL}{\alpha A G} \tag{b}$$

At the second floor the shear deflection is equal to the first floor deflection plus the relative deflection between floors, that is

$$y_{s2} = y_{s1} + \frac{2W(L/3)}{\alpha A G} = \frac{5WL}{3\alpha A G} \tag{c}$$

since the shear force of the second story is $V = 2W$, and at the third floor

$$y_{s3} = y_{s2} + \frac{W(L/3)}{\alpha A G} = \frac{6WL}{3\alpha A G} \tag{d}$$

The total deflection is then obtained by adding the flexural deflection determined in Example 6.7 to the above shear deflections. Hence,

$$y_1 = \frac{15}{162}\frac{WL^3}{EI} + \frac{WL}{\alpha A G}$$

$$y_2 = \frac{49}{162}\frac{WL^3}{EI} + \frac{5WL}{3\alpha A G}$$

$$y_3 = \frac{92}{162}\frac{WL^3}{EI} + \frac{6WL}{3\alpha A G} \tag{e}$$

We can see better the relative importance of the shear contribution to the total deflection by factoring the first terms in eqs. (e). Considering a rectangular wall for which $A = D \times t$, $E/G = 2.5$, $I = tD^3/12$, $\alpha = 1.2$ (t = thickness of the wall), we obtain

$$y_1 = \frac{15}{162}\frac{WL^3}{EI}\left[1 + 1.875\left(\frac{D}{L}\right)^2\right]$$

$$y_2 = \frac{49}{162}\frac{WL^3}{EI}\left[1 + 0.957\left(\frac{D}{L}\right)^2\right]$$

$$y_3 = \frac{92}{162}\frac{WL^3}{EI}\left[1 + 0.611\left(\frac{D}{L}\right)^2\right] \tag{f}$$

The next illustrative example presents a table showing the relative importance that shear deformation has in calculating the natural frequency for a series of values of the ratio D/L.

Example 6.9. For the structure modeled as shown in Fig. 6.15, study the relative importance of shear deformation in calculating the natural frequency.

Solution: In this study we will consider, for the wall, a range of values from 0 to 3.0 for the ratio D/L (depth-to-length ratio). The deflections y_1, y_2, y_3 at the floor levels are given by eqs. (f) of Example 6.8 and the natural frequency by eq. (6.63). The necessary calculations are conveniently shown in Table 6.2. It may be seen from the last column of Table 6.2, that for this example the natural frequency neglecting shear deformation $(D/L = 0)$ is $f = 4.72 \sqrt{EI/WL^3}$ cps. For short walls $(D/L > 0.5)$ the effect of shear deformation becomes increasingly important.

6.7   SUMMARY

The concept of generalized coordinate presented in this chapter permits the analysis of multiple interconnected rigid or elastic bodies with distributed properties as single degree-of-freedom systems. The analysis as one-degree-of-freedom systems can be made provided that by the specification of a single coordinate (the generalized coordinate) the configuration of the whole system is determined. Such a system may then be modeled as the simple oscillator with its various parameters of mass, stiffness, damping, and load, calculated to be dynamically equivalent to the actual system to be analyzed. The solution of this model provides the response in terms of the generalized coordinate.

TABLE 6.2   Calculation of the Natural Frequency for the Shear Wall Modeled as Shown in Fig. 6.15

D/L	y_1** (in)	y_2** (in)	y_3** (in)	ω*** (rad/sec)	f*** (cps)
0.00*	0.09259	0.30247	0.56790	29.66	4.72
0.50	0.13600	0.37483	0.65465	27.67	4.40
1.00	0.26620	0.59193	0.91489	23.30	3.71
1.50	0.48322	0.95376	1.34862	19.05	3.03
2.00	0.78704	1.46032	1.95585	15.71	2.50
2.50	1.17765	2.11161	2.73658	13.21	2.10
3.00	1.65509	2.90764	3.69079	11.33	1.80

*$D/L = 0$ is equivalent, to neglect shear deformations.
**Factor of WL^3/EI.
***Factor of $\sqrt{EI/WL^3}$.

The principle of virtual work which is applicable to systems in static or dynamic equilibrium is a powerful method for obtaining the equations of motion as an alternative to the direct application of Newton's law. The principle of virtual work states that for a system in equilibrium the summation of the work done by all its forces during any displacement compatible with the constraints of the system is equal to zero.

Rayleigh's Method for determining the natural frequency of a vibrating system is based on the principle of conservation of energy. In practice, it is applied by equating the maximum potential energy with the maximum kinetic energy of the system. To use Rayleigh's Method for the determining of the natural frequency of a discrete or a continuous system, it is necessary to assume a deformed shape. Often, this shape is selected as the one produced by gravitational loads acting in the direction of the expected displacements. This approach leads to the following formula for calculating the natural frequency:

$$\omega = \sqrt{\dfrac{g \displaystyle\sum_i W_i y_i}{\displaystyle\sum_i W_i y_i^2}}$$

where y_i is the deflection at coordinate i and W_i concentrated weight at this coordinate. Shear walls are structural walls designed to resist lateral forces in buildings. For short walls $(L/D \le 2)$ shear deformations are important and should be considered in the analysis in addition to the flexural deformations.

PROBLEMS

6.1   For the system shown in Fig. 6.17 determine the generalized mass M^*, damping C^*, stiffness K^*, and the generalized load $F^*(t)$. Select $Y(t)$ as the generalized coordinate.

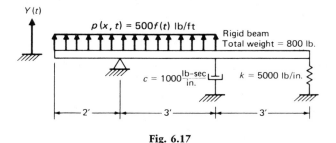

Fig. 6.17

6.2   Determine the generalized quantities M^*, C^*, K^*, and $F^*(t)$ for the structure shown in Fig. 6.18. Select $Y(t)$ as the generalized coordinate.

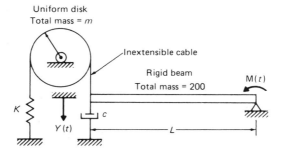

Fig. 6.18

6.3   Determine the generalized quantities M^*, C^*, K^*, and $F^*(t)$ for the structure shown in Fig. 6.19. Select $\theta(t)$ as the generalized coordinate.

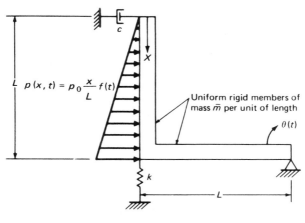

Fig. 6.19

6.4   For the elastic cantilever beam shown in Fig. 6.20, determine the generalized quantities M^*, K^*, and $F^*(t)$. Neglect damping. Assume that the deflected shape

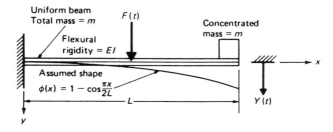

Fig. 6.20

is given by $\phi(x) = 1 - \cos(\pi x/2L)$ and select $Y(t)$ as the generalized coordinate as shown in Fig. 6.20. The beam is excited by a concentrated force $F(t) = F_0 f(t)$ at midspan.

6.5 Determine the generalized geometric stiffness K_G^* for the system in Fig. 6.20 if an axial tensile force N is applied at the free end of the beam along the x direction. What is the combined generalized stiffness K_c^*?

6.6 A concrete conical post of diameter d at the base and height L is shown in Fig. 6.21. It is assumed that the wind produces a dynamic pressure $p_0(t)$ per unit of projected area along a vertical plane. Determine the generalized quantities M^*, K^*, and $F^*(t)$. (Take modulus of elasticity $E_c = 3 \times 10^6$ psi; specific weight $\gamma = 150$ lb/ft^3 for concrete.)

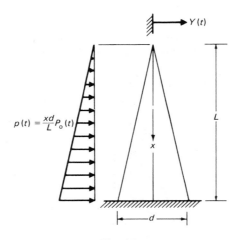

$$p(t) = \frac{xd}{L}P_o(t)$$

Fig. 6.21

6.7 A simply supported beam of total uniformly distributed mass m_b, flexural rigidity EI, and length L, carries a concentrated mass m at its center. Assume the deflection curve to be the deflection curve due to a concentrated force at the center of the beam and determine the natural frequency using Rayleigh's Method.

6.8 Determine the natural frequency of a simply supported beam with overhang which has a total uniformly distributed mass m_b, flexural rigidity EI, and dimensions shown in Fig. 6.22. Assume that during vibration the beam deflected curve is of the shape produced by a concentrated force applied at the free end of the beam.

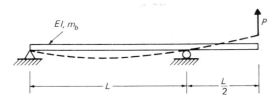

Fig. 6.22

6.9 Determine the natural frequency of the simply supported beam shown in Fig. 6.23 using Rayleigh's Method. Assume the deflection curve given by $\phi(x) = Y \sin \pi x/L$. The total mass of the beam is $m_b = 10$ lb $\cdot$ sec^2/in, flexural rigidity $EI = 10^8$ lb $\cdot$ in^2, and length $L = 100$ in. The beam carries a concentrated mass at the center $m = 5$ lb $\cdot$ sec^2/in.

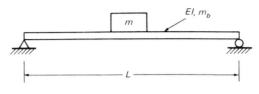

Fig. 6.23

6.10 A two-story building is modeled as the frame shown in Fig. 6.24. Use Rayleigh's Method to determine the natural frequency of vibration for the case in which only flexural deformation needs to be considered. Neglect the mass of the columns and assume rigid floors. [Hint: Use eq. (6.63)].

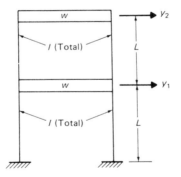

Fig. 6.24

6.11 Solve Problem 6.10 for the case in which the columns are short and only shear deformation needs to be considered. (The lateral force V for a fixed column of length L, cross-sectional area A, is approximately given by $V = AG\Delta/L$, where G is the shear modulus of elasticity and Δ the lateral deflection.)

6.12 Calculate the natural frequency of the shear wall carrying concentrated masses at the floor levels of a three-story building as shown in Fig. 6.25. Assume that the deflection shape of the shear wall is that resulting from a concentrated lateral force applied at its tip. Take flexural rigidity, $EI = 3.0 \times 10^{11}$ lb $\cdot$ in^2; length $L =$

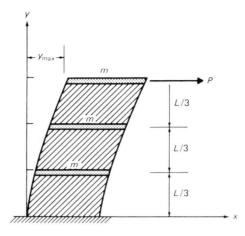

Fig. 6.25

36 ft; concentrated masses, $m = 100$ lb · sec²/in and mass per unit of length along the wall, $\bar{m} = 10$ lb · sec²/in².

6.13 Solve Problem 6.12 on the assumption that the deflection shape of the shear wall is that resulting from a lateral uniform load applied along its length.

7

Nonlinear Structural Response

In discussing the dynamic behavior of single degree-of-freedom systems, we assumed that in the model representing the structure, the restoring force was proportional to the displacement. We also assumed the dissipation of energy through a viscous damping mechanism in which the damping force was proportional to the velocity. In addition, the mass in the model was always considered to be unchanging with time. As a consequence of these assumptions, the equation of motion for such a system resulted in a linear, second-order ordinary differential equation with constant coefficients, namely,

$$m\ddot{y} + c\dot{y} + ky = F(t) \tag{7.1}$$

In the previous chapters it was illustrated that for particular forcing functions such as harmonic functions, it was relatively simple to solve this equation and that a general solution always existed in terms of Duhamel's integral. Equation (7.1) thus represents the dynamic behavior of many structures modeled as a single degree-of-freedom system. There are, however, physical situations for which this linear model does not adequately represent the dynamic characteristics of the structure. The analysis in such cases requires the introduction of a model in which the spring force or the damping force may not remain proportional, respectively, to the displacement or to the velocity. Consequently, the resulting equation of motion will no longer be linear and its mathematical solution, in general, will have a much greater complexity, often requiring a numerical procedure for its integration.

7.1   NONLINEAR SINGLE DEGREE-OF-FREEDOM MODEL

Consider in Fig. 7.1(a) the model for a single degree-of-freedom system and in Fig. 7.1(b) the corresponding free body diagram. The dynamic equilibrium in the system is established by equating to zero the sum of the inertial force $F_I(t)$, the damping force $F_D(t)$, the spring force $F_s(t)$, and the external force $F(t)$. Hence at time t_i the equilibrium of these forces is expressed as

$$F_I(t_i) + F_D(t_i) + F_s(t_i) = F(t_i) \tag{7.2}$$

and at short time Δt later as

$$F_I(t_i + \Delta t) + F_D(t_i + \Delta t) + F_s(t_i + \Delta t) = F(t_i + \Delta t) \tag{7.3}$$

Subtracting eq. (7.2) from eq. (7.3) results in the differential equation of motion in terms of increments, namely

$$\Delta F_I + \Delta F_D + \Delta F_s = \Delta F_i \tag{7.4}$$

where the incremental forces in this equation are defined as follows:

$$\Delta F_I = F_I(t_i + \Delta t) - F_I(t_i)$$

$$\Delta F_D = F_D(t_i + \Delta t) - F_D(t_i)$$

$$\Delta F_s = F_s(t_i + \Delta t) - F_s(t_i)$$

$$\Delta F_i = F(t_i + \Delta t) - F(t_i) \tag{7.5}$$

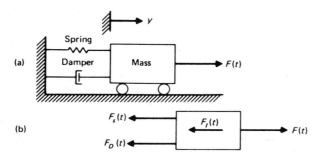

Fig. 7.1 (a) Model for a single degree-of-freedom system. (b) Free body diagram showing the inertial force, the damping force, the spring force, and the external force.

If we assume that the damping force is a function of the velocity and the spring force a function of displacement as shown graphically in Fig. 7.2, while the inertial force remains proportional to the acceleration, we may then express the incremental forces in eqs. (7.5) as

$$\Delta F_I = m\Delta\ddot{y}_i$$

$$\Delta F_D = c_i\Delta\dot{y}_i$$

$$\Delta F_s = k_i\Delta y_i \tag{7.6}$$

where the incremental displacement Δy_i, incremental velocity $\Delta\dot{y}_i$, and incremental acceleration $\Delta\ddot{y}$ are given by

$$\Delta y_i = y(t_i + \Delta t) - y(t_i) \tag{7.7}$$

$$\Delta\dot{y}_i = \dot{y}(t_i + \Delta t) - \dot{y}(t_i) \tag{7.8}$$

$$\Delta\ddot{y}_i = \ddot{y}(t_i + \Delta t) - \ddot{y}(t_i) \tag{7.9}$$

The coefficient k_i in eqs. (7.6) is defined as the current evaluation for the derivative of the spring force with respect to the displacement, namely,

$$k_i = \left(\frac{dF_s}{dy}\right)_{y=y_i} \tag{7.10}$$

Similarly, the coefficient c_i is defined as the current value of the derivative of the damping force with respect to the velocity, that is,

$$c_i = \left(\frac{dF_D}{d\dot{y}}\right)_{\dot{y}=\dot{y}_i} \tag{7.11}$$

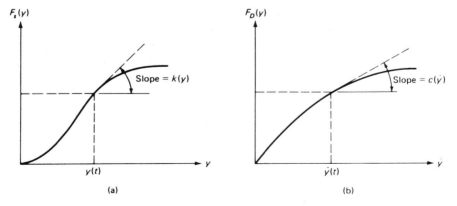

Fig. 7.2 (a) Nonlinear stiffness. (b) Nonlinear damping.

These two coefficients k_i and c_i are graphically depicted as the slopes of the corresponding curves shown in Fig. 7.2.

The substitution of eqs. (7.6) into eq. (7.4) results in a convenient form for the incremental equation, namely

$$m\Delta\ddot{y}_i + c_i\Delta\dot{y}_i + k_i\Delta y_i = \Delta F_i \tag{7.12}$$

where the coefficients c_i and k_i are calculated for values of velocity and displacement corresponding to time t_i and assumed to remain constant during the increment of time Δt. Since, in general, these two coefficients do not remain constant for that time increment, eq. (7.12) is an approximate equation.

7.2 INTEGRATION OF THE NONLINEAR EQUATION OF MOTION

Among the many methods available for the solution of the nonlinear equation of motion, probably one of the most effective is the step-by-step integration method. In this method, the response is evaluated at successive increments Δt of time, usually taken of equal lengths of time for computational convenience. At the beginning of each interval, the condition of dynamic equilibrium is established. Then, the response for a time increment Δt is evaluated approximately on the basis that the coefficients $k(y)$ and $c(\dot{y})$ remain constant during the interval Δt. The nonlinear characteristics of these coefficients are considered in the analysis by reevaluating these coefficients at the beginning of each time increment. The response is then obtained using the displacement and velocity calculated at the end of the time interval as the initial conditions for the next time step.

As we have said for each time interval, the stiffness coefficient $k(y)$ and the damping coefficient $c(\dot{y})$ are evaluated at the initiation of the interval but are assumed to remain constant until the next step; thus the nonlinear behavior of the system is approximated by a sequence of successively changing linear systems. It should also be obvious that the assumption of constant mass is unnecessary; it could just as well also be represented by a variable coefficient.

There are many procedures available for performing the step-by-step integration of eq. (7.12). Two of the most popular methods are the *constant acceleration method* and the *linear acceleration method*. As the names of these methods imply, in the first method the acceleration is assumed to remain constant during the time interval Δt, while in the second method, the acceleration is assumed to vary linearly during the interval. As may be expected, the constant acceleration method is simpler, but less accurate when compared with the linear acceleration method for the same value of the time increment. We shall present here in detail only the linear acceleration method.

This method has been found to yield excellent results with relatively little computational effort.

7.3   LINEAR ACCELERATION STEP-BY-STEP METHOD

In the linear acceleration method, it is assumed that the acceleration may be expressed by a linear function of time during the time interval Δt. Let t_i and $t_{i+1} = t_i + \Delta t$ be, respectively, the designation for the time at the beginning and at the end of the time interval Δt. In this type of analysis, the material properties of the system c_i and k_i may include any form of nonlinearity. Thus it is not necessary for the spring force to be only a function of displacement nor for the damping force to be specified only as a function of velocity. The only restriction in the analysis is that we evaluate these coefficients at an instant of time t_i and then assume that they remain constant during the increment of time Δt. When the acceleration is assumed to be a linear function of time for the interval of time t_i to $t_{i+1} = t_i + \Delta t$ as depicted in Fig. 7.3, we may express the acceleration as

$$\ddot{y}(t) = \ddot{y}_i + \frac{\Delta \ddot{y}_i}{\Delta t}(t - t_i) \tag{7.13}$$

where $\Delta \ddot{y}_i$ is given by eq. (7.9). Integrating eq. (7.13) twice with respect to time between the limits t_i and t yields

$$\dot{y}(t) = \dot{y}_i + \ddot{y}_i(t - t_i) + \frac{1}{2}\frac{\Delta \ddot{y}_i}{\Delta t}(t - t_i)^2 \tag{7.14}$$

and

$$y(t) = y_i + \dot{y}_i(t - t_i) + \frac{1}{2}\ddot{y}_i(t - t_i)^2 + \frac{1}{6}\frac{\Delta \ddot{y}_i}{\Delta t}(t - t_i)^3 \tag{7.15}$$

The evaluation of eqs. (7.14) and (7.15) at time $t = t_i + \Delta t$ gives

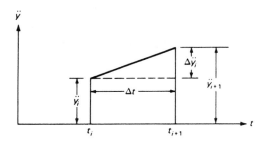

Fig. 7.3 Assumed linear variation of acceleration during time interval.

$$\Delta \dot{y}_i = \ddot{y}_i \Delta t + \tfrac{1}{2} \Delta \ddot{y}_i \Delta t \qquad (7.16)$$

and

$$\Delta y_i = \dot{y}_i \Delta t + \tfrac{1}{2} \ddot{y}_i \Delta t^2 + \tfrac{1}{6} \Delta \ddot{y}_i \Delta t^2 \qquad (7.17)$$

where Δy_i and $\Delta \dot{y}_i$ are defined in eqs. (7.7) and (7.8), respectively. Now to use the incremental displacement Δy as the basic variable in the analysis, eq. (7.17) is solved for the incremental acceleration $\Delta \ddot{y}_i$ and then substituted into eq. (7.16) to obtain

$$\Delta \ddot{y}_i = \frac{6}{\Delta t^2} \Delta y_i - \frac{6}{\Delta t} \dot{y}_i - 3\ddot{y}_i \qquad (7.18)$$

and

$$\Delta \dot{y}_i = \frac{3}{\Delta t} \Delta y_i - 3\dot{y}_i - \frac{\Delta t}{2} \ddot{y}_i \qquad (7.19)$$

The substitution of eqs. (7.18) and (7.19) into eq. (7.12) leads to the following form of the equation of motion:

$$m\left\{ \frac{6}{\Delta t^2} \Delta y_i - \frac{6}{\Delta t} \dot{y}_i - 3\ddot{y}_i \right\} + c_i \left\{ \frac{3}{\Delta t} \Delta y_i - 3\dot{y}_i - \frac{\Delta t}{2} \ddot{y}_i \right\} + k_i \Delta y_i = \Delta F_i \qquad (7.20)$$

Finally, transferring in eq. (7.20) all the terms containing the unknown incremental displacement Δy_i to the left-hand side gives

$$\bar{k}_i \Delta y_i = \Delta \bar{F}_i \qquad (7.21)$$

in which $\bar{k}_i$ is the effective spring constant, given by

$$\bar{k}_i = k_i + \frac{6m}{\Delta t^2} + \frac{3c_i}{\Delta t} \qquad (7.22)$$

and $\Delta \bar{F}_i$ is the effective incremental force, expressed by

$$\Delta \bar{F}_i = \Delta F_i + m\left\{ \frac{6}{\Delta t} \dot{y}_i + 3\ddot{y}_i \right\} + c_i \left\{ 3\dot{y}_i + \frac{\Delta t}{2} \ddot{y}_i \right\} \qquad (7.23)$$

It should be noted that eq. (7.21) is equivalent to the static incremental-equilibrium equation, and may be solved for the incremental displacement by simply dividing the effective incremental force $\Delta \bar{F}_i$ by the effective spring constant $\bar{k}_i$, that is,

$$\Delta y_i = \frac{\Delta \bar{F}_i}{\bar{k}_i} \qquad (7.24)$$

To obtain the displacement $y_{i+1} = y(t_i + \Delta t)$ at time $t_{i+1} = t_i + \Delta t$, this value of Δy_i is substituted into eq. (7.7) yielding

$$y_{i+1} = y_i + \Delta y_i \tag{7.25}$$

Then the incremental velocity $\Delta \dot{y}_i$ is obtained from eq. (7.19) and the velocity at time $t_{i+1} = t_i + \Delta t$ from eq. (7.8) as

$$\dot{y}_{i+1} = \dot{y}_i + \Delta \dot{y}_i \tag{7.26}$$

Finally, the acceleration $\ddot{y}_{i+1}$ at the end of the time step is obtained directly from the differential equation of motion, eq. (7.2), where the equation is written for time $t_{i+1} = t_i + \Delta t$. Hence, after setting $F_I = m\ddot{y}_{i+1}$ in eq. (7.2), it follows that

$$\ddot{y}_{i+1} = \frac{1}{m} \{F(t_{i+1}) - F_D(t_{i+1}) - F_S(t_{i+1})\} \tag{7.27}$$

where that damping force $F_D(t_{i+1})$ and the spring force $F_S(t_{i+1})$ are now evaluated at time $t_{i+1} = t_i + \Delta t$.

After the displacement, velocity and acceleration have been determined at time $t_{i+1} = t_i + \Delta t$, the outlined procedure is repeated to calculate these quantities at the following time step $t_{i+2} = t_{i+1} + \Delta t$, and the process is continued to any desired final value of time. The reader should, however, realize that this numerical procedure involves two significant approximations: (1) the acceleration is assumed to vary linearly during the time increment Δt; and (2) the damping and stiffness properties of the system are evaluated at the initiation of each time increment and assumed to remain constant during the time interval. In general, these two assumptions introduce errors which are small if the time step is short. However, these errors generally might tend to accumulate from step to step. This accumulation of errors should be avoided by imposing a total dynamic equilibrium condition at each step in the analysis. This is accomplished by expressing the acceleration at each step using the differential equation of motion in which the displacement and velocity as well as the stiffness and damping forces are evaluated at that time step.

There still remains the problem of the selection of the proper time increment Δt. As in any numerical method, the accuracy of the step-by-step integration method depends upon the magnitude of the time increment selected. The following factors should be considered in the selection of Δt: (1) the natural period of the structure; (2) the rate of variation of the loading function; and (3) the complexity of the stiffness and damping functions.

In general, it has been found that sufficiently accurate results can be obtained if the time interval is taken to be no longer than one-tenth of the natural period of the structure. The second consideration is that the interval should be small enough to represent properly the variation of the load with respect to time. The third point that should be considered is any abrupt variation in the rate of change of the stiffness or damping function. For example, in the usual assumption of elastoplastic materials, the stiffness suddenly changes from linear elastic to a yielding plastic phase. In this case, to obtain the best accuracy, it would be desirable to select smaller time steps in the neighborhood of such drastic changes.

7.4 ELASTOPLASTIC BEHAVIOR

If any structure modeled as a single degree-of-freedom system (spring-mass system) is allowed to yield plastically, then the restoring force exerted is likely to be of the form shown in Fig. 7.4(a). There is a portion of the curve in which linear elastic behavior occurs, whereupon, for any further deformation, plastic yielding takes place. When the structure is unloaded, the behavior is again elastic until further reverse loading produces compressive plastic yielding. The structure may be subjected to cyclic loading and unloading in this manner. Energy is dissipated during each cycle by an amount which is proportional to the area under the curve (hysteresis loop) as indicated in Fig. 7.4(a). This behavior is often simplified by assuming a definite yield point beyond which additional displacement takes place at a constant value for the restoring force without any further increase in the load. Such behavior is known as *elastoplastic* behavior; the corresponding force–displacement curve is shown in Fig. 7.4(b).

For the structure modeled as a spring-mass system, expressions of the restoring force for a system with elastoplastic behavior are easily written. These expressions depend on the magnitude of the restoring force as well as upon whether the motion is such that the displacement is increasing $(\dot{y} > 0)$ or decreasing $(\dot{y} < 0)$. Referring to Fig. 7.4(b) in which a general elastoplastic cycle is represented, we assume that the initial conditions are zero $(y_0 = 0,$

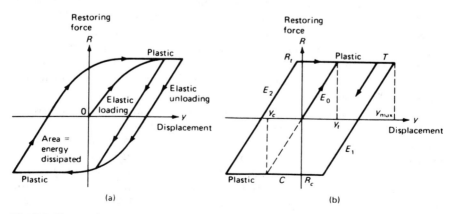

Fig. 7.4 Elastic-plastic structural models. (a) General plastic behavior. (b) Elastoplastic behavior.

$\dot{y}_0 = 0$) for the unloaded structure. Hence, initially, as the load is applied, the system behaves elastically along curve E_0. The displacement y_t, at which plastic behavior in tension may be initiated, and the displacement y_c, at which plastic behavior in compression may be initiated, are calculated, respectively, from

$$y_t = R_t/k \qquad (7.28)$$

and

$$y_c = R_c/k$$

where R_t and R_c are the respective values of the forces which produce yielding in tension and compression and k is the elastic stiffness of the structure. The system will remain on curve E_0 as long as the displacement y satisfies

$$y_c < y < y_t \qquad (7.29)$$

If the displacement y increases to y_t, the system begins to behave plastically in tension along curve T on Fig. 7.4(b); it remains on curve T as long as the velocity $\dot{y} > 0$. When $\dot{y} < 0$, the system reverses to elastic behavior on a curve such as E_1 with new yielding points given by

$$y_t = y_{max}$$
$$y_c = y_{max} - (R_t - R_c)/k \qquad (7.30)$$

in which y_{max} is the maximum displacement along curve T, which occurs when $\dot{y} = 0$.

Conversely, if y decreases to y_c, the system begins a plastic behavior in compression along curve C and it remains on this curve as long as $\dot{y} < 0$. The system returns to an elastic behavior when the velocity again changes direction and $\dot{y} > 0$. In this case, the new yielding limits are given by

$$y_c = y_{min}$$
$$y_t = y_{min} + (R_t - R_c)/k \qquad (7.31)$$

in which y_{min} is the minimum displacement along curve C, which occurs when $\dot{y} = 0$. The same condition given by eq. (7.29) is valid for the system to remain operating along any elastic segment such as E_0, E_1, E_2, ... as shown in Fig. 7.4(b).

We are now interested in calculating the restoring force at each of the possible segments of the elastoplastic cycle. The restoring force on an elastic phase of the cycle $(E_0, E_1, E_2, ...)$ may be calculated as

$$R = R_t - (y_t - y)k \qquad (7.32)$$

on a plastic phase in tension as

$$R = R_t \tag{7.33}$$

and on the plastic compressive phase as

$$R = R_c \tag{7.34}$$

The algorithm for the step-by-step linear acceleration method of a single degree-of-freedom system assuming an elastoplastic behavior is outlined in the following section.

7.5   ALGORITHM FOR STEP-BY-STEP SOLUTION FOR ELASTOPLASTIC SINGLE DEGREE-OF-FREEDOM SYSTEM

Initialize and input data:

(1) Input values for k, m, c, R_t, R_c, and a table giving the time t_i and magnitude of the excitation F_i.

(2) Set $y_0 = 0$ and $\dot{y}_0 = 0$.

(3) Calculate initial acceleration:

$$\ddot{y}_0 = \frac{F(t = 0)}{m} \tag{7.35}$$

(4) Select time step Δt and calculate constants:

$$a_1 = 3/\Delta t, \quad a_2 = 6/\Delta t, \quad a_3 = \Delta t/2, \quad a_4 = 6/\Delta t^2$$

(5) Calculate initial yield points:

$$y_t = R_t/k$$

$$y_c = R_c/k \tag{7.36}$$

For each time step:

(1) Use the following code to establish the elastic or plastic state of the system:

> KEY = 0 (elastic behavior)
>
> KEY = −1 (plastic behavior in compression)
>
> KEY = 1 (plastic behavior in tension)   (7.37)

(2) Calculate the displacement y and velocity $\dot{y}$ at the end of the time step and set the value of KEY according to the following conditions:

(a) When the system is behaving elastically at the beginning of the time step and

$$y_c < y < y_t \quad \text{KEY} = 0$$

$$y > y_t \qquad \text{KEY} = 1$$

$$y < y_c \qquad \text{KEY} = -1$$

(b) When the system is behaving plastically in tension at the beginning of the time step and

$$\dot{y} > 0 \quad \text{KEY} = 1$$

$$\dot{y} < 0 \quad \text{KEY} = 0$$

(c) When the system is behaving plastically in compression at the beginning of the time step and

$$\dot{y} < 0 \quad \text{KEY} = -1$$

$$\dot{y} > 0 \quad \text{KEY} = 0$$

(3) Calculate the effective stiffness:

$$\bar{k}_i = k_p + a_4 m + a_1 c_i \tag{7.38}$$

where

$$k_p = k \text{ for elastic behavior (KEY} = 0)$$

$$k_p = 0 \text{ for plastic behavior (KEY} = 1 \text{ or } -1) \tag{7.39}$$

(4) Calculate the incremental effective force:

$$\overline{\Delta F}_i = \Delta F_i + (a_2 m + 3 c_i) \dot{y}_i + (3m + a_3 c_i) \ddot{y}_i \tag{7.40}$$

(5) Solve for the incremental displacement:

$$\Delta y_i = \overline{\Delta F}_i / \bar{k}_i \tag{7.41}$$

(6) Calculate the incremental velocity:

$$\Delta \dot{y}_i = a_1 \Delta y_i - 3 \dot{y}_i - a_3 \ddot{y}_i \tag{7.42}$$

(7) Calculate displacement and velocity at the end of time interval:

$$y_{i+1} = y_i + \Delta y_i \tag{7.43}$$

$$\dot{y}_{i+1} = \dot{y}_i + \Delta \dot{y}_i \tag{7.44}$$

(8) Calculate acceleration $\ddot{y}_{i+1}$ at the end of time interval using the dynamic equation of equilibrium:

$$\ddot{y}_{i+1} = \frac{1}{m} \left[F(t_{i+1}) - c_{i+1}\dot{y}_{i+1} - R \right] \tag{7.45}$$

at which

$$R = R_t - (y_t - y_{i+1})k \quad \text{if KEY} = 0$$

$$R = R_t \quad\quad\quad\quad\quad \text{if KEY} = 1 \tag{7.46}$$

or

$$R = R_c \quad\quad\quad\quad\quad \text{if KEY} = -1$$

Example 7.1. To illustrate the hand calculations in applying the step-by-step integration method described above, consider the single degree-of-freedom system in Fig. 7.5 with elastoplastic behavior subjected to the loading history as shown. For this example, we assume that the damping coefficient remains constant ($\xi = 0.087$). Hence the only nonlinearities in the system arise from the changes in stiffness as yielding occurs.

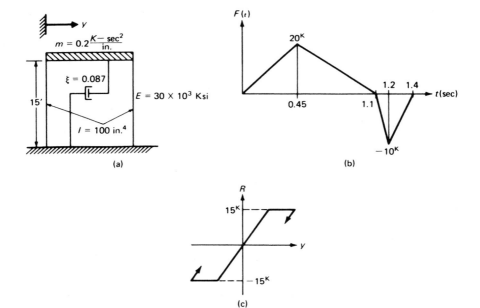

Fig. 7.5 Frame with elastoplastic behavior subjected to dynamic loading. (a) Frame. (b) Loading. (c) Elastoplastic behavior.

Solution: The stiffness of the system during elastic behavior is

$$k = \frac{12E(2I)}{L^3} = \frac{12 \times 30 \times 10^3 \times 2 \times 100}{(15 \times 12)^3} = 12.35 \text{ k/in}$$

and the damping coefficient

$$c = \xi c_{cr} = (0.087)(2)\sqrt{0.2 \times 12.35} = 0.274 \text{ k} \cdot \text{sec/in}$$

Initial displacement and initial velocity are $y_0 = \dot{y}_0 = 0$. Initial acceleration is

$$\ddot{y}_0 = \frac{F(0)}{k} = 0$$

Yield displacements are

$$y_t = \frac{R_t}{k} = \frac{15}{12.35} = 1.215 \text{ in}$$

and

$$y_c = -1.215 \text{ in}$$

The natural period is $T = 2\pi\sqrt{m/k} = 0.8$ sec (for the elastic system). For numerical convenience, we select $\Delta t = 0.1$ sec. The effective stiffness from eq. (7.38) is

$$\bar{k} = k_p + \frac{6}{0.1^2} 0.2 + \frac{3}{0.1} 0.274$$

or

$$\bar{k} = k_p + 128.22$$

where

$$k_p = k = 12.35 \text{ (elastic behavior)}$$

$$k_p = 0 \text{ (plastic behavior)}$$

The effective incremental loading from eq. (7.40) is

$$\overline{\Delta F} = \Delta F + \left(\frac{6}{\Delta t} m + 3c\right)\dot{y} + \left(3m + \frac{\Delta t}{2} c\right)\ddot{y}$$

$$\Delta \bar{F} = \Delta F + 12.822\dot{y} + 0.6137\ddot{y}$$

The velocity increment given by eq. (7.42) becomes

$$\Delta\dot{y} = 30\Delta y - 3\dot{y} - 0.05\ddot{y}$$

The necessary calculations may be conveniently arranged as illustrated in Table 7.1. In this example with elastoplastic behavior, the response changes abruptly as the yielding starts and stops. To obtain better accuracy, it would be desirable to subdivide the time step in the neighborhood of the change of state; however, an iterative procedure would be required to establish the length of the subintervals. This refinement has not been used in the present analysis nor in the computer program described in the next section. The stiffness computed at the initiation of the time step has been assumed to remain constant during the entire time increment. The reader is again cautioned that a significant error may arise during phase transitions unless the time step is selected relatively small.

7.6 PROGRAM 5—RESPONSE FOR ELASTOPLASTIC BEHAVIOR

The same as for the other programs presented in this book, Program 5 initially requests information about the data file: new, modify, or use existent data file. After the user has selected one of these options, the program requests the name of the file and the necessary input data. The program continues by reading and printing the data and by setting the initial values to the various constants and variables in the equations. Then by linear interpolation, values of the forcing function are computed at time increments equal to the selected time step Δt for the integration process. In the main body of the program, the displacement, velocity, and acceleration are computed at each time step. The nonlinear behavior of the restoring force is appropriately considered in the calculation by the variable KEY which is tested through a series of conditional statements in order to determine the correct expressions for the yield points and the magnitude of the restoring force in the system.

The output consists of a table giving the displacement, velocity, and acceleration at time increments Δt. The last column of the table shows the value of the index KEY which provides information about the state of the elastoplastic system. As indicated before, KEY = 0 for elastic behavior and KEY = 1 or KEY = −1 for plastic behavior, respectively, in tension or in compression.

Example 7.2. Using the computer Program 5, find the response of the structure in Example 7.1. Then repeat the calculation assuming elastic behavior. Plot and compare results for the elastoplastic behavior with the elastic response.

Solution:
Problem Data (from Problem 4.1):

Spring constant: $k = 12.35$ k/in

Damping coefficient: $c = 0.274$ k $\cdot$ sec/in

Mass: $m = 0.2$ (kip $\cdot$ sec^2/in)

Max. restoring force (tension): $R_t = 15$ kip

Max. restoring force (compression): $R_c = -15$ kip

Natural period: $T = 2\pi\sqrt{m/k} = 2\pi\sqrt{0.2/12.35} = 0.8$ sec

Select time step: $\Delta t = 0.1$ sec

Gravitational index: $G = 0$ (force on the mass)

The same computer program is used to obtain the response for a completely elastic behavior. It is only necessary to assign a large value to the maximum restoring forces in tension and in compression. These assigned values should be large enough in order for the structure to remain in the elastic range. For the present example, $R_t = 100$ kip and $R_c = -100$ kip were deemed adequate in this case. In order to visualize and facilitate a comparison between the elastic and inelastic responses for this example, the displacements are plotted in Fig. 7.6, with the response during yielding shown as dashed lines.

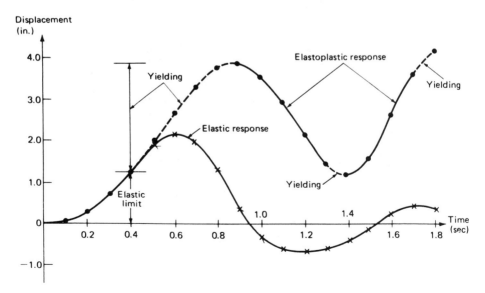

Fig. 7.6 Comparison of elastoplastic behavior with elastic response for Example 7.2.

TABLE 7.1 Nonlinear Response—Linear Acceleration Step-by-Step Method for Example 7.1

t (sec)	F (kip)	y (in)	KEY	$\dot{y}$ (in/sec)	R (kip)	$\ddot{y}$ (in/sec²)	$k_p/$ in kip/in	$\bar{k}$ kip/in	ΔF (kip)	$\Delta\bar{F}$ (kip)	Δy (in)	$\Delta\dot{y}$ (in/sec)
0	0	0	0	0	0	0	12.35	140.57	4.444	4.444	0.0316	0.9485
0.1	4.444	0.0316	0	0.9485	0.390	18.972	12.35	140.57	4.444	28.249	0.2010	2.2359
0.2	8.888	0.2326	0	3.1844	2.871	25.723	12.35	140.57	4.444	61.050	0.4343	2.193
0.3	13.333	0.6669	0	5.3760	8.233	18.134	12.35	140.57	4.444	84.510	0.6012	1.000
0.4	17.777	1.2681	1	6.3768	15.00	5.152	0	128.22	0.685	85.609	0.6677	0.6422
0.5	18.462	1.9358	1	7.0190	15.00	7.691	0	128.22	-3.077	91.641	0.7147	-0.0001
0.6	15.358	2.6505	1	7.0189	15.00	-7.693	0	128.22	-3.077	82.199	0.6409	-1.440
0.7	12.308	3.2916	1	5.5791	15.00	-21.105	0	128.22	-3.077	55.506	0.4329	-2.695
0.8	9.231	3.7244	1	2.8840	15.00	-32.797	0	128.22	-3.077	13.773	0.1074	-3.789
0.9	6.154	3.8319	0	-0.9054	15.00	-42.990	12.35	140.57	-3.077	-41.069	-0.2922	-3.899
1.0	3.077	3.5397	0	-4.8048	11.39	-34.998	12.35	140.57	-3.077	-86.162	-0.6130	-2.225
1.1	0	2.9268	0	-7.0295	3.825	-9.497	12.35	140.57	-10	-105.96	-0.7538	-1.051
1.2	-10	2.1729	0	-8.0806	-5.481	-11.525	12.35	140.57	5	-105.68	-0.7518	2.263
1.3	-5	1.4211	0	-5.8177	-14.76	56.784	12.35	140.57	5	-34.746	-0.2472	7.198
1.4	0	1.1739	-1	1.3860	-15.00	73.109	0	128.22	0	62.568	0.4880	6.842
1.5	0	1.6619	0	8.2227	-15.00	63.735	12.35	140.57	0	144.55	1.0283	2.995

Input Data and Output Results

```
PROGRAM 5: RESPONSE FOR ELASTOPLASTIC BEHAVIOR    DATA FILE:D5

    INPUT DATA:

NUMBER OF POINTS DEFINING THE EXCITATION          NE = 6
MASS                                              AM = .2
SPRING CONSTANT                                   AK = 12.35
DAMPING COEFFICIENT                                C = .274
TIME STEP INTEGRATION                              H = .1
MAX. RESTORING FORCE IN TENSION                   RT = 15
MIN. RESTORING FORCE COMPRESSION                  RC = -15
INDEX (GRAVITY OR ZERO)                            G = 0
```

TIME	EXCITATION	TIME	EXCITATION	TIME	EXCITATION	TIME	EXCITATION
0.000	0.00	0.450	20.00	1.100	0.00	1.200	-10.0
1.400	0.00	2.000	0.00				

```
    OUTPUT RESULTS:
```

TIME	DISPL.	VELOC.	ACC.	KEY
0.000	0.0000	0.0000	0.0000	0
0.100	0.0316	0.9485	18.9704	0
0.200	0.2326	3.1831	25.7221	0
0.300	0.6668	5.3755	18.1251	0
0.400	1.2679	6.3749	5.1553	1
0.500	1.9354	7.0173	7.6940	1
0.600	2.6500	7.0175	-7.6909	1
0.700	3.2909	5.5778	-21.1031	1
0.800	3.7237	2.8828	-32.7956	1
0.900	3.8310	-0.9064	-42.9890	0
1.000	3.5387	-4.8052	-34.9861	0
1.100	2.9258	-7.0283	-9.4766	0
1.200	2.1722	-8.0774	-11.5050	0
1.300	1.4208	-5.8129	56.7935	0
1.400	1.1741	1.3851	73.1025	-1
1.500	1.6625	8.2267	63.7295	0
1.600	2.6911	11.2188	-3.8869	0
1.700	3.6974	7.9471	-61.5465	0
1.800	4.1536	0.8688	-76.1903	1
1.900	3.8758	-6.2618	-66.4214	0
2.000	3.0147	-9.9897	-8.1384	0

7.7   SUMMARY

Structures are usually designed on the assumption that the structure is linearly elastic and that it remains linearly elastic when subjected to any expected dynamic excitation. However, there are situations in which the structure has to be designed for an eventual excitation of large magnitude such as the strong motion of an earthquake or the effects of nuclear explosion. In these cases, it is not realistic to assume that the structure will remain linearly elastic and it is then necessary to design the structure to withstand deformation beyond the elastic limit.

The simplest and most accepted assumption for the design beyond the elastic limit is to assume an elastoplastic behavior. In this type of behavior, the structure is elastic until the restoring force reaches a maximum value (tension or compression) at which it remains constant until the motion reverses its direction and returns to an elastic behavior.

There are many methods to solve numerically the differential equation of this type of motion. The step-by-step linear acceleration presented in this chapter provides satisfactory results with relatively simple calculations. However, these calculations are tedious and time consuming when performed by hand. The use of the computer and the availability of a computer program, such as the one described in this chapter, reduce the effort to a simple routine of data preparation.

PROBLEMS

7.1   The single degree-of-freedom of Fig. 7.7(a) is subjected to the foundation acceleration history in Fig. 7.7(b). Determine the maximum relative displacement of the columns. Assume elastoplastic behavior of Fig. 7.7(c).

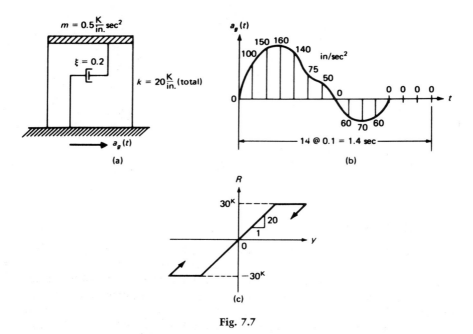

Fig. 7.7

7.2   Determine the displacement history for the structure in Fig. 7.7 when subjected to the impulse loading of Fig. 7.8 applied horizontally at the mass.

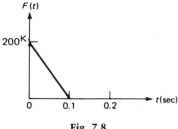

Fig. 7.8

7.3 Repeat Problem 7.2 for the impulse loading shown in Fig. 7.9 applied horizontally at the mass.

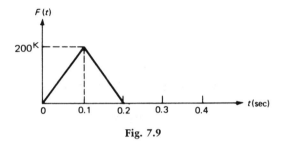

Fig. 7.9

7.4 Repeat Problem 7.2 for the acceleration history shown in Fig. 7.10 applied horizontally to the foundation.

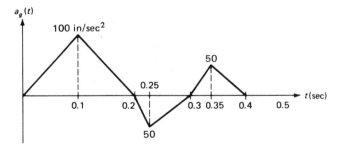

Fig. 7.10

7.5 Solve Problem 7.1 assuming elastic behavior of the structure. (Hint: Use computer Program 5 with $R_t = 200$ kip and $R_c = -200$ kip.)

7.6 Solve Problem 7.2 for elastic behavior of the structure. Plot the time-displacement response and compare with results from Problem 7.2.

7.7 Determine the ductility ratio from the results of Problem 7.2. (Ductility ratio is defined as the ratio of the maximum displacement to the displacement at the yield point.)

7.8   A structure modeled as spring-mass shown in Fig. 7.11(b) is subjected to the loading force depicted in Fig. 7.11(a). Assume elastoplastic behavior of Fig. 7.11(c).
Determine the response.

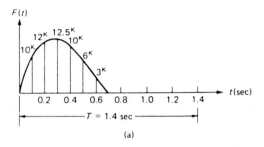

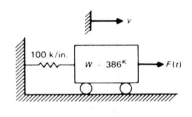

(a)

(b)

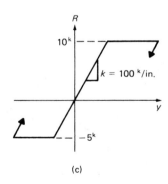

(c)

Fig. 7.11

7.9   Repeat Problem 7.8 assuming damping in the system equal to 20% of the critical
damping.

7.10   Solve Problem 7.8 assuming elastic behavior of the system. (Hint: Use Program
5 with $R_T = 1000$ kip and $R_C = -1000$ kip.)

7.11   Solve Problem 7.9 assuming elastic behavior of the system.

7.12 A structure modeled as the damped spring-mass system shown in Fig. 7.12(a) is subjected to the time-acceleration excitation acting at its support. The excitation function is expressed as $a(t) = a_0 f(t)$, where $f(t)$ is depicted in Fig. 7.12(b). Determine the maximum value that the factor a_0 may have for the structure to remain elastic. Assume that the structure has an elastoplastic behavior of Fig. 7.12(c).

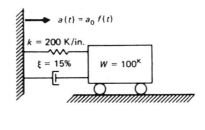

(a)

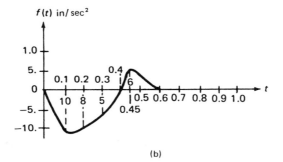

(b)

(c)

Fig. 7.12

8

Response Spectra

In this chapter, we introduce the concept of *response spectrum*, which in recent years has gained wide acceptance in structural dynamic practice, particularly in earthquake engineering design. Stated briefly, the response spectrum is a plot of the maximum response (maximum displacement, velocity, acceleration, or any other quantity of interest) to a specified load function for all possible single degree-of-freedom systems. The abscissa of the spectrum is the natural frequency (or period) of the system, and the ordinate the maximum response. A plot of this type is shown in Fig. 8.1, in which a one-story building is subjected to a ground displacement indicated by the function $y_s(t)$. The response spectral curve shown in Fig. 8.1(a) gives, for any single degree-of-freedom system, the maximum displacement of the mass m relative to the displacement at the support. Thus, to determine the response from an available spectral chart, for a specified excitation, we need only to know the natural frequency of the system.

8.1 CONSTRUCTION OF RESPONSE SPECTRUM

To illustrate the construction of a response spectral chart, consider in Fig. 8.2(a) the undamped oscillator subjected to one-half period of the sinusoidal

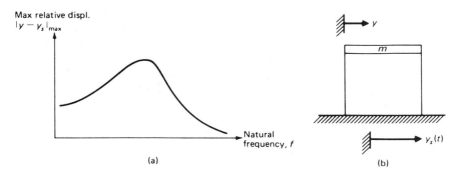

Max relative displ.
$|y - y_s|_{max}$

Natural
frequency, f

(a)

y

m

$y_s(t)$

(b)

Fig. 8.1 (a) Typical response spectrum. (b) Single degree-of-freedom system subjected to ground excitation.

exciting force shown in Fig. 8.2(b). The system is assumed to be initially at rest. The duration of the sinusoidal impulse is denoted by t_d. The differential equation of motion is obtained by equating to zero the sum of the forces in the corresponding free body diagram shown in Fig. 8.2(c), that is,

$$m\ddot{y} + ky = F(t) \tag{8.1}$$

in which

$$F(t) = \begin{cases} F_0 \sin \bar{\omega}t & \text{for } 0 \le t \le t_a \\ 0 & \text{for } t > t_d \end{cases} \tag{8.2}$$

and

$$\bar{\omega} = \frac{\pi}{t_d} \tag{8.3}$$

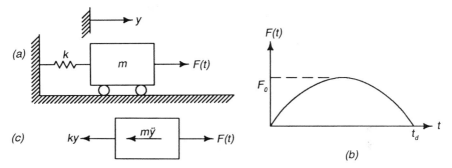

(a)

y

k

m

$F(t)$

(c)

$ky \leftarrow$ $\overset{m\ddot{y}}{\longleftarrow}$ $\rightarrow F(t)$

$F(t)$

F_0

t_d

t

(b)

Fig. 8.2 (a) Undamped simple oscillator subjected to load $F(t)$. (b) Loading function $F(t) = F_0 \sin \bar{\omega}t$ $(0 \le t \le t_d)$. (c) Free body diagram.

The solution of eq. (8.1) may be found by any of the methods studied in the preceding chapters such as the use of Duhamel's integral (Chapter 4) or the step-by-step linear acceleration method (Chapter 7). However, in this example, due to the simplicity of the exciting force, we can obtain the solution of eq. (8.1) by the direct method of integration of a linear differential equation, that is, the superposition of the complementary solution y_c and the particular solution y_p

$$y = y_c + y_p \tag{8.4}$$

The complementary solution of eq. (8.1) (right-hand side equals zero) is given by eq. (1.17) as

$$y_c = A \cos \omega t + B \sin \omega t \tag{8.5}$$

in which $\omega = \sqrt{k/m}$ is the natural frequency. The particular solution for the time interval $0 \le t \le t_d$ is suggested by the right-hand side of eq. (8.1) to be of the form

$$y_p = C \sin \bar{\omega} t \tag{8.6}$$

The substitution of eq. (8.6) into eq. (8.1) and solution of the resulting identity gives

$$C = \frac{F_0}{k - m\bar{\omega}^2} \tag{8.7}$$

Combining eqs. (8.4) through (8.7), we obtain the response for $0 \le t \le t_d$ as

$$y = A \cos \omega t + B \sin \omega t + \frac{F_0 \sin \bar{\omega} t}{k - m\bar{\omega}^2} \tag{8.8}$$

Introducing the initial conditions $y(0) = 0$ and $\dot{y}(0) = 0$ into eq. (8.8) and calculating the constants of integration A and B, we obtain

$$y = \frac{F_0/k}{1 - (\bar{\omega}/\omega)^2} [\sin \bar{\omega} t - (\bar{\omega}/\omega) \sin \omega t] \tag{8.9}$$

It is convenient to introduce the following notation:

$$y_{st} = \frac{F_0}{k}, \quad \bar{\omega} = \frac{\pi}{t_d}, \quad \omega = \frac{2\pi}{T}$$

Then eq. (8.9) becomes

$$\frac{y}{y_{st}} = \frac{1}{1 - \left(\dfrac{T}{2t_d}\right)^2} \left[\sin \pi \frac{t}{t_d} - \frac{T}{2t_d} \sin 2\pi \frac{t}{T} \right] \quad \text{for } 0 \le t \le t_d \tag{8.10a}$$

After a time t_d, the external force becomes zero and the system is then in free vibration. Therefore, the response for $t > t_d$ is of the form given by eq. (8.5) with the constants of integration determined from known values of displacement and velocity calculated from eq. (8.10a) at time $t = t_d$. The final expression for the response is then given by

$$\frac{y}{y_{st}} = \frac{T/t_d}{\left(\dfrac{T}{2t_d}\right)^2 - 1} \cos \pi \frac{t_d}{T} \sin 2\pi \left(\frac{t}{T} - \frac{t_d}{2T}\right) \quad \text{for } t \geq t_d \qquad (8.10b)$$

It may be seen from eq. (8.10) that the response in terms of y/y_{st} is a function of the ratio of the pulse duration to the natural period of the system (t_d/T) and of time expressed as t/T. Hence for any fixed value of the parameter t_d/T, we can obtain the maximum response from eq. (8.10). The plot in Fig. 8.3 of these maximum values as a function of t_d/T is the response spectrum for the half-sinusoidal force duration considered in this case. It can be seen from the response spectrum in Fig. 8.3 that the maximum value of the response (amplification factor) $y/y_{st} = 1.76$ occurs for this particular pulse when $t_d/T = 0.8$.

Due to the simplicity of the input force, it was possible in this case to obtain a closed solution and to plot the response spectrum in terms of dimensionless ratios, thus making this plot valid for any impulsive force described by one-half of the sine cycle. However, in general, for an arbitrary input load, we cannot expect to obtain such a general plot of the response spectrum and we normally have to be satisfied with the response spectrum plotted for a completely specified input excitation.

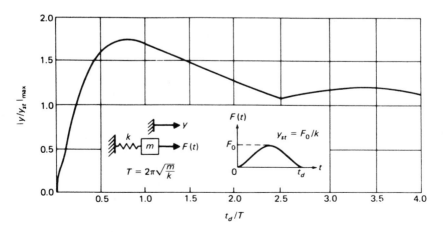

Fig. 8.3 Response spectrum for half-sinusoidal force of duration t_d.

8.2   RESPONSE SPECTRUM FOR SUPPORT EXCITATION

An important problem in structural dynamics is the analysis of a system subjected to excitation applied to the base or support of the structure. An example of such input excitation of the base acting on a damped oscillator which serves to model certain structures is shown in Fig. 8.4. The excitation in this case is given as an acceleration function which is represented in Fig. 8.5. The equation of motion which is obtained by equating to zero the sum of the forces in the corresponding free body diagram in Fig. 8.4(b) is

$$m\ddot{y} + c(\dot{y} - \dot{y}_s) + k(y - y_s) = 0 \tag{8.11}$$

or, with the usual substitution $\omega = \sqrt{k/m}$ and $\xi = c/c_{cr}$ $(c_{cr} = 2\sqrt{km})$,

$$\ddot{y} + 2\xi\omega\dot{y} + \omega^2 y = \omega^2 y_s(t) + 2\xi\omega\dot{y}_s(t) \tag{8.12}$$

Equation (8.12) is the differential equation of motion for the damped oscillator in terms of its absolute motion. A more useful formulation of this problem is to express eq. (8.12) in terms of the relative motion of the mass with respect to the motion of the support, that is, in terms of the spring deformation. The relative displacement u is defined as

$$u = y - y_s \tag{8.13}$$

Substitution into eq. (8.12) yields

$$\ddot{u} + 2\xi\omega\dot{u} + \omega^2 u = -\ddot{y}_s(t) \tag{8.14}$$

The formulation of the equation of motion in eq. (8.14) as a function of the relative motion between the mass and the support is particularly important since in design it is the deformation or stress in the "spring element" which

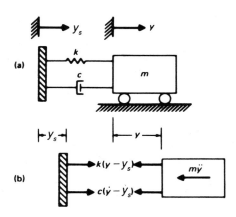

Fig. 8.4 (a) Damped simple oscillator subjected to support excitation. (b) Free body diagram.

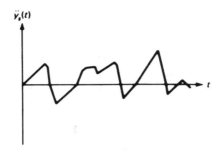

$\ddot{y}_s(t)$

Fig. 8.5 Acceleration function exciting the support of the oscillator in Fig. 8.4.

is required. Besides, the input motion at the base is usually specified by means of an acceleration function (e.g., earthquake accelerograph record); thus eq. (8.14) containing in the right-hand side the acceleration of the excitation is a more convenient form than eq. (8.12) which in the right-hand side has the support displacement and the velocity.

The solution of the differential equation, eq. (8.14), may be obtained by any of the methods presented in previous chapters for the solution of one-degree-of-freedom systems. In particular, the solution is readily expressed using Duhamel's integral as

$$u(t) = -\frac{1}{\omega} \int_0^t \ddot{y}_s(\tau) e^{-\xi\omega(t-\tau)} \sin \omega(t-\tau) \, d\tau \tag{8.15}$$

8.3   TRIPARTITE RESPONSE SPECTRA

It is possible to plot in a single chart using logarithmic scales the maximum response in terms of the acceleration, the relative displacement, and a third quantity known as the relative pseudovelocity. The pseudovelocity is not exactly the same as the actual velocity, but it is closely related and provides for a convenient substitute for the true velocity. These three quantities—the maximum absolute acceleration, the maximum relative displacement, and the maximum relative pseudovelocity—are known, respectively, as the spectral acceleration, spectral displacement, and spectral velocity.

It is significant that the spectral displacement S_D, that is, the maximum relative displacement, is proportional to the spectral acceleration S_a, the maximum absolute acceleration. To demonstrate this fact, consider the equation of motion, eq. (8.11), which, after using eq. (8.13), becomes for the damped system

$$m\ddot{y} + c\dot{u} + ku = 0 \tag{8.16}$$

and for the undamped system

$$m\ddot{y} + ku = 0 \qquad (8.17)$$

We observe from eq. (8.17) that the absolute acceleration is at all times proportional to the relative displacement. In particular, at maximum values, the spectral acceleration is proportional to the spectral displacement, that is, from eq. (8.17)

$$S_a = -\omega^2 S_D \qquad (8.18)$$

where $\omega = \sqrt{k/m}$ is the natural frequency of the system, $S_a = \ddot{y}_{max}$, and $S_D = u_{max}$.

When damping is considered in the system, it may be rationalized that the maximum relative displacement occurs when the relative velocity is zero ($\dot{u} = 0$). Hence we again obtain eq. (8.18) relating spectral acceleration and spectral displacement. This relationship is by mere coincidence the same as for simple harmonic motion. The fictitious velocity associated with the apparent harmonic motion is the pseudovelocity and, for convenience, its maximum value S_v is defined as the spectral velocity, that is

$$S_v = \omega S_D = \frac{S_a}{\omega} \qquad (8.19)$$

Dynamic response spectra for single degree-of-freedom elastic systems have been computed for a number of input motions. A typical example of displacement response spectrum for a single degree-of-freedom system subjected to support motion is shown in Fig. 8.6. This plot is the response for the input motion given by the recorded ground acceleration of the 1940 El Centro earthquake. The acceleration record of this earthquake has been used extensively in earthquake engineering investigations. A plot of the acceleration record for this earthquake is shown in Fig. 8.7. Until the time of the San Fernando, California earthquake of 1971, the El Centro record was one of the few records available for long and strong earthquake motions. In Fig. 8.8, the same type of data which was used to obtain the displacement response spectrum in Fig. 8.6 is plotted in terms of the spectral velocity, for several values of the damping coefficient, with the difference that the abscissa as well as the ordinate are in these cases plotted on a logarithmic scale. In this type of plot, because of eqs. (8.18) and (8.19), it is possible to draw diagonal scales for the displacement sloping 135° with the abscissa, and for the acceleration 45°, so that we can read from a single plot values of spectral acceleration, spectral velocity, and spectral displacements.

To demonstrate the construction of a tripartite diagram such as the one of Fig. 8.8, we write eq. (8.19) in terms of the natural frequency f in cycles per second (cps) and take the logarithm of the terms, so that

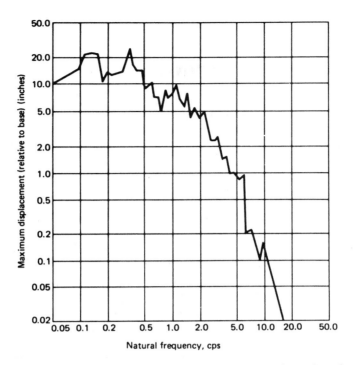

Fig. 8.6 Displacement response spectrum for elastic system subjected to the ground motion of 1940 El Centro earthquake. (from *Design of Multistory Reinforced Building for Earthquake Motions* by J. A. Blum, N. M. Newmark, and L. H. Corning, Portland Cement Association, 1961).

$$S_v = \omega S_D = 2\pi f S_D$$

$$\log S_v = \log f + \log (2\pi S_D) \tag{8.20}$$

For constant values of S_D, eq. (8.20) is the equation of a straight line of log S_v versus log f with a slope of 45°. Analogously, from eq. (8.19)

$$S_v = \frac{S_a}{\omega} = \frac{S_a}{2\pi f}$$

$$\log S_v = -\log f + \log \frac{S_a}{2\pi} \tag{8.21}$$

For a constant value of S_a, eq. (8.21) is the equation of a straight line of log S_v versus log f with a slope of 135°.

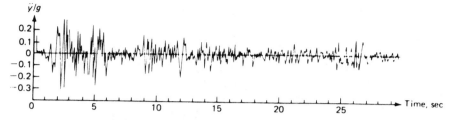

Fig. 8.7 Ground acceleration record for El Centro, California earthquake of May 18, 1940 north–south component.

8.4   RESPONSE SPECTRA FOR ELASTIC DESIGN

In general, response spectra are prepared by calculating the response to a specified excitation of single degree-of-freedom systems with various amounts of damping. Numerical integration with short time intervals are applied to calculate the response of the system. The step-by-step process is continued until the total earthquake record has been completed. The largest value of the function of interest is recorded and becomes the response of the system to that excitation. Changing the parameters of the system to change the natural frequency, we repeat the process and record a new maximum response. This

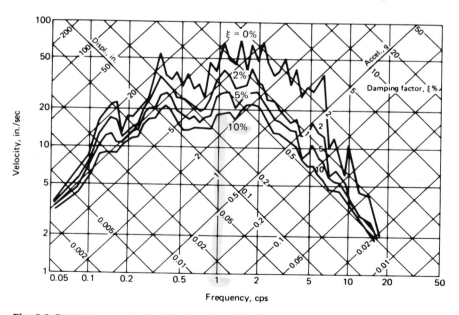

Fig. 8.8 Response spectra for elastic system for the 1940 El Centro earthquake (from Blume et al. 1961).

process is repeated until all frequencies of interest have been covered and the results plotted. Since no two earthquakes are alike, this process must be repeated for all earthquakes of interest.

Until the San Fernando, California earthquake of 1971, there were few recorded strong earthquake motions because there were few accelerometers emplaced to measure them; the El Centro, California earthquake of 1940 was the most severe earthquake recorded and was used as the basis for much analytical work. Recently, however, other strong earthquakes have been recorded. Maximum values of ground acceleration of about 0.32 g for the El Centro earthquake to values of more than 0.5 g for other earthquakes have been recorded. It can be expected that even larger values will be recorded as more instruments are placed closer to the epicenters of earthquakes.

Earthquakes consist of a series of essentially random ground motions. Usually the north–south, east–west, and vertical components of the ground acceleration are measured. Currently, no accurate method is available to predict the particular motion that a site can be expected to experience in future earthquakes. Thus it is reasonable to use a design response spectrum which incorporates the spectra for several earthquakes and which represents a kind of "average" response spectrum for design. Such a design response spectrum is shown in Fig. 8.9 normalized for a maximum ground acceleration of 1.0 g. This figure shows the design maximum ground motion and a series of response spectral plots corresponding to various values of the damping ratio in the system.

Details for the construction of the basic spectrum for design purposes are given by Newmark and Hall (1973), who have shown that smooth response spectra of idealized ground motion may be obtained by amplifying the ground motion by factors depending on the damping in the system. In general, for any given site, estimates might be made of the maximum ground acceleration, maximum ground velocity, and maximum ground displacement. The lines representing these maximum values are drawn on a tripartite logarithmic paper of which Fig. 8.10 is an example. The lines in this figure are shown for a maximum ground acceleration of 1.0 g, a velocity of 48 in/sec, and a displacement of 36 in. These values correspond to motions which are more intense than those generally expected in seismic design. They are, however, of proportional magnitudes which are generally correct for most practical applications. These maximum values normalized for a ground acceleration of 1.0 g are simply scaled down for other than 1.0 g acceleration of the ground. Recommended amplification factors to obtain the response spectra from maximum values of the ground motion are given in Table 8.1. For each value of the damping coefficient, the amplified displacement lines are drawn at the left, the amplified velocities at the top, and the amplified acceleration at the right of the chart. At a frequency of approximately 6 cps (Fig. 8.9), the amplified acceleration region line intersects a line sloping down toward the max-

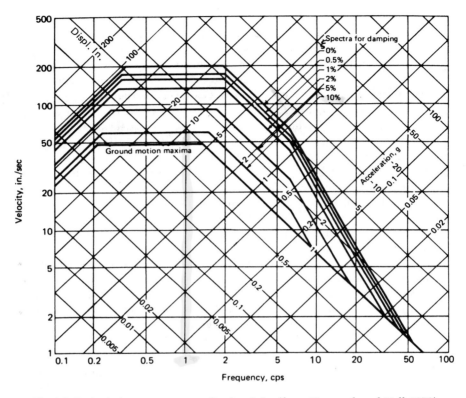

Fig. 8.9 Basic design spectra normalized to 1.0 g (from Newmark and Hall 1973).

imum ground acceleration value at a frequency of about 30 cps for a system with 2% damping. The lines corresponding to other values of damping are drawn parallel to the 2% damping line as shown in Fig. 8.9.

The amplification factors in Table 8.1 were developed on the basis of earthquake records available at the time. As new records of more recent earthquakes become available, these amplification factors have been recalculated. Table 8.2 shows the results of a statistical study based on a selection of 10 strong motion earthquakes. The table gives recommended amplification factors as well as the corresponding standard deviation values obtained in the study. The relatively large values shown in Table 8.2 for the standard deviation of the amplification factors provides further evidence on the uncertainties surrounding earthquake prediction and analysis. The response spectra for designs presented in Fig. 8.9 has been constructed using the amplification factors shown in Table 8.1.

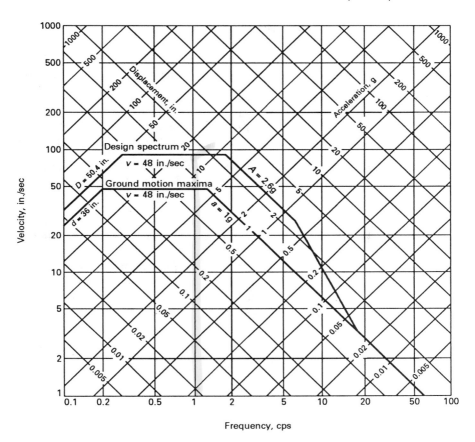

Fig. 8.10 Elastic design spectrum normalized to peak ground acceleration of 1.0 g for 5% damping (Newmark and Hall 1973).

TABLE 8.1   Relative Values Spectrum Amplification Factors

| Percent Damping | Amplification Factors | | |
	Displacement	Velocity	Acceleration
0	2.5	4.0	6.4
0.5	2.2	3.6	5.8
1	2.0	3.2	5.2
2	1.8	2.8	4.3
5	1.4	1.9	2.6
7	1.2	1.5	1.9
10	1.1	1.3	1.5
20	1.0	1.1	1.2

TABLE 8.2   Spectral Amplification Factors and Standard Deviation Values[*]

Percent Damping	Displacement		Velocity		Acceleration	
	Factor	Standard Deviation	Factor	Standard Deviation	Factor	Standard Deviation
2	1.691	0.828	2.032	0.853	3.075	0.738
5	1.465	0.630	1.552	0.605	2.281	0.502
10	1.234	0.481	1.201	0.432	1.784	0.321

[*]Newmark, N. M., and Riddell, R., Inelastic Spectra for Seismic Design: Seventh World Earthquake Conference, Istanbul, Turkey, Vol. 4, pp. 129–136, 1980.

Example 8.1.   A structure modeled as a single degree-of-freedom system has a natural period, $T = 1$ sec. Use the response spectral method to determine the maximum absolute acceleration, the maximum relative displacement, and the maximum relative pseudovelocity for: (a) a foundation motion equal to the El Centro earthquake of 1940, and (b) the design earthquake with a maximum ground acceleration equal to 0.32 g. Assume 10% of the critical damping.

Solution: (a) From the response spectra in Fig. 8.8 with $f = 1/T = 1.0$ cps, corresponding to the curve labeled $\xi = 0.10$, we read on the three scales the following values:

$$S_D = 3.3 \text{ in}$$

$$S_v = 18.5 \text{ in/sec}$$

$$S_a = 0.30 \, g$$

(b) From the basic design spectra in Fig. 8.9 with frequency $f = 1$ cps and 10% critical damping, we obtain after correcting for 0.32 g maximum ground acceleration in the following results:

$$S_D = 9.5 \times 0.32 = 3.04 \text{ in}$$

$$S_v = 60 \times 0.32 = 19.2 \text{ in/sec}$$

$$S_a = 0.95 \times 0.32 \, g = 0.304 \, g$$

8.5   RESPONSE SPECTRA FOR INELASTIC SYSTEMS

For certain types of extreme events such as nuclear blast explosions or strong motion earthquakes, it is sometimes necessary to design structures to withstand strains beyond the elastic limit. For example, in seismic design for an earthquake of moderate intensity, it is reasonable to assume elastic behavior for a well-designed and -constructed structure. However, for very strong motions, this is not a realistic assumption even for a well-designed structure. Although structures can be designed to resist severe earthquakes, it is not feasible economically to design buildings to elastically withstand earthquakes of the greatest foreseeable intensity. In order to design structures for strain levels beyond the linear range, the response spectrum has been extended to include the inelastic range (Newmark and Hall 1973). Generally, the elastoplastic relation between force and displacement, which was discussed in detail in Chapter 7, is used in structural dynamics. Such a force-displacement relationship is shown in Fig. 8.11. Because of the assumption of elastoplastic behavior, if the force is removed prior to the occurrence of yielding, the material will return along its loading line to the origin. However, when yielding occurs at a displacement y_t, the restoring force remains constant at a magnitude R_t. If the displacement is not reversed, the displacement may reach a maximum value y_{max}. If, however, the displacement is reversed, the elastic recovery follows along a line parallel to the initial line and the recovery proceeds elastically until a negative yield value R_c is reached in the opposite direction.

The preparation of response spectra for such an inelastic system is more difficult than that for elastic systems. However, response spectra have been prepared for several kinds of input disturbances. These spectra are usually plotted as a series of curves corresponding to definite values of the ductility ratio μ. The ductility ratio μ is defined as the ratio of the maximum displacement of the structure in the inelastic range to the displacement corresponding to the yield point y_y, that is,

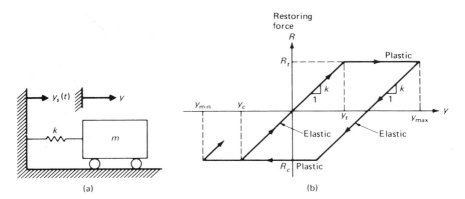

Fig. 8.11 Force-displacement relationship for an elastoplastic single degree-of-freedom system.

$$\mu = \frac{y_{max}}{y_y} \tag{8.22}$$

The response spectra for an undamped single degree-of-freedom system subjected to a support motion equal to the El Centro 1940 earthquake is shown
in Fig. 8.12 for several values of the ductility ratio. The tripartite logarithmic
scales used to plot these spectra give simultaneously for any single degree-
of-freedom system of natural period T and specified ductility ratio μ, the spectral values of displacement, velocity, and acceleration. Similarly, in Fig. 8.13,
are shown the response spectra for an elastoplastic system with 10% of critical damping. The spectral velocity and the spectral acceleration are read directly from the plots in Figs. 8.12 and 8.13, whereas the values obtained for
the spectral displacement must be multiplied by the ductility ratio in order
to obtain the correct value for the spectral displacement.

The concept of ductility ratio has been associated mainly with steel struc-

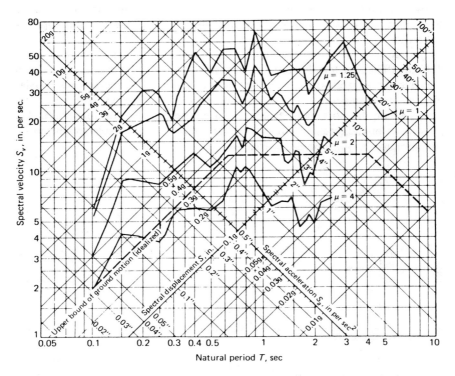

Fig. 8.12 Response spectra for undamped elastoplastic systems for the 1940 El Centro
earthquake (from Blume et al. 1961).

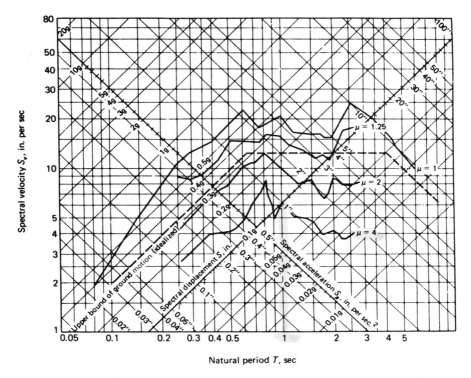

Fig. 8.13 Response spectra for elastoplastic systems with 10% critical damping for the 1940 El Centro earthquake (from Blume et al. 1961).

tures. These structures have a load–deflection curve that is often approximated as the elastoplastic curve shown in Fig. 8.11(b). For other types of structures, such as reinforced concrete structures or masonry shear walls, still, conveniently, the load–deflection curve is modeled as the elastoplastic curve. Although for steel structures, ductility factors as high as 6 are often used in collapse-level earthquake design, lower values for the ductility ratio are applicable to masonry shear walls. The selection of ductility values for seismic design must also be based on the design objectives and the loading criteria as well as the risk level acceptable for the structure as it relates to its use.

For reinforced concrete structures or masonry walls, a ductility factor of 1.0 to 1.5 seems appropriate for earthquake design where the objective is to limit damage. In other words, the objective of limit damage requires that structural members should be designed to undergo little if any yielding. When the design objective is to prevent collapse of the structure, ductilities of 2 to 3 are appropriate in this case.

8.6   RESPONSE SPECTRA FOR INELASTIC DESIGN

In the preceding section of this chapter, we discussed the procedure for calculating the seismic response spectra for elastic design. Figure 8.9 shows the elastic design spectra for several values of the damping ratio. The same procedure of constructing a basic response spectrum which consolidates the "average" effect of several earthquake records may also be applied to design in the inelastic range. The spectra for elastoplastic systems have the same appearance as the spectra for elastic systems, but the curves are displaced downward by an amount which is related to the ductility factor μ. Figure 8.14

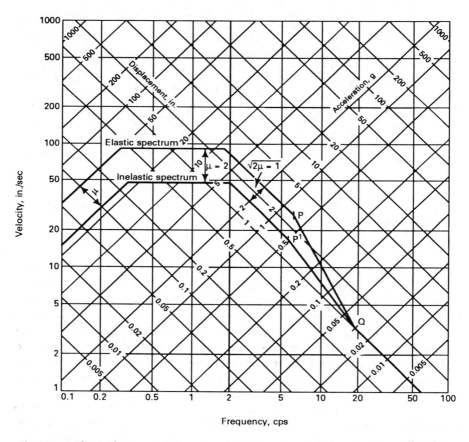

Fig. 8.14 Inelastic design spectrum normalized to peak ground acceleration of 1.0 g for 5% damping and ductility factor $\mu = 2.0$. (NOTE: Chart gives directly S_v and S_a; however, displacement values obtained from the chart should be multiplied by μ to obtain S_D.)

shows the construction of a typical design spectrum currently recommended (Newmark and Hall 1973) for use when inelastic action is anticipated.

The elastic spectrum for design from Fig. 8.9 corresponding to the desired damping ratio is copied in a tripartite logarithmic paper as shown in Fig. 8.14 for the spectra corresponding to 5% damping. Then lines reduced by the specified ductility factor are drawn parallel to elastic spectral lines in the displacement region (the left region) and in the velocity region (the central region). However, in the acceleration region (the right region) the recommended reducing factor is $\sqrt{2\mu - 1}$. This last line is extended up to a frequency of about 6 cps (point P' in Fig. 8.14). Then the inelastic design spectrum is completed by drawing a line from this last point P' to point Q, where the descending line from point P of the elastic spectrum intersects the line of constant acceleration as shown in Fig. 8.14.

The development of the reduction factors in the displacement and velocity regions is explained with the aid of Fig. 8.15(a). This figure shows the force–displacement curves for elastic and for elastoplastic behavior. At equal maximum displacement y_{max} for the two curves, we obtain from Fig. 8.15(a) the following relationship:

$$\mu = \frac{y_{max}}{y_y} = \frac{F_E}{F_y}$$

or

$$F_y = \frac{F_E}{\mu} \tag{8.23}$$

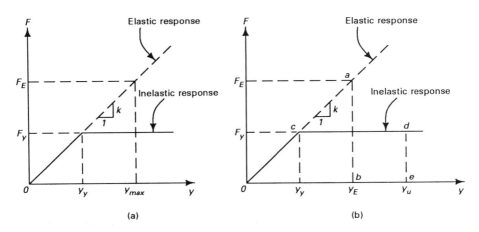

Fig. 8.15 Elastic and inelastic force–displacement curves. (a) Displacement and velocity regions. (b) Acceleration region.

where F_E is the force corresponding to the maximum displacement y_{max} in the elastic curve and F_y is the force at the yield condition. Equation (8.23) then shows that the force and consequently the acceleration in the elasto-plastic system is equal to the corresponding value in the elastic system reduced by the ductility factor. Therefore, the spectral acceleration S_a for elastoplastic behavior is related to the elastic spectral acceleration S_{aE} as

$$S_a = \frac{S_{aE}}{\mu} \tag{8.24}$$

However, in the acceleration region of the response spectrum, a ductility reduction factor μ does not result in close agreement with experimental data as does the recommended reduction factor $\sqrt{2\mu - 1}$. This factor may be rationally obtained by establishing the equivalence of the energy between the elastic and the inelastic system. In reference to Fig. 8.15(b) this equivalence is established by equating the area under elastic curve "Oab" with the area under the inelastic curve "$Ocde$." Namely,

$$\frac{F_E y_E}{2} = \frac{F_y y_y}{2} + F_y(y_\mu - y_y) \tag{8.25}$$

where F_E and F_y are the elastic and inelastic forces corresponding respectively to the maximum elastic displacement y_E and to the maximum inelastic displacement y_μ and where y_y is the yield displacement.

The substitution of $y_E = F_E/k$, $y_y = F_y/k$, and $y_\mu = \mu y_y$ into eq. (8.25) gives

$$\frac{F_E^2}{2k} = \frac{F_y^2}{2k}(2\mu - 1)$$

or

$$\frac{F_E}{F_y} = \sqrt{2\mu - 1}$$

Thus,

$$F_y = \frac{F_E}{\sqrt{2\mu - 1}}$$

and, consequently,

$$S_{a\mu} = \frac{S_{aE}}{\sqrt{2\mu - 1}}$$

Thus, the inelastic spectral acceleration in the acceleration region is obtained by reducing the elastic spectral by the factor $\sqrt{2\mu - 1}$.

The inelastic response spectrum thus constructed and shown in Fig. 8.14

gives directly the values for the spectral acceleration S_a and spectral velocity S_v. However, the values read from this chart in a displacement scale must be multiplied by the ductility factor μ to obtain the spectral displacement S_D. Inelastic response spectral charts for design developed by the procedure explained are presented in Figs. 8.16, 8.17, and 8.18 correspondingly to damping factors $\xi = 0$, 5%, and 10% and for ductility ratios $\mu = 1, 2, 5$, and 10.

Example 8.2.   Calculate the response of the single degree-of-freedom system of Example 8.1, assuming that the structure is designed to withstand

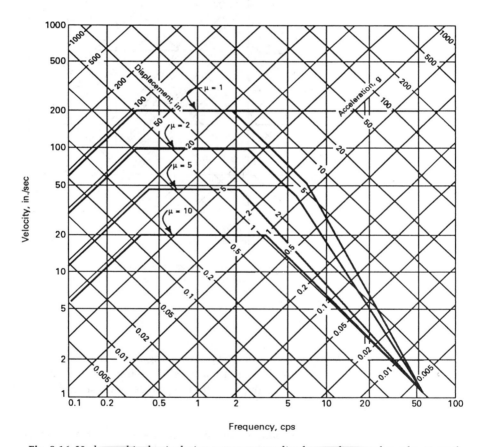

Fig. 8.16 Undamped inelastic design spectra normalized to peak ground acceleration of 1.0 g. (Spectral values S_a for acceleration and S_v for velocity are obtained directly from the graph. However, values of S_D for spectral displacement should be amplified by the ductility ratio μ.)

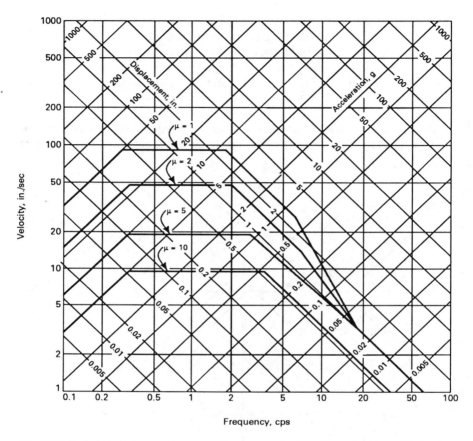

Fig. 8.17 Inelastic design spectra normalized to peak ground acceleration of 1.0 g for 5% damping. (Spectral values S_a for acceleration and S_v for velocity are directly obtained from the graph. However, values of S_D for spectral displacement should be amplified by the ductility ratio μ.)

seismic motions with an elastoplastic behavior having a ductility ratio μ = 4.0. Assume damping equal to 10% of the critical damping. (a) Use the response spectra of the El Centro earthquake. (b) Use the design response spectra.

Solution: (a) From the response spectrum corresponding to 10% of the critical damping (Fig. 8.13), we read for $T = 1$ sec and the curve labeled μ = 4.0

$$S_D = 1.0 \times 4.0 = 4.0 \text{ in}$$

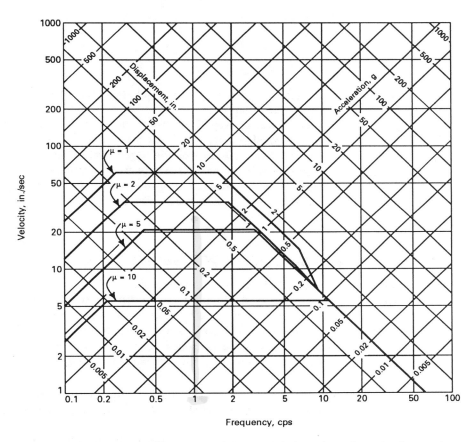

Fig. 8.18 Inelastic design spectra normalized to peak ground acceleration of 1.0 g' for 10% damping. (Spectral values S_a for acceleration and S_v for spectral velocity are directly obtained from the graph. However, values of S_D for spectral displacement should be amplified by the ductility ratio μ.)

$$S_v = 6.2 \text{ in/sec}$$

$$S_a = 0.1 \text{ g}$$

The factor 4.0 is required in the calculation of S_D since as previously noted the spectra plotted in Fig. 8.13 are correct for acceleration and for pseudovelocity, but for displacements it is necessary to amplify the values read from the chart by the ductility ratio.

(b) Using the inelastic design spectra with 10% damping for design in Fig. 8.18, corresponding to 1 cps, we obtain the following maximum values for the response:

$$S_D = 3.0 \times 0.32 \times 4.0 = 3.84 \text{ in}$$

$$S_v = 15.6 \times 0.32 = 5.00 \text{ in/sec}$$

$$S_a = 0.3 \times 0.32 = 0.096 \text{ g}$$

As can be seen, these spectral values based on the design spectrum are somewhat different than those obtained from the response spectrum of the El Centro earthquake of 1940. Also, if we compare these results for the elastoplastic behavior with the results in Example 8.1 for the elastic structure, we observe that the maximum relative displacement has essentially the same magnitude whereas the acceleration and the relative pseudovelocity are appreciably less. This observation is in general true for any structure when inelastic response is compared with the response based on elastic behavior.

8.7   PROGRAM 6—SEISMIC RESPONSE SPECTRA

The computer program described in this chapter serves to calculate elastic response spectra in terms of spectral displacement (maximum relative displacement), spectral velocity (maximum relative pseudovelocity), and spectral acceleration (maximum absolute acceleration) for any prescribed time–acceleration–seismic excitation. The response is calculated in the specified range of frequencies using the direct integration method presented in Chapter 4.

Example 8.3.   Use Program 6 to develop the response spectra for elastic systems subjected to the first ten seconds of the 1940 El Centro earthquake. Assume 10% damping.

The digitized values corresponding to the accelerations recorded for the first 10 sec of the El Centro earthquake are given in Table 8.3.

Solution:

```
PROGRAM 6:    SEISMIC RESPONSE SPECTRA      DATA FILE:D6

      INPUT DATA:

NUMBER OF POINTS DEFINING THE EXCITATION        NE= 186
INITIAL FREQUENCY (C.P.S.)                      FI= .05
FREQUENCY INCREMENT (C.P.S.)                     DF= .05
FINAL FREQUENCY C.P.S.                           FF= 20
DAMPING RATIO                                   XSI= .1
TIME STEP INTEGRATION                            H= .01
ACCELERATION GRAVITY                             G= 386
```

OUTPUT RESULTS

FREQ. (C.P.S.)	SPECT.DISPL. (IN)	SPECT.VELOC. (IN/SEC)	SPECT.ACC. (IN/SEC/SEC)
0.05	24.65	7.70	2.41
0.10	21.86	13.67	8.54
0.15	15.36	14.41	13.51
0.20	12.63	15.79	19.75
0.25	13.33	20.83	32.56
0.30	13.62	25.55	47.92
0.35	12.33	26.97	59.02
0.40	12.02	30.06	75.17
0.45	8.75	24.63	69.28
0.50	6.31	19.72	61.64
1.00	3.26	20.38	127.40
1.50	2.43	22.78	213.57
2.00	2.02	25.29	316.18
2.50	1.22	19.13	299.00
3.00	0.74	13.96	261.73
3.50	0.53	11.49	251.39
4.00	0.41	10.28	257.19
4.50	0.32	9.08	255.40
5.00	0.30	9.42	294.58
10.00	0.05	3.20	200.10
15.00	0.02	1.77	165.54
20.00	0.01	0.99	124.09

8.8 SUMMARY

Response spectra are plots which give the maximum response for a single degree-of-freedom system subjected to a specified excitation. The construction of these plots requires the solution of single degree-of-freedom systems for a sequence of values of the natural frequency and of the damping ratio in the range of interest. Every solution provides only one point (the maximum value) of the response spectrum. In solving the single degree-of-freedom systems, use is made of Duhamel's integral or of the direct method (Chapter 4) for elastic systems and of the step-by-step linear acceleration method for inelastic behavior (Chapter 7). Since a large number of systems must be analyzed in order to fully plot each response spectrum, the task is lengthy and time-consuming even with the use of the computer. However, once these curves are constructed and are available for the excitation of interest, the analysis for the design of structures subjected to dynamic loading is reduced to a simple calculation of the natural frequency of the system and the use of the response spectrum.

Table 8.3   Digitized Values of the Acceleration Recorded for the First Ten Seconds for the El Centro Earthquake of 1940

Time (Sec.)	Acc. (Acc.g')	Time (Sec.)	Acc. (Acc.g')	Time (Sec.)	Acc. (Acc.g')	Time (Sec.)	Acc. (Acc.g')
0.0000	0.0108	0.0420	0.0020	0.0970	0.0159	0.1610	-0.0001
0.2210	0.0189	0.2630	0.0001	0.2910	0.0059	0.3320	-0.0012
0.3740	0.0200	0.4290	-0.0237	0.4710	0.0076	0.5810	0.0425
0.6230	0.0094	0.6650	0.0138	0.7200	-0.0088	0.7400	-0.0256
0.7890	-0.0387	0.8290	-0.0568	0.8720	-0.0232	0.9020	-0.0343
0.9410	-0.0402	0.9610	-0.0603	0.9970	-0.0789	1.0660	-0.0666
1.0760	-0.0381	1.0940	-0.0429	1.1680	0.0897	1.3150	-0.1696
1.3840	-0.0828	1.4120	-0.0828	1.4400	-0.0945	1.4810	-0.0885
1.5090	-0.1080	1.5370	-0.1280	1.6280	0.1144	1.7030	0.2355
1.8550	0.1428	1.8800	0.1777	1.9240	-0.2610	2.0070	-0.3194
2.2150	0.2952	2.2700	0.2634	2.3200	-0.2984	2.3950	0.0054
2.4500	0.2865	2.5190	-0.0469	2.5750	0.1516	2.6520	0.2077
2.7080	0.1087	2.7690	-0.0325	2.8930	0.1033	2.9760	-0.0803
3.0680	0.0520	3.1290	-0.1547	3.2120	0.0065	3.2530	-0.2060
3.3860	0.1927	3.4190	-0.0937	3.5300	0.1708	3.5990	-0.0359
3.6680	0.0365	3.7380	-0.0736	3.8350	0.0311	3.9040	-0.1833
4.0140	0.0227	4.0560	-0.0435	4.1060	0.0216	4.2220	-0.1972
4.3140	-0.1762	4.4160	0.1460	4.4710	-0.0047	4.6180	0.2572
4.6650	-0.2045	4.7560	0.0608	4.8310	-0.2733	4.9700	0.1779
5.0390	0.0301	5.1080	0.2183	5.1990	0.0267	5.2330	0.1252
5.3020	0.1290	5.3300	0.1089	5.3430	-0.0239	5.4540	0.1723
5.5100	-0.1021	5.6060	0.0141	5.6900	-0.1949	5.7730	-0.2420

5.8000	-0.0050	5.8090	-0.0275	5.8690	-0.0573	5.8830	-0.0327
5.9250	0.0216	5.9800	0.0108	6.0130	0.0235	6.0850	-0.0665
6.1320	0.0014	6.1740	0.0493	6.1880	0.0149	6.1980	-0.0200
6.2290	-0.0381	6.2790	0.0207	6.3260	-0.0058	6.3680	-0.0603
6.3820	-0.0162	6.4090	0.0200	6.4590	-0.1760	6.4780	-0.0033
6.5200	0.0043	6.5340	-0.0040	6.5620	-0.0099	6.5750	-0.0017
6.6030	-0.0170	6.6450	0.0373	6.6860	0.0457	6.7140	0.0385
6.7280	0.0009	6.7490	-0.0288	6.7690	0.0016	6.8110	0.0113
6.8520	0.0022	6.9080	0.0092	6.9910	-0.0996	7.0740	0.0360
7.1210	0.0078	7.1430	-0.0277	7.1490	0.0026	7.1710	0.0272
7.2260	0.0576	7.2950	-0.0492	7.3700	0.0297	7.4060	0.0109
7.4250	0.0186	7.4610	-0.2530	7.5250	-0.0347	7.5720	0.0036
7.6000	-0.0628	7.6410	-0.0280	7.6690	-0.0196	7.6910	0.0068
7.7520	-0.0054	7.7940	-0.0603	7.8350	-0.0357	7.8770	-0.0716
7.9600	-0.0140	7.9870	-0.0056	8.0010	0.0222	8.0700	0.0468
8.1260	0.0260	8.1660	-0.0335	8.1950	-0.0128	8.2230	0.0661
8.2780	0.0305	8.3340	0.0246	8.4030	0.0347	8.4580	-0.0369
8.5330	-0.0344	8.5960	-0.0104	8.6380	-0.0260	8.7350	0.1534
8.8180	-0.0028	8.8600	0.0233	8.8820	-0.0261	8.9150	-0.0022
8.9560	-0.1849	9.0530	0.1260	9.0950	0.0320	9.1230	0.0955
9.1500	0.1246	9.2530	-0.0328	9.2890	-0.0451	9.4270	0.1301
9.4410	-0.1657	9.5100	0.0419	9.6350	-0.0936	9.7040	0.0816
9.8150	-0.0881	9.8980	0.0064	9.9390	-0.0006	9.9950	0.0586
10.0200	-0.0713	10.0500	-0.0448	10.0800	-0.0221	10.1000	0.0093
10.1500	0.0024	10.1900	0.0510				

In the following chapters dealing with structures which are modeled as systems with many degrees of freedom, it will be shown that the dynamic analysis of a system with n degrees of freedom can be transformed to the problem of solving n systems in which each one is a single degree-of-freedom system. Consequently, this transformation extends the usefulness of response spectra for single degree-of-freedom systems to the solution of systems of any number of degrees of freedom.

The reader should thus realize the full importance of a thorough understanding and mastery of the concepts and methods of solutions for single degree-of-freedom systems, since these same methods are also applicable to systems of many degrees of freedom after the problem has been transformed to independent single degree-of-freedom systems.

PROBLEMS

8.1   The steel frame shown in Fig. 8.19 is subjected to horizontal force at the girder level of (1000 sin 10t lb) for a time duration of half a cycle of the forcing sine function. Use the appropriate response spectral chart to obtain the maximum displacement. Neglect damping.

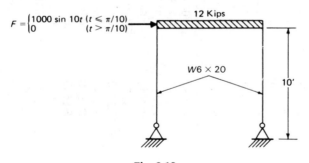

Fig. 8.19

8.2   Determine the maximum stresses in the columns of the frame of Problem 8.1.

8.3   Consider the frame shown in Fig. 8.20(a) subjected to a foundation excitation produced by a half cycle of the function $a_g = 200 \sin 10t$ in/sec^2 as shown in Fig. 8.20(b). Determine the maximum horizontal displacement of the girder relative to the motion of the foundation. Neglect damping.

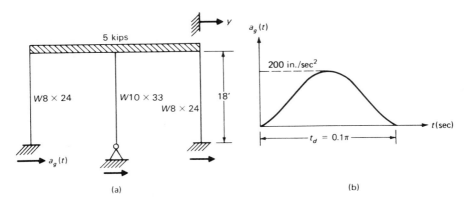

Fig. 8.20

8.4 Determine the maximum stress in the columns of the frame of Problem 8.3.

8.5 The frame shown in Fig. 8.19 is subjected to the excitation produced by the El Centro earthquake of 1940. Assume 10% damping and from the appropriate chart determine the spectral values for displacement, velocity, and acceleration. Assume elastic behavior.

8.6 Repeat Problem 8.5 using the basic design spectra given in Fig. 8.9 to determine the spectral values for acceleration, velocity, and displacement. (Scale down spectral values by factor 0.32.)

8.7 A structure modeled as the spring-mass system shown in Fig. 8.21 is assumed to be subjected to a support motion produced by the El Centro earthquake of 1940. Assuming elastic behavior and using the appropriate response spectral chart find the maximum relative displacement between the mass and the support. Also compute the maximum force acting on the spring. Neglect damping.

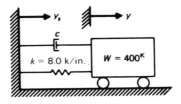

Fig. 8.21

8.8 Repeat Problem 8.7 assuming that the system has 10% of the critical damping.

8.9 Determine the force transmitted to the foundation for the system analyzed in Problem 8.8.

8.10 Consider the spring-mass system of Problem 8.7 and assume that the spring element follows, and that elastoplastic behavior with a maximum value for the restoring force in tension or in compression is equal to half the value of the elastic maximum force in the spring calculated in Problem 8.7. Determine the spectral value for the displacement. Neglect damping. (Hint: Start by assuming $\mu = 2$, find spectral value S_D, calculate μ, and find new spectral values, etc.)

8.11 Repeat Problem 8.10 for 10% damping.

8.12 A structure modeled as a single degree-of-freedom system has a natural period $T = 0.5$ sec. Use the response spectral method to determine in the elastic range the maximum absolute acceleration, the maximum relative displacement, and the maximum pseudorelative velocity for: (a) a foundation motion equal to the El Centro earthquake of 1940 and (b) the design spectrum with a maximum ground acceleration equal to 0.3 g. Neglect damping.

8.13 Solve Problem 8.12 assuming elastoplastic behavior of the system with ductility ratio $\mu = 4$.

8.14 Use Program 6 to develop a table having the spectral values for displacements, velocities, and accelerations in the range of frequency from 0.10 cps to 1.0 cps. The excitation is a constant acceleration of magnitude 0.01 g applied for 10 sec. Neglect damping.

8.15 Use Program 6 to develop the response spectra for elastic systems subjected to the first 10 sec of the 1940 El Centro earthquake. Neglect damping. The digitized values corresponding to the accelerations recorded for the first 10 sec of the El Centro earthquake are given in Table 8.3.

8.16 Repeat Problem 8.15 corresponding to a system with 10% of the critical damping.

PART II

Structures Modeled as Shear Buildings

9

The Multistory Shear Building

In Part I we analyzed and obtained the dynamic response for structures modeled as a single degree-of-freedom system. Only if the structure can assume a unique shape during its motion will the single degree model provide the exact dynamic response. Otherwise, when the structure takes more than one possible shape during motion, the solution obtained from a single degree model will be an approximation to the true dynamic behavior.

Structures cannot always be described by a single degree model and, in general, have to be represented by multiple degree models. In fact, structures are continuous systems and as such possess an infinite number of degrees of freedom. There are analytical methods to describe the dynamic behavior of continuous structures which have uniform material properties and regular geometry. These methods of analysis, though interesting in revealing information for the discrete modeling of structures, are rather complex and are applicable only to relatively simple actual structures. They require considerable mathematical analysis, including the solution of partial differential equations which will be presented in Chapter 20. For the present, we shall consider one of the most instructive and practical types of structure which involve many degrees of freedom, the multistory *shear building*.

9.1   STIFFNESS EQUATIONS FOR THE SHEAR BUILDING

A shear building may be defined as a structure in which there is no rotation of a horizontal section at the level of the floors. In this respect, the deflected building will have many of the features of a cantilever beam that is deflected by shear forces only, hence the name *shear building.* To accomplish such deflection in a building, we must assume that: (1) the total mass of the structure is concentrated at the levels of the floors; (2) the girders on the floors are infinitely rigid as compared to the columns; and (3) the deformation of the structure is independent of the axial forces present in the columns. These assumptions transform the problem from a structure with an infinite number of degrees of freedom (due to the distributed mass) to a structure which has only as many degrees as it has lumped masses at the floor levels. A three-story structure modeled as a shear building [Fig. 9.1(a)] will have three degrees of freedom, that is, the three horizontal displacements at the floor levels. The second assumption introduces the requirement that the joints between girders and columns are fixed against rotation. The third assumption leads to the condition that the rigid girders will remain horizontal during motion.

It should be noted that the building may have any number of bays and that it is only as a matter of convenience that we represent the shear building solely in terms of a single bay. Actually, we can further idealize the shear building as a single column [Fig. 9.2(a)], having concentrated masses at the floor levels with the understanding that only horizontal displacements of these masses are possible. Another alternative is to adopt a multimass spring system shown in Fig. 9.3(a) to represent the shear building. In any of the three

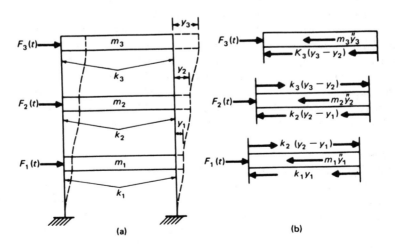

Fig. 9.1 Single bay model representation of a shear building.

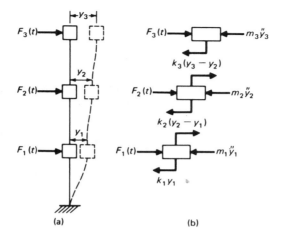

Fig. 9.2 Single column model representation of shear building.

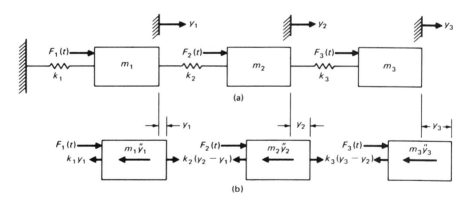

Fig. 9.3 Multimass spring model representation of a shear building.

representations depicted in these figures, the stiffness coefficient or spring constant k_i shown between any two consecutive masses is the force required to produce a relative unit displacement of the two adjacent floor levels.

For a uniform column with the two ends fixed against rotation, the spring constant is given by

$$k = \frac{12EI}{L^3} \tag{9.1a}$$

and for a column with one end fixed and the other pinned by

$$k = \frac{3EI}{L^3} \tag{9.1b}$$

where E is the material modulus of elasticity, I the cross-sectional moment of inertia, and L the length of the column.

It should be clear that all three representations shown in Figs. 9.1 to 9.3 for the shear building are equivalent. Consequently, the following equations of motion for the three-story shear building are obtained from any of the corresponding free body diagrams shown in these figures by equating to zero the sum of the forces acting on each mass. Hence

$$m_1\ddot{y}_1 + k_1y_1 - k_2(y_2 - y_1) - F_1(t) = 0$$

$$m_2\ddot{y}_2 + k_2(y_2 - y_1) - k_3(y_3 - y_2) - F_2(t) = 0$$

$$m_3\ddot{y}_3 + k_3(y_3 - y_2) - F_3(t) = 0 \tag{9.2}$$

This system of equations constitutes the *stiffness* formulation of the equations of motion for a three-story shear building. It may conveniently be written in matrix notation as

$$[M]\{\ddot{y}\} + [K]\{y\} = [F] \tag{9.3}$$

where $[M]$ and $[K]$ are, respectively, the mass and stiffness matrices given, respectively, by

$$[M] = \begin{bmatrix} m_1 & 0 & 0 \\ 0 & m_2 & 0 \\ 0 & 0 & m_3 \end{bmatrix} \tag{9.4}$$

$$[K] = \begin{bmatrix} k_1 + k_2 & -k_2 & 0 \\ -k_2 & k_2 + k_3 & -k_3 \\ 0 & -k_3 & k_3 \end{bmatrix} \tag{9.5}$$

and $\{y\}$, $\{\ddot{y}\}$, and $\{F\}$ are, respectively, the displacement, acceleration, and force vectors given by

$$\{y\} = \begin{Bmatrix} y_1 \\ y_2 \\ y_3 \end{Bmatrix}, \quad \{\ddot{y}\} = \begin{Bmatrix} \ddot{y}_1 \\ \ddot{y}_2 \\ \ddot{y}_3 \end{Bmatrix}, \quad \{F\} = \begin{Bmatrix} F_1(t) \\ F_2(t) \\ F_3(t) \end{Bmatrix} \tag{9.6}$$

It should be noted that the mass matrix, eq. (9.4), corresponding to the shear building is a diagonal matrix (the nonzero elements are only in the main diagonal). The elements of the stiffness matrix, eq. (9.5), are designated *stiffness coefficients*. In general, the stiffness coefficient k_{ij} is defined as the force at coordinate i when a unit displacement is given at j, all other coordinates being fixed. For example, the coefficient in the second row and second column of eq. (9.5) $k_{22} = k_2 + k_3$ is the force required at the second floor when a unit displacement is given to this floor.

9.2 FLEXIBILITY EQUATIONS FOR THE SHEAR BUILDING

An alternative approach in developing the equation of motion of a structure is the *flexibility* formulation. In this approach, the elastic properties of the structure are described by *flexibility coefficients*, which are defined as deflections produced by a unit load applied at one of the coordinates. Specifically, the flexibility coefficient f_{ij} is defined as the displacement at coordinate i when a unit static force is applied at coordinate j. Figure 9.4 depicts the flexibility coefficients corresponding to unit force applied at one of the story levels of a shear building. Using these coefficients and applying superposition, we may state that the displacement at any coordinate is equal to the sum of the products of flexibility coefficients at that coordinate multiplied by the corresponding forces. The forces acting on the three-story shear building (including the inertial forces) are shown in Fig. 9.5. Therefore, the displacements for the three-story building may be expressed in terms of the flexibility coefficients as

$$y_1 = (F_1(t) - m_1\ddot{y}_1)f_{11} + (F_2(t) - m_2\ddot{y}_2)f_{12} + (F_3(t) - m_3\ddot{y}_3)f_{13}$$

$$y_2 = (F_1(t) - m_1\ddot{y}_1)f_{21} + (F_2(t) - m_2\ddot{y}_2)f_{22} + (F_3(t) - m_3\ddot{y}_3)f_{23}$$

$$y_3 = (F_1(t) - m_1\ddot{y}_1)f_{31} + (F_2(t) - m_2\ddot{y}_2)f_{32} + (F_3(t) - m_3\ddot{y}_3)f_{33}$$

Rearranging the terms in these equations and using matrix notation, we obtain

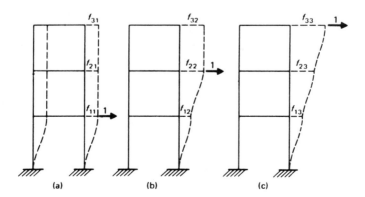

Fig. 9.4 Flexibility coefficients for a three-story shear building.

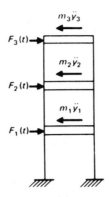

Fig. 9.5 Forces acting on a three-story shear building.

$$\{y\} = [f]\{F\} - [f][M]\{\ddot{y}\},\tag{9.7}$$

where $[M]$ is the mass matrix, eq. (9.4), $[f]$ is the flexibility matrix given by

$$[f] = \begin{bmatrix} f_{11} & f_{12} & f_{13} \\ f_{21} & f_{22} & f_{23} \\ f_{31} & f_{32} & f_{33} \end{bmatrix}\tag{9.8}$$

and $\{y\}$, $\{\ddot{y}\}$, and $\{F\}$ are, respectively, the displacement, acceleration, and force vectors given by eq. (9.6).

9.3   RELATIONSHIP BETWEEN STIFFNESS AND FLEXIBILITY MATRICES

The definitions given for either stiffness or flexibility coefficients are based on static considerations in which the displacements are produced by static forces. The relation between static forces and displacements may be obtained by equating to zero the acceleration vector $\{\ddot{y}\}$ in either eq. (9.3) or eq. (9.7). Hence

$$[K]\{y\} = \{F\}\tag{9.9}$$

$$[f]\{F\} = \{y\}\tag{9.10}$$

From these relations it follows that the stiffness matrix $[K]$ and the flexibility matrix $[f]$ are inverse matrices, that is

$$[K] = [f]^{-1}$$

or

$$[f] = [K]^{-1}\tag{9.11}$$

Consequently, the flexibility matrix $[f]$ for a shear building may be obtained either by calculating the inverse of the stiffness matrix or directly from the definition of the flexibility coefficients. Taking this last approach, we obtain, for the three-story shear building shown in Fig. 9.4(a),

$$k_1 f_{11} = 1$$

and

$$f_{11} = f_{21} = f_{31} = \frac{1}{k_1}$$

Analogously from Figs. 9.4(b) and 9.4(c) we obtain

$$f_{22} = f_{32} = \frac{1}{k_1} + \frac{1}{k_2}, \quad f_{12} = \frac{1}{k_1}$$

and

$$f_{33} = \frac{1}{k_1} + \frac{1}{k_2} + \frac{1}{k_3}, \quad f_{23} = \frac{1}{k_1} + \frac{1}{k_2}, \quad \text{and} \quad f_{13} = \frac{1}{k_1}$$

since the flexibility coefficients for springs in series are given by the summation of the reciprocal values of the spring constants.

Inserting these expressions for the flexibility coefficients into the flexibility matrix, eq. (9.8), results in

$$[f] = \begin{bmatrix} \dfrac{1}{k_1} & \dfrac{1}{k_1} & \dfrac{1}{k_1} \\[2mm] \dfrac{1}{k_1} & \dfrac{1}{k_1} + \dfrac{1}{k_2} & \dfrac{1}{k_1} + \dfrac{1}{k_2} \\[2mm] \dfrac{1}{k_1} & \dfrac{1}{k_1} + \dfrac{1}{k_2} & \dfrac{1}{k_1} + \dfrac{1}{k_2} + \dfrac{1}{k_3} \end{bmatrix} \tag{9.12}$$

The extension of the flexibility matrix for a three-story shear building to any number of stories is obvious from the pattern of eq. (9.12).

9.4 PROGRAM 7—MODELING STRUCTURES AS SHEAR BUILDINGS

The computer program presented in this chapter serves to model a structure as a shear building. Such modeling requires the development of the stiffness and mass matrices of the system. In Program 7, the elements of these matrices are stored in a file in preparation for dynamic analysis such as the calculation of natural frequencies and the determination of the response of the structure when subjected to dynamic forces or to seismic excitation.

Since the stiffness of the mass matrices are symmetric, only the elements in the upper triangular portion of these matrices need to be stored in files.

The notation used, in this case, consists of numbering consecutively those elements of the matrix located in the upper triangular part. For a matrix of order 5, the numbering of elements is as follows:

$$\begin{bmatrix} (1) & (3) & (6) & (10) & (15) \\ & (2) & (5) & (9) & (14) \\ & & (4) & (8) & (13) \\ & & & (7) & (12) \\ \text{symmetric} & & & & (11) \end{bmatrix}$$

Using this notation, we need to store only about half of the total number of coefficients in the stiffness matrix. As a matter of fact, we could save more memory in the computer by not storing the zero coefficients at the top of each column of the matrix. The implementation of such saving in computer memory requires special coding to locate each entry in the diagonal of the matrix and, therefore, each coefficient of the matrix. Such coding has not been implemented in the computer programs presented in this volume.

Example 9.1.   Determine the stiffness and mass matrices for the four-story shear building shown in Fig. 9.6. The modulus of elasticity is $E = 2 \times 10^6$ psi.

Solution:

Input Data and Output Results

```
        PROGRAM 7: SHEAR BUILDING          DATA FILE:D7

        INPUT DATA:

NUMBER OF STORIES (DEGREES OF FREEDOM)        ND = 4

            STORY DATA:

STORY #     STORY HEIGHT          STORY FLEXURAL         STORY MASS
                                STIFFNESS (TOTAL EI)

        1       180.00              0.1591E+09             1.000
        2       180.00              0.1591E+09             1.000
        3       180.00              0.1591E+09             1.000
        4       180.00              0.1591E+09             1.000

        OUTPUT RESULTS:
                **SYSTEM STIFFNESS MATRIX**

    6.5473E+02        -3.2737E+02         0.0000E+00        0.0000E+00
   -3.2737E+02         6.5473E+02        -3.2737E+02        0.0000E+00
```

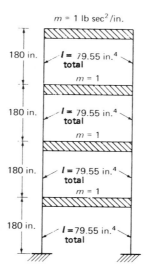

Fig. 9.6 Modeled structure for Example 9.1.

```
0.0000E+00      -3.2737E+02       6.5473E+02      -3.2737E+02
0.0000E+00       0.0000E+00      -3.2737E+02       3.2737E+02


                    **SYSTEM MASS MATRIX**

1.0000E+00       0.0000E+00       0.0000E+00       0.0000E+00
0.0000E+00       1.0000E+00       0.0000E+00       0.0000E+00
0.0000E+00       0.0000E+00       1.0000E+00       0.0000E+00
0.0000E+00       0.0000E+00       0.0000E+00       1.0000E+00
```

9.5 SUMMARY

The shear building idealization of structures provides a simple and useful mathematical model for the analysis of dynamic systems. This model permits the representation of the structure by lumped rigid masses interconnected by elastic springs. In obtaining the equations of motion, two different formulations are possible: (1) the stiffness method in which the equations of equilibrium are expressed in terms of stiffness coefficients; and (2) the flexibility method in which the equations of compatibility are written in terms of flexibility coefficients. The stiffness matrix and the flexibility matrix of a system, in reference to the same coordinates, are inverse matrices.

PROBLEMS

9.1 For the three-story shear building verify that the stiffness matrix, eq. (9.5), and the flexibility matrix, eq. (9.12), are inverse matrices.

9.2 For the two-story shear building shown in Fig. 9.7 determine the stiffness and flexibility matrices and then verify that these matrices are inverse matrices.

9.3 For the three-story shear building shown in Fig. 9.8 obtain the stiffness and flexibility matrices and show that these matrices are inverse matrices. All the columns are steel members W10 × 21.

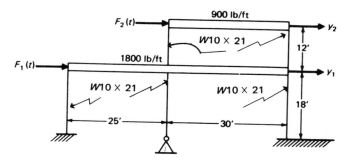

Fig. 9.7

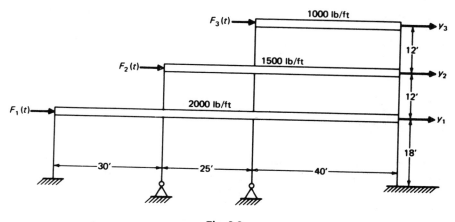

Fig. 9.8

9.4   Write the differential equation of motion using the stiffness formulation for the shear building in Fig. 9.7. Model the structure by a multimass-spring system.

9.5   Write the differential equation for the motion of the shear building in Fig. 9.8. Model the structure as a shear column with lumped masses as the floor levels.

9.6   The three-story shear building in Fig. 9.9 is subjected to a foundation motion which is given as an acceleration function $\ddot{y}_s(t)$. Obtain the stiffness differential equation of motion. Express the displacement of the floors relative to the foundation displacement (i.e., $u_i = y_i - y_s$).

9.7   Generalize the results of Problem 9.6 and obtain the equations of motion for a shear building of n stories.

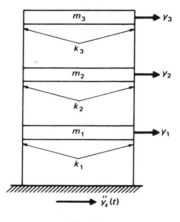

Fig. 9.9

9.8   Determine the stiffness and mass matrices for the shear building shown in Fig. 9.10. The modulus of elasticity is $E = 30 \times 10^6$ psi.

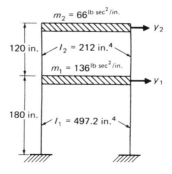

Fig. 9.10

9.9   Use Program 7 to determine the stiffness and mass matrices for the shear building shown in Fig. 9.11. The modulus of elasticity is $E = 30 \times 10^6$ psi.

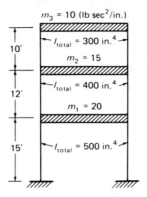

Fig. 9.11

10

Free Vibration of a Shear Building

When free vibration is under consideration, the structure is not subjected to any external excitation (force or support motion) and its motion is governed only by the initial conditions. There are occasionally circumstances for which it is necessary to determine the motion of the structure under conditions of free vibration, but this is seldom the case. Nevertheless, the analysis of the structure in free motion provides the most important dynamic properties of the structure which are the natural frequencies and the corresponding modal shapes. We begin by considering both formulations for the equations of motion, namely, the stiffness and the flexibility equations.

10.1 NATURAL FREQUENCIES AND NORMAL MODES

The problem of free vibration requires that the force vector $\{F\}$ be equal to zero in either the stiffness, eq. (9.3), or flexibility, eq. (9.7), formulations of the equations of motion. For the stiffness equation with $\{F\} = \{0\}$, we have

$$[M]\{\ddot{y}\} + [K]\{y\} = \{0\} \tag{10.1}$$

For free vibrations of the undamped structure, we seek solutions of eq. (10.1) in the form

$$y_i = a_i \sin (\omega t - \alpha), \quad i = 1, 2, \ldots, n$$

or in vector notation

$$\{y\} = \{a\} \sin (\omega t - \alpha) \tag{10.2}$$

where a_i is the amplitude of motion of the ith coordinate and n is the number of degrees of freedom. The substitution of eq. (10.2) into eq. (10.1) gives

$$-\omega^2 [M] \{a\} \sin (\omega t - \alpha) + [K] \{a\} \sin (\omega t - \alpha) = \{0\}$$

or rearranging terms

$$[[K] - \omega^2[M]] \{a\} = \{0\} \tag{10.3}$$

which, for the general case, is set for n homogeneous (right-hand side equal to zero) algebraic system of linear equations with n unknown displacements a_i and an unknown parameter ω^2. The formulation of eq. (10.3) is an important mathematical problem known as an *eigenproblem*. Its nontrivial solution, that is, the solution for which not all $a_i = 0$, requires that the determinant of the matrix factor of $\{a\}$ be equal to zero; in this case,

$$| [K] - \omega^2 [M] | = 0 \tag{10.4}$$

In general, eq. (10.4) results in a polynomial equation of degree n in ω^2 which should be satisfied for n values of ω^2. This polynomial is known as the *characteristic equation* of the system. For each of these values of ω^2 satisfying the characteristic eq. (10.4), we can solve eq. (10.3) for $a_1, a_2, \ldots, a_n$ in terms of an arbitrary constant.

Analogously, for the flexibility formulation, we have for free vibration from eq. (9.7) with $\{F\} = 0$,

$$\{y\} + [f] [M] \{\ddot{y}\} = \{0\} \tag{10.5}$$

We again assume harmonic motion as given by eq. (10.2) and substitute eq. (10.2) into eq. (10.5) to obtain

$$\{a\} = \omega^2 [f] [M] \{a\} \tag{10.6}$$

or

$$1/\omega^2 \{a\} = [D] \{a\} \tag{10.7}$$

where $[D]$ is known as the *dynamic matrix* and is defined as

$$[D] = [f] [M] \tag{10.8}$$

Equation (10.7) may also be written as

$$[[D] - 1/\omega^2 [I]] \{a\} = 0 \tag{10.9}$$

where $[I]$ is the unit matrix with ones in the main diagonal and zeros every-where else. For a nontrivial solution of eq. (10.9), it is required that the de-terminant of the coefficient matrix of $\{a\}$ be equal to zero, that is,

$$|[D] - 1/\omega^2 [I]| = 0 \qquad (10.10)$$

Equation (10.10) is a polynomial of degree n in $1/\omega^2$. This polynomial is the characteristic equation of the system for the flexibility formulation. For each one of the n solutions for $(1/\omega^2)$ of eq. (10.10), we can obtain from eq. (10.9) corresponding solutions for the amplitudes a_i in terms of an arbitrary con-stant. The necessary calculations are better explained with the use of a nu-merical example.

Example 10.1. The building to be analyzed is the simple steel rigid frame shown in Fig. 10.1. The weights of the floors and walls are indicated in the figure and are assumed to include the structural weight as well. The building consists of a series of frames spaced 15 ft apart. It is further assumed that the structural properties are uniform along the length of the building and, there-fore, the analysis to be made of an interior frame yields the response of the entire building.

Solution: The building is modeled as a shear building and, under the as-sumptions stated, the entire building may be represented by the spring-mass system shown in Fig. 10.2. The concentrated weights which are each taken as the total floor weight plus that of the tributary walls are computed as fol-lows:

$$W_1 = 100 \times 30 \times 15 + 20 \times 12.5 \times 15 \times 2 = 52{,}500 \text{ lb}$$

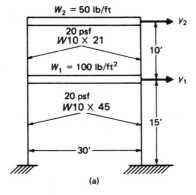

(a)

Fig. 10.1 Two-story shear building for Example 10.1.

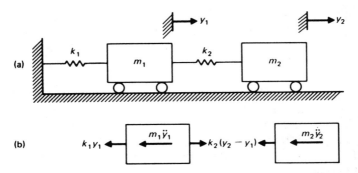

Fig. 10.2 Multimass-spring model for a two-story shear building. (a) Model. (b) Free body diagram.

$$m_1 = 136 \text{ lb} \cdot \text{sec}^2/\text{in}$$

$$W_2 = 50 \times 30 \times 15 + 20 \times 5 \times 15 \times 2 = 25{,}500 \text{ lb}$$

$$m_2 = 66 \text{ lb} \cdot \text{sec}^2/\text{in}$$

Since the girders are assumed to be rigid, the stiffness (spring constant) of each story is given by

$$k = \frac{12E(2I)}{L^3}$$

and the individual values for the steel column sections indicated are thus

$$k_1 = \frac{12 \times 30 \times 10^6 \times 248.6 \times 2}{(15 \times 12)^3} = 30{,}700 \text{ lb/in}$$

$$k_2 = \frac{12 \times 30 \times 10^6 \times 106.3 \times 2}{(10 \times 12)^3} = 44{,}300 \text{ lb/in}$$

The equations of motion for the system which are obtained by considering in Fig. 10.2(b) the dynamic equilibrium of each mass in free vibration are

$$m_1\ddot{y}_1 + k_1 y_1 - k_2(y_2 - y_1) = 0$$

$$m_2\ddot{y}_2 + k_2 (y_2 - y_1) = 0$$

In the usual manner, these equations of motion are solved for free vibration by substituting

$$y_1 = a_1 \sin (\omega t - \alpha)$$

$$y_2 = a_2 \sin (\omega t - \alpha) \tag{10.11}$$

for the displacements and

$$\ddot{y}_1 = -a_1\omega^2 \sin{(\omega t - \alpha)}$$

$$\ddot{y}_2 = -a_2\omega^2 \sin{(\omega t - \alpha)}$$

for the accelerations. In matrix notation, we obtain

$$\begin{bmatrix} k_1 + k_2 - m_1\omega^2 & -k_2 \\ -k_2 & k_2 - m_2\omega^2 \end{bmatrix} \begin{bmatrix} a_1 \\ a_2 \end{bmatrix} = \begin{bmatrix} 0 \\ 0 \end{bmatrix} \qquad (10.12)$$

For a nontrivial solution, we require that the determinant of the coefficient matrix be equal to zero, that is,

$$\begin{vmatrix} k_1 + k_2 - m_1\omega^2 & -k_2 \\ -k_2 & k_2 - m_2\omega^2 \end{vmatrix} = 0 \qquad (10.13)$$

The expansion of this determinant gives a quadratic equation in ω^2, namely,

$$m_1 m_2 \omega^4 - ((k_1 + k_2)m_2 + m_1 k_2)\omega^2 + k_1 k_2 = 0 \qquad (10.14)$$

or by introducing the numerical values for this example, we obtain

$$8976\omega^4 - 10{,}974{,}800\omega^2 + 1.36 \times 10^9 = 0 \qquad (10.15)$$

The roots of this quadratic are

$$\omega_1^2 = 140$$

$$\omega_2^2 = 1082$$

Therefore, the natural frequencies of the structure are

$$\omega_1 = 11.83 \text{ rad/sec}$$

$$\omega_2 = 32.89 \text{ rad/sec}$$

or in cycles per second

$$f_1 = \omega_1/2\pi = 1.88 \text{ cps}$$

$$f_2 = \omega_2/2\pi = 5.24 \text{ cps}$$

and the corresponding natural periods

$$T_1 = \frac{1}{f_1} = 0.532 \text{ sec}$$

$$T_2 = \frac{1}{f_2} = 0.191 \text{ sec}$$

To solve eq. (10.12) for the amplitudes a_1 and a_2, we note that by equating the determinant to zero in eq. (10.13), the number of independent equations is one less. Thus in the present case, the system of two equations is reduced to one independent equation. Considering the first equation in eq. (10.12) and substituting the first natural frequency, $\omega_1 = 11.8$ rad/sec, we obtain

$$55{,}960a_{11} - 44{,}300a_{21} = 0 \qquad (10.16)$$

We have introduced a second subindex in a_1 and a_2 to indicate that the value ω_1 has been used in this equation. Since in the present case there are two unknowns and only one equation, we can solve eq. (10.16) only for the relative value of a_{21} to a_{11}. This relative value is known as the normal mode or modal shape corresponding to the first frequency. For this example eq. (10.16) gives

$$\frac{a_{21}}{a_{11}} = 1.263$$

It is customary to describe the normal modes by assigning a unit value to one of the amplitudes; thus, for the first mode we set a_{11} equal to unity so that

$$a_{11} = 1.000$$

$$a_{21} = 1.263 \qquad (10.17)$$

Similarly, substituting the second natural frequency, $\omega_2 = 32.9$ rad/sec into eq. (10.12), we obtain the second normal mode as

$$a_{12} = 1.000$$

$$a_{22} = -1.629 \qquad (10.18)$$

It should be noted that although we obtained only ratios, the amplitudes of motion could, of course, be found from initial conditions.

We have now arrived at two possible simple harmonic motions of the structure which can take place in such a way that all the masses move in phase at the same frequency, either ω_1 or ω_2. Such a motion of an undamped system is called a *normal or natural mode of vibration*. The shapes (for this example a_{21}/a_{11} and a_{22}/a_{12}) are called normal mode shapes or simply modal shapes for the corresponding natural frequencies ω_1 and ω_2. The two modes which have been obtained for this example are depicted in Fig. 10.3.

We often use the phrase *first mode* or *fundamental mode* to refer to the mode associated with the lowest frequency. The other modes are sometimes called *harmonics* or *higher harmonics*. It is evident that the modes of vibration, each having its own frequency, behave essentially as single degree-of-freedom systems. The total motion of the system, that is, the total solution of the equations of motion, eq. (10.1), is given by the superposition of the modal harmonic vibrations which in terms of arbitrary constants of integration may be written as

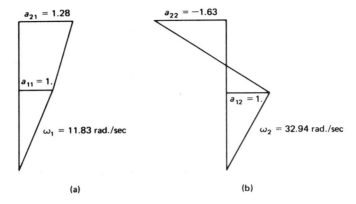

Fig. 10.3 Normal modes for Example 10.1. (a) First mode. (b) Second mode.

$$y_1(t) = C_1' a_{11} \sin(\omega_1 t - \alpha_1) + C_2' a_{12} \sin(\omega_2 t - \alpha_2)$$

$$y_2(t) = C_1' a_{21} \sin(\omega_1 t - \alpha_1) + C_2' a_{22} \sin(\omega_2 t - \alpha_2) \qquad (10.19)$$

Here C_1' and C_2' as well as α_1 and α_2 are four constants of integration to be determined from four initial conditions which are the initial displacement and velocity for each mass in the system. For a two-degree-of-freedom system, these initial conditions are

$$y_1(0) = y_{01}, \quad \dot{y}_1(0) = \dot{y}_{01}$$

$$y_2(0) = y_{02}, \quad \dot{y}_2(0) = \dot{y}_{02} \qquad (10.20)$$

For computational purposes, it is convenient to eliminate the phase angles α_1 and α_2 in eq. (10.19) in favor of other constants. Expanding the trigonometric functions and renaming the constants, we obtain

$$y_1(t) = C_1 a_{11} \sin \omega_1 t + C_2 a_{11} \cos \omega_1 t + C_3 a_{12} \sin \omega_2 t + C_4 a_{12} \cos \omega_2 t$$

$$y_2(t) = C_1 a_{21} \sin \omega_1 t + C_2 a_{21} \cos \omega_1 t + C_3 a_{22} \sin \omega_2 t + C_4 a_{22} \cos \omega_2 t \qquad (10.21)$$

in which C_1, C_2, C_3, and C_4 are new constants of integration. From the first two initial conditions in eq. (10.20), we have the following two equations:

$$y_{01} = C_2 a_{11} + C_4 a_{12}$$

$$y_{02} = C_2 a_{21} + C_4 a_{22} \qquad (10.22)$$

Since the modes are independent, these equations can always be solved for C_2 and C_4. Similarly, by expressing in eq. (10.21) the velocities at time equal to zero, we find

$$\dot{y}_{01} = \omega_1 C_1 a_{11} + \omega_2 C_3 a_{12}$$

$$\dot{y}_{02} = \omega_1 C_1 a_{21} + \omega_2 C_3 a_{22} \tag{10.23}$$

The solution of these two sets of equations allows us to express the motion of the system in terms of the two modal vibrations, each proceeding at its own frequency, completely independent of the other, the amplitudes and phases being determined by the initial conditions.

10.2   ORTHOGONALITY PROPERTY OF THE NORMAL MODES

We shall now introduce an important property of the normal modes, the orthogonality property. This property constitutes the basis of one of the most attractive methods for solving dynamic problems of multidegree-of-freedom systems. We begin by rewriting the equations of motion in free vibration, eq. (10.3), as

$$[K]\{a\} = \omega^2 [M]\{a\} \tag{10.24}$$

For the two-degree-of-freedom system, we obtain from eq. (10.12)

$$(k_1 + k_2)a_1 - k_2 a_2 = m_1 \omega^2 a_1$$

$$-k_2 a_1 + k_2 a_2 = m_2 \omega^2 a_2 \tag{10.25}$$

These equations are exactly the same as eq. (10.12) but written in this form they may be given a static interpretation as the equilibrium equations for the system acted on by forces of magnitude $m_1 \omega^2 a_1$ and $m_2 \omega^2 a_2$ applied to masses m_1 and m_2, respectively. The modal shapes may then be considered as the static deflections resulting from the forces on the right-hand side of eq. (10.25) for any of the two modes. This interpretation, as a static problem, allows us to use the results of the general static theory of linear structures. In particular, we may make use of Betti's theorem, which states: For a structure acted upon by two systems of loads and corresponding displacements, the work done by the first system of loads moving through the displacements of the second system is equal to the work done by this second system of loads undergoing the displacements produced by the first load system. The two systems of loading and corresponding displacements which we shall consider are as follows:
   System I:

$$\text{forces} \qquad \omega_1^2 a_{11} m_1, \quad \omega_1^2 a_{21} m_2$$

$$\text{and displacements} \quad a_{11}, \quad a_{21}$$

System II:

forces $\quad\quad \omega_2^2 a_{12} m_1, \quad \omega_2^2 a_{22} m_2$

and displacements $\quad a_{12}, \quad a_{22}$

The application of Betti's theorem for these two systems yields

$$\omega_1^2 m_1 a_{11} a_{12} + \omega_1^2 m_2 a_{21} a_{22} = \omega_2^2 m_1 a_{12} a_{11} + \omega_2^2 m_2 a_{22} a_{21}$$

or

$$(\omega_1^2 - \omega_2^2)(m_1 a_{11} a_{12} + m_2 a_{21} a_{22}) = 0 \tag{10.26}$$

If the natural frequencies are different $(\omega_1 \neq \omega_2)$, it follows from eq. (10.26) that

$$m_1 a_{11} a_{12} + m_2 a_{21} a_{22} = 0$$

which is the so-called orthogonality relation between modal shapes of a two-degree-of-freedom system. For an n-degree-of-freedom system in which the mass matrix is diagonal, the orthogonality condition between any two modes i and j may be expressed as

$$\sum_{k=1}^{m} m_k a_{ki} a_{kj} = 0, \quad \text{for } i \neq j \tag{10.27}$$

and in general for any n-degree-of-freedom system as

$$\{a_i\}^T [M] \{a_j\} = 0, \quad \text{for } i \neq j \tag{10.28}$$

in which $\{a_i\}$ and $\{a_j\}$ are any two modal vectors and $[M]$ is the mass matrix of the system.

As mentioned before, the amplitudes of vibration in a normal mode are only relative values which may be scaled or normalized to some extent as a matter of choice. The following is an especially convenient normalization for a general system:

$$\phi_{ij} = \frac{a_{ij}}{\sqrt{\{a_j\}^T [M] \{a_j\}}} \tag{10.29}$$

which, for a system having a diagonal mass matrix, may be written as

$$\phi_{ij} = \frac{a_{ij}}{\sqrt{\displaystyle\sum_{k=1}^{n} m_k a_{kj}^2}} \tag{10.30}$$

in which ϕ_{ij} is the normalized i component of the j modal vector. For normalized eigenvectors, the orthogonality condition is given by

$$\{\phi\}_i^T [M] \{\phi\}_j = 0 \quad \text{for } i \neq j$$

$$= 1 \quad \text{for } i = j \tag{10.31}$$

Another orthogonality condition is obtained by writing eq. (10.24) as

$$[K] \{\phi\}_{ij} = \omega^2 [M] \{\phi\}_j \tag{10.32}$$

Then premultiplying by $\{\phi\}_i^T$ we obtain, in view of the orthogonality condition of eq. (10.31), the following orthogonality condition between eigenvectors:

$$\{\phi\}_i^T [K] \{\phi\}_j = 0 \quad \text{for } i \neq j$$

$$= \omega^2 \quad \text{for } i = j \tag{10.33}$$

Example 10.2. For the two degree shear building of illustrative Example 10.1, determine the normalized modal shapes and verify the orthogonality condition between modes.

Solution: The substitution of eqs. (10.17) and (10.18) together with the values of the masses from Example 10.1 into the normalization factor required in eq. (10.30) gives

$$\sqrt{(136)(1.00)^2 + (66)(1.263)^2} = \sqrt{241.31}$$

$$\sqrt{(136)(1.00)^2 + (66)(-1.629)^2} = \sqrt{311.08}$$

Consequently, the normalized modes are

$$\phi_{11} = \frac{1.00}{\sqrt{241.31}} = 0.06437, \quad \phi_{12} = \frac{1.00}{\sqrt{311.08}} = 0.0567$$

$$\phi_{21} = \frac{1.263}{\sqrt{241.31}} = 0.0813, \quad \phi_{22} = \frac{-1.6287}{\sqrt{311.08}} = -0.0924$$

The normal modes may be conveniently arranged in the columns of a matrix known as the *modal matrix* of the system. For the general case of n degrees of freedom, the modal matrix is written as

$$[\Phi] = \begin{bmatrix} \phi_{11} & \phi_{12} \cdots & \phi_{1n} \\ \phi_{21} & \phi_{22} \cdots & \phi_{2n} \\ \phi_{n1} & \phi_{n2} \cdots & \phi_{nn} \end{bmatrix} \tag{10.34}$$

The orthogonality condition may then be expressed in general as

$$[\Phi]^T [M] [\Phi] = [I] \tag{10.35}$$

where $[\Phi]^T$ is the matrix transpose of $[\Phi]$ and $[M]$ the mass matrix of the system. For this example of two degrees of freedom, the modal matrix is

$$[\Phi] = \begin{bmatrix} 0.06437 & 0.0567 \\ 0.0813 & -0.0924 \end{bmatrix} \qquad (10.36)$$

To check the orthogonality condition, we simply substitute the normal modes from eq. (10.36) into eq. (10.35) and obtain

$$\begin{bmatrix} 0.06437 & 0.0813 \\ 0.0567 & -0.0924 \end{bmatrix} \begin{bmatrix} 136 & 0 \\ 0 & 66 \end{bmatrix} \begin{bmatrix} 0.06437 & 0.0567 \\ 0.0813 & -0.0924 \end{bmatrix} = \begin{bmatrix} 1 & 0 \\ 0 & 1 \end{bmatrix}$$

We have seen that to determine the natural frequencies and normal modes of vibration of a structural system, we have to solve an eigenvalue problem. The direct method of solution based on the expansion of the determinant and the solution of the characteristic equation is limited in practice to systems having only a few degrees of freedom. For a system of many degrees of freedom, the algebraic and numerical work required for the solution of an eigenproblem becomes so immense as to make the direct method impossible. However, there are many numerical methods available for the calculation of eigenvalues and eigenvectors of an eigenproblem. The discussion of these methods belongs in a mathematical text on numerical methods rather than in a text such as this on structural dynamics. For our purpose we have selected, among the various methods available for a numerical solution of an eigenproblem, the *Jacobi Method*, which is an iterative method to calculate the eigenvalues and eigenvectors of the system. The basic Jacobi solution method has been developed for the solution of standard eigenproblems (i.e., $[M]$ being the identity matrix). The method was proposed over a century ago and has been used extensively. This method can be applied to all symmetric matrices $[K]$ with no restriction on the eigenvalues. It is possible to transform the generalized eigenproblem, $[[K] - \omega^2 [M]] \{\Phi\} = \{0\}$ into the standard form and still maintain the symmetry required for the Jacobi Method. However, this transformation can be dispensed with by using a generalized Jacobi solution method (Bathe, K. J. 1982) which operates directly on $[K]$ and $[M]$.

10.3   PROGRAM 8—NATURAL FREQUENCIES AND NORMAL MODES

The program presented in this section uses the generalized Jacobi method to determine the natural frequencies and corresponding modal shapes for a structure modeled as a discrete system.

Example 10.3   Use Program 8 to solve the eigenproblem corresponding to a system having the following stiffness and mass matrices:

$$[K] = \begin{bmatrix} 3000 & -1500 & 0 \\ -1500 & 3000 & -1500 \\ 0 & -1500 & 1500 \end{bmatrix}$$

$$[M] = \begin{bmatrix} 1 & 0 & 0 \\ 0 & 1 & 0 \\ 0 & 0 & 1 \end{bmatrix}$$

Solution: The execution of Program 8 to calculate natural frequencies and modal shapes requires the previous preparation of a file containing the stiffness and mass matrices of the system. This file is created during the execution of one of the programs to model the structure or by execution of the auxiliary Program X1. This program accepts as input the stiffness and mass matrices of the structure and creates the file required to execute Program 8. In the solution of Example 10.3, the required file was created by executing Program X1.

Input Data and Output Results

```
PROGRAM 8: NATURAL FREQUENCIES AND NORMAL MODES     DATA FILE: SK

        INPUT DATA:

                  ***STIFFNESS MATRIX***

0.30000E+04          -.15000E+04          0.00000E+00
-.15000E+04          0.30000E+04          -.15000E+04
0.00000E+00          -.15000E+04          0.15000E+04

                  ***MASS MATRIX***

0.10000E+01          0.00000E+00          0.00000E+00
0.00000E+00          0.10000E+01          0.00000E+00
0.00000E+00          0.00000E+00          0.10000E+01


                  OUTPUT RESULTS:

        EIGENVALUES:

  297.1               2332.4               4870.5

        NATURAL FREQUENCIES (C.P.S.):

  2.74                7.69                 11.11
```

EIGENVECTORS BY ROWS:

```
0.32799   0.59101   0.73698
0.73698   0.32799  -0.59101
0.59101  -0.73698   0.32799
```

10.4 SUMMARY

The motion of an undamped dynamic system in free vibration is governed by a homogeneous system of differential equations which in matrix notation is

$$[M]\{\ddot{y}\} + [K]\{y\} = \{0\}$$

The process of solving this system of equations leads to a homogeneous system of linear algebraic equations of the form

$$([K] - \omega^2[M])\{a\} = \{0\}$$

which mathematically is known as an eigenproblem.

For a nontrivial solution of this problem, it is required that the determinant of the coefficients of the unknown $\{a\}$ be equal to zero, that is,

$$|[K] - \omega^2[M]| = 0$$

The roots ω_i^2 of this equation provide the natural frequencies ω_i. It is then possible to solve for the unknowns $\{a\}_i$ in terms of relative values. The vectors $\{a\}_i$ corresponding to the roots ω_i^2 are the modal shapes (eigenvectors) of the dynamic system. The arrangement in matrix format of the modal shapes constitutes the modal matrix $[\Phi]$ of the system. It is particularly convenient to normalize the eigenvectors to satisfy the following condition:

$$\{\phi\}_i^T[M]\{\phi\}_i = 1, \quad i = 1, 2, \dots, n$$

where the normalized modal vectors $\{\phi\}_i$ are obtained by dividing the components of the vector $\{a\}_i$ by $\sqrt{\{a\}_i^T[M]\{a\}_i}$.

The normalized modal vectors satisfy the following important conditions of orthogonality:

$$\{\phi\}_i^T[M]\{\phi\}_j = 0 \quad \text{for } i \neq j.$$

$$\{\phi\}_i[M]\{\phi\}_j = 1 \quad \text{for } i = j$$

and

$$\{\phi\}_i^T[K]\{\phi\}_j = 0 \quad \text{for } i \neq j$$

$$\{\phi\}_i^T[K]\{\phi\}_j = \omega_i^2 \quad \text{for } i = j$$

The above relations are equivalent to

$$[\Phi]^T [M] [\Phi] = [I]$$

and

$$[\Phi]^T [K] [\Phi] = [\Omega]$$

in which $[\Phi]$ is the modal matrix of the system and $[\Omega]$ is a diagonal matrix containing the eigenvalues ω_i^2 in the main diagonal.

For a dynamic system with only a few degrees of freedom, the natural frequencies and modal shapes may be determined by expanding the determinant and calculating the roots of the resulting characteristic equation. However, for a system with a large number of degrees of freedom, this direct method of solution becomes impractical. It is then necessary to resort to other numerical methods which usually require an iteration process. Among the various methods available for the solution of an eigenproblem, we have selected the Jacobi Method.

PROBLEMS

10.1 Determine the natural frequencies and normal modes for the two-story shear building shown in Fig. 10.4.

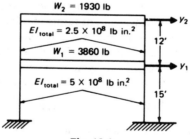

Fig. 10.4

10.2 A certain structure has been modeled as a three-degree-of-freedom system having the numerical values indicated in Fig. 10.5. Determine the natural frequencies

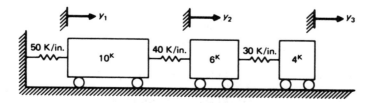

Fig. 10.5

of the structure and the corresponding normal modes. Check your answer using computer Program 8.

10.3 Assume a shear building model for the frame shown in Fig. 10.6 and determine the natural frequencies and normal modes.

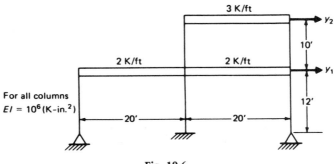

3 K/ft

2 K/ft   2 K/ft

For all columns
$EI = 10^6 (K\text{-}in.^2)$

←——— 20' ———→ ←——— 20' ———→

y_2

10'

y_1

12'

Fig. 10.6

10.4 A movable structural frame is supported on rollers as shown in Fig. 10.7. Determine natural periods and corresponding normal modes. Assume shear building model and check answers using Program 8.

10.5 Consider the uniform shear building in which the mass of each floor is m and the stiffness of each story is k. Determine the general form of the system of differential equations for a uniform shear building of N stories.

10.6 Write a main computer program which will use Subroutine JACOBI of Program 8 to calculate the natural frequencies and modal shapes for a uniform shear building as described in Problem 10.5. Express the natural frequencies in function of the quantity k/m.

10.7 Modify the computer program required in Problem 10.6 to incorporate the option

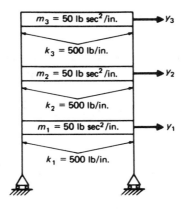

$m_3 = 50$ lb sec^2/in. y_3

$k_3 = 500$ lb/in.

$m_2 = 50$ lb sec^2/in. y_2

$k_2 = 500$ lb/in.

$m_1 = 50$ lb sec^2/in. y_1

$k_1 = 500$ lb/in.

Fig. 10.7

of setting the ratio k/m to a value which will result in the uniform shear building having a prescribed fundamental natural period T sec/cycle.

10.8 Write a main computer program which will use Subroutine JACOBI of Program 8 to calculate the natural frequencies and modal shapes for a shear building of N stories in which the stiffness of each story changes linearly from k at the first story to k_N at the Nth story. Assume equal mass at each floor.

10.9 Modify the computer program requested in Problem 10.8 for the case that the mass in each floor varies linearly from m at the first floor to m_N at the Nth floor.

10.10 Modify the computer program requested in Problem 10.8 to incorporate the option of setting the magnitude of the floor mass to a value which will result in the shear building having a prescribed fundamental natural period T sec/cycle.

10.11 Find the natural frequencies and modal shapes for the three-degree-of-freedom shear building in Fig. 10.8.

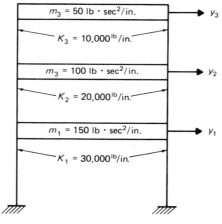

Fig. 10.8

10.12 Use the results of Problem 10.11 to write the expressions for the free vibration displacements y_1, y_2, and y_3 of the shear building in Fig. 10.8 in terms of constants of integration.

10.13 The stiffness matrix $[K]$ and the mass matrix $[M]$ of a structure modeled as a three-degree-of-freedom system are given respectively by

$$[K] = 10^6 \begin{bmatrix} 210 & 0 & -5 \\ 0 & 210 & 5 \\ -5 & 5 & 0.333 \end{bmatrix} \text{ (lb/in)}$$

and

$$[M] = \begin{bmatrix} 23{,}070 & 0 & -1{,}886 \\ 0 & 23{,}070 & 1{,}886 \\ -1{,}886 & 1{,}886 & 187{,}200 \end{bmatrix} \text{ (lb} \cdot \text{sec}^2/\text{in)}$$

Determine the natural frequencies and corresponding modal shapes of the structure.

10.14 Solve the eigenvalue problem corresponding to a structure for which the stiffness and mass matrices are

$$[K] = \begin{bmatrix} 18.8600 & -12.0000 & 5.1430 \\ -12.0000 & 15.0000 & -12.0000 \\ 5.1430 & -12.0000 & 18.8600 \end{bmatrix}$$

$$[M] = \begin{bmatrix} 0.8169 & 0.1286 & -0.0740 \\ 0.1286 & 0.8571 & 0.1286 \\ -0.0740 & 0.1286 & 0.8169 \end{bmatrix}$$

10.15 Solve the eigenproblem corresponding to a structure for which the stiffness and mass matrices are:

$$[K] = \begin{bmatrix} 1263.4 & 0 & -631.68 & 15192 & 0 & 0 \\ & 1052800 & -15792 & 263200 & 0 & 0 \\ & & 1263.4 & 0 & -631.68 & 15792 \\ & & & 1052800 & -15792 & 256300 \\ \text{Symmetric} & & & & 1263.4 & 0 \\ & & & & & 1052800 \end{bmatrix}$$

$$[M] = \begin{bmatrix} 3.7143 & 0 & 0.6429 & -7.7381 & 0 & 0 \\ & 238.10 & 7.7381 & -89.286 & 0 & 0 \\ & & 3.7143 & 0 & 0.6429 & -7.7381 \\ & & & 238.1 & 7.7381 & -89.286 \\ \text{Symmetric} & & & & 3.7143 & 0 \\ & & & & & 238.1 \end{bmatrix}$$

11

Forced Motion of Shear Buildings

In the preceding chapter, we have shown that the free motion of a dynamic system may be expressed in terms of free modal vibrations. Our present interest is to demonstrate that the forced motion of such a system may also be expressed in terms of the normal modes of vibration and that the total response may be obtained as the superposition of the solution of independent modal equations. In other words, our aim in this chapter is to show that the normal modes may be used to transform the system of *coupled* differential equations into a set of *uncoupled* differential equations in which each equation contains only one dependent variable. Thus the *modal superposition method* reduces the problem of finding the response of a multidegree-of-freedom system to the determination of the response of single degree-of-freedom systems.

11.1 MODAL SUPERPOSITION METHOD

We have shown that any free motion of a multidegree-of-freedom system may be expressed in terms of normal modes of vibration. It will now be demonstrated that the forced motion of such a system may also be expressed in terms of the normal modes of vibration. We return to the equations of motion, eq. (9.3), which for the particular case of a two-degree-of-freedom shear building are

$$m_1 \ddot{y}_1 + (k_1 + k_2) y_1 - k_2 y_2 = F_1(t)$$

$$m_2 \ddot{y}_2 - k_2 y_1 + k_2 y_2 = F_2(t) \tag{11.1}$$

We seek to transform this coupled system of equation into a system of independent or uncoupled equations in which each equation contains only one unknown function of time. It is first necessary to express the solution in terms of the normal modes multiplied by some factors determining the contribution of each mode. In the case of free motion, these factors were sinusoidal functions of time; in the present case, for forced motion, they are general functions of time which we designate as $z_i(t)$. Hence the solution of eq. (11.1) is assumed to be of the form

$$y_1(t) = a_{11} z_1(t) + a_{12} z_2(t)$$

$$y_2(t) = a_{21} z_1(t) + a_{22} z_2(t) \tag{11.2}$$

Upon substitution into eq. (11.1), we obtain

$$m_1 a_{11} \ddot{z}_1 + (k_1 + k_2) a_{11} z_1 - k_2 a_{21} z_1 + m_1 a_{12} \ddot{z}_2 + (k_1 + k_2) a_{12} z_2 - k_2 a_{22} z_2 = F_1(t)$$

$$m_2 a_{21} \ddot{z}_1 - k_2 a_{11} z_1 + k_2 a_{21} z_1 + m_2 a_{22} \ddot{z}_2 - k_2 a_{12} z_2 + k_2 a_{22} z_2 = F_2(t) \tag{11.3}$$

To determine the appropriate factors $z_1(t)$ and $z_2(t)$ which will uncouple eq. (11.3), it is advantageous to make use of the orthogonality relations to separate the modes. The orthogonality relations are used by multiplying the first of eq. (11.3) by a_{11} and the second by a_{21}. Addition of these equations after multiplication and simplification by using eq. (10.25) and (10.27) yields

$$(m_1 a_{11}^2 + m_2 a_{21}^2) \ddot{z}_1 + \omega_1^2 (m_1 a_{11}^2 + m_2 a_{21}^2) z_1 = a_{11} F_1(t) + a_{21} F_2(t) \tag{11.4a}$$

Similarly, multiplying the first of eqs. (11.3) by a_{12} and the second by a_{22}, we obtain

$$(m_1 a_{12}^2 + m_2 a_{22}^2) \ddot{z}_2 + \omega_2^2 (m_1 a_{12}^2 + m_2 a_{22}^2) z_2 = a_{12} F_1(t) + a_{22} F_2(t) \tag{11.4b}$$

The results obtained in eqs. (11.4) permit a simple physical interpretation. The force which is effective in exciting a mode is equal to the work done by the external force displaced by the modal shape in question. From the mathematical point of view, what we have accomplished is to separate or uncouple, by a change of variables, the original system of differential equations. Consequently, each of these equations, eq. (11.4a) or eq. (11.4b), corresponds to a single degree-of-freedom system which may be written as

$$M_1 \ddot{z}_1 + K_1 z_1 = P_1(t)$$

$$M_2 \ddot{z}_2 + K_2 z_2 = P_2(t) \tag{11.5}$$

where $M_1 = m_1 a_{11}^2 + m_2 a_{21}^2$ and $M_2 = m_1 a_{12}^2 + m_2 a_{22}^2$ are the modal masses; $K_1 = \omega_1^2 M_1$ and $K_2 = \omega_2^2 M_2$, the modal spring constants; and $P_1(t) = a_{11} F_1(t) + a_{21} F_2(t)$ and $P_2(t) = a_{12} F_1(t) + a_{22} F_2(t)$, the modal forces. Alternatively, if we had used the previous normalization, eqs. (10.29) or (10.30), these equations may be written simply as

$$\ddot{z}_1 + \omega_1^2 z_1 = P_1(t)$$

$$\ddot{z}_2 + \omega_2^2 z_2 = P_2(t) \tag{11.6}$$

where P_1 and P_2 are now given by

$$P_1 = \phi_{11} F_1(t) + \phi_{21} F_2(t)$$

$$P_2 = \phi_{12} F_1(t) + \phi_{22} F_2(t) \tag{11.7}$$

The solution for the uncoupled differential equations, eqs. (11.5) or eqs. (11.6), may now be found by any of the methods presented in the previous chapters. In particular, Duhamel's integral provides a general solution for these equations regardless of the functions describing the forces acting on the structure. Also, maximum values of the response for each modal equation may readily be obtained using available response spectra. However, the superposition of modal maximum responses presents a problem. The fact is that these modal maximum values will in general not occur simultaneously as the transformation of coordinates, eq. (11.2), requires. To obviate the difficulty, it is necessary to use an approximate method. An upper limit for the maximum response may be obtained by adding the absolute values of the maximum modal contributions, that is, by substituting z_1 and z_2 in eqs. (11.2) for the maximum modal responses $(z_{1\,\mathrm{max}}$ and $z_{2\,\mathrm{max}})$ and adding the absolute values of the terms in these equations, so that

$$y_{1\,\mathrm{max}} = |\phi_{11} z_{1\,\mathrm{max}}| + |\phi_{12} z_{2\,\mathrm{max}}|$$

$$y_{2\,\mathrm{max}} = |\phi_{21} z_{1\,\mathrm{max}}| + |\phi_{22} z_{2\,\mathrm{max}}| \tag{11.8}$$

The results obtained by this method will overestimate the maximum response. Another method, which is widely accepted and which gives a reasonable estimate of the maximum response from these spectral values, is the square root of the sum of the squares of the modal contributions (SRSS method). Thus the maximum displacements may be approximated by

$$y_{1\,\mathrm{max}} = \sqrt{(\phi_{11} z_{1\,\mathrm{max}})^2 + (\phi_{12} z_{2\,\mathrm{max}})^2}$$

and

$$y_{2\,\mathrm{max}} = \sqrt{(\phi_{21} z_{1\,\mathrm{max}})^2 + (\phi_{22} z_{2\,\mathrm{max}})^2} \tag{11.9}$$

The results obtained by application of the SRSS method (square root of the sum of the squares of modal contributions), may substantially underestimate or overestimate the total response when two or more modes are closely spaced. In this case, another method known as the Complete Quadratic Combination for combining modal responses to obtain the total response is recommended. The discussion of such a method is presented in Section 11.6.

The transformation from a system of two coupled differential equations, eq. (11.1), to a set of two uncoupled differential equations, eq. (11.6), may be extended to a system of N degrees of freedom. For such a system, it is particularly convenient to use matrix notation. With such notation, the equation of motion for a linear system of N degrees of freedom is given by eq. (9.3) as

$$[M]\{\ddot{y}\} + [K]\{y\} = \{F(t)\} \qquad (11.10)$$

where $[M]$ and $[K]$ are respectively the mass and the stiffness matrix of the system, $\{F(t)\}$ the vector of external forces, and $\{y\}$ the vector of unknown displacements at the nodal coordinates.

Introducing into eq. (11.10) the linear transformation of coordinates

$$\{y\} = [\Phi]\{z\} \qquad (11.11)$$

in which $[\Phi]$ is the modal matrix of the system, yields

$$[M][\Phi]\{\ddot{z}\} + [K][\Phi]\{z\} = \{F(t)\} \qquad (11.12)$$

The premultiplication of eq. (11.12) by the transpose of the ith modal vector, $\{\phi\}_i^T$, results in

$$\{\phi\}_i^T [M][\Phi]\{\ddot{z}\} + \{\phi\}_i^T [K][\Phi]\{z\} = \{\phi\}_i^T \{F(t)\} \qquad (11.13)$$

The orthogonality conditions between normalized modes, eqs. (10.31) and (10.33), imply that

$$\{\phi\}_i^T [M][\Phi] = 1 \qquad (11.14)$$

and

$$\{\phi\}_i^T [K][\Phi] = \omega_i^2 \qquad (11.15)$$

Consequently, eq. (11.13) may be written as

$$\ddot{z}_i + \omega_i^2 z_i = P_i(t) \quad (i = 1, 2, 3, \ldots, N) \qquad (11.16)$$

where the modal force $P_i(t)$ is given by

$$P_i(t) = \phi_{1i}F_1(t) + \phi_{2i}F_2(t) + \ldots + \phi_{Ni}F_N(t) \qquad (11.17)$$

Equation (11.16) constitutes a set of N uncoupled or independent equations of motion in terms of the modal coordinates z_i. These uncoupled equations, as may be observed, may readily be written after the natural frequencies ω_i and the modal vector $\{\phi\}_i$ have been determined in the solution of the corresponding eigenproblem as presented in Chapter 10.

Example 11.1.   The two-story frame of Example 10.1 is acted upon at the floor levels by triangular impulsive forces shown in Fig. 11.1. For this frame, determine the maximum floor displacements and the maximum shear forces in the columns.

Solution: The results obtained in Examples 10.1 and 10.2 for the free vibration of this frame gave the following values for the natural frequencies and normalized modes:

$$\omega_1 = 11.8 \text{ rad/sec,} \quad \omega_2 = 32.9 \text{ rad/sec}$$

$$\phi_{11} = 0.06437, \quad \phi_{12} = 0.0567$$

$$\phi_{21} = 0.08130, \quad \phi_{22} = -0.0924$$

The forces acting on the frame which are shown in Fig. 11.1(b) may be expressed by

$$F_1(t) = 10,000(1 - t/t_d) \text{ lb}$$

$$F_2(t) = 20,000(1 - t/t_d) \text{ lb,} \quad \text{for } t \le 0.1 \text{ sec}$$

in which $t_d = 0.1$ sec and

$$F_1(t) = F_2(t) = 0, \quad \text{for } t > 0.1 \text{ sec}$$

The substitution of these values into the uncoupled equations of motion, eqs. (11.6), gives

$$\ddot{z}_1 + 139.24z_1 = 2270 f(t)$$

$$\ddot{z}_2 + 1082.41z_2 = -1281 f(t)$$

in which $f(t) = 1 - t/t_d$ for $t \le 0.1$ and $f(t) = 0$ for $t > 0.1$. The maximum

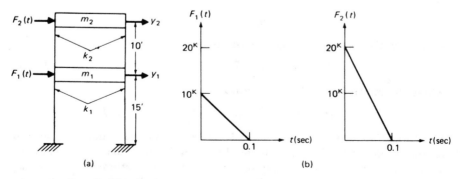

Fig. 11.1 Shear building with impulsive loadings. (a) Two-story shear building. (b) Impulsive loadings.

values for z_1 and z_2 are then obtained from available spectral charts such as the one shown in Fig. 4.5. For this example,

$$\frac{t_d}{T_1} = \frac{0.1}{0.532} = 0.188$$

and

$$\frac{t_d}{T_2} = \frac{0.1}{0.191} = 0.524$$

From Fig. 4.5, we obtain

$$(DLF)_{1_{max}} = \frac{z_{1_{max}}}{z_{1_{st}}} = 0.590$$

$$(DLF)_{2_{max}} = \frac{z_{2_{max}}}{z_{2_{st}}} = 1.22$$

where in this case the static deflections are calculated as

$$z_{1_{st}} = \frac{F_{01}}{\omega_1^2} = \frac{2270}{139.24} = 16.3, \quad z_{2_{st}} = \frac{F_{02}}{\omega_2^2} = \frac{1281}{1082.41} = 1.18$$

Then the maximum modal response is

$$z_{1_{max}} = 0.590 \times 16.3 = 9.62, \quad z_{2_{max}} = 1.22 \times 1.18 = 1.44$$

As indicated above these maximum modal values do not occur simultaneously and therefore cannot simply be superimposed to obtain the maximum response of the system. However, an upper limit for the absolute maximum displacement may be calculated with eqs. (11.8) as

$$y_{1_{max}} = |0.06437 \times 9.62| + |0.0567 \times 1.44| = 0.70 \text{ in}$$

$$y_{2_{max}} = |0.08130 \times 9.62| + |-0.0924 \times 1.44| = 0.92 \text{ in}$$

A second acceptable estimate of the maximum response is obtained by taking the square root of the sum of the squared modal contributions as indicated by eqs. (11.9). For this example, we have

$$y_{1_{max}} = \sqrt{(0.06437 \times 9.62)^2 + (0.0567 \times 1.44)^2} = 0.62 \text{ in}$$

$$y_{2_{max}} = \sqrt{(0.08130 \times 9.62)^2 + (-0.0924 \times 1.44)^2} = 0.79 \text{ in} \qquad (a)$$

The maximum shear force V_{max} in the columns is given by

$$V_{max} = k\Delta y \qquad (11.18)$$

in which k is the stiffness of the story and Δy the difference between the displacements at the two ends of the column. Since the maximum displacements calculated as in eq. (a) may have positive or negative values, the relative displacement Δy cannot be determined as the difference of the absolute displacements of the two ends of the column. The maximum possible value for Δy could be estimated as the sum of the absolute maximum displacements at the ends of the columns. However, this procedure will in most cases greatly overestimate the actual forces in the columns. The recommended procedure is first to calculate the shear forces in the columns for each mode separately and then combine these modal forces by a suitable method, such as the square root of the sum of the squares of modal contributions. This procedure is based on the fact that modal displacements are known with their correct relative sign and not as absolute values.

The maximum shear force V_{ij} at story i corresponding to mode j is given by

$$V_{ij} = z_{j_{max}} (\phi_{ij} - \phi_{i-1,j}) k_i \qquad (11.19)$$

where $z_{j_{max}}$ is the maximum modal response, $(\phi_{ij} - \phi_{i-1,j})$ the relative modal displacement of story i (with $\phi_{0j} = 0$), and k_i the stiffness of the story. For this example we have for the first story

$$k_1 = \frac{12EI_1}{L_1^3} = \frac{12 \times 30 \times 10^6 \times 248.6}{(15 \times 12)^3} = 15{,}345 \text{ lb/in}$$

$$V_{11} = 9.62 \times 0.06437 \times 15{,}345 = 9502 \text{ lb}$$

$$V_{12} = 1.44 \times 0.0567 \times 15{,}345 = 1253 \text{ lb}$$

$$V_{1 \text{ max}} = \sqrt{9502^2 + 1253^2} = 9584 \text{ lb}$$

and for the second story

$$k_2 = \frac{12EI_2}{L_2^3} = \frac{12 \times 30 \times 10^6 \times 106.3}{(10 \times 12)^3} = 22{,}146 \text{ lb/in}$$

$$V_{21} = 9.62 \times (0.08130 - 0.06437) \times 22{,}146 = 3607 \text{ lb}$$

$$V_{22} = 1.44 \times (-0.0924 - 0.0567) \times 22{,}146 = 4755$$

$$V_{2 \text{ max}} = \sqrt{3607^2 + 4755^2} = 5968 \text{ lb}$$

11.2   RESPONSE OF A SHEAR BUILDING TO BASE MOTION

The response of a shear building to the base or foundation motion is conveniently obtained in terms of floor displacements relative to the base motion.

For the two-story shear building of Fig. 11.2(a), which is modeled as shown in Fig. 11.2(b), the equations of motion obtained by equating to zero the sum of forces in the free body diagrams of Fig. 11.2(c) are the following:

$$m_1\ddot{y}_1 + k_1(y_1 - y_s) - k_2(y_2 - y_1) = 0$$

$$m_2\ddot{y}_2 + k_2(y_2 - y_1) = 0 \qquad (11.20)$$

where $y_s = y_s(t)$ is the displacement imposed at the foundation of the structure. Expressing the floor displacements relative to the base motion, we have

$$u_1 = y_1 - y_s$$

$$u_2 = y_2 - y_s \qquad (11.21)$$

Then differentiation yields

$$\ddot{y}_1 = \ddot{u}_1 + \ddot{y}_s$$

$$\ddot{y}_2 = \ddot{u}_2 + \ddot{y}_s \qquad (11.22)$$

Substitution of eqs. (11.21) and (11.22) into eqs. (11.20) results in

$$m_1\ddot{u}_1 + (k_1 + k_2)u_1 - k_2u_2 = -m_1\ddot{y}_s$$

$$m_2\ddot{u}_2 - k_2u_1 + k_2u_2 = -m_2\ddot{y}_s \qquad (11.23)$$

We note that the right-hand sides of eqs. (11.23) are proportional to the same function of time, $\ddot{y}_s(t)$. This fact leads to a somewhat simpler solution compared to the solution of eqs. (11.6) which may contain different functions of time in each equation. For the base motion of the shear building, eqs. (11.4) may be written as

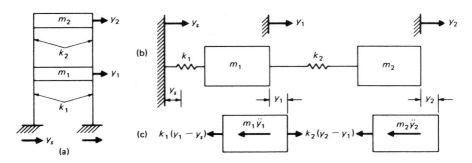

Fig. 11.2 Shear building with base motion. (a) Two-story shear building. (b) Mathematical model. (c) Free body diagram.

$$\ddot{z}_1 + \omega_1^2 z_1 = -\frac{m_1 a_{11} + m_2 a_{21}}{m_1 a_{11}^2 + m_2 a_{21}^2} \ddot{y}_s(t)$$

$$\ddot{z}_2 + \omega_2^2 z_2 = -\frac{m_1 a_{12} + m_2 a_{22}}{m_1 a_{12}^2 + m_2 a_{22}^2} \ddot{y}_s(t) \tag{11.24}$$

or

$$\ddot{z}_1 + \omega_1^2 z_1 = \Gamma_1 \ddot{y}_s(t)$$

$$\ddot{z}_2 + \omega_2^2 z_2 = \Gamma_2 \ddot{y}_s(t) \tag{11.25}$$

where Γ_1 and Γ_2 are called *participation factors* and are given by

$$\Gamma_1 = -\frac{m_1 a_{11} + m_2 a_{21}}{m_1 a_{11}^2 + m_2 a_{21}^2}$$

$$\Gamma_2 = -\frac{m_1 a_{12} + m_2 a_{22}}{m_1 a_{12}^2 + m_2 a_{22}^2} \tag{11.26}$$

The relation between the modal displacements z_1, z_2 and the relative displacement u_1, u_2 is given from eqs. (11.2) as

$$u_1 = a_{11} z_1 + a_{12} z_2$$

$$u_2 = a_{21} z_1 + a_{22} z_2 \tag{11.27}$$

In practice it is convenient to introduce a change of variables in eqs. (11.25) such that the second members of these equations equal $\ddot{y}_s(t)$. The required change of variables to accomplish this simplification is

$$z_1 = \Gamma_1 g_1$$

$$z_2 = \Gamma_2 g_2 \tag{11.28}$$

which when introduced into eqs. (11.25) gives

$$\ddot{g}_1 + \omega_1^2 g_1 = \ddot{y}_s(t)$$

$$\ddot{g}_2 + \omega_2^2 g_2 = \ddot{y}_s(t) \tag{11.29}$$

Finally, solving for $g_1(t)$ and $g_2(t)$ in the uncoupled eqs. (11.29) and substituting the solution into eqs. (11.27) and (11.28) give the response as

$$u_1(t) = \Gamma_1 a_{11} g_1(t) + \Gamma_2 a_{12} g_2(t)$$

$$u_2(t) = \Gamma_1 a_{21} g_1(t) + \Gamma_2 a_{22} g_2(t) \tag{11.30}$$

When the maximum modal response $g_{1\max}$ and $g_{2\max}$ are obtained from spectral charts, we may estimate the maximum values $u_{1\max}$ and $u_{2\max}$ from eqs. (11.9) as

$$u_{1_{max}} = \sqrt{(\Gamma_1 a_{11} g_{1_{max}})^2 + (\Gamma_2 a_{12} g_{2_{max}})^2}$$

$$u_{2_{max}} = \sqrt{(\Gamma_1 a_{21} g_{1_{max}})^2 + (\Gamma_1 a_{22} g_{2_{max}})^2} \tag{11.31}$$

The equations of motion for an N-story shear building [Fig. 11.3(a)] subjected to excitation motion at its base are obtained by equating to zero the sum of forces shown in the free body diagrams of Fig. 11.3(b), namely

$$m_1 \ddot{y}_1 + k_1(y_1 - y_s) - k_2(y_2 - y_1) = 0$$

$$m_2 \ddot{y}_2 + k_2(y_2 - y_1) - k_3(y_3 - y_2) = 0$$

$$\cdots$$

$$m_{N-1} \ddot{y}_{N-1} + k_{N-1}(y_{N-1} - y_{N-2}) - k_N(y_N - y_{N-1}) = 0$$

$$m_N \ddot{y}_N + k_N(y_N - y_{N-1}) = 0 \tag{11.32}$$

Introducing into eq. (11.32)

$$u_i = y_i - y_s \quad (i = 1, 2, \ldots, N) \tag{11.33}$$

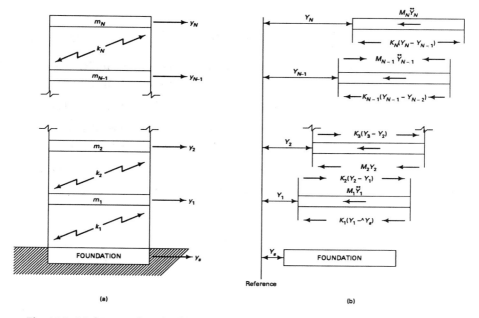

Fig. 11.3. Multistory shear building excited at the foundation. (a) Structural model. (b) Free body diagrams.

results in

$$m_1\ddot{u}_1 + k_1 u_1 - k_2(u_2 - u_1) = -m_1\ddot{y}_s$$

$$m_2\ddot{u}_2 + k_2 u_2 - k_3(u_3 - u_2) = -m_2\ddot{y}_s$$

$$\cdots$$

$$m_{N-1}\ddot{u}_{N-1} + k_{N-1}(u_{N-1} - u_{N-2}) - k_N(u_N - u_{N-1}) = -m_{N-1}\ddot{y}_s$$

$$m_N\ddot{u}_N + k_N(u_N - u_{N-1}) = -m_N\ddot{y}_s \qquad (11.34)$$

where $\ddot{y}_s = \ddot{y}_s(t)$ is the acceleration function exciting the base of the structure. Equations (11.34) may conveniently be written in matrix notation as

$$[M]\{\ddot{u}\} + [K]\{u\} = -[M]\{1\}\ddot{y}_s(t) \qquad (11.35)$$

in which $[M]$, the mass matrix, is a diagonal matrix, $[K]$, the stiffness matrix, is a symmetric matrix, $\{1\}$ is a vector with all its elements equal to 1, $\ddot{y}_s = \ddot{y}_s(t)$ is the applied acceleration at the foundation of the building, and $\{u\}$ and $\{\ddot{u}\}$ are, respectively, the displacement and acceleration vectors relative to the foundation.

As has been demonstrated, the system of differential equations (11.35) can be uncoupled through the transformation given by eq. (11.11) as

$$\{u\} = [\Phi]\{z\} \qquad (11.36)$$

where $[\Phi]$ is the modal matrix obtained in the solution of corresponding eigenproblem $[[K] - \omega^2 [M]]\{\phi\} = \{0\}$.

The substitution of eq. (11.36) into eq. (11.35) followed by premultiplication by the transpose of the ith eigenvector, $\{\phi\}_i^T$ (the ith modal shape), results in

$$\{\phi\}_i^T [M][\Phi]\{\ddot{z}\} + \{\phi\}_i^T [K][\Phi]\{z\} = -\{\phi\}_i^T [M]\{1\}\ddot{y}_s(t) \qquad (11.37)$$

which upon introduction of orthogonality property of the normalized eigenvectors [eqs. (11.14) and (11.15)] results in the modal equations

$$\ddot{z}_i + \omega_i^2 z_i = \Gamma_i \ddot{y}_s(t) \quad (i = 1, 2, \ldots, N) \qquad (11.38)$$

where the participation factor Γ_i is given by

$$\Gamma_i = -\frac{\displaystyle\sum_{j=1}^{N} m_j \phi_{ji}}{\displaystyle\sum_{j=1}^{N} m_j \phi_{ji}^2} \qquad (11.39)$$

or for normalized eigenvectors by

$$\Gamma_i = -\sum_{j=1}^{N} m_j \phi_{ji} \quad (i = 1, 2, 3, \ldots, N) \qquad (11.40)$$

The maximum response in terms of maximum values for displacements or for accelerations at the nodal coordinates calculated by the SRSS method is then given, respectively, by

$$u_{i_{max}} = \sqrt{\sum_{j=1}^{N} (\Gamma_i \phi_{ij} S_{Dj})^2} \qquad (11.41)$$

and

$$\ddot{u}_{i_{max}} = \sqrt{\sum_{j=1}^{N} (\Gamma_i \phi_{ij} S_{Aj})^2} \qquad (11.42)$$

where S_{Dj} and S_{Aj} are, respectively, the spectral displacement and spectral acceleration for the jth mode. The participation factors Γ_i indicated in eqs. (11.39) or (11.40) are the factors of the excitation function $\ddot{y}_s(t)$ in eq. (11.38). As presented in Chapter 8, response spectral charts are prepared as the solution eq. (11.38) (with $\Gamma_i = 1$) for values of the natural frequency ω_i in the range of interest. Therefore, the spectral values obtained from these charts S_{Di} or S_{Ai} should be multiplied as indicated in eqs. (11.41) and (11.42) by the participation factor Γ_i, which is omitted in solving eq. (11.38).

Example 11.2.   Determine the response of the frame of Example 11.1 shown in Fig. 11.2 when it is subjected to a suddenly applied constant acceleration $\ddot{y}_s = 0.28\ g$ at its base.

Solution: The natural frequencies and corresponding normal modes from calculations in Examples 10.1 and 10.2 are

$$\omega_1 = 11.83 \text{ rad/sec}, \quad \omega_2 = 32.89 \text{ rad/sec}$$

$$\phi_{11} = 0.06437, \qquad \phi_{12} = 0.0567$$

$$\phi_{21} = 0.08130, \qquad \phi_{22} = -0.0924$$

The acceleration acting at the base of this structure is

$$\ddot{y}_s = 0.28 \times 386 = 108.47 \text{ in/sec}^2$$

The participation factors are calculated from eqs. (11.39) with the denominators set equal to unity since the modes are normalized. These factors are then

$$\Gamma_1 = -(136 \times 0.06437 + 66 \times 0.08130) = -14.120$$

$$\Gamma_2 = -(136 \times 0.0567 - 66 \times 0.0924) = -1.613 \qquad \text{(a)}$$

The modal equations (11.29) are

$$\ddot{g}_1 + 140g_1 = 108.47$$

$$\ddot{g}_2 + 1082g_2 = 108.47 \qquad \text{(b)}$$

and their solution, assuming zero initial conditions for velocity and displacement, is given by eqs. (4.5) as

$$g_1(t) = \frac{108.47}{140}(1 - \cos 11.83t)$$

$$g_2(t) = \frac{108.47}{1082}(1 - \cos 32.89t) \qquad \text{(c)}$$

The response in terms of the relative motion of the stories at the floor levels with respect to the displacement of the base is given as a function of time by eqs. (11.27) as

$$u_1(t) = -14.120 \times 0.06437 \times 0.775 \,(1 - \cos 11.83t)$$

$$-1.613 \times 0.0567 \times 0.100 \,(1 - \cos 32.89t)$$

$$u_2(t) = -14.120 \times 0.08130 \times 0.775 \,(1 - \cos 11.83t)$$

$$+1.614 \times 0.0924 \times 0.100 \,(1 - \cos 32.89t)$$

or, upon simplification, as

$$u_1 = -0.7135 + 0.704 \cos 11.83t + 0.009 \cos 32.89t$$

$$u_2 = -0.874 + 0.900 \cos 11.83t - 0.015 \cos 32.89t \qquad \text{(d)}$$

In this example, due to the simple excitation function (a constant acceleration), it was possible to obtain a closed solution of the problem as a function of time. For a complex excitation function such as the one produced by an actual earthquake, it would be necessary either to resort to numerical integration to obtain the response or to use response spectra if available. The maximum modal response is obtained for the present example when the cosine functions in eqs. (c) are equal to -1. In this case the maximum modal response is then

$$g_{1\text{max}} = 1.55$$

$$g_{2\text{max}} = 0.20 \qquad \text{(e)}$$

and the maximum response, calculated from the approximate formulas (11.31), is

$$u_{1\text{max}} = 1.409 \text{ in}$$

$$u_{2\text{max}} = 1.800 \text{ in} \qquad \text{(f)}$$

The possible maximum values for the response calculated from eqs. (d) by setting the cosine functions to their maximum value result in

$$u_{1_{max}} = 1.426 \text{ in}$$

$$u_{2_{max}} = 1.789 \text{ in} \tag{g}$$

which for this particular example certainly compares very well with the approximate results obtained in eqs. (f) above.

11.3 PROGRAM 9—RESPONSE BY MODAL SUPERPOSITION

Program 9 calculates the response of a linear system by superposition of the solutions of the modal equations. Before one can use this program, it is necessary to solve an eigenproblem to determine the natural frequencies and modal shapes of the structure. The program determines the response of the structure excited either by time-dependent forces applied at nodal coordinates or a time-dependent acceleration at the support of the structure.

Example 11.3. Use computer Program 9 to find the response of the frame analyzed in Example 11.2.

Solution: The natural frequencies and the modal matrix for this structure as calculated in Example 10.1 are

$$\omega_1 = 11.83 \text{ rad/sec}$$

$$\omega_2 = 32.89 \text{ rad/sec}$$

and

$$[\Phi] = \begin{bmatrix} 0.0644 & 0.0567 \\ 0.0813 & -0.0924 \end{bmatrix}$$

The mass matrix is

$$[M] = \begin{bmatrix} 136 & 0 \\ 0 & 66 \end{bmatrix}$$

Previous to the execution of Program 9, it is necessary to execute one of the programs to model the structure followed by the execution of Program 8 to solve the eigenproblem. The execution of these programs will prepare the file required to execute Program 9. Alternately, when the eigensolution is known, we would execute the auxiliary Program X2 to prepare the required file. Since the eigensolution in the present example is known, we proceed to execute Program X2:

Input Data and Output Results

```
        PROGRAM X2: EIGENSOLUTION DATA      DATA FILE: DX2

        INPUT DATA:

NUMBER OF DEGREES OF FREEDOM              ND = 2

           EIGENVALUES:

       0.140E+03    0.108E+04

          MODE SHAPES BY ROWS:

0.06437  0.08130
0.05670  -.09240

PROGRAM 9: MODAL SUPERPOSITION       DATA FILE: D9

        INPUT DATA:

NUMBER OF DEGREES OF FREEDOM                          ND = 2
NUMBER OF EXTERNAL FORCES                             NF = 1
TIME STEP OF INTEGRATION                             H = .05
GRAVITATIONAL INDEX                                  G = 386
PRINT TIME HISTORY NPRT = 1; ONLY MAX. VALUES NPRT = 0    NPRT = 0

FORCE # ,   COORD. # WHERE FORCE IS APPLIED , NUM. OF POINTS DEFINING THE FORCE
     1              0                         2

          EXCITATION FUNCTION FOR FORCE # 1:

     TIME          EXCITATION

     0.00          0.280
     1.00          0.280

FLOOR MASSES OF THE SHEAR BUILDING:

   136                66

   MODAL DAMPING RATIOS:
0                    0

        MAXIMUM RESPONSE:

        COORD.    MAX. DISPL.    MAX. VELOC.    MAX. ACC.
          1          1.405          8.394         212.48
          2          1.728         10.650         239.71
```

Example 11.4. Solve Example 11.1 using Program 9.

Solution:

Input Data and Output Results

```
    PROGRAM 9: SHEAR BUILDING       DATA FILE: D11.4

    INPUT DATA:

NUMBER OF DEGREES OF FREEDOM                          ND = 2
NUMBER OF EXTERNAL FORCES                             NF = 2
TIME STEP OF INTEGRATION                             H = .01
GRAVITATIONAL INDEX                                   G = 0
PRINT TIME HISTORY NPRT = 1; ONLY MAX. VALUES NPRT = 0   NPRT = 0

FORCE #,  COORD. # WHERE FORCE IS APPLIED, NUM. OF POINTS DEFINING THE
FORCE
    1               1                 3
    2               2                 3

        EXCITATION FUNCTION FOR FORCE # 1:

    TIME           EXCITATION

    0.00          10000.00
    0.10              0.00
    0.20              0.00

        EXCITATION FUNCTION FOR FORCE # 2:

    TIME           EXCITATION

    0.00          20000.00
    0.10              0.00
    0.20              0.00

  MODAL DAMPING RATIOS:
0                   0

        MAXIMUM RESPONSE:

    COORD.    MAX. DISPL.    MAX. VELOC.    MAX. ACC.
       1         0.672          6.600         169.49
       2         0.673          8.871         302.89
```

11.4   Harmonic Forced Excitation

When the excitation, that is, the external forces or base motion, is harmonic (sine or cosine function), the analysis is quite simple and the response can readily be found without the use of modal analysis. Let us consider the two-story shear building as shown in Fig. 11.4 subjected to a single harmonic force $F = F_0 \sin \bar{\omega}t$ which is applied at the level of the second floor. In this case eqs. (11.1) with $F_1(t) = 0$ and $F_2 = F_0 \sin \bar{\omega}t$ become

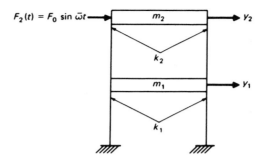

Fig. 11.4 Shear building with harmonic loading.

$$m_1\ddot{y}_1 + (k_1 + k_2)y_1 - k_2y_2 = 0$$

$$m_2\ddot{y}_2 - k_2y_1 + k_2y_2 = F_0 \sin \bar{\omega}t \tag{11.43}$$

For the steady-state response we seek a solution of the form

$$y_1 = Y_1 \sin \bar{\omega}t$$

$$y_2 = Y_2 \sin \bar{\omega}t \tag{11.44}$$

After substitution of eqs. (11.44) into eqs. (11.43) and cancellation of the common factor $\sin \bar{\omega}t$, we then obtain

$$(k_1 + k_2 - m_1\bar{\omega}^2)Y_1 - k_2Y_2 = 0$$

$$-k_2Y_1 + (k_2 - m_2\bar{\omega}^2)Y_2 = F_0 \tag{11.45}$$

which is a system of two equations in two unknowns, Y_1 and Y_2. This system always has a unique solution except in the case when the determinant formed by the coefficients of the unknowns is equal to zero. The reader should remember that in this case the forced frequency $\bar{\omega}$ would equal one of the natural frequencies, since this determinant when equated to zero is precisely the condition used for determining the natural frequencies. In other words, unless the structure is forced to vibrate at one of the resonant frequencies, the algebraic system of eqs. (11.43) has a unique solution for Y_1 and Y_2.

Example 11.5. Determine the steady-state response of the two-story shear building of Example 10.1 when a force $F_2(t) = 10{,}000 \sin 20t$ is applied to the second story as shown in Fig. 11.4.

Solution: The natural frequencies for this frame were determined in Example 10.1 to be

$$\omega_1 = 11.83 \text{ rad/sec}$$

$$\omega_2 = 32.89 \text{ rad/sec}$$

Since the forcing frequency is 20 rad/sec, the system is not at resonance. The steady-state response is then given by solving eqs. (11.45) for Y_1 and Y_2. Substituting numerical values in this system of equations, we have

$$(75,000 - 136 \times 20^2)Y_1 - 44,300 \, Y_2 = 0$$

$$-44,300 \, Y_1 + (44,300 - 66 \times 20^2)Y_2 = 10,000$$

Solving these equations simultaneously results in

$$Y_1 = -0.28 \text{ in}, \quad Y_2 = -0.13 \text{ in}$$

Therefore, according to eqs. (11.44), the steady-state response is

$$y_1 = -0.28 \sin 20t \text{ in}$$

$$y_2 = -0.13 \sin 20t \text{ in} \qquad \text{(Ans.)}$$

Damping may be considered in the analysis by simply including damping elements in the model as it is shown in Fig. 11.5 for a two-story shear building. The equations of motion which are obtained by equating to zero the sum of the forces in the free body diagram shown in Fig. 11.5(c) are

$$m_1 \ddot{y}_1 + (c_1 + c_2)\dot{y}_1 + (k_1 + k_2)y_1 - c_2\dot{y}_2 - k_2 y_2 = F_1(t)$$

$$m_2 \ddot{y}_2 - c_2\dot{y}_1 - k_2 y_1 + c_2\dot{y}_2 + k_2 y_2 = F_2(t) \qquad (11.46)$$

Now, considering the general case of applied forces of the form given by

$$F(t) = F_c \cos \bar{\omega}t + F_s \sin \bar{\omega}t \qquad (11.47)$$

we, conveniently, express such force in complex form as

$$F(t) = (F_c - iF_s) e^{i\bar{\omega}t} \qquad (11.48)$$

with the tacit understanding that only the real part of the eq. (11.48) is the applied. We show that the real part of the complex force in eq. (11.48) is precisely the force in (11.47). Using Euler's formula $e^{i\bar{\omega}t} = \cos \bar{\omega}t + i \sin \bar{\omega}t$, we obtain

$$\text{Real } \{(F_c - iF_s) e^{i\bar{\omega}t}\} = \text{Real } \{(F_c - iF_s)(\cos \bar{\omega}t + i \sin \bar{\omega}t)\}$$

$$= F_c \cos \bar{\omega}t + F_s \sin \bar{\omega}t \qquad (11.49)$$

which is the expression in eq. (11.47).

Assuming that the forces $F_1(t)$ and $F_2(t)$ in eq. (11.46) are in the form given by eq. (11.47), we substitute eq. (11.48) into eq. (11.46) to obtain

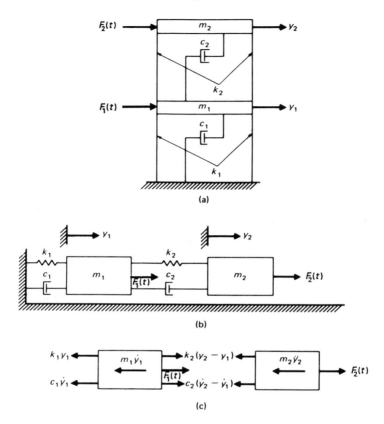

Fig. 11.5 (a) Damped shear building with harmonic load. (b) Multidegree mass-spring model. (c) Free body diagram.

$$m_1\ddot{y}_1 + (c_1 + c_2)\dot{y}_1 + (k_1 + k_2)y_1 - c_2\dot{y}_2 - k_2y_2 = (F_{c1} - iF_{s1})\,e^{i\bar{\omega}t}$$

$$m_2\ddot{y}_2 - c_2\dot{y}_1 - k_2y_1 + c_2\dot{y}_2 + k_2y_2 = (F_{c2} - iF_{s2})\,e^{i\bar{\omega}t} \quad (11.50)$$

The solution of the complex system of eqs. (11.50) will, in general, be of the form $y(t) = (Y_c + iY_s)\,e^{i\bar{\omega}t}$. Since only the real parts of the forces in eqs. (11.50) are applied, the solution is also only the real part of $y(t)$. Then, analogously to eq. (11.49),

$$\text{Real}\,\{Y_c + iY_s)\,e^{i\bar{\omega}t}\} = Y_c\cos\bar{\omega}t - Y_s\sin\bar{\omega}t \quad (11.51)$$

For the steady-state response of eqs. (11.50), we seek solutions in the form of

$$y_1 = Y_1\,e^{i\bar{\omega}t}$$

$$y_2 = Y_2\,e^{i\bar{\omega}t} \quad (11.52)$$

The substitution of eqs. (11.52) and the first and second derivatives of y_1 and y_2, namely

$$\dot{y}_1 = i\bar{\omega}Y_1\,e^{i\bar{\omega}t}, \quad \ddot{y}_1 = -\bar{\omega}^2 Y_1\,e^{i\bar{\omega}t},$$

$$\dot{y}_2 = i\bar{\omega}Y_2\,e^{i\bar{\omega}t}, \quad \ddot{y}_2 = -\bar{\omega}^2 Y_2\,e^{i\bar{\omega}t}$$

into eqs. (11.50) results in the following system of complex algebraic equations:

$$\{(k_1 + k_2 - m_1\bar{\omega}^2) + i\bar{\omega}(c_1 + c_2)\}Y_1 - (k_2 + i\bar{\omega}c_2)Y_2 = F_{c1} - iF_{s1}$$

$$- (k_2 + i\bar{\omega}c_2)Y_1 + \{(k_2 - m_2\bar{\omega}^2) + i\bar{\omega}c_2\}Y_2 = F_{c2} - iF_{s2} \quad (11.53)$$

As already stated, the response is then found by solving the complex system of equations (11.53) and considering the real part. Hence, from eq. (11.51), we have

$$y_1(t) = Y_{c1}\cos\bar{\omega}t - Y_{s1}\sin\bar{\omega}t$$

$$y_2(t) = Y_{c2}\cos\bar{\omega}t - Y_{s2}\sin\bar{\omega}t \quad (11.54)$$

in which $Y_1 = Y_{c1} + iY_{s1}$, $Y_2 = Y_{c2} + iY_{s2}$ is the solution of the complex equations (11.53). The necessary calculations are better explained through the use of a numerical example.

Example 11.6.   Determine the steady-state response for the two-story shear building of Example 11.5 in which damping is considered in the analysis (Fig. 11.5). Assume for this example that the damping constants c_1 and c_2 are, respectively, proportional to the magnitude of spring constants k_1 and k_2 in which the factor of proportionality, $a_0 = 0.01$.

Solution: The damping constants are calculated as

$$c_1 = a_0 k_1 = 307\text{ lb} \cdot \text{sec/in}$$

$$c_2 = a_0 k_2 = 443 \quad \text{(a)}$$

The substitution of numerical values for this example into eqs. (11.53) gives the following system of equations:

$$(20600 + 15000i)Y_1 - (44300 + 8860i)Y_2 = 0$$

$$-(44300 + 8860i)Y_1 + (17900 + 8860i)Y_2 = -10,000i \quad \text{(b)}$$

The solution of this system of equations is

$$Y_1 = 0.0006814 + 0.26865i$$

$$Y_2 = -0.06309 + 0.13777i$$

Therefore, by the relation established in eq. (11.53), the steady-state response is given by

$$y_1(t) = 0.0006814 \cos 20t - 0.26865 \sin 20t$$

$$y_2(t) = -0.06309 \cos 20t - 0.137775 \sin 20t$$

which also may be written as

$$y_1 = 0.2686 \sin (20t + 3.139) \text{ in}$$

$$y_2 = 0.1516 \sin (20t + 3.571) \tag{Ans.}$$

When these results are compared with those obtained for the undamped structure in Example 11.5, we note only a small change in the amplitude of motion. This is always the case for systems lightly damped and subjected to harmonic excitation of a frequency which is not close to one of the natural frequencies of the system. For this example, the forced frequency $\bar{\omega} = 20$ rad/sec is relatively far from the natural frequencies $\omega_1 = 11.83$ rad/sec or $\omega_2 = 32.89$ rad/sec which were calculated in Example 10.1.

11.5   PROGRAM 10—HARMONIC RESPONSE

Program 10 calculates the response to harmonic excitations of a structural system for which the stiffness and mass matrices have been determined in modeling the structure. Damping in the system is assumed to be proportional to the stiffness and mass coefficients, that is, the damping matrix is calculated as

$$[C] = a_0 [M] + a_1 [K] \tag{11.55}$$

in which a_0 and a_1 are constants specified in the input data. The program calculates the steady-state response for structures subjected to harmonic forces applied at the nodal coordinates or a harmonic acceleration applied at the base of the structure.

Example 11.7.   Obtain the response of the damped two-degree-of-freedom shear building of Example 11.6 using computer Program 10.

Solution:

Input Data and Output Results

```
PROGRAM 10: HARMONIC RESPONSE      DATA FILE: D10

    INPUT DATA:
```

```
DAMPING STIFFNESS FACTOR                                    KFAC = .01
DAMPING MASS FACTOR                                         MFAC = 0
FORCED FREQUENCY (RAD/SEC)                                  W = 20

      ***SYSTEM STIFFNESS MATRIX***

    7.5000E+04          -4.4300E+04
   -4.4300E+04           4.4300E+04

      ***SYSTEM MASS MATRIX***

  1.3600E+02             0.0000E+00
  0.0000E+00             6.6000E+01

  FORCE COEFFICIENTS OF [FC*COS(W*T)+FI*SIN(W*T)] FOR EACH COORD.

      COORD.        FC COMPONENT              FI COMPONENT

        1                 0                        0
        2                 0                      10000

  THE STEADY-STATE SOLUTION IS

      [FC*COS(W*T)+FI*SIN(W*T)] FOR EACH COORD.

      COORD.        FC COMPONENT              FI COMPONENT

        1             6.814E-04               -2.6865E-01
        2            -6.309E-02               -1.3777E-01
```

As given by the output results of computer Program 10, the response for this two-degree-of-freedom system is

$$y_1(t) = 0.00068 \cos 20t - 0.2686 \sin 20t$$

$$y_2(t) = -0.0631 \cos 20t - 0.1378 \sin 20t$$

or

$$y_1(t) = 0.2686 \sin (20t + 3.139)$$

$$y_2(t) = 0.1516 \sin (20t + 3.571)$$

The results given by the computer, as expected, are the same as the values calculated in Example 11.6.

Example 11.8. For the structure modeled as a four-story shear building shown in Fig. 11.6, determine the steady-state response when subjected to the force $F(t) = 10000 \sin 20t$ (lb) applied at the top floor of the building.

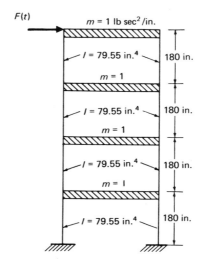

Fig. 11.6 Modeled structure for Example 11.8

Assume damping proportionate to stiffness (factor of proportionality $a_1 = 0.01$. Modulus of elasticity $E = 30 \times 10^6$ psi.

Solution: The solution of this problem is indicated by executing Program 7 to model the structure as a shear building followed by the execution of Program 10 to obtain the response to harmonic forces. Modeling of this structure has been undertaken in the solution of Example 9.1.

Input Data and Output Results

```
PROGRAM 10: HARMONIC RESPONSE      DATA FILE: D10.8

    INPUT DATA:

DAMPING STIFFNESS FACTOR                        KFAC = .01
DAMPING MASS FACTOR                             MFAC = 0
FORCED FREQUENCY (RAD/SEC)                      W = 20

    ***SYSTEM STIFFNESS MATRIX***

   6.5473E+02      -3.2737E+02       0.0000E+00        0.0000E+00
  -3.2737E+02       6.5473E+02      -3.2737E+02        0.0000E+00
   0.0000E+00      -3.2737E+02       6.5473E+02       -3.2737E+02
   0.0000E+00       0.0000E+00      -3.2737E+02        3.2737E+02
```

SYSTEM MASS MATRIX

1.0000E+00	0.0000E+00	0.0000E+00	0.0000E+00
0.0000E+00	1.0000E+00	0.0000E+00	0.0000E+00
0.0000E+00	0.0000E+00	1.0000E+00	0.0000E+00
0.0000E+00	0.0000E+00	0.0000E+00	1.0000E+00

FORCE COEFFICIENTS OF [FC*COS(W*T)+FI*SIN(W*T)] FOR EACH COORD.

COORD.	FC COMPONENT	FI COMPONENT
1	0	0
2	0	0
3	0	0
4	0	10000

THE STEADY-STATE SOLUTION IS

[FC*COS(W*T)+FI*SIN(W*T)] FOR EACH COORD.

COORD.	FC COMPONENT	FI COMPONENT
1	2.038E+01	2.6492E+01
2	2.304E+01	1.7070E+01
3	2.643E+00	-1.7821E+01
4	-2.505E+01	-3.2396E+01

11.6   COMBINING MAXIMUM VALUES OF MODAL RESPONSE

The square root of the sum of squared contributions (SRSS), to estimate the total response from calculated maximum modal values, may be expressed in general, from eq. (11.41) or eq. (11.42), as

$$R = \sqrt{\sum_{i=1}^{N} R_i^2} \qquad (11.56)$$

where R is the estimated response (force, displacement, etc.) at a specified coordinate and R_i is the corresponding maximum response of the ith mode at that coordinate.

Application of the SRSS method for combining modal response generally provides an acceptable estimation of the total maximum response. However, when some of the modes are closely spaced, the use of the SRSS method may result in grossly underestimating or overestimating the maximum response. In particular, large errors have been found in the analysis of three-dimensional structures in which torsional effects are significant. The term "closely spaced" may arbitrarily define the case when the difference between two natural frequencies is within 10% of the smallest of the two frequencies.

A formulation known as the complete quadratic combination (CQC), which is based on the theory of random vibrations, has been proposed by Kiureghian (1980) and by Wilson et al. (1981). The CQC method, which may be considered as an extension of the SRSS method, is given by the following equations:

$$R = \sqrt{\sum_{i=1}^{N} \sum_{j=1}^{N} R_i \rho_{ij} R_j} \qquad (11.57)$$

in which the cross-modal coefficient ρ_{ij} may be approximated by

$$\rho_{ij} = \frac{8(\xi_i \xi_j)^{1/2} (\xi_i + r\xi_j) r^{3/2}}{(1 - r^2)^2 + 4\xi_i \xi_j r (1 + r^2) + 4(\xi_i^2 + \xi_j^2) r^2} \qquad (11.58)$$

where $r = \omega_j / \omega_i$ is the ratio of the natural frequencies or order i and j and ξ_i and ξ_j the corresponding damping ratios for modes i and j. For constant modal damping ξ, eq. (11.58) reduces to

$$\rho_{ij} = \frac{8\xi^2 (1 + r) r^{3/2}}{(1 - r^2)^2 + 4\xi^2 r (1 + r)^2} \qquad (11.59)$$

It is important to note that, for $i = j$, eq. (11.58) or eq. (11.59) yields $\rho_{ii} = 1$ for any value of the damping ratio, including $\xi = 0$. Thus, for an undamped structure, the CQC method [eq. (11.57)] is identical to the SRSS method [eq. (11.56)].

11.7   SUMMARY

For the solution of linear equations of motion, we may employ either the modal superposition method of dynamic analysis or a step-by-step numerical integration procedure. The modal superposition method is restricted to the analysis of structures governed by linear systems of equations whereas the step-by-step methods of numerical integration are equally applicable to systems with linear or nonlinear behavior. We have deferred the presentation of the step-by-step integration method to Chapter 19 on nonlinear structural response of multidegree-of-freedom systems.

In the present chapter, we have introduced the modal superposition method in obtaining the response of a shear building subjected to either force excitation or to base motion and have demonstrated that the use of the normal modes of free vibration for transforming the coordinates leads to a set of uncoupled differential equations. The solution of these equations may then be obtained by any of the methods presented in Part I for the single degree-of-freedom system.

When use is made of response spectra to determine maximum values for modal response, these values are usually combined by the square root of the sum of squares (SRSS) method. However, the SRSS method could seriously overestimate or underestimate the total response when some frequencies are closely spaced. A more precise method of combining maximum values of the modal response is the complete quadratic combination (CQC). This method has been strongly recommended in lieu of the SRSS method. In the particular case of harmonic excitation, the response may be obtained in closed form by simply solving a system of algebraic equations in which the unknowns are the amplitudes of the response at the various coordinates.

PROBLEMS

11.1 Determine the response as a function of time for the two-story shear building of Problem 10.1 when a constant force of 5000 lb is suddenly applied at the level of the second floor as shown in Fig. 11.7. Bays are 15 ft apart.

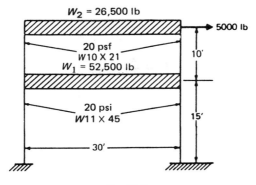

$W_2 = 26,500$ lb

5000 lb

20 psf
W10 X 21
$W_1 = 52,500$ lb

10'

20 psi
W11 X 45

15'

30'

Fig. 11.7

11.2 Repeat Problem 11.1 if the excitation is applied to the base of the structure in the form of a suddenly applied acceleration of magnitude 0.5 g.

11.3 Determine the maximum displacement at the floor levels of the three-story shear building [Fig. 11.8(a)] subjected to an impulsive triangular load as shown in Fig. 11.8(b). The total stiffness of the columns of each story is $k = 1500$ lb/in and the mass at each floor level is $m = 0.3886$ lb · sec^2/in.

11.4 Solve Problem 11.3 using Program 9. Set time increment $\Delta t = 0.01$ sec and total time of integration $T_{max} = 0.3$ sec.

11.5 Determine the maximum shear force in the columns of the second story of Problem 11.3. (Hint: Calculate modal shear forces and combine contributions using method of square root sum of squares.)

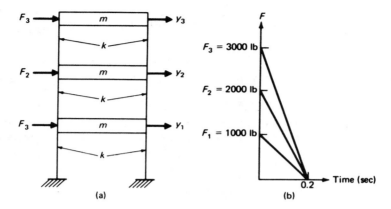

Fig. 11.8

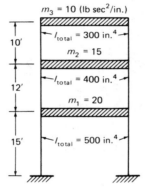

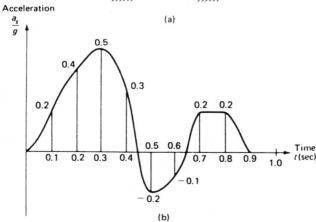

Fig. 11.9

11.6 Use computer Program 9 to obtain the time history response of the three-story building in Fig. 11.9(a) subjected to the support acceleration plotted in Fig. 11.9(b). Determine the response for a total time of 1.0 sec using time step $\Delta t = 0.05$ sec, and modal damping coefficient of 10% for all the modes ($E = 30 \times 10^6$ psi).

11.7 Find the steady-state response of the shear building shown in Fig. 11.10 subjected to the harmonic forces indicated in the figure. Neglect damping.

11.8 Solve Problem 11.7 assuming damping coefficients proportional to the story stiffness, $c_i = 0.05 K_i$.

11.9 For the structure (shear building) shown in Fig. 11.11 determine the steady-state motion for the following load systems (loads in pounds):

(a) $F_1(t) = 1000 \sin t,$ $F_2(t) = 2000 \sin t,$ $F_3(t) = 1500 \sin t$
(b) $F_1(t) = 2000 \cos t,$ $F_2(t) = 3000 \cos t,$ $F_3(t) = 4000 \cos t$

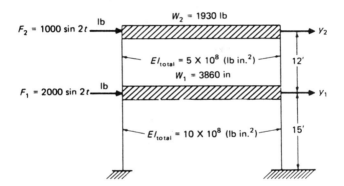

Fig. 11.10

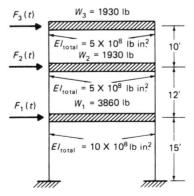

Fig. 11.11

Also load the structure simultaneously with load systems (a) and (b) and verify the superposition of results.

11.10 For the structure modeled as a four-story shear building shown in Fig. 11.12, determine the steady-state response when it is subjected to the force $F = 10000 \sin 20t$ (lb) applied at the top floor of the building. The modulus of elasticity is $E = 2.0 \times 10^6$ psi. Assume damping in the system proportional to the stiffness coefficient ($c_0 = 0.01$).

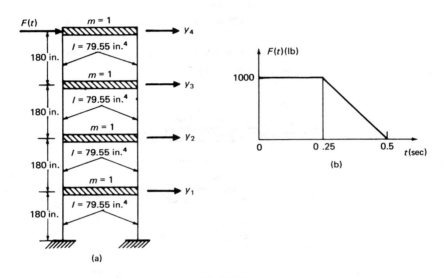

Fig. 11.12

12

Damped Motion of Shear Buildings

In the previous chapters, we determined the natural frequencies and modal shapes for undamped structures when modeled as shear buildings. We also determined the response of these structures using the modal superposition method. In this method, as we have seen, the differential equations of motion are uncoupled by means of a transformation of coordinates which incorporates the orthogonality property of the modal shapes.

The consideration of damping in the dynamic analysis of structures complicates the problem. Not only will the differential equations of motion have additional terms due to damping forces, but the uncoupling of the equations will only be possible by imposing some restrictions or conditions on the functional expression for the damping coefficients.

The damping normally present in structures is relatively small and practically does not affect the calculation of natural frequencies and modal shapes of the system. Hence, the effect of damping is neglected in determining the natural frequencies and modal shapes of the structural systems. Therefore, in practice, the eigenproblem for the damped structure is solved by using the same methods employed for undamped structures.

12.1   EQUATIONS FOR DAMPED SHEAR BUILDING

For a viscously damped shear building, such as the three-story building shown in Fig. 12.1, the equations of motion obtained by summing forces in the corresponding free body diagrams are

$$m_1\ddot{y}_1 + c_1\dot{y}_1 + k_1y_1 - c_2(\dot{y}_2 - \dot{y}_1) - k_2(y_2 - y_1) = F_1(t)$$

$$m_2\ddot{y}_2 + c_2(\dot{y}_2 - \dot{y}_1) + k_2(y_2 - y_1) - c_3(\dot{y}_3 - \dot{y}_2) - k_3(y_3 - y_2) = F_2(t)$$

$$m_3\ddot{y}_3 + c_3(\dot{y}_3 - \dot{y}_2) + k_3(y_3 - y_2) = F_3(t) \qquad (12.1)$$

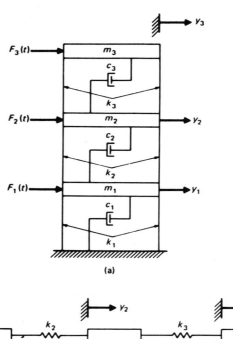

(a)

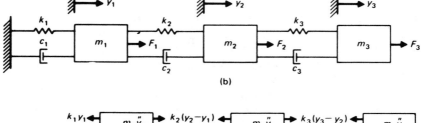

(b)

(c)

Fig. 12.1 (a) Damped shear building. (b) Mathematical model. (c) Free body diagram.

These equations may be conveniently written in matrix notation as

$$[M]\{\ddot{y}\} + [C]\{\dot{y}\} + [K]\{y\} = \{F(t)\} \tag{12.2}$$

where the matrices and vectors are as previously defined, except for the damping matrix $[C]$, which is given by

$$[C] = \begin{bmatrix} c_1 + c_2 & -c_2 & 0 \\ -c_2 & c_2 + c_3 & -c_3 \\ 0 & -c_3 & c_3 \end{bmatrix} \tag{12.3}$$

In the next section, we shall establish the conditions under which the damped equations of motion may be transformed to an uncoupled set of independent equations.

12.2   UNCOUPLED DAMPED EQUATIONS

To solve the differential equations of motion, eq. (12.2), we seek to uncouple these equations. We, therefore, introduce the transformation of coordinates

$$\{y\} = [\Phi]\{z\} \tag{12.4}$$

where $[\Phi]$ is the modal matrix obtained in the solution of the undamped free-vibration system. The substitution of eq. (12.4) and its derivatives into eq. (12.2) leads to

$$[M][\Phi]\{\ddot{z}\} + [C][\Phi]\{\dot{z}\} + [K][\Phi]\{z\} = \{F(t)\} \tag{12.5}$$

Premultiplying eq. (12.5) by the transpose of the nth modal vector $\{\phi\}_n^T$ yields

$$\{\phi\}_n^T [M][\Phi]\{\ddot{z}\} + \{\phi\}_n^T [C][\Phi]\{\dot{z}\} + \{\phi\}_n^T [K][\Phi]\{z\} = \{\phi\}_n^T \{F(t)\} \tag{12.6}$$

It is noted that the orthogonality property of the modal shapes,

$$\{\phi\}_n^T [M]\{\phi\}_m = 0, \quad m \neq n$$
$$\{\phi\}_n^T [K]\{\phi\}_m = 0 \tag{12.7}$$

causes all components except the nth mode in the first and third terms of eq. (12.6) to vanish. A similar reduction is *assumed* to apply to the damping term in eq. (12.6), that is, if it is assumed that

$$\{\phi\}_n^T [C]\{\phi\}_m = 0, \quad n \neq m \tag{12.8}$$

then the coefficient of the damping term in eq. (12.6) will reduce to

$$\{\phi\}_n^T [C]\{\phi\}_n$$

In this case eq. (12.6) may be written as

$$M_n \ddot{z}_n + C_n \dot{z}_n + K_n z_n = F_n(t)$$

or alternatively as

$$\ddot{z}_n + 2\xi_n\omega_n\dot{z}_n + \omega_n^2 z_n = F_n(t)/M_n \tag{12.9}$$

in which case

$$M_n = \{\phi\}_n^T [M] \{\phi\}_n \tag{12.10a}$$

$$K_n = \{\phi\}_n^T [K] \{\phi\}_n = \omega_n^2 M_n \tag{12.10b}$$

$$C_n = \{\phi\}_n^T [C] \{\phi\}_n = 2\xi\omega_n M_n \tag{12.10c}$$

$$F_n(t) = \{\phi\}_n^T \{F(t)\} \tag{12.10d}$$

The normalization discussed previously (section 10.2)

$$\{\phi\}_n^T [M] \{\phi\}_n = 1 \tag{12.11}$$

will result in

$$M_n = 1$$

so that eq. (12.9) reduces to

$$\ddot{z}_n + 2\xi_n\omega_n\dot{z}_n + \omega_n^2 z_n = F_n(t) \tag{12.12}$$

which is a set of N uncoupled differential equations $(n = 1, 2, \ldots, N)$.

12.3   CONDITIONS FOR DAMPING UNCOUPLING

In the derivation of the uncoupled damped equation (12.12), it has been assumed that the normal coordinate transformation, eq. (12.4), that serves to uncouple the inertial and elastic forces also uncouples the damping forces. It is of interest to consider the conditions under which this uncoupling will occur, that is, the form of the damping matrix $[C]$ to which eq. (12.8) applies.

When the damping matrix is of the form

$$[C] = a_0 [M] + a_1 [K] \tag{12.13}$$

in which a_0 and a_1 are arbitrary proportionality factors, the orthogonality condition will be satisfied. This may be demonstrated by applying the orthogonality condition to eq. (12.13), that is, premultiplying both sides of this equation by the transpose of the nth mode $\{\phi\}_n^T$ and postmultiplying by the modal matrix $[\Phi]$. We obtain

$$\{\phi\}_n^T [C] [\Phi] = a_0 \{\phi\}_n^T [M] [\Phi] + a_1 \{\phi\}_n^T [K] [\Phi] \tag{12.14}$$

The orthogonality conditions, eqs. (12.7), then reduce eq. (12.14) to

$$\{\phi\}_n^T [C] [\Phi] = a_0 \{\phi\}_n^T [M] \{\phi\}_n + a_1 \{\phi\}_n^T [K] \{\phi\}_n$$

or, by eqs. (12.10), to

$${\phi}_n^T [C] [\Phi]_n = a_0 M_n + a_1 M_n \omega_n^2$$

$${\phi}_n^T [C] [\Phi]_n = (a_0 + a_1 \omega_n^2) M_n$$

which shows that, when the damping matrix is of the form of eq. (12.13), the damping forces are also uncoupled with the transformation eq. (12.4). However, it can be shown that there are other matrices formed from the mass and stiffness matrices which also satisfy the orthogonality condition. In general, the damping matrix may be of the form

$$[C] = [M] \sum_i a_i ([M]^{-1} [K])^i \qquad (12.15)$$

where i can be anywhere in the range $-\infty < i < \infty$ and the summation may include as many terms as desired. The damping matrix, eq. (12.13), can obviously be obtained as a special case of eq. (12.15). By taking two terms corresponding to $i = 0$ and $i = 1$ in eq. (12.15), we obtain the damping matrix expressed by eq. (12.13). With this form of the damping matrix it is possible to compute the damping coefficients necessary to provide uncoupling of a system having any desired damping ratios in any specified number of modes. For any mode n, the modal damping is given by eq. (12.10c), that is

$$C_n = {\phi}_n^T [C] {\phi}_n = 2\xi_n \omega_n M_n$$

If $[C]$ as given by eq. (12.15) is substituted in the expression for C_n, we obtain

$$C_n = {\phi}_n^T [M] \sum_i a_i ([M]^{-1} [K])^i {\phi}_n \qquad (12.16)$$

Now, using eq. (10.24) $(K{\phi}_n = \omega_n^2 M{\phi}_n)$ and performing several algebraic operations, we can show (Clough and Penzien 1975, p. 195) that the damping coefficient associated with any mode n may be written as

$$C_n = \sum_i a_i \omega_n^{2i} M_n = 2\xi_n \omega_n M_n \qquad (12.17)$$

from which

$$\xi_n = \frac{1}{2\omega_n} \sum_i a_i \omega_n^{2i} \qquad (12.18)$$

Equation (12.18) may be used to determine the constants a_i for any desired values of modal damping ratios corresponding to any specified numbers of modes. For example, to evaluate these constants specifying the first four modal damping ratios $\xi_1, \xi_2, \xi_3, \xi_4$, we may choose $i = 1, 2, 3, 4$. In this case eq. (12.18) gives the following system of equations:

$$\begin{Bmatrix} \xi_1 \\ \xi_2 \\ \xi_3 \\ \xi_4 \end{Bmatrix} = \frac{1}{2} \begin{pmatrix} \omega_1 & \omega_1^3 & \omega_1^5 & \omega_1^7 \\ \omega_2 & \omega_2^3 & \omega_2^5 & \omega_2^7 \\ \omega_3 & \omega_3^3 & \omega_3^5 & \omega_3^7 \\ \omega_4 & \omega_4^3 & \omega_4^5 & \omega_4^7 \end{pmatrix} \begin{Bmatrix} a_1 \\ a_2 \\ a_3 \\ a_4 \end{Bmatrix} \qquad (12.19)$$

In general, eq. (12.19) may be written symbolically as

$$\{\xi\} = \tfrac{1}{2} [Q] \{a\} \qquad (12.20)$$

where $[Q]$ is a square matrix having different powers of the natural frequencies. The solution of eq. (12.20) gives the constants $\{a\}$ as

$$\{a\} = 2[Q]^{-1} \{\xi\} \qquad (12.21)$$

Finally the damping matrix is obtained after the substitution of eq. (12.21) into eq. (12.15).

It is interesting to observe from eq. (12.18) that in the special case when the damping matrix is proportional to the mass $\{C\} = a_0 [M]$ $(i = 0)$, the damping ratios are inversely proportional to the natural frequencies; thus the higher modes of the structures will be given very little damping. Analogously, when the damping is proportional to the stiffness matrix $([C] = a_1 [K])$, the damping ratios are directly proportional to the corresponding natural frequencies, as can be seen from eq. (12.18) evaluated for $i = 1$; and in this case the higher modes of the structure will be very heavily damped.

Example 12.1.   Determine the absolute damping coefficients for the structure presented in Example 10.1. Assume 10% of the critical damping for each mode.

Solution: From Example 10.1, we have the following information.
Natural frequencies:

$$\omega_1 = 11.83 \text{ rad/sec}$$
$$\omega_2 = 32.89 \text{ rad/sec} \qquad (a)$$

Modal matrix:

$$[\Phi] = \begin{bmatrix} 1.00 & 1.00 \\ 1.26 & -1.63 \end{bmatrix} \qquad (b)$$

Mass matrix:

$$[M] = \begin{bmatrix} 136 & 0 \\ 0 & 66 \end{bmatrix}$$

Stiffness matrix:

$$[K] = \begin{bmatrix} 75000 & -44300 \\ -44300 & 44300 \end{bmatrix}$$

Using eq. (12.18) with $i = 0, 1$ to calculate the constants a_i needed in eq. (12.15), we obtain the following system of equations:

$$\begin{Bmatrix} 0.1 \\ 0.1 \end{Bmatrix} = \frac{1}{2} \begin{bmatrix} 11.83 & (11.83)^3 \\ 32.89 & (32.89)^3 \end{bmatrix} \begin{Bmatrix} a_1 \\ a_2 \end{Bmatrix}$$

Solving this system of equations gives

$$a_1 = 0.01851$$

$$a_2 = -0.00001146$$

We also calculate

$$[M]^{-1} = \begin{bmatrix} 0.007353 & 0 \\ 0 & 0.01515 \end{bmatrix}$$

and

$$[M]^{-1}[K] = \begin{bmatrix} 551.475 & -325.738 \\ -671.145 & 671.145 \end{bmatrix}$$

Then

$$\sum_{i=1}^{2} a_i([M]^{-1}[K])^i = 0.01851 \begin{bmatrix} 551.475 & -325.738 \\ -671.145 & 671.145 \end{bmatrix}$$

$$- 0.00001146 \begin{bmatrix} 551.475 & -325.738 \\ -671.145 & 671.145 \end{bmatrix}^2$$

$$= \begin{bmatrix} 4.2172 & -1.4654 \\ -3.0193 & 4.7556 \end{bmatrix}$$

Finally, substituting this matrix into eq. (12.15) yields the damping matrix as

$$[C] = \begin{bmatrix} 136 & 0 \\ 0 & 66 \end{bmatrix} \begin{bmatrix} 4.2172 & -1.4654 \\ -3.0193 & 4.7556 \end{bmatrix} = \begin{bmatrix} 573.5 & -199.3 \\ -199.3 & 313.9 \end{bmatrix}$$

There is yet a second method for evaluating the damping matrix corresponding to any set of specified modal damping ratio. The method may be explained starting with the relationship

$$[A] = [\Phi]^T[C][\Phi] \tag{12.22}$$

where the matrix $[A]$ is defined as

$$[A] = \begin{bmatrix} 2\xi_1\omega_1 M_1 & 0 & 0 & - \\ 0 & 2\xi_2\omega_2 M_2 & 0 & - \\ 0 & 0 & 2\xi_3\omega_3 M_3 & - \\ - & - & - & - \end{bmatrix} \tag{12.23}$$

in which the modal masses M_1, M_2, M_3, ... are equal to one if the modal matrix $\{\Phi\}$ has been normalized. It is evident that the damping matrix $[C]$ may be evaluated by postmultiplying and premultiplying eq. (12.22) by the inverse of the modal matrix $[\Phi]^{-1}$ and its inverse transpose $[\Phi]^{-T}$, such that

$$[C] = [\Phi]^{-T} [A] [\Phi]^{-1} \qquad (12.24)$$

Therefore, for any specified set of modal damping ratios $\{\xi\}$, matrix $[A]$ can be evaluated from eq. (12.23) and the damping matrix $[C]$ from eq. (12.24). However, in practice, the inversion of the modal matrix is a large computational effort. Instead, taking advantage of orthogonality properties of the mode shapes, we can deduce the following expression for the system damping matrix:

$$[C] = [M] \left(\sum_{n=1}^{N} \frac{2\xi_n \omega_n}{M_n} \{\Phi\}_n \{\Phi\}_n^T \right) [M] \qquad (12.25)$$

Equation (12.25) may be obtained from the condition of orthogonality of the normal modes given by eq. (10.35) as

$$[I] = [\Phi]^T [M] [\Phi] \qquad (12.26)$$

Postmultiplying eq. (12.26) by $[\Phi]^{-1}$, we obtain

$$[\Phi]^{-1} = [\Phi]^T [M] \qquad (12.27)$$

Applying the transpose operation to eq. (12.27) results in

$$[\Phi]^{-T} = [M] [\Phi] \qquad (12.29)$$

in which $[M] = [M]^T$ since the mass matrix $[M]$ is a symmetric matrix.

Finally, the substitution into eq. (12.24) of eqs. (12.27) and (12.28) gives

$$[C] = [M] [\Phi] [A] [\Phi]^T [M]$$

which results in eq. (12.25) after substituting matrix $[A]$ from eq. (12.23). The damping matrix $[C]$ obtained from eq. (12.25) will satisfy the orthogonality property and, therefore, the damping term in the differential equation (12.2) will be uncoupled with the same transformation, eq. (12.4), which serves to uncouple the inertial and elastic forces.

It is of interest to note in eq. (12.25) that the contribution to the damping matrix of each mode is proportional to the modal damping ratio; thus any undamped mode will contribute nothing to the damping matrix.

We should mention at this point the circumstances under which it will be desirable to evaluate the elements of the damping matrix, as eq. (12.15) or eq. (12.25). It has been stated that absolute structural damping is a rather difficult quantity to determine or even to estimate. However, modal damping ratios may be estimated on the basis of past experience. This past experience includes laboratory determination of damping in different materials, as well as damping values obtained from vibration tests in existing buildings and other structures. Numerical values for damping ratios in structures are generally in the range of 1% to 10%. These values depend on the type of structure, materials utilized, nonstructural elements, etc. They also depend on the period and the magnitude of vibration. It has also been observed that damping ratios corresponding to higher modes have increasing values. Figure 12.2 shows the values of damping ratios measured in existing buildings as reported by H. Aoyama (1980) [in Wakabayashi (1986)]. It may be observed from this figure that experimental values for damping scatter over a wide range, and that it is difficult to give definite recommendations. The scatter observed in Fig. 12.2 is typical for experiments conducted to determine damping. On this basis, the obvious conclusion should be that the assumption of viscous damping to represent damping does not describe the real mechanism of energy dissipation in structural dynamics. However, for analytical expediency and also because of the uncertainties involved in attempting other formulation of damping, we still accept the assumption of viscous damping. At this time, the best recommendation that can be given in regard to damping is to use conservative values, 1% to 2% for steel buildings and 3% to 5% for reinforced concrete buildings for the fundamental frequency, and to assume damping ratios for

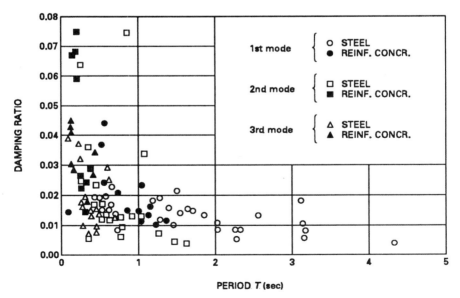

Fig. 12.2 Damping ratios measured in existing buildings [H. Aoyama in Wakabayashi (1986)].

the higher modes to increase in proportion to the natural frequencies. Thus, on the basis of giving some consideration to the type of structure, we assign numerical values to the damping ratios in all the modes of interest. These values are then used directly in the modal equations or they are used to determine the damping matrix which is needed when the dynamic response is obtained by some analytical method other than modal superposition, e.g., the time history response of a linear or nonlinear system.

Example 12.2 Determine the damping matrix of Example 12.1 using the method based on eq. (12.22).

Solution: From Example 10.2 we have the natural frequencies

$$\omega_1 = 11.83 \text{ rad/sec}$$

$$\omega_2 = 32.89 \text{ rad/sec}$$

and the mass matrix

$$[M] = \begin{bmatrix} 136 & 0 \\ 0 & 66 \end{bmatrix}$$

To determine $[C]$, we could use either eq. (12.24) or eq. (12.25). From Example 10.2, the normalized modal matrix is

$$[\Phi] = \begin{bmatrix} 0.06437 & 0.0567 \\ 0.0813 & -0.0924 \end{bmatrix}$$

and its inverse by eq. (12.27) is

$$[\Phi]^{-1} = \begin{bmatrix} 8.752 & 5.370 \\ 7.700 & -6.097 \end{bmatrix}$$

Substituting into eq. (12.23), we obtain

$$2\xi_1\omega_1 M_1 = (2)(0.1)(11.83)(1) = 2.366$$

$$2\xi_2\omega_2 M_2 = (2)(0.1)(32.89)(1) = 6.578$$

Then by eq. (12.24)

$$[C] = \begin{bmatrix} 8.752 & 7.700 \\ 5.370 & -6.097 \end{bmatrix} \begin{bmatrix} 2.366 & 0 \\ 0 & 6.578 \end{bmatrix} \begin{bmatrix} 8.752 & 5.370 \\ 7.700 & -6.097 \end{bmatrix}$$

$$[C] = \begin{bmatrix} 572 & -198 \\ -198 & 313 \end{bmatrix} \qquad \text{(Ans.)}$$

which in this case of equal damping ratios in all the models checks with the damping matrix obtained in Example 12.1 for the same structure using eq. (12.15).

12.4   PROGRAM 11—ABSOLUTE DAMPING FROM MODAL DAMPING RATIOS

Program 11, "DAMPING," serves to calculate the absolute damping coefficients, that is, the elements of the damping matrix from known values of modal damping ratios.

Example 12.3.   Use Program 11 to calculate the damping matrix for a structure with three degrees of freedom for which the squares of the natural frequencies (eigenvalues) are

$$\omega_1^2 = 1.9618, \quad \omega_2^2 = 15.3927, \quad \omega_3^2 = 60.7968$$

The modal and mass matrices are

$$[\Phi] = \begin{bmatrix} 0.4330 & -0.7421 & 0.7228 \\ 0.7967 & 0.0000 & -0.7719 \\ 0.4330 & 0.7421 & 0.7228 \end{bmatrix}$$

and

$$[M] = \begin{bmatrix} 0.8169 & 0.1286 & -0.0740 \\ 0.1286 & 0.8571 & 0.1286 \\ -0.0740 & 0.1286 & 0.8169 \end{bmatrix}$$

Assume damping ratio of 10% in all the modes.

Solution: The execution of Program 11 requires a file which is prepared during the execution of Program 8 to solve the eigenproblem of a structure previously modeled. Alternatively, the required file is prepared by the auxiliary Program X2. In the solution of Example 12.3, the auxiliary Program X2 is executed, followed by the execution of Program 11.

Input Data and Output Results

```
PROGRAM 11: DAMPING MATRIX      DATA FILE:D11
   NATURAL FREQUENCIES (C.P.S.):

0.223      0.624      1.241

MODAL SHAPES BY ROW (EIGENVECTORS):

 0.4330     0.7967     0.4330
-0.7421     0.0000     0.7421
 0.7228    -0.7719 ,   0.7228

 INPUT DATA:

NUMBER OF DEGREES OF FREEDOM      ND = 3
```

```
          ***SYSTEM MASS MATRIX***

   8.1690E-01    1.2860E-01   -7.4000E-02
   1.2860E-01    8.5710E-01    1.2860E-01
  -7.4000E-02    1.2860E-01    8.1690E-01

MODAL DAMPING RATIOS:

  .1               .1             .1

             OUTPUT RESULTS:
       ***SYSTEM DAMPING MATRIX***

   6.9214E-01   -2.3033E-01    6.1736E-03
  -2.3033E-01    5.2958E-01   -2.3033E-01
   6.1736E-03   -2.3033E-01    6.9214E-01
```

12.5 SUMMARY

The most common method of taking into account the dissipation of energy in structural dynamics is to assume in the mathematical model the presence of damping forces of magnitudes which are proportional to the relative velocity and of directions opposite to the motion. This type of damping is known as viscous damping because it is the kind of damping that will be developed by motion in an ideal viscous fluid. The inclusion of this type of damping in the equations does not alter the linearity of the differential equations of motion. Since the amount of damping commonly presented in structural systems is relatively small, its effect is neglected in the calculation of natural frequencies and mode shapes. However, to uncouple the damped differential equations of motion, it is necessary to impose some restrictions on the values of damping coefficients in the system. These restrictions are of no consequence due to the fact that in practice it is easier to determine or to estimate modal damping ratios rather than absolute damping coefficients. In addition, when solving the equations of motion by the modal superposition method, only damping ratios are required. When the solution is sought by other methods, the absolute value of the damping coefficients may be calculated from modal damping ratios by the two methods presented in this chapter.

PROBLEMS

12.1 The stiffness and mass matrices for a certain two-degree-of-freedom structure are

$$[K] = \begin{bmatrix} 400 & -200 \\ -200 & 200 \end{bmatrix}, \quad [M] = \begin{bmatrix} 2 & 0 \\ 0 & 1 \end{bmatrix}$$

Determine the damping matrix for this system corresponding to 20% of the critical damping for the first mode and 10% for the second mode. Use the method based on eqs. (12.16) and (12.17).

12.2 Repeat Problem 12.1 using the method based on eqs. (12.22) and (12.25).

12.3 The natural frequencies and corresponding normal modes (arranged in the modal

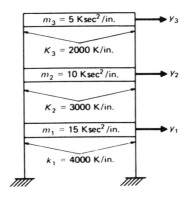

Fig. 12.3

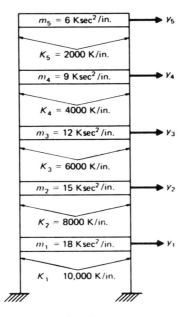

Fig. 12.4

matrix) for the three-story shear building shown in Fig. 12.3 are $\omega_1 = 9.31$ rad/sec, $\omega_2 = 20.94$ rad/sec, $\omega_3 = 29.00$ rad/sec, and

$$[\Phi] = \begin{bmatrix} 0.1114 & -0.1968 & -0.1245 \\ 0.2117 & -0.0277 & 0.2333 \\ 0.2703 & 0.2868 & -0.2114 \end{bmatrix}$$

Determine the damping matrix for the system corresponding to damping ratios of 10% for all the modes.

12.4 Repeat Problem 12.3 for damping ratios of 20% for all the modes.

12.5 Repeat Problem 12.3 for the following value of modal damping ratios:

$$\xi_1 = 0.2, \quad \xi_2 = 0.1, \quad \xi_3 = 0.0$$

12.6 Use Program 7 and 8 to model and determine the natural frequencies and normal modes for the five-story shear building shown in Fig. 12.4; then use Program 11 to determine the damping matrix corresponding to an 8% damping ratio in all the modes.

12.7 Repeat Problem 12.6 for the following values of the modal damping ratios:

$$\xi_1 = 0.20, \quad \xi_2 = 0.15, \quad \xi_3 = 0.10, \quad \xi_4 = 0.05, \quad \xi_5 = 0$$

13

Reduction of Dynamic Matrices

In the discretization process, it is sometimes necessary to divide a structure into a large number of elements because of changes in geometry, loading, or material properties. When the elements are assembled for the entire structure, the number of unknown displacements, that is, the number of degrees of freedom, may be quite large. As a consequence, the stiffness, mass, and damping matrices will be of large dimensions. The solution of the corresponding eigenproblem to determine natural frequencies and modal shapes will be difficult and, in addition, expensive. In such cases it is desirable to reduce the size of these matrices in order to make the solution of the eigenproblem more manageable and economical. Such reduction is referred to as *condensation.*

A popular method of reduction is the *Static Condensation Method.* This method, though simple to apply, is only approximate and may produce relatively large errors in the results when applied to dynamic problems. An improved method for condensation of dynamic problems, which gives virtually exact results, has recently been proposed. This new method, called the *Dynamic Condensation Method,* will be presented in this chapter after the introduction of the Static Condensation Method and some earlier modifications.

13.1   STATIC CONDENSATION

A practical method of accomplishing the reduction of the stiffness matrix is to identify those degrees of freedom to be condensed as *dependent* or *secondary* degrees of freedom, and express them in terms of the remaining *independent* or *primary* degrees of freedom. The relation between the secondary and primary degrees of freedom is found by establishing the static relation between them, hence the name *Static Condensation Method* (Guyan, R. J., 1965). This relation provides the means to reduce the stiffness matrix. This method is also used in static problems to eliminate unwanted degrees of freedom such as the internal degrees of freedom of an element used with the Finite Element Method. In order to describe the Static Condensation Method, let us assume that those (secondary) degrees of freedom to be reduced or condensed are arranged as the first s coordinates, and the remaining (primary) degrees of freedom are the last p coordinates. With this arrangement, the stiffness equation for the structure may be written using partition matrices as

$$\left[\begin{array}{c|c} [K_{ss}] & [K_{sp}] \\ \hline [K_{ps}] & [K_{pp}] \end{array}\right] \left\{\begin{array}{c} \{y_s\} \\ \{y_p\} \end{array}\right\} = \left\{\begin{array}{c} \{0\} \\ \{F_p\} \end{array}\right\} \tag{13.1}$$

where $\{y_s\}$ is the displacement vector corresponding to the s degrees of freedom to be reduced and $\{y_p\}$ is the vector corresponding to the remaining p independent degrees of freedom. In eq. (13.1), it was assumed that the external forces were zero at the dependent (i.e., secondary) degrees of freedom; this assumption is not mandatory (Gallagher, R. H., 1975), but serves to simplify explanations without affecting the final results. A simple multiplication of the matrices on the left side of eq. (13.1) expands this equation into two matrix equations, namely,

$$[K_{ss}]\{y_s\} + [K_{sp}]\{y_p\} = \{0\} \tag{13.2}$$

$$[K_{ps}]\{y_s\} + [K_{pp}]\{y_p\} = \{F_p\} \tag{13.3}$$

Equation (13.2) is equivalent to

$$\{y_s\} = [\overline{T}]\{y_p\} \tag{13.4}$$

where $[\overline{T}]$ is the *transformation matrix* given by

$$[\overline{T}] = -[K_{ss}]^{-1}[K_{sp}] \tag{13.5}$$

Substituting eq. (13.4) and using eq. (13.5) in eq. (13.3) results in the *reduced stiffness equation* relating forces and displacements at the primary coordinates, that is,

$$[\overline{K}]\{y_p\} = \{F_p\} \tag{13.6}$$

where $[\overline{K}]$ is the *reduced stiffness matrix* given by

$$[\overline{K}] = [K_{pp}] - [K_{ps}][K_{ss}]^{-1}[K_{sp}] \qquad (13.7)$$

Equation (13.4), which expresses the static relation between the secondary coordinates $\{y_s\}$ and primary coordinates $\{y_p\}$, may also be written using the identity $\{y_p\} = [I]\{y_p\}$ as

$$\left\{\begin{array}{c} \{y_s\} \\ \hline \{y_p\} \end{array}\right\} = \left[\begin{array}{c} [\overline{T}] \\ \hline [I] \end{array}\right]\{y_p\}$$

or

$$\{y\} = [T]\{y_p\} \qquad (13.8)$$

where

$$\{y\} = \left\{\begin{array}{c} \{y_s\} \\ \hline \{y_p\} \end{array}\right\} \quad \text{and} \quad [T] = \left[\begin{array}{c} [\overline{T}] \\ \hline [I] \end{array}\right] \qquad (13.9)$$

Substituting eqs. (13.8) and (13.9) into eq. (13.1) and premultiplying by the transpose of $[T]$ results in

$$[T]^T[K][T]\{y_p\} = [[\overline{T}]^T[I]]\left\{\begin{array}{c} \{0\} \\ \{F_p\} \end{array}\right\}$$

or

$$[T]^T[K][T]\{y_p\} = \{F_p\}$$

and using eq. (13.6)

$$[\overline{K}] = [T]^T[K][T] \qquad (13.10)$$

thus showing that the reduced stiffness matrix $[\overline{K}]$ can be expressed as a transformation of the system stiffness matrix $[K]$.

It may appear that the calculation of the reduced stiffness matrix $[\overline{K}]$ given by eq. (13.7) requires the inconvenient calculation of the inverse matrix $[K_{ss}]^{-1}$. However, the practical application of the Static Condensation Method does not require a matrix inversion. Instead the standard Gauss–Jordan elimination process is applied systematically on the system's stiffness matrix $[K]$ up to the elimination of the secondary coordinates $\{y_s\}$. At this stage of the elimination process the stiffness equation (13.1) has been reduced to

$$\left[\begin{array}{c|c} [I] & -[\overline{T}] \\ \hline [0] & [\overline{K}] \end{array}\right]\left\{\begin{array}{c} \{y_s\} \\ \hline \{y_p\} \end{array}\right\} = \left\{\begin{array}{c} \{0\} \\ \hline \{F_p\} \end{array}\right\} \qquad (13.11)$$

It may be seen by expanding eq. (13.11) that the partition matrices $[\overline{T}]$ and $[\overline{K}]$ are precisely the transformation matrix and the reduced stiffness matrix defined by eqs. (13.4) and (13.6), respectively. In this way, the Gauss–Jordan elimination process yields both the transformation matrix $[\overline{T}]$ and the reduced

stiffness matrix $[\overline{K}]$. There is thus no need to calculate $[K_{ss}]^{-1}$ in order to reduce the secondary coordinates of the system.

Example 13.1.   Consider the two-degree-of-freedom system represented by
the model shown in Fig. 13.1 and use static condensation to reduce the first
coordinate.

Solution: For this system the equations of equilibrium are readily obtained as

$$\begin{bmatrix} 2k & -k \\ -k & k \end{bmatrix} \begin{Bmatrix} y_1 \\ y_2 \end{Bmatrix} = \begin{Bmatrix} 0 \\ F_2 \end{Bmatrix} \tag{13.12}$$

The reduction of y_1 using Gauss elimination leads to

$$\begin{bmatrix} 1 & -\frac{1}{2} \\ 0 & k/2 \end{bmatrix} \begin{Bmatrix} y_1 \\ y_2 \end{Bmatrix} = \begin{Bmatrix} 0 \\ F_2 \end{Bmatrix} \tag{13.13}$$

Comparing eq. (13.13) with eq. (13.11), we identify in this example

$$[\overline{T}] = \tfrac{1}{2}$$

$$[\overline{K}] = k/2 \tag{13.14}$$

Consequently, from eq. (13.9) the transformation matrix is

$$[T] = \begin{bmatrix} \frac{1}{2} \\ 1 \end{bmatrix} \tag{13.15}$$

We can now check eq. (13.10) by simply performing the indicated multiplications, namely

$$[\overline{K}] = \begin{bmatrix} \frac{1}{2} & 1 \end{bmatrix} \begin{bmatrix} 2k & -k \\ -k & k \end{bmatrix} \begin{bmatrix} \frac{1}{2} \\ 1 \end{bmatrix} = \frac{k}{2} \tag{13.16}$$

which agrees with the result given in eqs. (13.14).

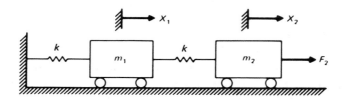

Fig. 13.1 Mathematical model for a two-degree-of-freedom system.

13.2   STATIC CONDENSATION APPLIED TO DYNAMIC PROBLEMS

In order to reduce the mass and the damping matrices, it is assumed that the same static relation between the secondary and primary degrees of freedom remains valid in the dynamic problem. Hence the same transformation based on static condensation for the reduction of the stiffness matrix is also used in reducing the mass and damping matrices. In general this method of reducing the dynamic problem is not exact and introduces errors in the results. The magnitude of these errors depends on the relative number of degrees of freedom reduced as well as on the specific selection of these degrees of freedom for a given structure.

We consider first the case in which the discretization of the mass results in a number of massless degrees of freedom. In this case it is only necessary to carry out the static condensation of the stiffness matrix and to delete from the mass matrix the rows and columns corresponding to the massless degrees of freedom. The Static Condensation Method in this case does not alter the original problem, and thus results in an equivalent eigenproblem without introducing any error.

In the general case, that is, the case involving the condensation of degrees of freedom to which the discretization process has allocated mass, the reduced mass and damping matrices are obtained using transformations analogous to eq. (13.10). Specifically, if $[M]$ is the mass matrix of the system, then the *reduced mass matrix* is given by

$$[\overline{M}] = [T]^T [M] [T] \tag{13.17}$$

where $[T]$ is the transformation matrix defined by eq. (13.9). Analogously, for a damped system, the *reduced damping matrix* is given by

$$[\overline{C}] = [T]^T [C] [T] \tag{13.18}$$

where $[C]$ is the damping matrix of the system.

This method of reducing the mass and damping matrices may be justified as follows: The potential elastic energy V and the kinetic energy KE of the structure may be written, respectively, as

$$V = \tfrac{1}{2} \{y\}^T [K] \{y\} \tag{13.19}$$

$$KE = \tfrac{1}{2} \{\dot{y}\}^T [M] \{\dot{y}\} \tag{13.20}$$

Analogously, the work δW_d done by the damping forces $F_d = [C] \{\dot{y}\}$ corresponding to displacement $\{\delta y\}$ may be expressed as

$$\delta W_d = \{\delta y\}^T [C] \{\dot{y}\} \tag{13.21}$$

Introduction of the transformation eq. (13.8) in the above equations results in

$$V = \frac{1}{2} \{y_p\}^T [T]^T [K] [T] \{y_p\} \tag{13.22}$$

$$KE = \frac{1}{2} \{\dot{y}_p\}^T [T]^T [M] [T] \{\dot{y}_p\} \tag{13.23}$$

$$\delta W_d = \{\delta y_p\}^T [T]^T [C] [T] \{\dot{y}_p\} \tag{13.24}$$

The respective substitution of $[\bar{K}]$, $[\bar{M}]$, and $[\bar{C}]$ from eqs. (13.10), (13.17), and (13.18) for the product of the three central matrices in eqs. (13.22), (13.23), and (13.24) yield

$$V = \frac{1}{2} \{y_p\}^T [\bar{K}] \{y_p\} \tag{13.25}$$

$$KE = \frac{1}{2} \{\dot{y}_p\}^T [\bar{M}] \{\dot{y}_p\} \tag{13.26}$$

$$\delta W_d = \{\delta y_p\}^T [\bar{C}] \{\dot{y}_p\} \tag{13.27}$$

These last three equations express the potential energy, the kinetic energy, and the virtual work of the damping forces in terms of the independent coordinates $\{y_p\}$. Hence the matrices $[\bar{K}]$, $[\bar{M}]$, and $[\bar{C}]$ may be interpreted, respectively, as the stiffness, mass, and damping matrices of the structure corresponding to the independent degrees of freedom $\{y_p\}$.

Example 13.2. Find the natural frequencies and modal shapes for the three-degree-of-freedom shear building shown in Fig. 13.2; then condense the first degree of freedom and compare the resulting values obtained for natural frequencies and modal shapes. The stiffness of each story and the mass at each floor level are indicated in the figure.

Solution:

Calculation of Natural Frequencies and Modal Shapes   The equation of motion in free vibration for this structure is given by eq. (9.3) with the force vector $\{F\} = \{0\}$, namely,

$$[M] \{\ddot{y}\} + [K] \{y\} = \{0\} \tag{a}$$

where the matrices $[M]$ and $[K]$ are given, respectively, by eqs. (9.4) and (9.5). Substituting the corresponding numerical values in eq. (a) yields

$$\begin{bmatrix} 25 & 0 & 0 \\ 0 & 50 & 0 \\ 0 & 0 & 100 \end{bmatrix} \begin{Bmatrix} \ddot{y}_1 \\ \ddot{y}_2 \\ \ddot{y}_3 \end{Bmatrix} + \begin{bmatrix} 40,000 & -10,000 & 0 \\ -10,000 & 20,000 & -10,000 \\ 0 & -10,000 & 10,000 \end{bmatrix} \begin{Bmatrix} y_1 \\ y_2 \\ y_3 \end{Bmatrix} = \begin{Bmatrix} 0 \\ 0 \\ 0 \end{Bmatrix}$$

Upon substituting $y_i = Y_i \sin \omega t$ and canceling the factor $\sin \omega t$, we obtain

$$\begin{bmatrix} 40,000 - 25\omega^2 & -10,000 & 0 \\ -10,000 & 20,000 - 50\omega^2 & -10,000 \\ 0 & -10,000 & 10,000 - 100\omega^2 \end{bmatrix} \begin{Bmatrix} Y_1 \\ Y_2 \\ Y_3 \end{Bmatrix} = \begin{Bmatrix} 0 \\ 0 \\ 0 \end{Bmatrix} \tag{b}$$

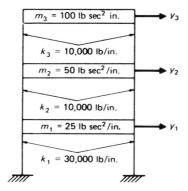

Fig. 13.2 Shear building for Example 13.2.

for which a nontrivial solution requires that the determinant of the coefficients be equal to zero, that is,

$$\begin{vmatrix} 40{,}000 - 25\omega^2 & -10{,}000 & 0 \\ -10{,}000 & 20{,}000 - 50\omega^2 & -10{,}000 \\ 0 & -10{,}000 & 10{,}000 - 100\omega^2 \end{vmatrix} = 0$$

The expansion of this determinant results in a third degree equation in terms of ω^2 having the following roots:

$$\omega_1^2 = 36.1$$

$$\omega_2^2 = 400.0$$

$$\omega_3^2 = 1664.0 \qquad\qquad (c)$$

The natural frequencies are calculated as $f = \omega/2\pi$, so that

$$f_1 = 0.96 \text{ cps}$$

$$f_2 = 3.18 \text{ cps}$$

$$f_3 = 264.8 \text{ cps}$$

The modal shapes are then determined by substituting in turn each of the values for the natural frequencies into eq. (b), deleting a redundant equation, and solving the remaining two equations for two of the unknowns in terms of the third. As we mentioned previously, in solving for these unknowns it is expedient to set the first nonzero unknown equal to one. Performing these operations, we obtain from eqs. (b) and (c) the following values for the modal shapes:

$$Y_{11} = 1.00, \quad Y_{12} = 1.00, \quad Y_{13} = 1.00$$

$$Y_{21} = 3.91, \quad Y_{22} = 3.00, \quad Y_{23} = 3.338$$

$$Y_{31} = 6.11, \quad Y_{32} = -1.00, \quad Y_{33} = -2.025$$

Condensation of Coordinate y_1   The stiffness matrix for this structure is

$$\begin{bmatrix} 40,000 & -10,000 & 0 \\ -10,000 & 20,000 & -10,000 \\ 0 & -10,000 & 10,000 \end{bmatrix}$$

Gauss elimination of the first unknown gives

$$\begin{bmatrix} 1 & -0.25 & 0 \\ 0 & 17,500 & -10,000 \\ 0 & -10,000 & 10,000 \end{bmatrix} \tag{d}$$

A comparison of eq. (d) with eq. (13.11) indicates that

$$[\bar{T}] = [0.25 \quad 0]$$

$$[\bar{K}] = \begin{bmatrix} 17,500 & -10,000 \\ -10,000 & 10,000 \end{bmatrix} \tag{e}$$

and from eq. (13.9),

$$[T] = \begin{bmatrix} 0.25 & 0 \\ 1 & 0 \\ 0 & 1 \end{bmatrix} \tag{f}$$

To check, we use eq. (13.10) to compute $[\bar{K}]$. Hence

$$[\bar{K}] = \begin{bmatrix} 0.25 & 1 & 0 \\ 0 & 0 & 1 \end{bmatrix} \begin{bmatrix} 40,000 & -10,000 & 0 \\ -10,000 & 20,000 & -10,000 \\ 0 & -10,000 & 10,000 \end{bmatrix} \begin{bmatrix} 0.25 & 0 \\ 1 & 0 \\ 0 & 1 \end{bmatrix}$$

$$[\bar{K}] = \begin{bmatrix} 17,500 & -10,000 \\ -10,000 & 10,000 \end{bmatrix}$$

which checks with eqs. (e). The reduced mass matrix is calculated by substituting matrix $[T]$ and its transpose from eq. (f) into eq. (13.17), so that

$$[\bar{M}] = \begin{bmatrix} 0.25 & 1 & 0 \\ 0 & 0 & 1 \end{bmatrix} \begin{bmatrix} 25 & 0 & 0 \\ 0 & 50 & 0 \\ 0 & 0 & 100 \end{bmatrix} \begin{bmatrix} 0.25 & 0 \\ 1 & 0 \\ 0 & 1 \end{bmatrix}$$

which results in

$$[\bar{M}] = \begin{bmatrix} 51.6 & 0 \\ 0 & 100 \end{bmatrix}$$

The condensed dynamic problem is then

$$\begin{bmatrix} 51.6 & 0 \\ 0 & 100 \end{bmatrix} \begin{Bmatrix} \ddot{y}_2 \\ \ddot{y}_3 \end{Bmatrix} + \begin{bmatrix} 17,500 & -10,000 \\ -10,000 & 10,000 \end{bmatrix} \begin{Bmatrix} y_2 \\ y_3 \end{Bmatrix} = \begin{Bmatrix} 0 \\ 0 \end{Bmatrix}$$

The natural frequencies and modal shapes are determined from the solution of the following reduced eigenproblem:

$$\begin{bmatrix} 17,500 - 51.6\omega^2 & -10,000 \\ -10,000 & 10,000 - 100\omega^2 \end{bmatrix} \begin{Bmatrix} Y_2 \\ Y_3 \end{Bmatrix} = \begin{Bmatrix} 0 \\ 0 \end{Bmatrix} \tag{g}$$

Equating to zero the determinant of the coefficient matrix in eq. (g) and solving the resulting quadratic equation in ω^2 gives

$$\omega_1^2 = 36.1$$

$$\omega_2^2 = 403.3 \tag{h}$$

from which

$$f_1 = \sqrt{36.1}/2\pi = 0.95 \text{ cps}$$

$$f_2 = \sqrt{403.3}/2\pi = 3.20 \text{ cps}$$

Corresponding modal shapes are obtained from eq. (g) after substituting the numerical values for ω_1^2 or ω_2^2 and solving the first equation for Y_3 with $Y_2 = 1$. We then obtain

$$Y_{21} = 1.00, \quad Y_{22} = 1.00$$

$$Y_{31} = 1.56, \quad Y_{32} = -0.33$$

Application of eq. (13.8) for the first mode gives

$$\begin{Bmatrix} Y_1 \\ Y_2 \\ Y_3 \end{Bmatrix}_1 = \begin{bmatrix} 0.25 & 0 \\ 1 & 0 \\ 0 & 1 \end{bmatrix} \begin{Bmatrix} 1.00 \\ 1.56 \end{Bmatrix} = \begin{Bmatrix} 0.25 \\ 1.00 \\ 1.56 \end{Bmatrix}$$

or, after normalizing so that the first component is 1,

$$Y_{11} = 1.00, \quad Y_{21} = 4.00, \quad Y_{31} = 6.24$$

and analogously for the second mode

$$Y_{12} = 1.00, \quad Y_{22} = 4.00, \quad Y_{32} = -1.32$$

For this system of only three degrees of freedom, the reduction of one coordinate gives natural frequencies that compare rather well for the first two modes [eqs. (h) and (c)]. However, experience shows that static condensation may produce large errors in the calculation of eigenvalues and eigenvectors obtained from the reduced system. A general recommendation given by users of this method is to assume that the static condensation process results in an eigenproblem which provides acceptable approximations of only about a third of the calculated eigenvalues (natural frequencies) and eigenvectors (modal shapes).

Example 13.3.   Figure 13.3 shows a uniform four-story shear building. For this structure, determine the following: (a) the natural frequencies and corresponding modal shapes as a four-degree-of-freedom system, (b) the natural frequencies and modal shapes after static condensation of coordinates y_1 and y_3.

Solution:

(a) Natural Frequencies and Modal Shapes as a Four-Degree-of-Freedom System   The stiffness and the mass matrices for this structure are respectively

$$[K] = 327.35 \begin{bmatrix} 2 & -1 & 0 & 0 \\ -1 & 2 & -1 & 0 \\ 0 & -1 & 2 & -1 \\ 0 & 0 & -1 & 1 \end{bmatrix} \qquad \text{(a)}$$

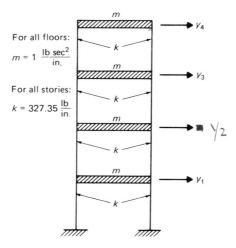

For all floors:

$m = 1 \dfrac{\text{lb sec}^2}{\text{in.}}$

For all stories:

$k = 327.35 \dfrac{\text{lb}}{\text{in.}}$

Fig. 13.3
Uniform four-story shear building for Example 13.3.

and

$$[M] = \begin{bmatrix} 1 & 0 & 0 & 0 \\ 0 & 1 & 0 & 0 \\ 0 & 0 & 1 & 0 \\ 0 & 0 & 0 & 1 \end{bmatrix}$$

(b)

Substituting eqs. (a) and (b) into eq. (10.3) and solving the corresponding eigenvalue problem (using Program 8) yields

$$\omega_1^2 = 39.48, \quad \omega_2^2 = 327.35, \quad \omega_3^2 = 768.3, \quad \text{and} \quad \omega_4^2 = 1156.00$$

corresponding to the natural frequencies

$$f_1 = \frac{\omega_1}{2\pi} = 1.00 \text{ cps}$$

$$f_2 = \frac{\omega_2}{2\pi} = 2.88 \text{ cps}$$

(c)

$$f_3 = \frac{\omega_3}{2\pi} = 4.41 \text{ cps}$$

$$f_4 = \frac{\omega_4}{2\pi} = 5.41 \text{ cps}$$

and the normalized modal matrix (see Section 10.2)

$$[\Phi] = \begin{bmatrix} 0.2280 & 0.5774 & -0.6565 & 0.4285 \\ 0.4285 & 0.5774 & 0.2280 & -0.6565 \\ 0.5774 & 0 & 0.5774 & 0.5774 \\ 0.6565 & -0.5774 & -0.4285 & -0.2280 \end{bmatrix}$$

(d)

(b) Natural Frequencies and Modal Shapes after Reduction to Two-Degree-of-Freedom System   To reduce coordinates y_1 and y_3, we first, for convenience, rearrange the stiffness matrix in eq. (a) to have the coordinates in order y_1, y_3, y_2, y_4:

$$[K] = 327.35 \begin{bmatrix} 2 & 0 & -1 & 0 \\ 0 & 2 & -1 & -1 \\ -1 & -1 & 2 & 0 \\ 0 & -1 & 0 & 1 \end{bmatrix}$$

(e)

Applying Gauss-Jordan elimination to the first two rows of the matrix in eq. (e) results in

$$\begin{bmatrix} 1 & 0 & \vdots & -0.5 & 0 \\ 0 & 1 & \vdots & -0.5 & -0.5 \\ \cdots & \cdots & \cdots & \cdots & \cdots \\ 0 & 0 & \vdots & 327.35 & -163.70 \\ 0 & 0 & \vdots & -163.70 & 163.70 \end{bmatrix} \qquad \text{(f)}$$

A comparison of eq. (f) with eq. (13.11) reveals that

$$[\bar{T}] = \begin{bmatrix} 0.5 & 0 \\ 0.5 & 0.5 \end{bmatrix} \qquad \text{(g)}$$

and

$$[\bar{K}] = \begin{bmatrix} 327.35 & -163.70 \\ -163.70 & 163.70 \end{bmatrix} \qquad \text{(h)}$$

Use of eq. (13.9) gives

$$[T] = \begin{bmatrix} 0.5 & 0 \\ 0.5 & 0.5 \\ 1 & 0 \\ 0 & 1 \end{bmatrix}$$

The reduced mass matrix can now be calculated by eq. (13.17) as

$$[\bar{M}] = [T]^T [M] [T] = \begin{bmatrix} 1.5 & 0.25 \\ 0.25 & 1.25 \end{bmatrix} \qquad \text{(i)}$$

The condensed eigenproblem is then

$$\begin{bmatrix} 327.35 - 1.5\omega^2 & -163.70 - 0.25\omega^2 \\ -163.70 - 0.25\omega^2 & 163.70 - 1.25\omega^2 \end{bmatrix} \begin{Bmatrix} Y_2 \\ Y_4 \end{Bmatrix} = \begin{Bmatrix} 0 \\ 0 \end{Bmatrix} \qquad \text{(j)}$$

and its solution is

$$\omega_1^2 = 40.39, \quad \omega_2^2 = 365.98 \qquad \text{(k)}$$

$$[Y]_p = \begin{bmatrix} 0.4380 & 0.7056 \\ 0.6723 & -0.6128 \end{bmatrix} \qquad \text{(l)}$$

where $[Y]_p$ is the modal matrix corresponding to the primary degrees of freedom. The eigenvectors for the four-degree-of-freedom system are calculated for the first mode by eq. (13.8) as

$$\begin{Bmatrix} Y_1 \\ Y_3 \\ Y_2 \\ Y_4 \end{Bmatrix}_1 = \begin{bmatrix} 0.5 & 0 \\ 0.5 & 0.5 \\ 1 & 0 \\ 0 & 1 \end{bmatrix} \begin{Bmatrix} 0.4380 \\ 0.6723 \end{Bmatrix} = \begin{Bmatrix} 0.2190 \\ 0.5552 \\ 0.4380 \\ 0.6723 \end{Bmatrix}$$

or

$$\begin{Bmatrix} Y_1 \\ Y_2 \\ Y_3 \\ Y_4 \end{Bmatrix}_1 = \begin{Bmatrix} 0.2190 \\ 0.4380 \\ 0.5552 \\ 0.6723 \end{Bmatrix} \tag{m}$$

and for the second mode

$$\begin{Bmatrix} Y_1 \\ Y_3 \\ Y_2 \\ Y_4 \end{Bmatrix}_2 = \begin{bmatrix} 0.5 & 0 \\ 0.5 & 0.5 \\ 1 & 0 \\ 0 & 1 \end{bmatrix} \begin{Bmatrix} 0.7056 \\ -0.6128 \end{Bmatrix} = \begin{Bmatrix} 0.3528 \\ 0.0464 \\ 0.7056 \\ -0.6128 \end{Bmatrix}$$

or

$$\begin{Bmatrix} Y_1 \\ Y_2 \\ Y_3 \\ Y_4 \end{Bmatrix}_2 = \begin{Bmatrix} 0.3528 \\ 0.7056 \\ 0.0464 \\ -0.6128 \end{Bmatrix} \tag{n}$$

Example 13.4.   The shear building of Example 13.3 is subjected to an earthquake motion at its foundation. For design purposes, use the response spectrum of Fig. 8.10 (Section 8.4) and determine the maximum horizontal displacements of the structure at the level of the floors.

Solution: The participation factor of a shear building with N stories is given by eq. (11.40) as

$$\text{sum over} \qquad \Gamma_i = - \sum_{j=1}^{N} (m_j \phi_{ji}) \tag{a}$$
$$\text{floors}$$

where m_j is the mass at jth floor and ϕ_{ji} the jth element of the normalized ith eigenvector.

(a) Response Considering Four Degrees of Freedom   The substitution into eq. (a) of the corresponding numerical results from Example 13.3 gives

$$\Gamma_1 = -1.890, \quad \Gamma_2 = -0.5775, \quad \Gamma_3 = -0.2797, \quad \text{and} \quad \Gamma_4 = -0.1213 \tag{b}$$

The spectral displacements corresponding to the values of the natural frequencies of this building [eq. (c) of Example 13.3] are obtained from the response spectrum, Fig. 8.10, as

$$S_{D1} = 14.32, \quad S_{D2} = 3.240, \quad S_{D3} = 1.433, \quad \text{and} \quad S_{D4} = 0.969 \qquad \text{(c)}$$

The maximum displacements at the floor levels relative to the displacement at the base of the building are calculated using eq. (11.41), namely,

sum over modes $\qquad u_{i\max} = \sqrt{\sum_{j=1}^{N} (\Gamma_j S_{Dj} \phi_{ij})^2} \qquad$ (d)

to obtain

$$u_{1\max} = 6.274 \text{ in}, \quad u_{2\max} = 11.65 \text{ in}, \quad u_{3\max} = 15.64 \text{ in}, \quad \text{and}$$
$$u_{4\max} = 17.81 \text{ in}$$

(b) Response Considering the System Reduced to Two Degrees of Freedom The natural frequencies, calculated from eq. (k) in Example 13.3, are

$$f_1 = \sqrt{40.39}/2\pi = 1.011 \text{ cps}$$

and

$$f_2 = \sqrt{365.98}/2\pi = 3.044 \text{ cps} \qquad \text{(e)}$$

Upon introducing, into eq. (a), the corresponding eigenvectors given in eqs. (m) and (n) of Example 13.3, we obtain the participation factors

$$\Gamma_1 = -1.884$$

$$\Gamma_2 = -0.492$$

The values of spectral displacements corresponding to the frequencies in eq. (e) can be read from Fig. 8.10:

$$S_{D1} = 14.16$$

$$S_{D2} = 2.913$$

Use of eq. (d) gives the relative maximum displacements at the level of the floors as

$$u_{1\max} = \sqrt{(1.884 \times 14.16 \times 0.2190)^2 + (0.4920 \times 2.913 \times 0.3528)^2} = 5.864 \text{ in}$$

$$u_{2\max} = \sqrt{(1.884 \times 14.16 \times 0.4380)^2 + (0.4920 \times 2.913 \times 0.7056)^2} = 11.73 \text{ in}$$

$$u_{3\max} = \sqrt{(1.884 \times 14.16 \times 0.5552)^2 + (0.4920 \times 2.913 \times 0.0464)^2} = 14.81 \text{ in}$$

$$u_{4\max} = \sqrt{(1.884 \times 14.16 \times 0.6723)^2 + (0.4920 \times 2.913 \times 0.6128)^2} = 17.97 \text{ in.}$$

13.3 DYNAMIC CONDENSATION

A method of reduction which may be considered an extension of the Static Condensation Method has recently been proposed (Paz 1984). The algorithm for the proposed method starts by assigning an approximate value (e.g., zero) to the first eigenvalue ω_1^2, applying dynamic condensation to the dynamic matrix of the system $[D_1] = [K] - \omega_1^2[M]$, and then solving the reduced eigenproblem to determine the first and second eigenvalues ω_1^2 and ω_2^2. Next, dynamic condensation is applied to the dynamic matrix $[D_2] = [K] - \omega_2^2[M]$ to reduce the problem and calculate the second and third eigenvalues, ω_2^2 and ω_3^2. The process continues in this manner, with one virtually exact eigenvalue and an approximation of the next order eigenvalue calculated at each step.

The Dynamic Condensation Method requires neither matrix inversion nor series expansion. To demonstrate this fact, consider the eigenvalue problem of a discrete structural system for which it is desired to reduce the secondary degrees of freedom $\{y_s\}$ and retain the primary degrees of freedom $\{y_p\}$. In this case, the equations of free motion may be written in partitioned matrix form as

$$\begin{bmatrix} [M_{ss}] & [M_{sp}] \\ [M_{ps}] & [M_{pp}] \end{bmatrix} \begin{Bmatrix} \{\ddot{y}_s\} \\ \{\ddot{y}_p\} \end{Bmatrix} + \begin{bmatrix} [K_{ss}] & [K_{sp}] \\ [K_{ps}] & [K_{pp}] \end{bmatrix} \begin{Bmatrix} \{y_s\} \\ \{y_p\} \end{Bmatrix} = \begin{Bmatrix} \{0\} \\ \{0\} \end{Bmatrix} \quad (13.28)$$

The substitution of $\{y\} = \{Y\} \sin \omega_i t$ in eq. (13.28) results in the generalized eigenproblem

$$\begin{bmatrix} [K_{ss}] - \omega_i^2[M_{ss}] & [K_{sp}] - \omega_i^2[M_{sp}] \\ [K_{ps}] - \omega_i^2[M_{ps}] & [K_{pp}] - \omega_i^2[M_{pp}] \end{bmatrix} \begin{Bmatrix} \{Y_s\} \\ \{Y_p\} \end{Bmatrix} = \begin{Bmatrix} \{0\} \\ \{0\} \end{Bmatrix} \quad (13.29)$$

where ω_i^2 is the approximation of the ith eigenvalue which was calculated in the preceding step of the process. To start the process one takes an approximate or zero value for the first eigenvalue ω_1^2.

The following three steps are executed to calculate the ith eigenvalue ω_i^2 and the corresponding eigenvector $\{Y\}_i$ as well as an approximation of the eigenvalue of the next order ω_{i+1}^2:

Step 1. The approximation of ω_i^2 is introduced in eq. (13.29); Gauss–Jordan elimination of the secondary coordinates $\{Y_s\}$ is then used to reduce eq. (13.29) to

$$\begin{bmatrix} [I] & -[\bar{T}_i] \\ [0] & [\bar{D}_i] \end{bmatrix} \begin{Bmatrix} \{Y_s\} \\ \{Y_p\} \end{Bmatrix} = \begin{Bmatrix} \{0\} \\ \{0\} \end{Bmatrix} \quad (13.30)$$

The first equation in eq. (13.30) can be written as

$$\{Y_s\} = [\bar{T}_i] \{Y_p\} \quad (13.31)$$

Consequently, the ith modal shape $\{Y\}_i$ can be expressed as

$$\{Y\}_i = [T_i] \{Y_p\} \quad (13.32)$$

where

$$[T_i] = \left[\frac{[\bar{T_i}]}{[I]}\right] \quad \text{and} \quad \{Y\}_i = \left\{\frac{\{Y_s\}}{\{Y_p\}}\right\} \tag{13.33}$$

Step 2. The reduced mass matrix $[\bar{M_i}]$ and the reduced stiffness matrix $[\bar{K_i}]$ are calculated as

$$[\bar{M_i}] = [T_i]^T [M] [T_i] \tag{13.34}$$

and

$$[\bar{K_i}] = [\bar{D_i}] + \omega_i^2 [\bar{M_i}] \tag{13.35}$$

where the transformation matrix $[T_i]$ is given by eq. (13.33) and the reduced dynamic matrix $[\bar{D_i}]$ is defined in eq. (13.30).

Step 3. The reduced eigenproblem

$$[[\bar{K_i}] - \omega^2 [\bar{M_i}]] \{Y_p\} = \{0\} \tag{13.36}$$

is solved to obtain an improved eigenvalue ω_i^2, its corresponding eigenvector $\{Y_p\}_i$, and also an approximation for the next order eigenvalue ω_{i+1}^2.

This three-step process may be applied iteratively. That is, the value of ω_i^2 obtained in step 3 may be used as an improved approximate value in step 1 to obtain a further improved value of ω_i^2 in step 3. Experience has shown that one or two such iterations will produce virtually exact eigensolutions. Once an eigenvector $\{Y_p\}_i$ for the reduced system is found, the ith modal shape of the system is determined as $\{Y\}_i = [T_i] \{Y_p\}_i$ using eq. (13.32).

Example 13.5.   Repeat Example 13.3 of Section 13.2 using the Dynamic Condensation Method.

Solution: The stiffness matrix and the mass matrix with the coordinates in the order y_1, y_3, y_2, y_4 are given, respectively, by eqs. (e) and (b) of Example 13.3. Substitution of these matrices into eq. (13.29) results in the dynamic matrix for the system:

$$[D] = \left[\begin{array}{cc:cc} 654.70 - \omega_i^2 & 0 & -327.35 & 0 \\ 0 & 654.70 - \omega_i^2 & -327.35 & -327.35 \\ \hdashline -327.35 & -327.35 & 654.70 - \omega_i^2 & 0 \\ 0 & -327.35 & 0 & 327.35 - \omega_i^2 \end{array}\right] \tag{a}$$

Step 1.   Assuming we have no initial approximation of ω_i^2, we start step 1 by setting $\omega_i^2 = 0$ and substituting this value into eq. (a):

$$[D_1] = \left[\begin{array}{cc:cc} 654.70 & 0 & -327.35 & 0 \\ 0 & 654.70 & -327.35 & -327.35 \\ \hdashline -327.35 & -327.35 & 654.70 & 0 \\ 0 & -327.35 & 0 & 327.35 \end{array}\right] \tag{b}$$

Application of the Gauss–Jordan elimination process to the first two rows gives

$$\begin{bmatrix} 1 & 0 & \vdots & -0.5 & 0.0 \\ 0 & 1 & \vdots & -0.5 & -0.5 \\ \hdashline 0 & 0 & \vdots & 327.35 & -163.67 \\ 0 & 0 & \vdots & -163.67 & 163.67 \end{bmatrix}$$

from which, by eqs. (13.30) and (13.33)

$$[T_1] = \begin{bmatrix} 0.5 & 0.0 \\ 0.5 & 0.5 \\ 1 & 0 \\ 0 & 1 \end{bmatrix}$$

and

$$[\bar{D}_1] = \begin{bmatrix} 327.35 & -163.67 \\ -163.67 & 163.67 \end{bmatrix}$$

Step 2. The reduced mass and stiffness matrices, eqs. (13.34) and (13.35), are

$$[\bar{M}_1] = [T_1]^T [M] [T_1] = \begin{bmatrix} 1.5 & 0.25 \\ 0.25 & 1.25 \end{bmatrix}$$

and

$$[\bar{K}_1] = [\bar{D}_1] + \omega_1^2 [\bar{M}_1] = \begin{bmatrix} 327.35 & -163.67 \\ -163.67 & 163.67 \end{bmatrix}$$

Step 3. The solution of the reduced eigenproblem $[[\bar{K}_1] - \omega^2 [\bar{M}_1]] \{Y_p\} = \{0\}$ yields

$$\omega_1^2 = 40.39 \quad \text{and} \quad \omega_2^2 = 365.98$$

These values for ω_1^2 and ω_2^2 may be improved by iterating the calculations, that is, by introducing $\omega_1^2 = 40.39$ into eq. (13.29). This substitution results in

$$[D_1] = \begin{bmatrix} 614.31 & 0 & \vdots & -327.35 & 0 \\ 0 & 614.31 & \vdots & -327.35 & -327.35 \\ \hdashline -327.35 & -327.35 & \vdots & 614.31 & 0 \\ 0 & -327.35 & \vdots & 0 & 286.96 \end{bmatrix}$$

Application of Gauss–Jordan elimination to the first two rows gives

$$\begin{bmatrix} 1 & 0 & \vdots & -0.533 & 0.0 \\ 0 & 1 & \vdots & -0.533 & -0.533 \\ \hdashline 0 & 0 & \vdots & 265.44 & -174.44 \\ 0 & 0 & \vdots & -174.44 & 112.53 \end{bmatrix}$$

from which

$$[T_1] = \begin{bmatrix} 0.533 & 0.0 \\ 0.533 & 0.533 \\ 1 & 0 \\ 0 & 1 \end{bmatrix}$$

and

$$[\bar{D}_1] = \begin{bmatrix} 265.44 & -174.44 \\ -174.44 & 112.53 \end{bmatrix}$$

The reduced mass and stiffness matrices are then

$$[\bar{M}_1] = [T_1]^T [M] [T_1] = \begin{bmatrix} 1.568 & 0.284 \\ 0.284 & 1.284 \end{bmatrix}$$

and

$$[\bar{K}_1] = \bar{D}_1 + \omega_1^2 [\bar{M}_1] = \begin{bmatrix} 328.76 & -162.97 \\ -162.67 & 164.39 \end{bmatrix}$$

The solution of the reduced eigenproblem,

$$[[\bar{K}_1] - \omega^2[\bar{M}_1]] \{Y_p\} = \{0\}$$

yields the eigenvalues

$$\omega_1^2 = 39.48, \quad \omega_2^2 = 360.21 \tag{c}$$

and corresponding eigenvectors

$$\{Y_p\}_1 = \begin{bmatrix} 0.4283 \\ 0.6562 \end{bmatrix}, \quad \{Y_p\}_2 = \begin{bmatrix} 0.6935 \\ -0.6171 \end{bmatrix} \tag{d}$$

The same process is now applied to the second mode, starting by substituting into eq. (13.29) the approximate eigenvalue $\omega_2^2 = 360.21$ calculated for the second mode in eq. (c). In this case we obtain

$$[D_2] = \left[\begin{array}{cc|cc} 294.49 & 0 & -327.35 & 0 \\ 0 & 294.49 & -327.35 & -327.35 \\ \hline -327.35 & -327.35 & 294.49 & 0 \\ 0 & -327.35 & 0 & -32.86 \end{array} \right]$$

Gauss–Jordan elimination of the first two rows yields

$$\left[\begin{array}{cc|cc} 1 & 0 & -1.112 & 0.0 \\ 0 & 1 & -1.112 & -1.112 \\ \hline 0 & 0 & -433.27 & -363.88 \\ 0 & 0 & -363.88 & -396.74 \end{array} \right]$$

from which

$$[T_2] = \begin{bmatrix} 1.112 & 0.0 \\ 1.112 & 1.112 \\ \hline 1 & 0 \\ 0 & 1 \end{bmatrix}$$

and

$$[\bar{D}_2] = \begin{bmatrix} -433.27 & -363.88 \\ -363.88 & -396.74 \end{bmatrix}$$

and the reduced mass and stiffness matrices are

$$[\bar{M}_2] = [T_2]^T [M] [T_2] = \begin{bmatrix} 3.471 & 1.236 \\ 1.236 & 2.236 \end{bmatrix}$$

$$[\bar{K}_2] = [\bar{D}_2] + \omega_2^2 [\bar{M}_2] = \begin{bmatrix} 817.12 & 81.21 \\ 81.21 & 408.56 \end{bmatrix}$$

The solution is of the reduced eigenproblem $[[\bar{K}_2] - \omega^2[\bar{M}_2]] \{Y_p\} = \{0\}$ yields for the second mode

$$\omega_2^2 = 328.61$$

An iteration is performed by introducing $\omega_2^2 = 328.61$ into eq. (13.29) to obtain the following:

$$[D_2] = \begin{bmatrix} 326.09 & 0 & -327.35 & 0 \\ 0 & 326.09 & -327.35 & -327.35 \\ \hline -327.35 & -327.35 & 326.09 & 0 \\ 0 & -327.35 & 0 & -1.26 \end{bmatrix}$$

Applying Gauss–Jordan elimination to the first two rows yields

$$\begin{bmatrix} 1 & 0 & -1.004 & 0.0 \\ 0 & 1 & -1.004 & -1.004 \\ \hline 0 & 0 & -331.14 & -328.62 \\ 0 & 0 & -328.62 & -329.88 \end{bmatrix}$$

from which

$$[T_2] = \begin{bmatrix} 1.004 & 0.0 \\ 1.004 & 1.004 \\ \hline 1 & 0 \\ 0 & 1 \end{bmatrix}$$

and

$$[\overline{D}_2] = \begin{bmatrix} -331.14 & -328.62 \\ -328.62 & -329.88 \end{bmatrix}$$

and the reduced mass and stiffness matrices are

$$[\overline{M}_2] = [T_2]^T [M] [T_2] = \begin{bmatrix} 3.015 & 1.008 \\ 1.008 & 2.008 \end{bmatrix}$$

and

$$[\overline{K}_2] = [\overline{D}_2] + \omega_2^2 [\overline{M}_2] = \begin{bmatrix} 659.78 & 2.54 \\ 2.54 & 329.89 \end{bmatrix}$$

The solution of the reduced eigenproblem $[[\overline{K}_2] - \omega^2 [\overline{M}_2]] \{Y_p\} = \{0\}$ now gives for the second mode

$$\omega_2^2 = 327.35, \qquad \{Y_p\} = \begin{Bmatrix} 0.5766 \\ -0.5766 \end{Bmatrix} \qquad \text{(e)}$$

Therefore, from eqs. (c), (d), and (e) we have obtained for the first two eigenvalues

$$\omega_1^2 = 39.48 \qquad \text{and} \qquad \omega_2^2 = 327.35 \qquad \text{(f)}$$

and corresponding eigenvectors

$$\{Y_p\}_1 = \begin{Bmatrix} 0.4283 \\ 0.6562 \end{Bmatrix}, \qquad \{Y_p\}_2 = \begin{Bmatrix} 0.5766 \\ -0.5766 \end{Bmatrix} \qquad \text{(g)}$$

The eigenvectors of the system are then computed using eq. (13.32) as follows:

$$\begin{Bmatrix} Y_1 \\ Y_3 \\ Y_2 \\ Y_4 \end{Bmatrix}_1 = \begin{bmatrix} 0.533 & 0.0 \\ 0.533 & 0.533 \\ 1 & 0 \\ 0 & 1 \end{bmatrix} \begin{Bmatrix} 0.4283 \\ 0.6562 \end{Bmatrix} = \begin{Bmatrix} 0.2283 \\ 0.5780 \\ 0.4283 \\ 0.6562 \end{Bmatrix}$$

Hence

$$\begin{Bmatrix} Y_1 \\ Y_2 \\ Y_3 \\ Y_4 \end{Bmatrix}_1 = \begin{Bmatrix} 0.2283 \\ 0.4283 \\ 0.5780 \\ 0.6562 \end{Bmatrix} \qquad \text{(h)}$$

and

$$\begin{Bmatrix} Y_1 \\ Y_3 \\ Y_2 \\ Y_4 \end{Bmatrix}_2 = \begin{bmatrix} 1.004 & 0.0 \\ 1.004 & 1.004 \\ 1 & 0 \\ 0 & 1 \end{bmatrix} \begin{Bmatrix} 0.5766 \\ -0.5766 \end{Bmatrix} = \begin{Bmatrix} 0.5789 \\ 0.0 \\ 0.5766 \\ -0.5766 \end{Bmatrix}$$

Hence

$$\begin{Bmatrix} Y_1 \\ Y_2 \\ Y_3 \\ Y_4 \end{Bmatrix}_2 = \begin{Bmatrix} 0.5789 \\ 0.5766 \\ 0.0 \\ -0.5766 \end{Bmatrix} \qquad\qquad (i)$$

The eigenvalues and eigenvectors [eqs. (f), (h), and (i)] calculated for this example using dynamic condensation are virtually identical to the exact solution determined in eqs. (c) and (d) of Example 13.3.

It should be noted that normalization of the eigenvectors is not needed in eqs. (h) and (i) if the reduced vectors are normalized with respect to the reduced mass of the system, that is, if a reduced eigenvector $\{Y_p\}$ satisfies the normalizing equation

$$\{Y_p\}^T [\overline{M}] \{Y_p\} = 1$$

then by eq. (13.34)

$$\{Y_p\}^T [T]^T [M] [T] \{Y_p\} = 1$$

and since by eq. (13.32)

$$\{Y\}_i = [T] \{Y_p\}$$

we see that

$$\{Y\}_i^T [M] \{Y\}_i = 1$$

thus demonstrating that the eigenvector $\{Y\}_i$ is normalized with respect to the mass matrix of the system $[M]$ if $\{Y_p\}$ is normalized with respect to $[\overline{M}]$.

13.4 MODIFIED DYNAMIC CONDENSATION

The dynamic condensation method essentially requires the application of elementary operations, as is routinely done to solve a linear system of algebraic equations, using the Gauss–Jordan elimination process. The elementary operations are required to transform eq. (13.29) to the form given by eq. (13.30). However, the method also requires the calculation of the reduced mass matrix by eq. (13.34). This last equation involves the multiplication of three matrices of dimensions equal to the total number of coordinates in the system.

Thus, for a system defined with many degrees of freedom, the calculation of the reduced mass matrix $[\bar{M}]$ requires a large number of numerical operations. A modification (Paz 1989), recently proposed, obviates such large number of numerical operations. This modification consists of calculating the reduced stiffness matrix $[\bar{K}]$ only once by simple elimination of s displacements in eq. (13.29) after setting $\omega^2 = 0$, thus making unnecessary the repeated calculation of $[\bar{K}]$ for each mode using eq. (13.35). Furthermore, it also eliminates the time consumed in calculating the reduced mass matrix $[\bar{M}]$ using eq. (13.34). In the modified method, the reduced mass matrix for any mode i is calculated from eq. (13.35) as

$$[\bar{M}_i] = \frac{1}{\omega_i^2} [[\bar{K}] - [\bar{D}_i]] \tag{13.37}$$

where $[\bar{K}]$ = the reduced stiffness matrix, already calculated, and $[\bar{D}_i]$ = dynamic matrix given in the partitioned matrix of eq. (13.30).

As can be seen, the modified algorithm essentially requires, for each eigenvalue calculated, only the application of the Gauss–Jordan process to eliminate s unknowns in a linear system of equations such as the system in eq. (13.29).

Example 13.6.   Repeat Example 13.5 using the modified dynamic condensation method.

Solution: The initial calculations of the modified method are the same as those in Example 13.5. Thus, from Example 13.5 we have

$$[\bar{K}_1] = \begin{bmatrix} 327.35 & -163.67 \\ -163.67 & 163.67 \end{bmatrix} \tag{a}$$

$$[\bar{M}_1] = \begin{bmatrix} 1.5 & 0.25 \\ 0.25 & 1.25 \end{bmatrix} \tag{b}$$

$$\omega_1^2 = 40.39, \quad \omega_2^2 = 365.98 \tag{c}$$

$$[\bar{D}_1] = \begin{bmatrix} 265.44 & -174.44 \\ -174.44 & 112.53 \end{bmatrix} \tag{d}$$

and

$$[\bar{T}_1] = \begin{bmatrix} 0.533 & 0 \\ 0.533 & 0.533 \end{bmatrix} \tag{e}$$

Now, the reduced mass matrix $[\bar{M}_1]$ is calculated from eq. (13.37), after substitution in this equation of $[\bar{K}_1]$ from eq. (a) and $[\bar{D}_1]$ from eq. (d), as

$$[\bar{M}_1] = \frac{1}{\omega_1^2} [[\bar{K}_1] - [\bar{D}_1]] = \begin{bmatrix} 1.530 & 0.267 \\ 0.267 & 1.266 \end{bmatrix}$$ (f)

Then the reduced stiffness and mass matrix from eqs. (a) and (f) are used to solve the reduced eigenproblem

$$[[\bar{K}_1] - \omega^2 [\bar{M}_1]] \{Y_p\} = \{0\}$$

to obtain eigenvalues

$$\omega_1^2 = 39.46, \quad \omega_2^2 = 363.67$$ (g)

and the eigenvector for the first mode

$$\begin{Bmatrix} Y_2 \\ Y_4 \end{Bmatrix} = \begin{Bmatrix} 0.43359 \\ 0.66424 \end{Bmatrix}$$ (h)

The eigenvector for the first mode, in terms of the original four coordinates, is then obtained from eq. (13.31) as

$$\begin{Bmatrix} Y_1 \\ Y_3 \end{Bmatrix} = \begin{bmatrix} 0.533 & 0 \\ 0.533 & 0.533 \end{bmatrix} \begin{Bmatrix} 0.43359 \\ 0.66424 \end{Bmatrix} = \begin{Bmatrix} 0.23110 \\ 0.58514 \end{Bmatrix}$$ (i)

The combination of eqs. (h) and (i) gives the eigenvector for the first mode as

$$\begin{Bmatrix} Y_1 \\ Y_2 \\ Y_3 \\ Y_4 \end{Bmatrix}_1 = \begin{Bmatrix} 0.23110 \\ 0.43359 \\ 0.58514 \\ 0.66424 \end{Bmatrix}$$ (j)

Analogously, for the second mode, we substitute $\omega_2^2 = 363.67$ from eq. (g) into eq. (a) of Example 13.5, to obtain the following matrices after reducing the first two coordinates:

$$[\bar{D}_2] = \begin{bmatrix} -445.43 & -368.20 \\ -368.20 & -404.54 \end{bmatrix}$$

$$[\bar{T}_2] = \begin{bmatrix} 1.1248 & 0 \\ 1.1248 & 1.1248 \end{bmatrix}$$

The reduced mass matrix $[\bar{M}_2]$ is then calculated from eq. (13.37) as

$$\bar{M}_2 = \frac{1}{363.67} \left\{ \begin{bmatrix} 327.35 & -163.67 \\ -163.67 & 163.67 \end{bmatrix} - \begin{bmatrix} -445.43 & -368.20 \\ 368.20 & -404.54 \end{bmatrix} \right\}$$

$$[\bar{M}_2] = \begin{bmatrix} 2.1250 & 0.5624 \\ 0.5624 & 1.5624 \end{bmatrix}$$

Then, for the second mode, the solution of the corresponding reduced eigenproblem gives

$$\omega_2^2 = 319.41, \qquad \begin{Bmatrix} Y_2 \\ Y_4 \end{Bmatrix} = \begin{Bmatrix} 0.61894 \\ -0.63352 \end{Bmatrix}$$

$$\begin{Bmatrix} Y_1 \\ Y_3 \end{Bmatrix} = \begin{bmatrix} 1.1248 & 0 \\ 1.1248 & 1.1248 \end{bmatrix} \begin{Bmatrix} 0.61894 \\ -0.63352 \end{Bmatrix} = \begin{Bmatrix} 0.69618 \\ -0.01640 \end{Bmatrix}$$

and

$$\begin{Bmatrix} Y_1 \\ Y_2 \\ Y_3 \\ Y_4 \end{Bmatrix}_2 = \begin{Bmatrix} 0.69618 \\ 0.61894 \\ -0.01640 \\ -0.63352 \end{Bmatrix}$$

The results obtained for this example using the modified method, although sufficiently approximate, are not as close to the exact solution as those obtained in Example 13.5 by the direct application of the dynamic condensation method. Table 13.1 shows and compares eigenvalues calculated in Examples 13.5 and 13.6 with the exact solution obtained previously in Example 13.3.

13.5 PROGRAM 12—REDUCTION OF THE DYNAMIC PROBLEM

The computer program described in this section reduces by static condensation or by dynamic condensation the stiffness and mass matrices of a structural system and solves the reduced eigenproblem. The user has the option of selecting either of these two methods.

Example 13.7. For the four-story shear building shown in Fig. 13.3, use Program 12 to determine the natural frequencies and modal shapes after reducing the system to coordinates y_2 and y_4 using the following methods:

TABLE 13.1 Comparison of Results in Examples 13.5 and 13.6 Using Dynamic Condensation and Modified Dynamic Condensation

	EIGENVALUE				
Mode	Exact Solution	Dynamic Condensation	Error %	Modified Method	Error %
1	39.48	39.48	0.00	39.46	0.05
2	327.35	327.35	0.00	319.41	2.42

(a) Static condensation,

(b) Dynamic condensation,

(c) Exact solution as a system with four degrees of freedom.

Solution: The execution of "Program 12: Condensation," requires the previous preparation of a file modeling the structure, that is, a file storing the stiffness and mass matrix of the system. For this example, this file can be prepared by either executing Program 7 to Model the structure as a shear building or executing the Auxiliary Program X1, which directly stores in a file the stiffness and mass matrices from input data.

Input Data and Output Results

SYSTEM STIFFNESS MATRIX

```
+.65470E+03      -.32735E+03      +.00000E+00      +.00000E+00
-.32735E+03      +.65470E+03      -.32735E+03      +.00000E+00
+.00000E+00      -.32735E+03      +.65470E+03      -.32735E+03
+.00000E+00      +.00000E+00      -.32735E+03      +.32735E+03
```

SYSTEM MASS MATRIX

```
+.10000E+01      +.00000E+00      +.00000E+00      +.00000E+00
+.00000E+00      +.10000E+01      +.00000E+00      +.00000E+00
+.00000E+00      +.00000E+00      +.10000E+01      +.00000E+00
+.00000E+00      +.00000E+00      +.00000E+00      +.10000E+01
```

(a)   STATIC CONDENSATION

    PROGRAM 12: CONDENSATION   DATA FILE:D12

    INPUT DATA:

```
NUMBER OF DEGREES OF FREEDOM                      ND=  4
NUMBER OF PRIMARY COORDINATES                     NL=  2
CONDENSATION INDEX                             INDEX=-1
  PRIMARIES COORDINATES:
```

    2   4

    OUTPUT RESULTS:

```
FREQ. #        EIGENVALUE          FREQUENCY (C.P.S.)
   1           4.039E+0.1               1.011
```

EIGENVECTOR # 1

```
0.218979    0.437958    0.555152    0.672345
```

```
FREQ. #        EIGENVALUE           FREQUENCY (C.P.S.)
    2          3.660E+02                 3.045

EIGENVECTOR # 2

0.352792    0.705584    0.046386    -.612812
```

(b) DYNAMIC CONDENSATION

        PROGRAM 12: CONDENSATION   DATA FILE:D12

    INPUT DATA:

```
NUMBER OF DEGREES OF FREEDOM                       ND= 4
NUMBER OF PRIMARY COORDINATES                      NL= 2
CONDENSATION INDEX                              INDEX= 1
  PRIMARIES COORDINATES:

    2     4
```

        OUTPUT RESULTS:

```
FREQ. #        EIGENVALUE           FREQUENCY (C.P.S.)

    1          3.948E+0.1                1.000

EIGENVECTOR # 1

0.228219    0.428282    0.577871    0.656167

FREQ. #        EIGENVALUE           FREQUENCY (C.P.S.)

    2          3.289E+02                 2.885

EIGENVECTOR # 2

0.618702    0.556594    0.002393    -.554441
```

(c) EXACT SOLUTION

        PROGRAM 8: NATURAL FREQUENCIES   DATA FILE: SK

    INPUT DATA:

                        STIFFNESS MATRIX

```
0.65473E+03    -.32737E+03    0.00000E+00    0.00000E+00
-.32737E+03    0.65473E+03    -.32737E+03    0.00000E+00
0.00000E+00    -.32737E+03    0.65473E+03    -.32737E+03
0.00000E+00    0.00000E+00    -.32737E+03    0.32737E+03
```

MASS MATRIX

```
0.10000E+01      0.00000E+00      0.00000E+00      0.00000E+00
0.00000E+00      0.10000E+01      0.00000E+00      0.00000E+00
0.00000E+00      0.00000E+00      0.10000E+01      0.00000E+00
0.00000E+00      0.00000E+00      0.00000E+00      0.10000E+01
```

OUTPUT RESULTS:

EIGENVALUES:

```
3.948E+01        3.274E+02        7.684E+02        1.156E+03
```

NATURAL FREQUENCIES (C.P.S.):

```
1.000            2.880            4.411            5.412
```

EIGENVECTORS BY ROWS:

```
 0.22801  0.42853  0.57735  0.65654
-0.57735-0.57735  0.00000  0.57735
-0.65654  0.22801  0.57735-0.42853
 0.42853-0.65654  0.57735-0.22801
```

As expected, results given by the computer for Example 13.7 agree with calculations for the same structure obtained previously using static condensation in Example 13.3 and dynamic condensation in Example 13.5.

13.6   SUMMARY

The reduction of unwanted or secondary degrees of freedom is usually accomplished in practice by the Static Condensation Method. This method consists of determining, by a partial Gauss–Jordan elimination, the reduced stiffness matrix corresponding to the primary degrees of freedom and the transformation matrix relating the secondary and primary degrees of freedom. The same transformation matrix is used in an orthogonal transformation to reduce the mass and damping matrices of the system. Static condensation introduces errors when applied to the solution structural dynamics problems. However, as is shown in this chapter, the application of the Dynamic Condensation Method substantially reduces or eliminates these errors. Furthermore, the Dynamic Condensation Method converges rapidly to the exact solution when iteration is applied.

PROBLEMS

13.1 The stiffness and mass matrices of a certain structure are given by

$$[K] = \begin{bmatrix} 10 & -2 & -1 & 0 \\ -2 & 6 & -3 & -2 \\ -1 & -3 & 12 & -1 \\ 0 & -2 & -1 & 8 \end{bmatrix}, \quad [M] = \begin{bmatrix} 0 & 0 & 0 & 0 \\ 0 & 0 & 0 & 0 \\ 0 & 0 & 3 & 0 \\ 0 & 0 & 0 & 2 \end{bmatrix}$$

(a) Use the Static Condensation Method to determine the transformation matrix and the reduced stiffness and mass matrices corresponding to the elimination of the first two degrees of freedom (the massless degrees of freedom).

(b) Determine the natural frequencies and corresponding normal modes for the reduced system.

13.2 Repeat (a) and (b) of Problem 13.1 for a structure having stiffness matrix as indicated in that problem, but mass matrix given by

$$[M] = \begin{bmatrix} 1 & 0 & 0 & 0 \\ 0 & 1 & 0 & 0 \\ 0 & 0 & 3 & 0 \\ 0 & 0 & 0 & 2 \end{bmatrix}$$

13.3 Determine the natural frequencies and modal shape of the system in Problem 13.2 in terms of its four original coordinates; find the errors in the two modes obtained in Part (b) of Problem 13.2.

13.4 Consider the shear building shown in Fig. 13.4.

(a) Determine the transformation matrix and the reduced stiffness and mass matrices corresponding to the static condensation of the coordinates y_1, y_3, and y_4 as indicated in the figure.

(b) Determine the natural frequencies and normal modes for the reduced system obtained in Part (a).

(c) Use the results of Part (b) to determine the modal shapes, described in the five original coordinates, corresponding to the two lowest frequencies.

13.5 Use the results obtained in Problem 13.4 to determine the maximum displacements relative to the foundation for the structure shown in Fig. 13.4 when subjected to an earthquake for which the response spectrum is given in Fig. 8.10 of Section 8.4.

13.6 Use the results in Problem 13.4 to determine the maximum shear forces in the stories of the building in Fig. 13.4 when subjected to an earthquake for which the response spectrum is given in Fig. 8.10 of Section 8.4.

13.7 Repeat Problem 13.2 using the Dynamic Condensation Method.

13.8 Repeat Problem 13.4 using the Dynamic Condensation Method and compare results with the exact solution.

13.9 Consider the five-story shear building of Fig. 13.4 subjected at its foundation to

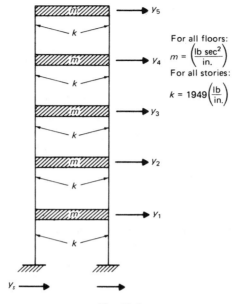

For all floors:
$$m = \left(\frac{\text{lb sec}^2}{\text{in.}}\right)$$
For all stories:
$$k = 1949 \left(\frac{\text{lb}}{\text{in.}}\right)$$

Fig. 13.4

the time-acceleration excitation depicted in Fig. 13.5. Use static condensation of the coordinates y_1, y_3, and y_4 and determine:

(a) The two natural frequencies and corresponding modal shapes of the reduced system.

(b) The displacements at the floor levels considering two modes.

(c) The shear forces in the columns of the structure also considering two modes.

13.10 Solve Problem 13.9 using the Dynamic Condensation Method.

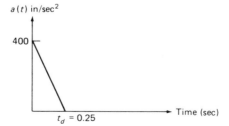

Fig. 13.5

13.11 The stiffness and mass matrices for a certain structure are

$$[K] = 10^6 \begin{bmatrix} 0.906 & 0.294 & 0.424 \\ 0.294 & 0.318 & 0.176 \\ 0.424 & 0.176 & 80.000 \end{bmatrix}$$

$$[M] = \begin{bmatrix} 288 & -8 & 1556 \\ -8 & 304 & 644 \\ 1556 & 644 & 80,000 \end{bmatrix}$$

Calculate the fundamental natural frequency of the system after reduction of the first coordinate by the following methods:

(a) Static condensation.

(b) Dynamic condensation.

Also obtain the natural frequencies as a three-degrees-of-freedom system and compare results for the fundamental frequency.

13.12 Repeat Problem 13.11 using the Modified Dynamic Condensation Method and compare results with the solution obtained with no condensation.

PART III

Framed Structures Modeled as Discrete Multidegree-of-Freedom Systems

14

Dynamic Analysis of Beams

In this chapter, we shall study the dynamic behavior of structures designated as beams, that is, structures which carry loads which are mainly transverse to the longitudinal direction, thus producing flexural stresses and lateral displacements. We begin by establishing the static characteristics for a beam segment; we then introduce the dynamic effects produced by the inertial forces. Two approximate methods are presented to take into account the inertial effect in the structure: (1) the lumped mass method in which the distributed mass is assigned to point masses, and (2) the consistent mass method in which the assignment to point masses includes rotational effects. The latter method is consistent with the static elastic deflections of the beam.

In Chapters 20 and 21, the exact theory for dynamics of beams considering the elastic and inertial distributed properties will be presented. In these chapters, the mathematical relationship between the exact solution and the stiffness and the consistent mass coefficients will be shown.

14.1   STATIC PROPERTIES FOR A BEAM SEGMENT

Consider a uniform beam segment of cross-sectional moment of inertia I, length L, and material modulus of elasticity E as shown in Fig. 14.1. We shall establish the relation between static forces and moments designated as P_1, P_2, P_3, and P_4 and the corresponding linear and angular displacements δ_1, δ_2, δ_3, and δ_4 at the ends of the beam segment as indicated in Fig. 14.1. The relation thus obtained is the stiffness matrix for a beam segment. The forces P_i and the displacements δ_i are said to be at the *nodal coordinates* defined for the beam segment.

The differential equation for small transverse displacements of a beam, which is well known from elementary studies of strength of materials, is given by

$$EI \frac{d^2y}{dx^2} = M(x) \tag{14.1}$$

in which $M(x)$ is the bending moment at a section of the beam and y is the transverse displacement.

The differential equation (14.1) for a uniform beam segment is equivalent to

$$EI \frac{d^4y}{dx^4} = p(x) \tag{14.2}$$

since

$$\frac{dM(x)}{dx} = V(x) \tag{14.3}$$

and

$$\frac{dV(x)}{dx} = p(x)$$

in which $p(x)$ is the beam load per unit length and $V(x)$ is the shear force.

We state first the general definition of stiffness coefficient which is designated by k_{ij}, that is, k_{ij} is the force at nodal coordinate i due to a unit displacement at nodal coordinate j while all other nodal coordinates are maintained at zero displacement. Figure 14.2 shows the displacement curves

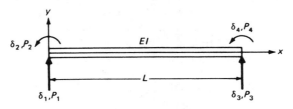

Fig. 14.1 Beam segment showing forces and displacements at the nodal coordinates.

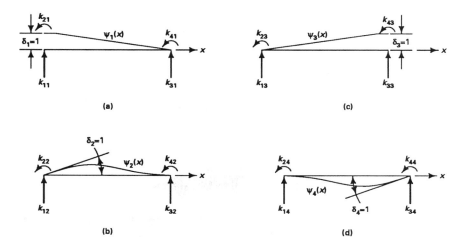

Fig. 14.2 Static deflection curves due to a unit displacement at one of the nodal co-ordinates.

corresponding to a unit displacement at each one of the four nodal coordinates for a beam segment indicating the corresponding stiffness coefficients. To determine the expressions for the stiffness coefficients k_{ij}, we begin by finding the equations for displaced curves shown in Fig. 14.2. We consider the beam segment in Fig. 14.1 free of loads $[p(x) = 0]$, except for the forces P_1, P_2, P_3, P_4 applied at the nodal coordinates. In this case, eq. (14.2) is reduced to

$$\frac{d^4y}{dx^4} = 0 \tag{14.4}$$

Successive integrations of eq. (14.4) yields

$$\frac{d^3y}{dx^3} = C_1$$

$$\frac{d^2y}{dx^2} = C_1x + C_2$$

$$\frac{dy}{dx} = \frac{1}{2} C_1x^2 + C_2x + C_3 \tag{14.5}$$

$$y = \frac{1}{6} C_1x^3 + \frac{1}{2} C_2x^2 + C_3x + C_4 \tag{14.6}$$

in which C_1, C_2, C_3, and C_4 are constants of integration to be evaluated using boundary conditions. For example, to determine the function $\psi_1(x)$ for the curve shown in Fig. 14.2(a), we make use of the following boundary conditions:

$$\text{at } x = 0 \quad y(0) = 1 \quad \text{and} \quad \frac{dy(0)}{dx} = 0 \tag{14.7}$$

$$\text{at } x = L \quad y(L) = 0 \quad \text{and} \quad \frac{dy(L)}{dx} = 0 \tag{14.8}$$

Use of these conditions in eqs. (14.5) and (14.6) results in an algebraic system of four equations to determine the constants C_1, C_2, C_3, and C_4.

The substitution of these constants into eq. (14.6) results in the equation of the deflected curve for the beam segment in Fig. 14.2(a) as

$$\psi_1(x) = 1 - 3\left(\frac{x}{L}\right)^2 + 2\left(\frac{x}{L}\right)^3 \tag{14.9a}$$

in which $\psi_1(x)$ is used instead of $y(x)$ to correspond to the condition $\delta_1 = 1$ imposed on the beam segment. Proceeding in analogous fashion, we obtain for the equations of the deflected curves in the other cases depicted in Fig. 14.2 the following equations:

$$\psi_2(x) = x\left(1 - \frac{x}{L}\right)^2 \tag{14.9b}$$

$$\psi_3(x) = 3\left(\frac{x}{L}\right)^2 - 2\left(\frac{x}{L}\right)^3 \tag{14.9c}$$

$$\psi_4(x) = \frac{x^2}{L}\left(\frac{x}{L} - 1\right) \tag{14.9d}$$

Since $\psi_1(x)$ is the deflection corresponding to a unit displacement $\delta_1 = 1$, the displacement resulting from an arbitrary displacement δ_1 is $\psi_1(x)\delta_1$. Analogously, the deflection resulting from nodal displacements δ_2, δ_3, and δ_4 are, respectively, $\psi_2(x)\delta_2$, $\psi_3(x)\delta_3$, and $\psi_4(x)\delta_4$. Therefore, the total deflection $y(x)$ at coordinate x due to arbitrary displacements at the nodal coordinates of the beam segment is given by superposition as

$$y(x) = \psi_1(x)\delta_1 + \psi_2(x)\delta_2 + \psi_3(x)\delta_3 + \psi_4(x)\delta_4 \tag{14.10}$$

The deflection equations which are given by eqs. (14.9) and which correspond to unit displacements at the nodal coordinates of a beam segment may be used to determine expressions for the stiffness coefficients. For example, consider the beam in Fig. 14.2(b) which is in equilibrium with the forces producing the displacement $\delta_2 = 1.0$. For this beam in the equilibrium position, we assume that a virtual displacement equal to the deflection curve shown in Fig. 14.2(a) takes place. We then apply the principle of virtual work which states that, for an elastic system in equilibrium, the work done by the external forces is equal to the work of the internal forces during the virtual displacement. In order to apply this principle, we note that the external work W_E is equal to the product of the force k_{12} displaced by $\delta_1 = 1$, that is

$$W_E = k_{12}\delta_1 \qquad (14.11)$$

This work, as stated above, is equal to the work performed by the elastic forces during the virtual displacement. Considering the work performed by the bending moment, we obtain for the internal work

$$W_I = \int_0^L M(x)\, d\theta \qquad (14.12)$$

in which $M(x)$ is the bending moment at section x of the beam and $d\theta$ is the relative angular displacement of this section.

For the virtual displacement under consideration, the transverse deflection of the beam is given by eq. (14.9b) which is related to the bending moment through the differential equation (14.1). Substitution of the second derivative $\psi_2''(x)$ of eq. (14.9b) into eq. (14.1) results in

$$EI\psi_2''(x) = M(x) \qquad (14.13)$$

The angular deflection $d\theta$ produced during this virtual displacement is related to the resulting transverse deflection of the beam $\psi_1(x)$ by

$$\frac{d\theta}{dx} = \frac{d^2\psi_1(x)}{dx^2} = \psi_1''(x)$$

or

$$d\theta = \psi_1''(x)\, dx \qquad (14.14)$$

Equating the external virtual work W_E from eq. (14.11) with the internal virtual work W_I from eq. (14.12) after using $M(x)$ and $d\theta$ from eqs. (14.13) and (14.14), respectively, finally gives the stiffness coefficient as

$$k_{12} = \int_0^L EI\psi_1''(x)\, \psi_2''(x)\, dx \qquad (14.15)$$

In general, any stiffness coefficient associated with beam flexure, therefore, may be expressed as

$$k_{ij} = \int_0^L EI\psi_i''(x)\, \psi_j''(x)\, dx \qquad (14.16)$$

It may be seen from eq. (14.16) that $k_{ij} = k_{ji}$ since the interchange of indices requires only an interchange of the two factors $\psi_i''(x)$ and $\psi_j''(x)$ in eq. (14.16). The equivalence of $k_{ij} = k_{ji}$ is a particular case of Betti's theorem, but it is better known as *Maxwell's reciprocal theorem*.

It should be pointed out that although the deflection functions, eqs. (14.9), were obtained for a uniform beam, in practice they are nevertheless also used in determining the stiffness coefficients for nonuniform beams.

Considering the case of a uniform beam segment of length L and cross-sectional moment of inertia I, we may calculate any stiffness coefficient from eq. (14.16) and the use of eqs. (14.9). In particular, the stiffness coefficient k_{12} is calculated as follows. From eq. (14.9a), we obtain

$$\psi_1''(x) = -\frac{6}{L^2} + \frac{12x}{L^3}$$

and from eq. (14.9b)

$$\psi_2''(x) = -\frac{4}{L} + \frac{6x}{L^2}$$

Substitution in eq. (14.15) gives

$$k_{12} = EI \int_0^L \left(\frac{-6}{L^2} + \frac{12x}{L^3}\right)\left(\frac{-4}{L} + \frac{6x}{L^2}\right) dx$$

and integration gives

$$k_{12} = \frac{6EI}{L^2}$$

Since the stiffness coefficient k_{1j} is defined as the force at the nodal coordinate 1 due to unit displacement at the coordinate j, the forces at coordinate 1 due to successive displacement δ_1, δ_2, δ_3, and δ_4 at the four nodal coordinates of the beam segment are given, respectively, by $k_{11}\delta_1$, $k_{12}\delta_2$, $k_{13}\delta_3$, and $k_{14}\delta_4$. Therefore, the total force P_1 at coordinate 1 resulting from these nodal displacements is obtained by the superposition of the resulting forces, that is,

$$P_1 = k_{11}\delta_1 + k_{12}\delta_2 + k_{13}\delta_3 + k_{14}\delta_4$$

Analogously, the forces at the other nodal coordinates are

$$P_2 = k_{21}\delta_1 + k_{22}\delta_2 + k_{23}\delta_3 + k_{24}\delta_4$$

$$P_3 = k_{31}\delta_1 + k_{32}\delta_2 + k_{33}\delta_3 + k_{34}\delta_4$$

$$P_4 = k_{41}\delta_1 + k_{42}\delta_2 + k_{43}\delta_3 + k_{44}\delta_4 \qquad (14.17)$$

The above equations are written conveniently in matrix notation as

$$\begin{bmatrix} P_1 \\ P_2 \\ P_3 \\ P_4 \end{bmatrix} = \begin{bmatrix} k_{11} & k_{12} & k_{13} & k_{14} \\ k_{21} & k_{22} & k_{23} & k_{24} \\ k_{31} & k_{32} & k_{33} & k_{34} \\ k_{41} & k_{42} & k_{43} & k_{44} \end{bmatrix} \begin{bmatrix} \delta_1 \\ \delta_2 \\ \delta_3 \\ \delta_4 \end{bmatrix} \qquad (14.18)$$

or symbolically as

$$\{P\} = [k]\{\delta\} \tag{14.19}$$

in which $\{P\}$ and $\{\delta\}$ are, respectively, the force and the displacement vectors at the nodal coordinates of the beam element and $[k]$ is the beam element stiffness matrix.

The use of eq. (14.16) in the manner shown above to determine the coefficient k_{12} will result in the evaluation of all the coefficients of the stiffness matrix. This result for a uniform beam segment is

$$\begin{Bmatrix} P_1 \\ P_2 \\ P_3 \\ P_4 \end{Bmatrix} = \frac{EI}{L^3} \begin{bmatrix} 12 & 6L & -12 & 6L \\ 6L & 4L^2 & -6L & 2L^2 \\ -12 & -6L & 12 & -6L \\ 6L & 2L^2 & -6L & 4L^2 \end{bmatrix} \begin{Bmatrix} \delta_1 \\ \delta_2 \\ \delta_3 \\ \delta_4 \end{Bmatrix} \tag{14.20}$$

or in condensed notation

$$\{P\} = [k]\{\delta\} \tag{14.21}$$

14.2   SYSTEM STIFFNESS MATRIX

Thus far we have established in eq. (14.20) the stiffness equation for a uniform beam segment, that is, we have obtained the relation between nodal displacements (linear and angular) and nodal forces (forces and moments). Our next objective is to obtain the same type of relation between the nodal displacements and the nodal forces, but now for the entire structure (system stiffness equation). Furthermore, our aim is to obtain the system stiffness matrix from the stiffness matrix of each element of the system. The procedure is perhaps better explained through a specific example such as the cantilever beam shown in Fig. 14.3.

The first step in obtaining the system stiffness matrix is to divide the structure into elements. The beam in Fig. 14.3 has been divided into three ele-

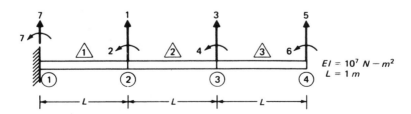

Fig. 14.3 Cantilever beam divided into three beam segments with numbered system nodal coordinates.

ments which are numbered sequentially for identification. The second step is to identify the nodes or joints between elements and to number consecutively those nodal coordinates which are not constrained. The constrained or fixed nodal coordinates are the last to be labeled. All the fixed coordinates may be given the same label as shown in Fig. 14.3. In the present case, we consider only two possible displacements at each node, a vertical deflection and an angular displacement. The cantilever beam in Fig. 14.3 with its three elements results in a total of six free nodal coordinates and two fixed nodal coordinates, the latter being labeled with the number seven. The third step is to obtain systematically the stiffness matrix for each element in the system and to add the element stiffness coefficients appropriately to obtain the system stiffness matrix. This method of assembling the system stiffness matrix is called the *direct method*. In effect, any stiffness coefficient k_{ij} of the system may be obtained by adding together the corresponding stiffness coefficients associated with those nodal coordinates. Thus, for example, to obtain the system stiffness coefficient k_{33}, it is necessary to add the stiffness coefficients of beam segments $\triangle$ and $\triangle$ corresponding to node three. These coefficients are designated as $k_{33}^{(2)}$ and $k_{11}^{(3)}$, respectively. The upper indices serve to identify the beam segment, and the lower indices to locate the appropriate stiffness coefficients in the corresponding element stiffness matrices.

Proceeding with the example in Fig. 14.3 and using eq. (14.20), we obtain the following expression for the stiffness matrix of beam segment $\triangle$, namely

$$[k^{(2)}] = 10^7 \begin{array}{cccc} & \begin{array}{cccc} 1 & 2 & 3 & 4 \end{array} & \\ & \left[\begin{array}{cccc} 12 & 6 & -12 & 6 \\ 6 & 4 & -6 & 2 \\ -12 & -6 & 12 & -6 \\ 6 & 2 & -6 & 4 \end{array}\right] & \begin{array}{c} 1 \\ 2 \\ 3 \\ 4 \end{array} \end{array} \qquad (14.22)$$

For the beam segment $\triangle$, the element nodal coordinates numbered one to four coincide with the assignment of system nodal coordinates also numbered 1 to 4 as may be seen in Fig. 14.3. However, for the beam segments $\triangle$ and $\triangle$ of this beam, the assignment of element nodal coordinates numbered 1 to 4 does not coincide with the assigned system coordinates. For example, for element $\triangle$ the assigned system coordinates as seen in Fig. 14.3 are 7, 7, 1, 2; for element $\triangle$, 3, 4, 5, 6. In the process of assembling the system stiffness, coefficients for element $\triangle$ will be correctly allocated to coordinates 1, 2, 3, 4; for element $\triangle$ to coordinates 7, 7, 1, 2; and for element $\triangle$ to coordinates 3, 4, 5, 6. A simple way to indicate this allocation of coordinates, when working by hand, is to write at the top and on the right of the element stiffness matrix the coordinate numbers corresponding to the system nodal coordinates for the element as it is indicated in eq. (14.22) for element $\triangle$. The stiffness matrices for elements ⚠ and ⚠ with the corresponding indication of system nodal coordinates are, respectively,

$$[k^{(1)}] = 10^7 \begin{array}{cccc} 7 & 7 & 1 & 2 \\ \begin{bmatrix} 12 & 6 & -12 & 6 \\ 6 & 4 & -6 & 2 \\ -12 & -6 & 12 & -6 \\ 6 & 2 & -6 & 4 \end{bmatrix} & & & \begin{array}{c} 7 \\ 7 \\ 1 \\ 2 \end{array} \end{array} \tag{14.23}$$

and

$$[k^{(3)}] = 10^7 \begin{array}{cccc} 3 & 4 & 5 & 6 \\ \begin{bmatrix} 12 & 6 & -12 & 6 \\ 6 & 4 & -6 & 2 \\ -12 & -6 & 12 & -6 \\ 6 & 2 & -6 & 4 \end{bmatrix} & & & \begin{array}{c} 3 \\ 4 \\ 5 \\ 6 \end{array} \end{array} \tag{14.24}$$

Proceeding systematically to assemble the system stiffness matrix, we translate each entry in the element stiffness matrices, eqs. (14.22), (14.23), and (14.24), to the appropriate location in the system stiffness matrix. For instance, the stiffness coefficient for element ⚠, $k^{(3)}_{13} = -12 \times 10^7$ should be translated to location at row 3 and column 5 since these are the coordinates indicated at right and top of matrix eq. (14.24) for this entry. Every element stiffness coefficient translated to its appropriate location in the system stiffness matrix is added to the other coefficients accumulated at that location. The stiffness coefficients corresponding to columns or rows carrying a label of a fixed system nodal coordinate (seven in the present example) are simply disregarded since the constrained nodal coordinates are not unknown quantities. The assemblage of the system matrix in the manner described results for this example in a 6×6 matrix, namely

$$[k] = 10^7 \begin{bmatrix} 24 & 0 & -12 & 6 & 0 & 0 \\ 0 & 8 & -6 & 2 & 0 & 0 \\ -12 & -6 & 24 & 0 & -12 & 6 \\ 6 & 2 & 0 & 8 & -6 & 2 \\ 0 & 0 & -12 & -6 & 12 & -6 \\ 0 & 0 & 6 & 2 & -6 & 4 \end{bmatrix} \tag{14.25}$$

Equation (14.25) is thus the system stiffness matrix for the cantilever beam shown in Fig. 14.3 which has been segmented into three elements. As such, the system stiffness matrix relates the forces and the displacements at the nodal system coordinates in the same manner as the element stiffness matrix relates forces and displacements at the element nodal coordinates.

14.3   INERTIAL PROPERTIES—LUMPED MASS

The simplest method for considering the inertial properties for a dynamic system is to assume that the mass of the structure is lumped at the nodal coordinates where translational displacements are defined, hence the name *lumped mass method*. The usual procedure is to distribute the mass of each element to the nodes of the element. This distribution of the mass is determined by statics. Figure 14.4 shows, for beam segments of length L and distributed mass $\bar{m}(x)$ per unit of length, the nodal allocation for uniform, triangular, and general mass distribution along the beam segment. The assemblage of the mass matrix for the entire structure will be a simple matter of adding the contributions of lumped masses at the nodal coordinates defined as translations.

In this method, the inertial effect associated with any rotational degree of freedom is usually assumed to be zero, although a finite value may be associated with rotational degrees of freedom by calculating the mass moment of inertia of a fraction of the beam segment about the nodal points. For example, for a uniform beam, this calculation would result in determining the mass moment of inertia of half of the beam segment about each node, that is

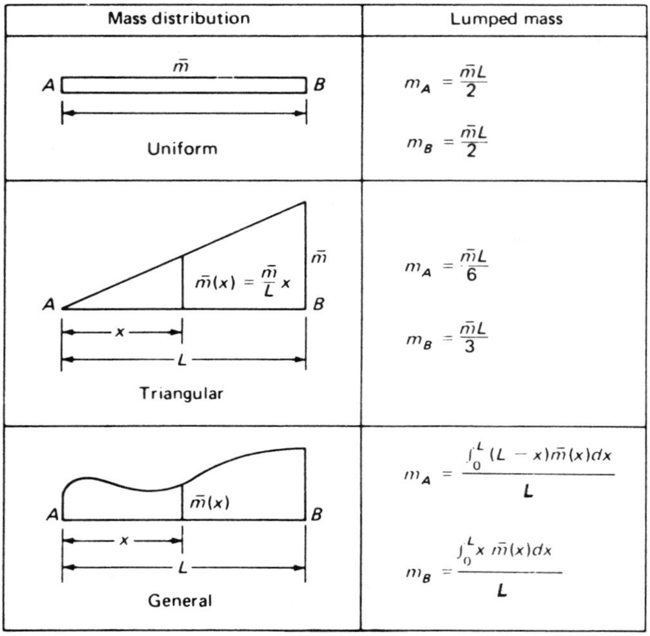

Mass distribution	Lumped mass
$\bar{m}$ A ▭ B Uniform	$m_A = \dfrac{\bar{m}L}{2}$ $m_B = \dfrac{\bar{m}L}{2}$
$\bar{m}(x) = \dfrac{\bar{m}}{L}x$ A ◿ B Triangular	$m_A = \dfrac{\bar{m}L}{6}$ $m_B = \dfrac{\bar{m}L}{3}$
$\bar{m}(x)$ A ▭ B General	$m_A = \dfrac{\int_0^L (L - x)\bar{m}(x)dx}{L}$ $m_B = \dfrac{\int_0^L x\,\bar{m}(x)dx}{L}$

Fig. 14.4 Lumped masses for beam segments with distributed mass.

$$I_A = I_B = \frac{1}{3}\left(\frac{\overline{m}L}{2}\right)\left(\frac{L}{2}\right)^2$$

where $\overline{m}$ is the mass per unit length along the beam. For the cantilever beam shown in Fig. 14.3 in which only translational mass effects are considered, the mass matrix of the system would be the diagonal matrix, namely

$$[M] = \begin{matrix} & \begin{matrix} 1 & 2 & 3 & 4 & 5 & 6 \end{matrix} \\ \begin{bmatrix} m_1 & & & & & \\ & 0 & & & & \\ & & m_3 & & & \\ & & & 0 & & \\ & & & & m_5 & \\ & & & & & 0 \end{bmatrix} & \begin{matrix} 1 \\ 2 \\ 3 \\ 4 \\ 5 \\ 6 \end{matrix} \end{matrix} \qquad (14.26)$$

in which

$$m_1 = \frac{\overline{m}L_1}{2} + \frac{\overline{m}L_2}{2}$$

$$m_3 = \frac{\overline{m}L_2}{2} + \frac{\overline{m}L_3}{2}$$

$$m_5 = \frac{\overline{m}L_3}{2}$$

Using a special symbol ($\lceil \ \rfloor$) for diagonal matrices, we may write eq. (14.26) as

$$[M] = \lceil m_1 \quad 0 \quad m_3 \quad 0 \quad m_5 \quad 0 \rfloor \qquad (14.27)$$

14.4 INERTIAL PROPERTIES—CONSISTENT MASS

It is possible to evaluate the mass coefficients corresponding to the nodal coordinates of a beam element by a procedure similar to the determination of element stiffness coefficients. First, we define the mass coefficient m_{ij} as the force at nodal coordinate i due to a unit acceleration at nodal coordinate j while all other nodal coordinates are maintained at zero acceleration.

Consider the beam segment shown in Fig. 14.5(a) which has distributed mass $\overline{m}(x)$ per unit of length. In the consistent mass method, it is assumed that the deflections resulting from unit dynamic displacements at the nodal coordinates of the beam element are given by the same functions $\psi_1(x)$, $\psi_2(x)$, $\psi_3(x)$, and $\psi_4(x)$ of eqs. (14.9) which were obtained from static considerations

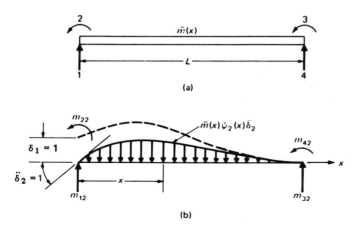

Fig. 14.5 (a) Beam element with distributed mass showing four nodal coordinates. (b) Beam element supporting inertial load due to acceleration $\ddot{\delta}_2 = 1$, undergoing virtual displacement $\delta_1 = 1$.

If the beam segment is subjected to a unit nodal acceleration at one of the nodal coordinates, say $\ddot{\delta}_2 = 1$, the transverse acceleration developed along the length of the beam is given by the second derivative with respect to time of eq. (14.10). In this case, with $\ddot{\delta}_1 = \ddot{\delta}_3 = \ddot{\delta}_4 = 0$, we obtain

$$\ddot{y}_2(x) = \psi_2(x)\ddot{\delta}_2 \tag{14.28}$$

The inertial force $f_I(x)$ per unit of length along the beam due to this acceleration is then given by

$$f_I(x) = \overline{m}(x)\ddot{y}_2(x)$$

or using eq. (14.28) by

$$f_I(x) = \overline{m}(x)\psi_2(x)\ddot{\delta}_2$$

or, since $\ddot{\delta}_2 = 1$,

$$f_I(x) = \overline{m}(x)\psi_2(x) \tag{14.29}$$

Now to determine the mass coefficient m_{12}, we give to the beam in Fig. 14.5(b) a virtual displacement corresponding to a unit displacement at coordinate 1, $\delta_1 = 1$ and proceed to apply the principle of virtual work for an elastic system (external work equal to internal virtual work). The virtual work of the external force is simply

$$W_E = m_{12}\delta_1 = m_{12} \tag{14.30}$$

since the only external force undergoing virtual displacement is the inertial force reaction $m_{12}\delta_1$ with $\delta_1 = 1$. The virtual work of the internal forces per unit of length along the beam segment is

$$\delta W_I = f_I(x)\,\psi_1(x)$$

or by eq. (14.29),

$$\delta W_I = \overline{m}(x)\,\psi_2(x)\,\psi_1(x)$$

and for the entire beam

$$W_I = \int_0^L \overline{m}(x)\,\psi_2(x)\,\psi_1(x)\,dx \tag{14.31}$$

Equating the external and internal virtual work given, respectively, by eqs. (14.30) and (14.31) results in

$$m_{12} = \int_0^L \overline{m}(x)\,\psi_2(x)\,\psi_1(x)\,dx \tag{14.32}$$

which is the expression for the consistent mass coefficient m_{12}. In general, a consistent mass coefficient may be calculated from

$$m_{ij} = \int_0^L \overline{m}(x)\,\psi_i(x)\,\psi_j(x)\,dx \tag{14.33}$$

It may be seen from eq. (14.33) that $m_{ij} = m_{ji}$ since the interchange of the subindices only results in an interchange of the order of the factors $\psi_i(x)$ and $\psi_j(x)$ under the integral.

In practice, the cubic equations (14.9) are used in calculating the mass coefficients of any straight beam element. For the special case of the beam with uniformly distributed mass, the use of eq. (14.33) gives the following relation between inertial forces and acceleration at the nodal coordinates:

$$\begin{Bmatrix} P_1 \\ P_2 \\ P_3 \\ P_4 \end{Bmatrix} = \frac{\overline{m}L}{420} \begin{bmatrix} 156 & 22L & 54 & -13L \\ 22L & 4L^2 & 13L & -3L^2 \\ 54 & 13L & 156 & -22L \\ -13L & -3L^2 & -22L & 4L^2 \end{bmatrix} \begin{Bmatrix} \ddot{\delta}_1 \\ \ddot{\delta}_2 \\ \ddot{\delta}_3 \\ \ddot{\delta}_4 \end{Bmatrix} \tag{14.34}$$

When the mass matrix, eq. (14.34), has been evaluated for each beam element of the structure, the mass matrix for the entire system is assembled by exactly the same procedure (direct method) as described in developing the stiffness matrix for the system. The resulting mass matrix will in general have the same arrangement of nonzero terms as the stiffness matrix.

The dynamic analysis using the lumped mass matrix requires considerably less computational effort than the analysis using the consistent mass method for the following reasons. The lumped mass matrix for the system results in a diagonal mass matrix whereas the consistent mass matrix has many off-diagonal terms which are called mass coupling. Also, the lumped mass matrix contains zeros in its main diagonal due to assumed zero rotational inertial forces. This fact permits the elimination by static condensation (Chapter 13) of the rotational degrees of freedom, thus reducing the dimension of the dynamic problem. Nevertheless, the dynamic analysis using the consistent mass matrix gives results which approximate better to the exact solution compared to the lumped mass method for the same element discretization.

Example 14.1.   Determine the lumped mass and the consistent mass matrices for the cantilever beam in Fig. 14.6. Assume uniform mass, $\overline{m} = 420$ kg/m.

Solution: (a) Lumped Mass Matrix. The lumped mass at each node of any of the three beam segments, into which the cantilever beam has been divided, is simply half of the mass of the segment. In the present case, the lumped mass at each node is 210 kg as shown in Fig. 14.6. The lumped mass matrix $[M_L]$ for this structure is a diagonal matrix of dimension 6×6, namely,

$$[M_L] = \lceil 420 \quad 0 \quad 420 \quad 0 \quad 210 \quad 0 \rfloor$$

(b) Consistent Mass Matrix. The consistent mass matrix for a uniform beam segment is given by eq. (14.34). The substitution of numerical values for this example $L = 1m$, $\overline{m} = 420$ kg/m into eq. (14.34) gives the consistent mass matrix $[M_c^{(2)}]$ for any of three beam segments as

$$[M_c^{(2)}] = \begin{matrix} & 1 & 2 & 3 & 4 & \\ & \begin{bmatrix} 156 & 22 & 54 & -13 \\ 22 & 4 & 13 & -3 \\ 54 & 13 & 156 & -22 \\ -13 & -3 & -22 & 4 \end{bmatrix} & \begin{matrix} 1 \\ 2 \\ 3 \\ 4 \end{matrix} \end{matrix} \qquad \text{(a)}$$

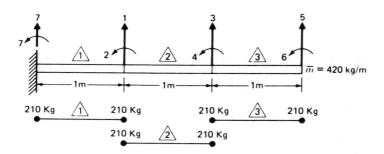

Fig. 14.6 Lumped masses for Example 14.1.

The assemblage of the system mass matrix from the element mass matrices is carried out in exactly the same manner as the assemblage of the system stiffness matrix from the element stiffness matrices, that is, the element mass matrices are allocated to appropriate entries in the system mass matrix. For the second beam segment, this allocation corresponds to the first four coordinates as indicated above and on the right of eq. (a). For the beam segment ⚠, the appropriate allocation is 3, 4, 5, 6 and for the beam segment ⚠, 7, 7, 1, 2 since these are the system nodal coordinates for these beam segments as indicated in Fig. 14.6. The consistent mass matrix $[M_c]$ for this example obtained in this manner is given by

$$[M_c] = \begin{array}{c} \\ \end{array}
\begin{array}{cccccc}
1 & 2 & 3 & 4 & 5 & 6 \\
\end{array}
\left[\begin{array}{cccccc}
312 & 0 & 54 & -13 & 0 & 0 \\
0 & 8 & 13 & -3 & 0 & 0 \\
54 & 13 & 312 & 0 & 54 & -13 \\
-13 & -3 & 0 & 8 & 13 & -3 \\
0 & 0 & 54 & 13 & 156 & -22 \\
0 & 0 & -13 & -3 & -22 & 4 \\
\end{array}\right]
\begin{array}{c}
1 \\ 2 \\ 3 \\ 4 \\ 5 \\ 6
\end{array} \qquad \text{(b)}$$

We note that the mass matrix $[M_c]$ is symmetric and also banded as in the case of the stiffness matrix for this system. These facts are of great importance in developing computer programs for structural analysis, since it is possible to perform the necessary calculations storing in the computer only the diagonal elements and the elements to one of the sides of the main diagonal. The maximum number of nonzero elements in any row which are required to be stored is referred to as the bandwidth of the matrix. For the matrix eq. (b) the bandwidth is equal to four (NBW = 4). In this case, it is necessary to store a total of $6 \times 4 = 24$ coefficients, whereas if the square matrix were to be stored, it would require $6 \times 6 = 36$ storage spaces. This economy in storing spaces becomes more dramatic for structures with a large number of nodal coordinates. The dimension of the bandwidth is directly related to the largest difference of the nodal coordinates assigned to any of the elements of the structure. Therefore, it is important to number the system nodal coordinates so as to minimize this difference.

14.5   DAMPING PROPERTIES

Damping coefficients are defined in a manner entirely parallel to the definition of the stiffness coefficient or the mass coefficient. Specifically, the damping coefficient c_{ij} is defined as the force developed at coordinate i due to a unit velocity at j. If the damping forces distributed in the structure could be determined, the damping coefficients of the various structural elements would then be used in obtaining the damping coefficient corresponding to the system. For example, the damping coefficient c_{ij} for an element might be of the form

$$c_{ij} = \int_0^L c(x)\,\psi_i(x)\,\psi_j(x)\,dx \tag{14.35}$$

where $c(x)$ represents the distributed damping coefficient per unit length. If the element damping matrix could be calculated, the damping matrix for the entire structure could be assembled by a superposition process equivalent to the direct stiffness matrix. In practice, the evaluation of the damping property $c(x)$ is impracticable. For this reason, the damping is generally expressed in terms of damping ratios obtained experimentally rather than by a direct evaluation of the damping matrix using eq. (14.35). These damping ratios are evaluated or estimated for each natural mode of vibration. If the explicit expression of the damping matrix $[C]$ is needed, it may be computed from the specified relative damping coefficients by any of the methods described in Chapter 12.

14.6   EXTERNAL LOADS

When the dynamic loads acting on the structure consist of concentrated forces and moments applied at defined nodal coordinates, the load vector can be written directly. In general, however, loads are applied at points other than nodal coordinates. In addition, the external load may include the action of distributed forces. In this case, the load vector corresponding to the nodal coordinates consists of the equivalent or generalized forces. The procedure to determine the equivalent nodal forces which is consistent with the derivation of the stiffness matrix and the consistent mass matrix is to assume the validity of the static deflection functions, eqs. (14.9), for the dynamic problem and to use the principle of virtual work.

Consider the beam element in Fig. 14.7 when subjected to an arbitrary

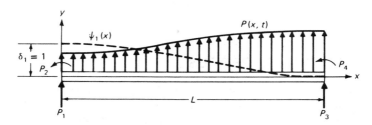

Fig. 14.7 Beam element supporting arbitrary distributed load undergoing virtual displacement $\delta_1 = 1$.

distributed force $p(x, t)$ which is a function of position along the beam as well as a function of time. The equivalent force P_1 at coordinate 1 may be found by giving a virtual displacement $\delta_1 = 1$ at this coordinate and equating the resulting external work and internal work during this virtual displacement. In this case, the external work is

$$W_E = P_1\delta_1 = P_1 \tag{14.36}$$

since $\delta_1 = 1$. The internal work per unit of length along the beam is $p(x, t)\psi_1(x)$ and the total internal work is then

$$W_I = \int_0^L p(x, t)\,\psi_1(x)\,dx \tag{14.37}$$

Equating external work, eq. (14.36), and internal work, eq. (14.37), gives the equivalent nodal force as

$$P_1(t) = \int_0^L p(x, t)\,\psi_1(x)\,dx \tag{14.38}$$

Thus the element equivalent nodal forces can be expressed in general as

$$P_i(t) = \int_0^L p(x, t)\,\psi_i(x)\,dx \tag{14.39}$$

Example 14.2.   Consider the beam segment in Fig. 14.8 and determine the element nodal forces for a uniform distributed force along the length of the beam given by

$$p(x, t) = 200 \sin 10t \text{ N/m}$$

Solution: Introduction of numerical values into the displacements functions, eqs. (14.9), and substitution in eq. (14.39) yield

$$P_1(t) = 200 \int_0^1 (1 - 3x^2 + 2x^3)\,dx \sin 10t = 100 \sin 10t$$

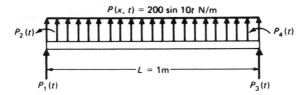

Fig. 14.8 Beam segment subjected to external distributed load showing equivalent nodal forces.

$$P_2(t) = 200 \int_0^1 x(1 - x)^2 \, dx \sin 10t = 16.67 \sin 10t$$

$$P_3(t) = 200 \int_0^1 (3x^2 - 2x^3) \, dx \sin 10t = 100 \sin 10t$$

$$P_4(t) = 200 \int_0^1 x^2(x - 1) \, dx \sin 10t = -16.67 \sin 10t$$

14.7 GEOMETRIC STIFFNESS

When a beam element is subjected to an axial force in addition to flexural loading, the stiffness coefficients are modified by the presence of the axial force. The modification corresponding to the stiffness coefficient k_{ij} is known as the geometric stiffness k_{Gij}, which is defined as the force corresponding to the nodal coordinate i due to a unit displacement at coordinate j and resulting from the axial forces in the structure. These coefficients may be evaluated by application of the principle of virtual work. Consider a beam element as used previously but now subjected to a distributed axial force per unit of length $N(x)$, as depicted in Fig. 14.9(a). In the sketch in Fig. 14.9(b), the beam segment is subjected to a unit rotation of the left end, $\delta_2 = 1$. By definition, the nodal

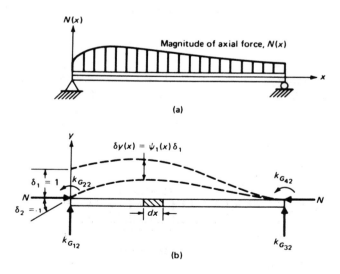

Fig. 14.9 (a) Beam element loaded with arbitrary distributed axial force. (b) Beam element acted on by nodal forces resulting from displacement $\delta_2 = 1$, undergoing a virtual displacement $\delta_1 = 1$.

forces due to this displacement are the corresponding geometric stiffness coefficients; for example, k_{G12} is the vertical force at the left end. If we now give to this deformed beam a unit displacement $\delta_1 = 1$, the resulting external work is

$$W_e = k_{G12}\delta_1$$

or

$$W_e = k_{G12} \tag{14.40}$$

since $\delta_1 = 1$.

The internal work during this virtual displacement is found by considering a differential element of length dx taken from the beam in Fig. 14.9(b) and shown enlarged in Fig. 14.10. The work done by the axial force $N(x)$ during the virtual displacement is

$$dW_I = N(x)\delta_e \tag{14.41}$$

where δ_e represents the relative displacement experienced by the normal force $N(x)$ acting on the differential element during the virtual displacement. From Fig. 14.10, by similar triangles (triangles I and II), we have

$$\frac{\delta_e}{d\psi_1(x)} = \frac{d\psi_2(x)}{dx}$$

or

$$\delta_e = \frac{d\psi_1}{dx} \cdot \frac{d\psi_2}{dx}\, dx$$

$$\delta_e = \psi_1'(x) \cdot \psi_2'(x)\, dx$$

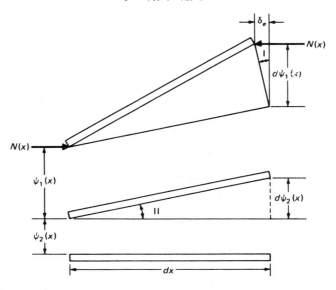

Fig. 14.10 Differential segment of deflected beam in Fig. 14.9.

in which $\psi_1'(x)$ and $\psi_2'(x)$ are the derivatives with respect to x of the corresponding displacement functions defined in eqs. (14.9).

Now, substituting δ_e into eq. (14.41), we have

$$dW_I = N(x)\,\psi_1'(x)\,\psi_2'(x)\,dx \qquad (14.42)$$

Then integrating this expression and equating the result to the external work, eq. (14.40), finally give

$$k_{G12} = \int_0^L N(x)\,\psi_1'(x)\,\psi_2'(x)\,dx \qquad (14.43)$$

In general, any geometric stiffness coefficient may be expressed as

$$k_{Gij} = \int_0^L N(x)\,\psi_i'(x)\,\psi_j'(x)\,dx \qquad (14.44)$$

In the derivation of eq. (14.44), it is assumed that the normal force $N(x)$ is independent of time. When the displacement functions, eqs. (14.9), are used in eq. (14.44) to calculate the geometric stiffness coefficients, the result is called the *consistent geometric stiffness* matrix. In the special case where the axial force is constant along the length of the beam, use of eqs. (14.44) and (14.9) gives the geometric stiffness equation as

$$\begin{Bmatrix} P_1 \\ P_2 \\ P_3 \\ P_4 \end{Bmatrix} = \frac{N}{30L} \begin{bmatrix} 36 & 3L & -36 & 3L \\ 3L & 4L^2 & -3L & -L^2 \\ -36 & -3L & 36 & -3L \\ 3L & -L^2 & -3L & 4L^2 \end{bmatrix} \begin{Bmatrix} \delta_1 \\ \delta_2 \\ \delta_3 \\ \delta_4 \end{Bmatrix} \qquad (14.45)$$

The assemblage of the system geometric stiffness matrix can be carried out exactly in the same manner as for the assemblage of the elastic stiffness matrix. The resulting geometric stiffness matrix will have the same configuration as the elastic stiffness matrix. It is customary to define the geometric stiffness matrix for a compressive axial force. In this case, the combined stiffness matrix $[K_c]$ for the structure is given by

$$[K_c] = [K] - [K_G] \qquad (14.46)$$

in which $[K]$ is the assembled elastic stiffness matrix for the structure and $[K_G]$ the corresponding geometric stiffness matrix.

Example 14.3. For the cantilever beam in Fig. 14.11, determine the system geometric matrix when an axial force of magnitude 30 N is applied at the free end as shown in this figure.

Solution: The substitution of numerical values into eq. (14.45) for any of

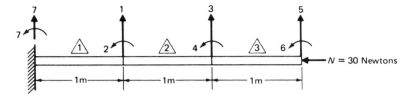

Fig. 14.11 Cantilever beam subjected to constant axial force (Example 14.3).

the three beam segments in which the cantilever beam has been divided gives the element geometric matrix

$$[K_G] = \begin{bmatrix} 36 & 3 & -36 & 3 \\ 3 & 4 & -3 & -1 \\ -36 & -3 & 36 & -3 \\ 3 & -1 & -3 & 4 \end{bmatrix}$$

since in this example $L = 1m$ and $N = 30$ N. Use of the direct method gives the assembled system geometric matrix as

$$[K_G] = \begin{bmatrix} 72 & 0 & -36 & 3 & 0 & 0 \\ 0 & 8 & -3 & -1 & 0 & 0 \\ -36 & -3 & 72 & 0 & -36 & 3 \\ 3 & -1 & 0 & 8 & -3 & -1 \\ 0 & 0 & -36 & -3 & 36 & -3 \\ 0 & 0 & 3 & -1 & -3 & 4 \end{bmatrix}$$

14.8   EQUATIONS OF MOTION

In the previous sections of this chapter, the distributed properties of a beam and its load were expressed in terms of discrete quantities at the nodal co-ordinates. The equations of motion as functions of these coordinates may then be established by imposing conditions of dynamic equilibrium between the inertial forces $\{F_I(t)\}$, damping forces $\{F_D(t)\}$, elastic forces $\{F_s(t)\}$, and the external forces $\{F(t)\}$, that is,

$$\{F_I(t)\} + \{F_D(t)\} + \{F_s(t)\} = \{F(t)\} \tag{14.47}$$

The forces on the left-hand side of eq. (14.47) are expressed in terms of the system mass matrix, the system damping matrix, and the system stiffness matrix as

$${F_I(t)} = [M]{\ddot{y}} \tag{14.48}$$

$${F_D(t)} = [C]{\dot{y}} \tag{14.49}$$

$${F_s(t)} = [K]{y} \tag{14.50}$$

Substitution of these equations into eq. (14.47) gives the differential equation of motion for a linear system as

$$[M]{\ddot{y}} + [C]{\dot{y}} + [K]{y} = {F(t)} \tag{14.51}$$

In addition, if the effect of axial forces is considered in the analysis, eq. (14.51) is modified so that

$$[M]{\ddot{y}} + [C]{\dot{y}} + [K_c]{y} = {F(t)} \tag{14.52}$$

in which

$$[K_c] = [K] - [K_G] \tag{14.53}$$

In practice, the solution of eq. (14.51) or eq. (14.52) is accomplished by standard methods of analysis and the assistance of appropriate computer programs as those described in this and the following chapters. We illustrate these methods by presenting here some simple problems for hand calculation.

Example 14.4.   Consider in Fig. 14.12 a uniform beam with the ends fixed against translation or rotation. In preparation for analysis, the beam has been divided into four equal segments. Determine the first three natural frequencies and corresponding modal shapes. Use the lumped mass method in order to simplify the numerical calculations.

Solution: We begin by numbering sequentially the nodal coordinates starting with the rotational coordinates which have to be condensed in the lumped mass method (no inertial effect associated with rotational coordinates), continuing to number the coordinates associated with translation, and assigning

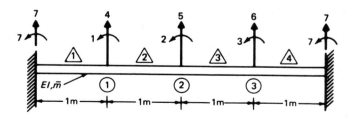

Fig. 14.12 Fixed beam divided in four elements with indication of system nodal coordinates (Example 14.4).

the dummy last number 7 to any fixed nodal coordinates as shown in Fig. 14.12. The stiffness matrix for any of the beam segments for this example is obtained from eq. (14.20) as

$$
[K] = EI
\begin{array}{cccc}
4 & 1 & 5 & 2 \\
7 & 7 & 4 & 1 \\
\begin{bmatrix}
12 & 6 & -12 & 6 \\
6 & 4 & -6 & 2 \\
-12 & -6 & 12 & -6 \\
6 & 2 & -6 & 4
\end{bmatrix}
\begin{array}{cc}
7 & 4 \\
7 & 1 \\
4 & 5 \\
1 & 2
\end{array}
\end{array}
\qquad (14.54)
$$

With the aid of the system nodal coordinates for each beam segment written at the top and on the right of the stiffness matrix, eq. (14.54), we proceed to assemble the system stiffness matrix using the direct method. For the beam segment △, the corresponding labels are 7, 7, 4, 1. Since the label 7 which corresponds to fixed coordinates should be ignored, we need to translate only the lowest 2 × 2 submatrix on the right to locations given by the combination of row indices 4, 1 and column indices 4, 1; for the beam segment △, we translate the 4 × 4 elements of matrix eq. (14.54) to the system stiffness matrix to rows and columns designated by combination of indices 4, 1, 5, 2 as labeled for this element and so forth for the other two beam segments. The assembled system stiffness matrix obtained in this manner is

$$
[K] = EI
\begin{bmatrix}
8 & 2 & 0 & 0 & -6 & 0 \\
2 & 8 & 2 & 6 & 0 & -6 \\
0 & 2 & 8 & 0 & 6 & 0 \\
0 & 6 & 0 & 24 & -12 & 0 \\
-6 & 0 & 6 & -12 & 24 & -12 \\
0 & -6 & 0 & 0 & -12 & 24
\end{bmatrix}
\qquad (14.55)
$$

The reduction or condensation of eq. (14.55) is accomplished as explained in Chapter 13 by simply performing the Gauss–Jordan elimination of the first three rows since, in this case, we should condense these first three coordinates. This elimination reduces eq. (14.55) to the following matrix:

$$
[A] =
\left[
\begin{array}{ccc:ccc}
1 & 0 & 0 & -0.214 & -0.750 & 0.214 \\
0 & 1 & 0 & 0.858 & 0 & -0.858 \\
0 & 0 & 1 & -0.214 & 0.750 & 0.214 \\
\hdashline
0 & 0 & 0 & 18.86EI & -12.00EI & 5.14EI \\
0 & 0 & 0 & -12.00EI & 15.00EI & -12.00EI \\
0 & 0 & 0 & 5.14EI & -12.00EI & 18.86EI
\end{array}
\right]
\qquad (14.56)
$$

Comparison of eq. (14.56) in partition form with eq. (13.11) permits the identification of the reduced stiffness matrix $[\bar{K}]$ and the transformation matrix $[\bar{T}]$, so that

$$[\bar{K}] = EI \begin{bmatrix} 18.86 & -12.00 & 5.14 \\ -12.00 & 15.00 & -12.00 \\ 5.14 & -12.00 & 18.86 \end{bmatrix} \tag{14.57}$$

and

$$[\bar{T}] = \begin{bmatrix} 0.214 & 0.750 & -0.214 \\ -0.858 & 0 & 0.858 \\ 0.214 & -0.750 & -0.214 \end{bmatrix} \tag{14.58}$$

The general transformation matrix, eq. (13.9), is then

$$[T] = \begin{bmatrix} 0.214 & 0.750 & -0.214 \\ -0.858 & 0 & 0.858 \\ 0.214 & -0.750 & -0.214 \\ 1 & 0 & 0 \\ 0 & 1 & 0 \\ 0 & 0 & 1 \end{bmatrix} \tag{14.59}$$

As an exercise, the reader may check eq. (13.7) for this example by simply performing the matrix multiplications

$$[\bar{K}] = [T]^T [K] [T]$$

The lumped mass method applied to this example gives three equal masses of magnitude $\bar{m}$ at each of the three translatory coordinates as indicated in Fig. 14.13. Therefore, the reduced lumped mass matrix is

$$[\bar{M}] = \bar{m} \begin{bmatrix} 1 & 0 & 0 \\ 0 & 1 & 0 \\ 0 & 0 & 1 \end{bmatrix} \tag{14.60}$$

The natural frequencies and modal shapes are found by solving the undamped free vibration problem, that is,

$$[\bar{M}] \{\ddot{y}\} + [\bar{K}] \{y\} = \{0\} \tag{14.61}$$

Assuming the harmonic solution ($\{y\} = \{a\} \sin \omega t$), we obtain

$$([\bar{K}] - \omega^2 [\bar{M}]) \{a\} = \{0\} \tag{14.62}$$

requiring for a nontrivial solution that the determinant

$$\|[\bar{K}] - \omega^2 [\bar{M}]\| = 0 \tag{14.63}$$

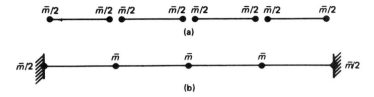

Fig. 14.13 (a) Lumped masses for uniform beam segments. (b) Lumped masses at system nodal coordinates.

Substitution in this last equation $[\bar{K}]$ and $[\bar{M}]$, respectively, from eqs. (14.57) and (14.60) yields

$$
\begin{vmatrix}
18.86 - \lambda & -12.00 & 5.14 \\
-12.00 & 15.00 - \lambda & -12.00 \\
5.14 & -12.00 & 18.86 - \lambda
\end{vmatrix} = 0
\qquad (14.64)
$$

in which

$$
\lambda = \frac{\bar{m}\omega^2}{EI}
\qquad (14.65)
$$

The roots of the cubic equation (14.64) are found to be

$$
\lambda_1 = 1.943, \quad \lambda_2 = 13.720, \quad \text{and} \quad \lambda_3 = 37.057
\qquad (14.66)
$$

Then from eq. (14.65)

$$
\omega_1 = 1.393 \sqrt{EI/\bar{m}}
$$

$$
\omega_2 = 3.704 \sqrt{EI/\bar{m}}
$$

$$
\omega_3 = 6.087 \sqrt{EI/\bar{m}}
\qquad (14.67)
$$

The first three natural frequencies for a uniform fixed beam of length $L = 4m$ determined by the exact analysis (Chapter 20) are

$$
\omega_1 \ (\text{exact}) = 1.398 \sqrt{EI/\bar{m}}
$$

$$
\omega_2 \ (\text{exact}) = 3.854 \sqrt{EI/\bar{m}}
$$

$$
\omega_3 \ (\text{exact}) = 7.556 \sqrt{EI/\bar{m}}
\qquad (14.68)
$$

The first two natural frequencies determined using the three-degrees-of-freedom reduced system compare very well with the exact values. A practical rule in condensing degrees of freedom is to reduce the system to a dimension of twice the number of required natural frequencies for the dynamic analysis.

The modal shapes are determined by solving two of the equations in eq. (14.62) after substituting successively values of ω_1, ω_2, ω_3 from eqs. (14.67) and letting the first element for each modal shape be equal to one. The resulting modal shapes are

$$\{a\}_1 = \begin{bmatrix} 1.00 \\ 1.84 \\ 1.00 \end{bmatrix}, \quad \{a\}_2 = \begin{bmatrix} 1.00 \\ 0 \\ -1.00 \end{bmatrix}, \quad \{a\}_3 = \begin{bmatrix} 1.00 \\ -1.08 \\ 1.00 \end{bmatrix} \qquad (14.69)$$

The normalized modal shapes which are obtained by division of the elements of eqs. (14.69) by corresponding values of $\sqrt{\Sigma m_i a_{ij}^2}$ ($m_i = \bar{m} = 100$ kg/m for this example) are arranged in the columns of the modal matrix as

$$[\Phi]_p = \begin{bmatrix} 0.0431 & 0.0707 & 0.0562 \\ 0.0793 & 0 & -0.0607 \\ 0.0431 & -0.0707 & 0.0562 \end{bmatrix} \qquad (14.70a)$$

The modal shapes in terms of the six original coordinates are then obtained by eq. (13.8) as

$$[\Phi] = [T][\Phi]_p$$

$$[\Phi] = \begin{bmatrix} 0.0594 & 0.0301 & -0.0457 \\ 0 & -0.1212 & 0 \\ -0.0594 & 0.0301 & 0.0457 \\ 0.0431 & 0.0707 & 0.0562 \\ 0.0793 & 0 & -0.0607 \\ 0.0431 & 0 & 0.0562 \end{bmatrix} \qquad (14.70b)$$

Example 14.5. Determine the steady-state response for the beam of Example 14.4 when subjected to the harmonic forces

$$F_1 = F_{01} \sin \bar{\omega} t$$

$$F_2 = F_{02} \sin \bar{\omega} t$$

and

$$F_3 = F_{03} \sin \bar{\omega} t$$

acting, respectively, at joints 1, 2, and 3 of the beam in Fig. 14.12. Neglect damping and let $EI = 10^8 (\text{N} \cdot \text{m}^2)$, $\bar{m} = 100$ kg/m, $\bar{\omega} = 3000$ rad/sec, $F_{01} = 2000$ N, $F_{02} = 3000$ N, and $F_{03} = 1000$ N.

Solution: The modal equations (uncoupled equations) can readily be written using the results of Example 14.4. In general the nth normal equation is given by

$$\ddot{z}_n + \omega_n^2 z_n = P_n \sin \bar{\omega} t \qquad (14.71)$$

in which

$$P_n = \sum_{i=1}^{N} \phi_{in} F_{0i}$$

The steady-state solution of eq. (14.71) is given by eq. (3.4) as

$$z_n = Z_n \sin \bar{\omega} t = \frac{P_n \sin \bar{\omega} t}{\omega_n^2 - \bar{\omega}^2} \qquad (14.72)$$

The calculations required in eq. (14.72) are conveniently arranged in Table 14.1.

The deflections at the nodal coordinates are found from the transformation

$$\{y\} = [\Phi]\{z\} \qquad (14.73)$$

in which $[\Phi]$ is the modal matrix and

$$\{y\} = \{Y\} \sin \bar{\omega} t$$

$$\{z\} = \{Z\} \sin \bar{\omega} t$$

The substitution of the modal matrix from eq. (14.70) and the values of $\{Z\}$ from the last column of Table 14.1 into eq. (14.73) gives the amplitudes at nodal coordinates as

$$\begin{Bmatrix} Y_1 \\ Y_2 \\ Y_3 \\ Y_4 \\ Y_5 \\ Y_6 \end{Bmatrix} = \begin{bmatrix} 0.0594 & 0.0301 & -0.0457 \\ 0 & -0.1212 & 0 \\ -0.0594 & 0.0301 & 0.0457 \\ 0.0431 & 0.0707 & 0.0562 \\ 0.0793 & 0 & -0.0607 \\ 0.0431 & -0.0707 & 0.0562 \end{bmatrix} \begin{Bmatrix} -5.200 \\ 1.500 \\ -0.048 \end{Bmatrix} 10^{-5}$$

or

$$Y_1 = -2.616 \times 10^{-6} \text{ rad}$$

$$Y_2 = -1.818 \times 10^{-6} \text{ rad}$$

$$Y_3 = 3.524 \times 10^{-6} \text{ rad}$$

TABLE 14.1   Modal Response for Example 14.5

Mode n	ω_n^2	$P_n = \sum_i \phi_{in} F_{0i}$	$Z_n = \dfrac{P_n}{\omega_n^2 - \bar{\omega}^2}$
1	1.943×10^6	367.2	$-5.200 \ 10^{-5}$
2	13.720×10^6	70.7	$1.500 \ 10^{-5}$
3	37.057×10^6	-13.5	$-0.048 \ 10^{-5}$

$$Y_4 = -1.207 \times 10^{-6} \text{ m}$$

$$Y_5 = -4.094 \times 10^{-6} \text{ m}$$

$$Y_6 = -3.329 \times 10^{-6} \text{ m}$$

Therefore, the motion at these nodal coordinates is given by

$$y_1 = -2.616 \times 10^{-6} \sin 3000t \text{ rad}$$

$$y_2 = -1.818 \times 10^{-6} \sin 3000t \text{ rad}$$

$$y_3 = 3.524 \times 10^{-6} \sin 3000t \text{ rad}$$

$$y_4 = -1.207 \times 10^{-6} \sin 3000t \text{ m}$$

$$y_5 = -4.094 \times 10^{-6} \sin 3000t \text{ m}$$

$$y_6 = -3.329 \times 10^{-6} \sin 3000t \text{ m} \qquad (14.74)$$

The minus sign in the resulting amplitudes of motion simply indicates that the motion is 180° out of phase with the applied harmonic forces.

14.9   ELEMENT FORCES AT NODAL COORDINATES

The central problem to be solved using the dynamic stiffness method is to determine the displacements at the nodal coordinates. Once these displacements have been determined, it is a simple matter of substituting the appropriate displacements in the condition of dynamic equilibrium for each element to calculate the forces at the nodal coordinates. The nodal element forces $\{P\}$ may be obtained by adding the inertial force $\{P_I\}$, the damping force $\{P_D\}$, the elastic force $\{P_S\}$, and subtracting the nodal equivalent forces $\{P_E\}$. Therefore, we may write

$$\{P\} = \{P_I\} + \{P_D\} + \{P_S\} - \{P_E\}$$

or

$$\{P\} = [m]\{\ddot{\delta}\} + [c]\{\dot{\delta}\} + [k]\{\delta\} - \{P_E\} \qquad (14.75)$$

In eq. (14.75) the inertial force, the damping force, and the elastic force are, respectively,

$$\{P_I\} = [m]\{\ddot{\delta}\}$$

$$\{P_D\} = [c]\{\dot{\delta}\}$$

$$\{P_S\} = [k]\{\delta\} \qquad (14.76)$$

where $[m]$ is the element mass matrix; $[c]$ the element damping matrix; $[k]$ the element stiffness matrix; and $\{\delta\}$, $\{\dot{\delta}\}$, $\{\ddot{\delta}\}$ represents, respectively, the displacement, velocity, and acceleration vectors at the nodal coordinates of the element.

The determination of the element nodal forces is illustrated in the following example.

Example 14.6. Determine the element nodal forces and moments for the four beam segments of Example 14.5.

Solution: Since in this example damping is neglected and there are no external forces applied except those at the nodal coordinates, eq. (14.75) reduces to

$$\{P\} = [m]\{\ddot{\delta}\} + [k]\{\delta\} \tag{14.77}$$

The displacement functions for the six nodal coordinates of the beam in Fig. 14.12 are given by eq. (14.74). These displacements are certainly also the displacements of the element nodal coordinates. The identification for this example of corresponding nodal coordinates between beam segments and system nodal coordinates is

$$\{\delta\}_1 = \begin{bmatrix} 0 \\ 0 \\ y_4 \\ y_1 \end{bmatrix}, \quad \{\delta\}_2 = \begin{Bmatrix} y_4 \\ y_1 \\ y_5 \\ y_2 \end{Bmatrix}, \quad \{\delta\}_3 = \begin{Bmatrix} y_5 \\ y_2 \\ y_6 \\ y_3 \end{Bmatrix}, \quad \{\delta\}_4 = \begin{Bmatrix} y_6 \\ y_3 \\ 0 \\ 0 \end{Bmatrix} \tag{14.78}$$

where $\{\delta\}_i$ is the vector of nodal displacement for i beam segment. The substitution of appropriate quantities into eq. (14.77) for the first beam segment results in

$$\begin{Bmatrix} P_1 \\ P_2 \\ P_3 \\ P_4 \end{Bmatrix} = \frac{\bar{m}}{2} \begin{bmatrix} 1 & 0 & 0 & 0 \\ 0 & 0 & 0 & 0 \\ 0 & 0 & 1 & 0 \\ 0 & 0 & 0 & 0 \end{bmatrix} \begin{Bmatrix} 0 \\ 0 \\ y_4 \\ y_1 \end{Bmatrix} (-\bar{\omega}^2) \sin \bar{\omega} t$$

$$+ EI \begin{bmatrix} 12 & 6 & -12 & 6 \\ 6 & 4 & -6 & 2 \\ -12 & -6 & 12 & -6 \\ 6 & 2 & -6 & 4 \end{bmatrix} \begin{Bmatrix} 0 \\ 0 \\ y_4 \\ y_1 \end{Bmatrix} \sin \bar{\omega} t$$

To complete this example, we substitute the numerical values of $\bar{\omega} = 3000$ rad/sec, $\bar{m} = 100$ kg/m, and $EI = 10^8$ (N·m^2) and obtain

$$\begin{Bmatrix} P_1 \\ P_2 \\ P_3 \\ P_4 \end{Bmatrix} = \begin{bmatrix} 7.5 & 6 & -12 & 6 \\ 6 & 4 & -6 & 2 \\ -12 & -6 & 7.5 & -6 \\ 6 & 2 & -6 & 4 \end{bmatrix} \begin{Bmatrix} 0 \\ 0 \\ -120.7 \\ -261.6 \end{Bmatrix} \sin 3000t$$

which then gives

$$P_1 = -121.2 \sin 3000t \text{ N}$$

$$P_2 = 201.0 \sin 3000t \text{ N} \cdot \text{m}$$

$$P_3 = 664.3 \sin 3000t \text{ N}$$

$$P_4 = -322.2 \sin 3000t \text{ N} \cdot \text{m}$$

The nodal element forces found in this manner for all of the four beam segments in this example are given in Table 14.2.

The results in Table 14.2 may be used to check that the dynamic conditions of equilibrium are satisfied in each beam segment. The free body diagrams of the four elements of this beam are shown in Fig. 14.14 with inclusion of nodal inertial forces. These forces are computed by multiplying the nodal mass by the corresponding nodal acceleration.

14.10 PROGRAM 13—MODELING STRUCTURES AS BEAMS

The computer program presented in this section calculates the stiffness and mass matrices for a beam and stores the coefficients of these matrices in a file, for future use. Since the stiffness and mass matrices are symmetric, only the upper triangular portion of these matrices needs to be stored. The program also stores in another file, named by the user, the general information on the beam. The information stored in these files is needed by programs which perform dynamic analysis such as calculation of natural frequencies or determination of the response of the structure subjected to external excitation.

Example 14.7. Determine the stiffness and mass matrices for the uniform fixed-ended beam shown in Fig. 14.15. The following are the properties of the beam:

TABLE 14.2 Element Nodal Forces (Amplitudes) for Example 14.6

Force	\multicolumn{4}{c}{Beam Segment}	Units
	1	2	3	4	
P_1	−121.2	1347.1	1947.9	−382.3	N
P_2	201.0	322.2	−481.4	−587.0	N · m
P_3	664.3	1038.3	1392.4	1880.4	N
P_4	−322.2	481.4	587.0	−1292.6	N · m

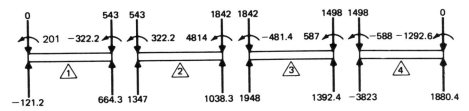

Fig. 14.14 Dynamic equilibrium for beam segments of Example 14.6.

Length: $L = 200$ in
Cross-section moment of inertia: $I = 100$ in^4
Modulus of elasticity: $E = 6.58$ E6 lb/in^2
Mass per unit length: $\bar{m} = 0.10$ (lb·sec^2/in/in)

Solution: The beam is divided in four segments of equal length as shown in Fig. 14.15. This division results in a total of five joints of which two are fixed, thus giving a total of four fixed coordinates (the two ends of the beam are fixed for translation and for rotation).

Input Data and Output Results

```
PROGRAM 13: MODELING BEAMS          DATA FILE:D13

    GENERAL DATA:

NUMBER OF JOINTS                                    NJ= 5
NUMBER OF BEAM ELEMENTS                             NE= 4
NUMBER OF FIXED COORDINATES                         NC= 4
MODULUS OF ELASTICITY                                E= 6580000

    JOINT DATA: (ONE LINE PER JOINT)

JOINT #          X-COORDINATE          CONCENTRATED JOINT MASS

    1               0.00                      0.00
    2              50.00                      0.00
    3             100.00                      0.00
    4             150.00                      0.00
    5             200.00                      0.00
```

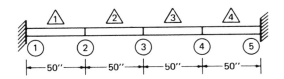

Fig. 14.15 Fixed beam modeled for Example 14.7.

```
BEAM ELEMENT DATA: (ONE LINE FOR EACH BEAM ELEMENT)
      (NEGATIVE JOINT NUMBERS TO RELEASED JOINTS)

BEAM-ELEMENT #,LEFT JOINT #,    RIGHT JOINT #,    MASS/LENGTH,  MOMENT OF INERTIA
         1               1               2            0.10           100.00
         2               2               3            0.10           100.00
         3               3               4            0.10           100.00
         4               4               5            0.10           100.00

FIXED COORDINATE DATA:

JOINT #,  DIRECTION # (VERTICAL = 1, ROTATION = 2)

  1          1
  1          2
  5          1
  5          2
```

OUTPUT RESULTS:

SYSTEM STIFFNESS MATRIX

```
 1.2634E+05    0.0000E+00   -6.3168E+04    1.5792E+06    0.0000E+00    0.0000E+00
 0.0000E+00    1.0528E+08   -1.5792E+06    2.6320E+07    0.0000E+00    0.0000E+00
-6.3168E+04   -1.5792E+06    1.2634E+05    0.0000E+00   -6.3168E+04    1.5792E+06
 1.5792E+06    2.6320E+07    0.0000E+00    1.0528E+08   -1.5792E+06    2.6320E+07
 0.0000E+00    0.0000E+00   -6.3168E+04   -1.5792E+06    1.2634E+05    0.0000E+00
 0.0000E+00    0.0000E+00    1.5792E+06    2.6320E+07    0.0000E+00    1.0528E+08
```

SYSTEM MASS MATRIX

```
 3.7143E+00    0.0000E+00    6.4286E-01   -7.7381E+00    0.0000E+00    0.0000E+00
 0.0000E+00    2.3810E+02    7.7381E+00   -8.9286E+01    0.0000E+00    0.0000E+00
 6.4286E-01    7.7381E+00    3.7143E+00    0.0000E+00    6.4286E-01    7.7381E+00
-7.7381E+00   -8.9286E+01    0.0000E+00    2.3810E+02    7.7381E+00   -8.9286E+01
 0.0000E+00    0.0000E+00    6.4286E-01    7.7381E+00    3.7143E+00    0.0000E+00
 0.0000E+00    0.0000E+00   -7.7381E+00   -8.9286E+01    0.0000E+00    2.3810E+02
```

Example 14.8. For the fixed beam modeled in Example 14.7 determine: (a) the natural frequencies and modal shapes, (b) the response to concentrated force of 1000 lb suddenly applied at the center of the beam for 0.1 sec and removed linearly as shown in Fig. 14.16. Use time step of integration $\Delta t = 0.01$ sec.

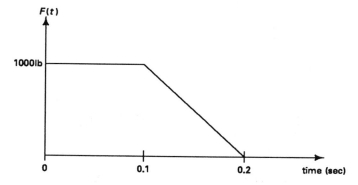

Fig. 14.16 Force applied to the beam of Example 14.8

Solution: (a) During the execution of Program 13, to model the beam in Problem 14.7, a file named "SK" has been created to store the data necessary to execute Program 8 to calculate natural frequencies and corresponding modal shapes. The execution of Program 8 gives the following output:

Output Results

PROGRAM 8: NATURAL FREQUENCIES   DATA FILE: SK

    OUTPUT RESULTS

EIGENVALUES:

2.064E+03 1.593E+04 6.271E+04 2.245E+05 6.140E+05 1.594E+06

NATURAL FREQUENCIES (C.P.S.):

7.23   20.09   39.86   75.40   124.71   200.93

EIGENVECTORS BY ROWS:

0.19352	0.00542	0.35606	−0.00000	0.19352	−0.00542
−0.32920	−0.00355	0.00000	0.01299	0.32920	−0.00355
0.32051	−0.00754	−0.33013	−0.00000	0.32051	0.00754
−0.15157	0.02892	−0.00000	−0.03233	0.15157	0.02892
0.06682	−0.04865	0.31435	−0.00000	−0.06682	0.04865
0.18579	0.04758	0.00000	0.07737	−0.18579	0.04758

(b) The execution of Program 8 "Jacobi" creates a file named "EA" in which the eigenvalues ω_i^2 and the eigenvectors $\{\phi\}_i$ are stored for subsequent use by other programs such as Program 9 which serve to calculate the response of the structure to externally applied forces. The following is the computer output for execution of Program 9:

Input Data and Output Results

    PROGRAM 9: MODAL   DATA FILE:D14.88

    INPUT DATA:

NUMBER OF DEGREES OF FREEDOM	ND= 6
NUMBER OF EXTERNAL FORCES	NF= 1
TIME STEP OF INTEGRATION	H= .01
GRAVITATIONAL INDEX	G= 0
PRINT TIME HISTORY NPRT = 1; ONLY MAX. VALUES NPRT = 0	NPRT= 0

FORCE #, COORD. # WHERE FORCE IS APPLIED, NUM. OF POINTS DEFINING THE FORCE

1   3   3

EXCITATION FUNCTION FOR FORCE # 1:

TIME	EXCITATION
0.0000	10000.00
0.1000	10000.00
0.2000	0.00

MODAL DAMPING RATIOS:

0	0	0	0	0
0				

MAXIMUM RESPONSE:

COORD.	MAX. DISPL.	MAX. VELOC.	MAX. ACC.
1	0.654	17.13	1740.37
2	0.019	0.70	177.90
3	1.243	30.15	3345.84
4	0.000	0.00	0.00
5	0.654	17.13	1740.37
6	0.019	0.70	177.90

14.11   SUMMARY

In this chapter, we have formulated the dynamic equations for beams in reference to a discrete number of nodal coordinates. These coordinates are translational and rotational displacements defined at joints between structural elements of the beam (beam segments). The dynamic equations for a linear system are conveniently written in matrix notation as

$$[M]\{\ddot{y}\} + [C]\{\dot{y}\} + [K]\{y\} = \{F(t)\}$$

where $F(t)$ is the force vector and $[M]$, $[C]$, and $[K]$ are, respectively, the mass, damping, and stiffness matrices of the structure. These matrices are assembled by the appropriate superposition (direct method) of the matrices determined for each beam segment of the structure.

The solution of the dynamic equations (i.e., the response) of a linear system may be found by the modal superposition method. This method requires the determination of the natural frequencies ω_n $(n = 1, 2, 3, ..., N)$ and the corresponding normal modes which are conveniently written as the columns of the modal matrix $[\Phi]$. The linear transformation $\{y\} = [\Phi]\{z\}$ applied to the dynamic equations reduces them to a set of independent equations (uncoupled equations) of the form

$$\ddot{z}_n + 2\omega_n\xi_n\dot{z}_n + \omega_n^2 z_n = P_n(t)$$

where ξ_n is the modal damping ratio and $P_n(t) = \Sigma_i\phi_{in}F_i(t)$ is the modal force.

An alternate method for determining the response of linear systems (also valid for nonlinear systems) is the numerical integration of the dynamic equations. Chapter 19 presents the step-by-step linear acceleration method (with a modification introduced by Wilson) which is an efficient method for solving the dynamic equations.

A computer program is also described for the dynamic analysis of beams. This program performs the task of assembling and storing in a file the stiffness and mass matrices of the system. These matrices are subsequently used by other programs to calculate natural frequencies or the response of the beam to external excitation.

PROBLEMS

14.1 A uniform beam of flexural stiffness $EI = 10^9$ (lb · in²) and length 300 in has one end fixed and the other simply supported. Determine the system stiffness matrix considering three beam segments and the nodal coordinates indicated in Fig. 14.17.

14.2 Assuming that the beam shown in Fig. 14.17 carries a uniform weight per unit length $q = 3.86$ lb/in, determine the system mass matrix corresponding to the lumped mass formulation.

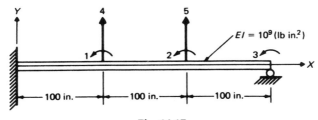

Fig. 14.17

14.3 Determine the system mass matrix for Problem 14.2 using the consistent mass method.

14.4 For the beam in Problems 14.1 and 14.2, use static condensation to eliminate the massless degrees of freedom. Find the transformation matrix and the reduced stiffness and mass matrices.

14.5 For the beam in Problems 14.1 and 14.3 use static condensation to eliminate the rotational degrees of freedom. Find the transformation matrix and the reduced stiffness and mass matrices.

14.6 Determine the natural frequencies and corresponding normal modes using the reduced stiffness and mass matrices obtained in Problem 14.4.

14.7 Determine the natural frequencies and corresponding normal modes using the reduced stiffness and mass matrices obtained in Problem 14.3.

14.8 Determine the geometric stiffness matrix for the beam of Problem 14.1 when it is subjected to a constant tensile force of 10,000 lb as shown in Fig. 14.18.

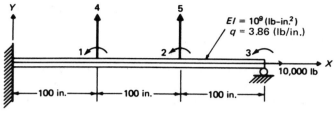

Fig. 14.18

14.9 Perform static condensation to reduce the geometric stiffness matrix obtained in Problem 14.8. Eliminate the rotational coordinates.

14.10 Use results from Problem 14.4 and 14.9 and determine the natural frequencies and corresponding normal modes for the beam shown in Fig. 14.18.

14.11 Use the results of Problems 14.5 and 14.9 and determine the natural frequencies and corresponding normal modes for the beam shown in Fig. 14.18.

14.12 Determine the stiffness matrix for a beam segment in which the flexural stiffness has a linear variation as shown in Fig. 14.19.

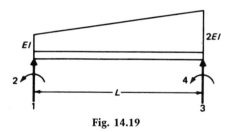

Fig. 14.19

14.13 Determine the lumped mass matrix for a beam segment in which the mass has a linear distribution as shown in Fig. 14.20.

14.14 Determine the consistent mass matrix for the beam segment shown in Fig. 14.20.

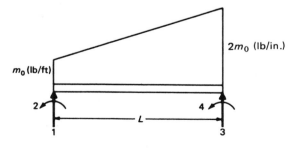

Fig. 14.20

14.15 The uniform beam shown in Fig. 14.21 is subjected to a constant force of 5000 lb suddenly applied along the nodal coordinate 4. Use the results obtained in Problem 14.6 to determine the response by the modal superposition method. (Use only the two modes left by the static condensation.)

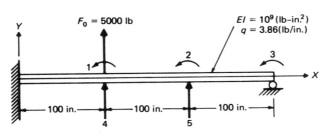

Fig. 14.21

14.16 Solve Problem 14.15 using the results obtained in Problem 14.7 which are based on the consistent mass formulation.

14.17 Solve Problem 14.15 using the results obtained in Problem 14.9 which includes the effect of the axial force in the stiffness of the system.

14.18 Determine the steady-state response for the beam shown in Fig. 14.22 which is acted upon by a harmonic force $F(t) = 5000 \sin 30t$ (lb) as shown in the figure. Eliminate the rotational coordinates by static condensation (Problem 14.5). Neglect damping in the system.

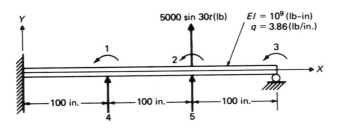

Fig. 14.22

14.19 Determine the natural frequencies and corresponding normal modes for the beam shown in Fig. 14.17: (a) condensing the three rotational nodal coordinates; (b) no condensing coordinates. (Use the consistent mass method.)

14.20 Determine the response for the beam shown in Fig. 14.21. Neglect damping. (a) Do not condense coordinates; (b) condense the three rotational coordinates.

14.21 Repeat Problem 14.20 assuming 10% damping in all the modes.

14.22 Determine the steady-state response for the beam shown in Fig. 14.22 when subjected to a harmonic force as shown in the figure. Do not condense coordinates, and neglect damping in the system.

14.23 Repeat Problem 14.22 assuming that the damping is proportional to stiffness of the system where the constant of proportionality $a_0 = 0.2$.

14.24 Solve Problem 14.22 after condensing the three rotational coordinates.

14.25 Repeat Problem 14.24 assuming 15% damping in all the modes.

14.26 Determine the steady-state response for the beam shown in Fig. 14.21. Do not condense coordinates, and neglect damping in the system.

15

Dynamic Analysis of Plane Frames

The dynamic analysis using the stiffness matrix method for structures modeled as beams was presented in Chapter 14. This method of analysis when applied to beams requires the calculation of element matrices (stiffness, mass, and damping matrices), the assemblage from these matrices of the corresponding system matrices, the formation of the force vector, and the solution of the resultant equations of motion. These equations, as we have seen, may be solved in general by the modal superposition method or by numerical integration of the differential equations of motion. In this chapter and in the following chapters, the dynamic analysis of structures modeled as frames is presented.

We begin in this chapter with the analysis of structures modeled as plane frames and with the loads acting in the plane of the frame. The dynamic analysis of such structures requires the inclusion of the axial effects in the stiffness and mass matrices. It also requires a coordinate transformation of the nodal coordinates from element or local coordinates to system or global coordinates. Except for the consideration of axial effects and the need to transform these coordinates, the dynamic analysis by the stiffness method when applied to frames is identical to the analysis of beams as discussed in Chapter 14.

15.1   ELEMENT STIFFNESS MATRIX FOR AXIAL EFFECTS

The inclusion of axial forces in the stiffness matrix of a flexural beam segment requires the determination of the stiffness coefficients for axial loads. To derive the stiffness matrix for an axially loaded member, consider in Fig. 15.1 a beam segment acted on by the axial forces P_1 and P_2 producing axial displacements δ_1 and δ_2 at the nodes of the element. For a prismatic and uniform beam segment of length L and cross-sectional A, it is relatively simple to obtain the stiffness relation for axial effects by the application of Hooke's law. In relation to the beam shown in Fig. 15.1, the displacements δ_1 produced by the force P_1 acting at node 1 while node 2 is maintained fixed ($\delta_2 = 0$) is given by

$$\delta_1 = \frac{P_1 L}{AE} \tag{15.1}$$

From eq. (15.1) and the definition of the stiffness coefficient k_{11} (force at node 1 to produce a unit displacement, $\delta_1 = 1$), we obtain

$$k_{11} = \frac{P_1}{\delta_1} = \frac{AE}{L} \tag{15.2a}$$

The equilibrium of the beam segment acted upon by the force k_{11} requires a force k_{21} at the other end, namely

$$k_{21} = -k_{11} = -\frac{AE}{L} \tag{15.2b}$$

Analogously, the other stiffness coefficients are

$$k_{22} = \frac{AE}{L} \tag{15.2c}$$

and

$$k_{12} = -\frac{AE}{L} \tag{15.2d}$$

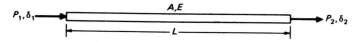

Fig. 15.1 Beam element with nodal axial loads P_1, P_2, and corresponding nodal displacements δ_1, δ_2.

The stiffness coefficients as given by eqs. (15.2) are the elements of the stiffness matrix relating axial forces and displacements for a prismatic beam segment, that is,

$$\begin{bmatrix} P_1 \\ P_2 \end{bmatrix} = \frac{AE}{L}\begin{bmatrix} 1 & -1 \\ -1 & 1 \end{bmatrix}\begin{bmatrix} \delta_1 \\ \delta_2 \end{bmatrix} \qquad (15.3)$$

The stiffness matrix corresponding to the nodal coordinates for the beam segment shown in Fig. 15.2 is obtained by combining in a single matrix the stiffness matrix for axial effects, eq. (15.3), and the stiffness matrix for flexural effects, eq. (14.20). The matrix resulting from this combination relates the forces P_i and the displacements δ_i at the coordinates indicated in Fig. 15.2 as

$$\begin{Bmatrix} P_1 \\ P_2 \\ P_3 \\ P_4 \\ P_5 \\ P_6 \end{Bmatrix} = \frac{EI}{L^3}\begin{bmatrix} AL^2/I & & & & \text{symmetric} & \\ 0 & 12 & & & & \\ 0 & 6L & 4L^2 & & & \\ -AL^2/I & 0 & 0 & AL^2/I & & \\ 0 & -12 & -6L & 0 & 12 & \\ 0 & 6L & 2L^2 & 0 & -6L & 4L^2 \end{bmatrix}\begin{Bmatrix} \delta_1 \\ \delta_2 \\ \delta_3 \\ \delta_4 \\ \delta_5 \\ \delta_6 \end{Bmatrix} \qquad (15.4)$$

or, in concise notation,

$$\{P\} = [K]\{\delta\} \qquad (15.5)$$

15.2 ELEMENT MASS MATRIX FOR AXIAL EFFECTS

The determination of mass influence coefficients for axial effects of a beam element may be carried out by any of two methods indicated previously for the flexural effects: (1) the lumped mass method and (2) the consistent mass method. In the lumped mass method, the mass allocation to the nodes of the beam element is found from static considerations which for a uniform beam gives half of the total mass of the beam segment allocated at each node. Then for a prismatic beam segment, the relation between nodal axial forces and nodal accelerations is given by

$$\begin{Bmatrix} P_1 \\ P_2 \end{Bmatrix} = \frac{\bar{m}L}{2}\begin{bmatrix} 1 & 0 \\ 0 & 1 \end{bmatrix}\begin{Bmatrix} \ddot{\delta}_1 \\ \ddot{\delta}_2 \end{Bmatrix} \qquad (15.6)$$

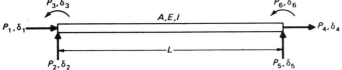

Fig. 15.2 Beam element showing flexural and axial nodal forces and displacements.

where $\overline{m}$ is the mass per unit of length. The combination of the flexural lumped mass coefficient and axial mass coefficients gives, in reference to the nodal coordinates in Fig. 15.2, the following diagonal matrix:

$$\begin{Bmatrix} P_1 \\ P_2 \\ P_3 \\ P_4 \\ P_5 \\ P_6 \end{Bmatrix} = \frac{\overline{m}L}{2} \begin{bmatrix} 1 & & & & & \\ & 1 & & & & \\ & & 0 & & & \\ & & & 1 & & \\ & & & & 1 & \\ & & & & & 0 \end{bmatrix} \begin{Bmatrix} \ddot{\delta}_1 \\ \ddot{\delta}_2 \\ \ddot{\delta}_3 \\ \ddot{\delta}_4 \\ \ddot{\delta}_5 \\ \ddot{\delta}_6 \end{Bmatrix} \qquad (15.7)$$

To calculate the coefficients for the consistent mass matrix, it is necessary first to determine the displacement functions corresponding to a unit axial displacement at one of the nodal coordinates. Consider in Fig. 15.3 an axial unit displacement $\delta_1 = 1$ of node 1 while the other node 2 is kept fixed so that $\delta_2 = 0$. If $u = u(x)$ is the displacement at section x, the displacement at section $x + dx$ will be $u + du$. It is evident then that the element dx in the new position has changed in length by an amount du, and thus, the strain is du/dx. Since from Hooke's law, the ratio of stress to strain is equal to the modulus of elasticity E, we can write

$$\frac{du}{dx} = \frac{P}{AE} \qquad (15.8)$$

Integration with respect to x yields

$$u = \frac{P}{AE} x + C \qquad (15.9)$$

in which C is a constant of integration. Introducing the boundary conditions, $u = 1$ at $x = 0$ and $u = 0$ at $x = L$, we obtain the displacement function $u_1(x)$ corresponding to a unit displacement δ_1 as

$$u_1(x) = 1 - \frac{x}{L} \qquad (15.10)$$

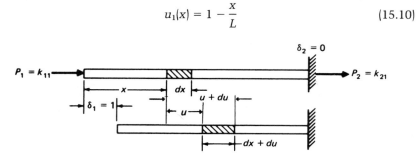

Fig. 15.3 Displacement at node 1 ($\delta_1 = 1$) of a beam element.

Analogously, the displacement function $u_2(x)$ corresponding to a unit displacement $\delta_2 = 1$ is

$$u_2(x) = \frac{x}{L} \tag{15.11}$$

The application of the principle of virtual work results in a general expression for the calculation of the stiffness coefficients. For example, consider the beam in Fig. 15.3, which is in equilibrium with the forces $P_1 = k_{11}$ and $P_2 = k_{21}$ at its two ends. Assume that a virtual displacement $\delta_2 = 1$ takes place. Then, according to the principle of virtual work, during this virtual displacement, the work of the external and internal forces are equal. The external force k_{21} performs the work

$$W_E = k_{21}\delta_2$$

or

$$W_E = k_{21} \tag{15.12}$$

since $\delta_2 = 1$. The internal force $P(x)$ at any section x is obtained from eq. (15.8) as

$$P(x) = AE\, u_1'(x) \tag{15.13}$$

in which $u_1'(x) = du_1/dx$. The relative displacement of element dx during this virtual displacement is

$$du_2 = \frac{du_2}{dx}\, dx \tag{15.14}$$

as shown in Fig. 15.4. Hence the internal work for element dx is obtained from eqs. (15.13) and (15.14) as

$$dW_I = AE\, u_1'(x)u_2'(x)\, dx$$

and for the beam segment of length L

$$W_I = \int_0^L AE\, u_1'(x)\, u_2'(x)\, dx \tag{15.15}$$

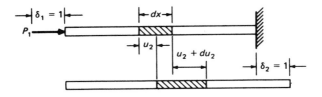

Fig. 15.4 Displacement at node 2 $(\delta_2 = 1)$ of a beam element subjected to axial displacement at node 1 $(\ddot{\delta}_1 = 1)$.

Finally, equating $W_E = W_I$ from eqs. (15.12) and (15.15) gives the stiffness coefficient

$$k_{21} = \int_0^L AE\, u_1'(x)\, u_2'(x)\, dx \qquad (15.16)$$

In general, the stiffness coefficient k_{ij} for axial effects may be obtained from

$$k_{ij} = \int_0^L AE\, u_i'(x)\, u_j'(x)\, dx \qquad (15.17)$$

Using eq. (15.17), the reader may check the results obtained in eq. (15.3) for a uniform beam. However, eq. (15.17) could as well be used for nonuniform beams in which AE would in general be a function of x. In practice, the same displacement $u_1(x)$ and $u_2(x)$ obtained for a uniform beam are also used in eq. (15.17) for a nonuniform member. The displacement $y(x, t)$ at any section x of a beam element due to dynamic nodal displacements, $\delta_1(t)$ and $\delta_2(t)$, is obtained by superposition. Hence

$$y(x, t) = u_1(x)\, \delta_1(t) + u_2(x)\, \delta_2(t) \qquad (15.18)$$

in which $u_1(x)$ and $u_2(x)$ are given by eqs. (15.10) and (15.11).

Now consider the beam of Fig. 15.5 while undergoing a unit acceleration, $\ddot{\delta}_1(t) = 1$ which by eq. (15.18) results in an acceleration at x given by

$$\ddot{u}_1(x, t) = u_1(x)\, \ddot{\delta}_1(t)$$

or

$$\ddot{u}_1(x, t) = u_1(x)$$

since $\ddot{\delta}_1(t) = 1$. The inertial force per unit length along the beam resulting from this unit acceleration is

$$f_I = \overline{m}(x) u_1(x) \qquad (15.19)$$

where $\overline{m}(x)$ is the mass per unit length along the beam. Now, to determine the mass coefficient m_{21}, we give to the beam shown in Fig. 15.5 a virtual

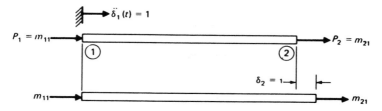

Fig. 15.5 Displacement at node 2 $(\delta_2 = 1)$ of a beam element undergoing axial acceleration at node 1 $(\ddot{\delta}_1(t) = 1)$.

displacement $\delta_2 = 1$. The only external force doing work during this virtual displacement is the reaction m_{21}. This work is then

$$W_E = m_{21}\, \delta_2$$

or

$$W_E = m_{21} \tag{15.20}$$

since $\delta_2 = 1$. The internal work per unit length along the beam performed by the inertial force f_I during this virtual displacement is

$$\delta W_I = f_I(x)\, u_2(x)$$

or, from eq. (15.19),

$$\delta W_I = \overline{m}(x)\, u_1(x)\, u_2(x)$$

Hence the total internal work is

$$W_I = \int_0^L \overline{m}(x)\, u_1(x)\, u_2(x)\, dx \tag{15.21}$$

Finally, equating eqs. (15.20) and (15.21) yields

$$m_{21} = \int_0^L \overline{m}(x)\, u_1(x)\, u_2(x)\, dx \tag{15.22}$$

or, in general,

$$m_{ij} = \int_0^L \overline{m}(x)\, u_i(x)\, u_j(x)\, dx \tag{15.23}$$

The application of eq. (15.23) to the special case of a uniform beam results in

$$m_{11} = \int_0^L \overline{m}\left(1 - \frac{x}{L}\right)^2 dx = \frac{\overline{m}L}{3} \tag{15.24}$$

Similarly,

$$m_{22} = \frac{\overline{m}L}{3}$$

and

$$m_{12} = m_{21} = \int_0^L \overline{m}\left(1 - \frac{x}{L}\right)\left(\frac{x}{L}\right) dx = \frac{\overline{m}L}{6} \tag{15.25}$$

In matrix form, the axial inertial force relationship for a uniform beam may be written as

$$\begin{Bmatrix} P_1 \\ P_2 \end{Bmatrix} = \frac{\overline{m}L}{6} \begin{bmatrix} 2 & 1 \\ 1 & 2 \end{bmatrix} \begin{Bmatrix} \ddot{\delta}_1 \\ \ddot{\delta}_2 \end{Bmatrix} \tag{15.26}$$

Finally, combining the mass matrix eq. (14.34) for flexural effects with eq. (15.26) for the axial effects, we obtain the consistent mass matrix for a uniform beam element in reference to the nodal coordinates as shown in Fig. 15.2 as

$$\begin{Bmatrix} P_1 \\ P_2 \\ P_3 \\ P_4 \\ P_5 \\ P_6 \end{Bmatrix} = \frac{\overline{m}L}{420} \begin{bmatrix} 140 & & & & \text{Symmetric} & \\ 0 & 156 & & & & \\ 0 & 22L & 4L^2 & & & \\ 70 & 0 & 0 & 140 & & \\ 0 & 54 & 13L & 0 & 156 & \\ 0 & -13L & -3L^2 & 0 & -22L & 4L^2 \end{bmatrix} \begin{Bmatrix} \ddot{\delta}_1 \\ \ddot{\delta}_2 \\ \ddot{\delta}_3 \\ \ddot{\delta}_4 \\ \ddot{\delta}_5 \\ \ddot{\delta}_6 \end{Bmatrix} \quad (15.27)$$

or, in condensed notation,

$$\{P\} = [M_c] \{\ddot{\delta}\}$$

in which $[M_c]$ is the consistent mass matrix.

15.3   COORDINATE TRANSFORMATION

The stiffness matrix for the beam element in eq. (15.4) as well as the mass matrix in eq. (15.27) are in reference to nodal coordinates defined by coordinate axes fixed on the beam element. These axes are called *local* or *element axes* while the coordinate axes for the whole structure are known as *global* or *system axes*. Figure 15.6 shows a beam element with nodal forces P_1, P_2,

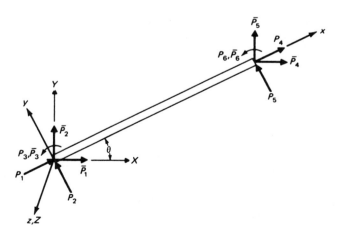

Fig. 15.6 Beam element showing nodal forces in local (x, y, z) and global coordinate axes (X, Y, Z).

. . ., P_6 referred to the local coordinate axes x, y, z, and $\bar{P}_1$, $\bar{P}_2$, . . ., $\bar{P}_6$ referred to global coordinate set of axes X, Y, Z. The objective is to transform the element matrices (stiffness, mass, etc.) from the reference of local coordinate axes to the global coordinate axes. This transformation is required in order that the matrices for all the elements refer to the same set of coordinates; hence, the matrices become compatible for assemblage into the system matrices for the structure. We begin by expressing the forces (P_1, P_2, P_3) in terms of the forces $(\bar{P}_1, \bar{P}_2, \bar{P}_3)$. Since these two sets of forces are equivalent, we obtain from Fig. 15.6 the following relations:

$$P_1 = \bar{P}_1 \cos \theta + \bar{P}_2 \sin \theta$$

$$P_2 = -\bar{P}_1 \sin \theta + \bar{P}_2 \cos \theta$$

$$P_3 = \bar{P}_3 \qquad (15.28)$$

The first two equations of eq. (15.28) may be written in matrix notation as

$$\begin{Bmatrix} P_1 \\ P_2 \end{Bmatrix} = \begin{bmatrix} \cos \theta & \sin \theta \\ -\sin \theta & \cos \theta \end{bmatrix} \begin{Bmatrix} \bar{P}_1 \\ \bar{P}_2 \end{Bmatrix} \qquad (15.29)$$

Analogously, we obtain for the forces on the other node the relations:

$$P_4 = \bar{P}_4 \cos \theta + \bar{P}_5 \sin \theta$$

$$P_5 = -\bar{P}_4 \sin \theta + \bar{P}_5 \cos \theta$$

$$P_6 = \bar{P}_6 \qquad (15.30)$$

Equations (15.28) and (15.30) may conveniently be arranged in matrix form as

$$\begin{Bmatrix} P_1 \\ P_2 \\ P_3 \\ P_4 \\ P_5 \\ P_6 \end{Bmatrix} = \begin{bmatrix} \cos \theta & \sin \theta & 0 & 0 & 0 & 0 \\ -\sin \theta & \cos \theta & 0 & 0 & 0 & 0 \\ 0 & 0 & 1 & 0 & 0 & 0 \\ 0 & 0 & 0 & \cos \theta & \sin \theta & 0 \\ 0 & 0 & 0 & -\sin \theta & \cos \theta & 0 \\ 0 & 0 & 0 & 0 & 0 & 1 \end{bmatrix} \begin{Bmatrix} \bar{P}_1 \\ \bar{P}_2 \\ \bar{P}_3 \\ \bar{P}_4 \\ \bar{P}_5 \\ \bar{P}_6 \end{Bmatrix} \qquad (15.31)$$

or in condensed notation

$$\{P\} = [T]\{\bar{P}\} \qquad (15.32)$$

in which $\{P\}$ and $\{\bar{P}\}$ are, respectively, the vectors of the element nodal forces in local and global coordinates and $[T]$ is the transformation matrix given by the square matrix in eq. (15.31).

Repeating the same procedure, we obtain the relation between nodal displacements $(\delta_1, \delta_2, . . ., \delta_6)$ in local coordinates and the components of the nodal displacements in global coordinates $(\bar{\delta}_1, \bar{\delta}_2, . . ., \bar{\delta}_6)$, namely

$$\begin{Bmatrix} \delta_1 \\ \delta_2 \\ \delta_3 \\ \delta_4 \\ \delta_5 \\ \delta_6 \end{Bmatrix} = \begin{bmatrix} \cos\theta & \sin\theta & 0 & 0 & 0 & 0 \\ -\sin\theta & \cos\theta & 0 & 0 & 0 & 0 \\ 0 & 0 & 1 & 0 & 0 & 0 \\ 0 & 0 & 0 & \cos\theta & \sin\theta & 0 \\ 0 & 0 & 0 & -\sin\theta & \cos\theta & 0 \\ 0 & 0 & 0 & 0 & 0 & 1 \end{bmatrix} \begin{Bmatrix} \bar{\delta}_1 \\ \bar{\delta}_2 \\ \bar{\delta}_3 \\ \bar{\delta}_4 \\ \bar{\delta}_5 \\ \bar{\delta}_6 \end{Bmatrix} \qquad (15.33)$$

or

$$\{\delta\} = [T]\{\bar{\delta}\} \qquad (15.34)$$

Now, the substitution of $\{P\}$ from eq. (15.32) and $\{\delta\}$ from eq. (15.34) into the stiffness equation referred to local axes $\{P\} = [K]\{\delta\}$ results in

$$[T]\{\bar{P}\} = [K][T]\{\bar{\delta}\}$$

or

$$\{\bar{P}\} = [T]^{-1}[K][T]\{\bar{\delta}\} \qquad (15.35)$$

where $[T]^{-1}$ is the inverse of matrix $[T]$. However, as the reader may verify, the transformation matrix $[T]$ in eq. (15.31) is an orthogonal matrix, that is, $[T]^{-1} = [T]^T$. Hence

$$\{\bar{P}\} = [T]^T[K][T]\{\bar{\delta}\} \qquad (15.36)$$

or, in a more convenient notation,

$$\{\bar{P}\} = [\bar{K}]\{\bar{\delta}\} \qquad (15.37)$$

in which

$$[\bar{K}] = [T]^T[K][T] \qquad (15.38)$$

is the stiffness matrix for a beam segment in reference to the global system of coordinates.

Repeating the procedure of transformation as applied to the stiffness matrix for the lumped mass, eq. (15.7), or consistent mass matrix, eq. (15.27), we obtain in a similar manner

$$\{\bar{P}\} = [\bar{M}]\{\ddot{\bar{\delta}}\}$$

in which

$$[\bar{M}] = [T]^T[M][T] \qquad (15.39)$$

is the mass matrix for a beam segment referred to global coordinates and $[T]$ is the matrix of the transformation given by the square matrix in eq. (15.33).

Example 15.1. Consider in Fig. 15.7 a plane frame having two prismatic beam elements and three degrees of freedom as indicated in the figure. Using

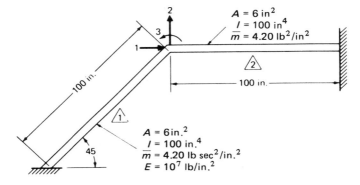

Fig. 15.7 Plane frame for Example 15.1.

the consistent mass formulation, determine the three natural frequencies and normal modes corresponding to this discrete model of the frame.

Solution: The stiffness matrix for element ▲ or ▲ in local coordinates by eq. (15.4) is

$$[K_1] = [K_2] = 1000 \begin{bmatrix} 600 & & & & & \text{Symmetric} \\ 0 & 12 & & & & \\ 0 & 600 & 40{,}000 & & & \\ -600 & 0 & 0 & 600 & & \\ 0 & -12 & -600 & 0 & 12 & \\ 0 & 600 & 20{,}000 & 0 & -600 & 40{,}000 \end{bmatrix}$$

The transformation matrix for element ▲ by eq. (15.31) with $\theta = 45°$ is

$$[T_1] = \frac{1}{\sqrt{2}} \begin{bmatrix} 1 & 1 & 0 & 0 & 0 & 0 \\ -1 & 1 & 0 & 0 & 0 & 0 \\ 0 & 0 & \sqrt{2} & 0 & 0 & 0 \\ 0 & 0 & 0 & 1 & 1 & 0 \\ 0 & 0 & 0 & -1 & 1 & 0 \\ 0 & 0 & 0 & 0 & 0 & \sqrt{2} \end{bmatrix}$$

and for element ▲ with $\theta = 0°$ is the identity matrix

$$[T_2] = [I]$$

The mass matrix in local coordinates for either of the two elements of this frame from eq. (15.27) is

$$[M_1] = [M_2] = \begin{bmatrix} 140 & & & & \text{Symmetric} & \\ 0 & 156 & & & & \\ 0 & 2200 & 40{,}000 & & & \\ 70 & 0 & 0 & 140 & & \\ 0 & 54 & 1300 & 0 & 156 & \\ 0 & -1300 & -30{,}000 & 0 & -2200 & 40{,}000 \end{bmatrix}$$

The element stiffness and mass matrices in reference to the global system of coordinates are, respectively, calculated by eqs. (15.38) and (15.39). For element $\triangle$ the stiffness matrix is

$$[\bar{K}_1] = 10^6 \begin{matrix} \\ \\ \\ \\ \\ \end{matrix} \begin{bmatrix} 0.306 & & & & \text{Symmetric} & \\ 0.294 & 0.306 & & & & \\ -0.424 & 0.424 & 40.000 & & & \\ -0.306 & -0.294 & 0.424 & 0.306 & & \\ -0.294 & -0.306 & -0.424 & 0.294 & 0.306 & \\ -0.424 & 0.424 & 20.000 & 0.424 & -0.424 & 40.000 \end{bmatrix} \begin{matrix} 4 \\ 4 \\ 4 \\ 1 \\ 2 \\ 3 \end{matrix}$$

$$\begin{matrix} \quad 4 \qquad 4 \qquad 4 \qquad 1 \qquad 2 \qquad 3 \end{matrix}$$

and the mass matrix

$$[\bar{M}_1] = \begin{bmatrix} 148 & & & & \text{Symmetric} & \\ -8 & 148 & & & & \\ -1556 & 1556 & 40{,}000 & & & \\ 62 & 8 & -919 & 148 & & \\ 8 & 62 & 919 & -8 & 148 & \\ 919 & -919 & -30{,}000 & 1556 & -1556 & 40{,}000 \end{bmatrix} \begin{matrix} 4 \\ 4 \\ 4 \\ 1 \\ 2 \\ 3 \end{matrix}$$

$$\begin{matrix} \quad 4 \qquad 4 \qquad 4 \qquad 1 \qquad 2 \qquad 3 \end{matrix}$$

For element $\triangle$,

$$[\bar{K}_2] = 10^6 \begin{bmatrix} 0.600 & & & & \text{Symmetric} & \\ 0 & 0.012 & & & & \\ 0 & 0.600 & 40.000 & & & \\ -0.600 & 0 & 0 & 0.600 & & \\ 0 & -0.012 & -0.600 & 0 & 0.012 & \\ 0 & 0.600 & 20.000 & 0 & -0.600 & 40.000 \end{bmatrix} \begin{matrix} 1 \\ 2 \\ 3 \\ 4 \\ 4 \\ 4 \end{matrix}$$

$$\begin{matrix} \quad 1 \qquad 2 \qquad 3 \qquad 4 \qquad 4 \qquad 4 \end{matrix}$$

and

$$[\bar{M}_2] = \begin{bmatrix} 140 & & & \text{Symmetric} & & \\ 0 & 156 & & & & \\ 0 & 2200 & 40{,}000 & & & \\ 70 & 0 & 0 & 140 & & \\ 0 & 54 & 1300 & 0 & 156 & \\ 0 & -1300 & -30{,}000 & 0 & -2200 & 40{,}000 \end{bmatrix} \begin{matrix} 1 \\ 2 \\ 3 \\ 4 \\ 4 \\ 4 \end{matrix}$$

with column labels $\begin{matrix}1 & 2 & 3 & 4 & 4 & 4\end{matrix}$

The system stiffness and mass matrices are assembled by the direct method. As was mentioned before, it is expedient for hand calculation of these matrices to indicate the corresponding system nodal coordinates at the top and right of the element matrices. We thus obtain the system stiffness matrix as

$$[\bar{K}] = 10^6 \begin{bmatrix} 0.906 & 0.294 & 0.424 \\ 0.294 & 0.318 & 0.176 \\ 0.424 & 0.176 & 80.000 \end{bmatrix}$$

and the system mass matrix as

$$[\bar{M}] = \begin{bmatrix} 288 & -8 & 1556 \\ -8 & 304 & 644 \\ 1556 & 644 & 80{,}000 \end{bmatrix}$$

The natural frequencies are found as the roots of the characteristic equation

$$\|[K] - \omega^2 [M]\| = 0$$

which, upon substituting the values given for this example, yields

$$10^3 \begin{vmatrix} 906 - 0.288\omega^2 & 294 + 0.008\omega^2 & 424 - 1.556\omega^2 \\ 294 + 0.008\omega^2 & 318 - 0.304\omega^2 & 176 - 0.644\omega^2 \\ 424 - 1.556\omega^2 & 176 - 0.644\omega^2 & 80{,}000 - 80\omega^2 \end{vmatrix} = 0$$

The roots then are

$$\omega_1^2 = 638.5, \quad \omega_2^2 = 976.6, \quad \omega_3^2 = 4211.6$$

and the natural frequencies are

$$\omega_1 = 25.26 \text{ rad/sec}, \quad \omega_2 = 31.24 \text{ rad/sec}, \quad \text{and} \quad \omega_3 = 64.90 \text{ rad/sec}$$

or

$$f_1 = 4.02 \text{ cps}, \quad f_2 = 4.97 \text{ cps}, \quad \text{and} \quad f_3 = 10.33 \text{ cps}$$

The normal modes are given as the nontrivial solution of the eigenproblem

$$([K] - \omega^2 [M]) \{a\} = \{0\}$$

Substituting $\omega_1^2 = 638.5$ and setting $a_{11} = 1.0$, we obtain the first mode shape as

$$\{a_1\} = \begin{Bmatrix} a_{11} \\ a_{21} \\ a_{31} \end{Bmatrix} = \begin{Bmatrix} 1.00 \\ -2.38 \\ 0 \end{Bmatrix}$$

which is normalized with the factor

$$\sqrt{\{a_1\}^T [M] \{a_1\}} = 45.81$$

The normalized eigenvector is then

$$\{\phi_1\} = \begin{Bmatrix} \phi_{11} \\ \phi_{21} \\ \phi_{31} \end{Bmatrix} = \begin{Bmatrix} 0.0218 \\ -0.0527 \\ 0 \end{Bmatrix}$$

Analogously, for the other two modes, we obtain

$$\{\phi_2\} = \begin{Bmatrix} \phi_{12} \\ \phi_{22} \\ \phi_{32} \end{Bmatrix} = \begin{Bmatrix} 0.00498 \\ 0.00206 \\ 0.00341 \end{Bmatrix} \quad \text{and} \quad \{\phi_3\} = \begin{Bmatrix} \phi_{13} \\ \phi_{23} \\ \phi_{33} \end{Bmatrix} = \begin{Bmatrix} 0.0583 \\ 0.0241 \\ -0.0016 \end{Bmatrix}$$

From these vectors, we obtain the modal matrix

$$[\phi] = \begin{bmatrix} 0.0218 & 0.00498 & 0.0583 \\ -0.0527 & 0.00206 & 0.0241 \\ 0 & 0.00341 & -0.0016 \end{bmatrix}$$

Example 15.2.   Determine the maximum displacement at the nodal coordinates of the frame in Fig. 15.7 when a force of magnitude 100,000 lb is suddenly applied at nodal coordinate 1. Neglect damping.

Solution: From Example 15.1, the natural frequencies are $\omega_1 = 25.26$ rad/sec, $\omega_2 = 31.24$ rad/sec, and $\omega_3 = 64.90$ rad/sec; and the modal matrix is

$$[\phi] = \begin{bmatrix} 0.0218 & 0.00498 & 0.0583 \\ -0.0527 & 0.00206 & 0.0241 \\ 0 & 0.00341 & -0.0016 \end{bmatrix}$$

The modal equations have the form of

$$\ddot{z}_i + \omega_i^2 z_i = P_i \tag{a}$$

where

$$P_i = \sum_j \phi_{ji} F_j \tag{b}$$

In this example, the nodal applied forces are

$$F_1 = 100,000 \text{ lb}, \qquad F_2 = 0, \qquad F_3 = 0$$

We thus obtain the modal equations as

$$\ddot{z}_1 + 638.5z_1 = 2180$$

$$\ddot{z}_2 + 976.6z_2 = 498$$

$$\ddot{z}_3 + 4211.6z_3 = 5830 \qquad (c)$$

The solutions of these equations are of the form

$$z_i = \frac{P_i}{\omega_i^2}(1 - \cos \omega_i t)$$

Substitution for P_i and ω_i yields

$$z_1 = 3.414(1 - \cos 25.2685t)$$

$$z_2 = 0.510(1 - \cos 31.2506t)$$

$$z_3 = 1.384(1 - \cos 64.8970t) \qquad (d)$$

The nodal displacements are obtained from

$$\{y\} = [\Phi]\{z\}$$

which results in

$$y_1 = 0.1577 - 0.0744 \cos 25.26t - 0.00254 \cos 31.25t - 0.0807 \cos 64.9t$$

$$y_2 = -0.1455 + 0.1800 \cos 25.26t - 0.00105 \cos 31.25t - 0.0333 \cos 64.9t$$

$$y_3 = -0.000475 + 0 \cos 25.26t - 0.00174 \cos 31.25t + 0.0022 \cos 64.9t$$

$$(e)$$

The maximum possible displacements at the nodal coordinates may then be estimated as the summation of the absolute values of the coefficients in the above expressions. Hence

$$y_{1_{max}} = 0.3177 \text{ in}$$

$$y_{2_{max}} = 0.3598 \text{ in}$$

$$Y_{3_{max}} = 0.0044 \text{ rad} \qquad (f)$$

15.4   PROGRAM 14—MODELING STRUCTURES AS PLANE FRAMES

Program 14 serves to determine the stiffness and the mass matrices for the plane frame and to store the coefficients of these matrices in a file. Since the stiffness and mass matrices are symmetric matrices, only the upper triangular portion of these matrices needs to be stored. The program also stores in another file, named by the user, the general information on the frame. The information stored in these files is needed by programs which the user may call to perform dynamic analysis of the frame, such as determination of natural frequencies or calculation of the response of the structure subject to external excitation.

Example 15.3.   Use Program 14 to determine the stiffness and mass matrices for the plane frame shown in Fig. 15.7.

Solution:

Input Data and Output Results

```
PROGRAM 14: MODELING PLANE FRAMES      DATA FILE:D14

   GENERAL DATA:

NUMBER OF JOINTS                          NJ= 3
NUMBER OF BEAM ELEMENTS                    NE= 2
NUMBER OF FIXED COORDINATES                NC= 6
MODULUS OF ELASTICITY                       E= 1E+07
      JOINT DATA:
```

JOINT #	X-COORDINATE	Y-COORDINATE	CONCENTRATED JOINT MASS
1	0.00	0.00	0.000
2	70.71	70.71	0.000
3	170.71	70.71	0.000

```
      BEAM ELEMENT DATA:
```

BEAM-ELEMENT#	FIRST JT.#	SECOND JT.#	MASS/LENGTH	SECT.INERTIA	SECT.AREA
1	1	2	4.200	100.00	6.00
2	2	3	4.200	100.00	6.00

```
      FIXED COORDINATES DATA:

      JOINT #, DIRECTION # (HORIZONTAL = 1 VERTICAL = 2, ROTATION = 3)
      1          1
      1          2
      1          3
      3          1
      3          2
      3          3

      OUTPUT RESULTS:

                              **SYSTEM STIFFNESS MATRIX**
         9.0600E+05           2.9400E+05          4.2427E+05
         2.9400E+05           3.1800E+05          1.7573E+05
         4.2427E+05           1.7573E+05          8.0000E+07

                              **SYSTEM MASS MATRIX**
         2.8800E+02          -7.9999E+00          1.5556E+03
        -7.9999E+00           3.0400E+02          6.4440E+02
         1.5556E+03           6.4440E+02          7.9999E+04
```

Example 15.4. For the structure shown in Fig. 15.7 which was modeled in Example 15.3, determine: (a) natural frequencies and modal shapes; (b) the response to a force of magnitude 100,000 lb suddenly applied at nodal coordinate 2. *The solution of these programs requires the use of files prepared during the execution of Program 14 to model the structure as a plane frame followed by the execution of Programs 8 and 9 to determine natural frequencies and to obtain the response using the modal superposition method.*

Solution:

(a) Output Results

```
PROGRAM 8: NATURAL FREQUENCIES        DATA FILE: SK

                          *OUTPUT RESULTS*
        EIGENVALUES:
        6.385E+02      9.766E+02      4.212E+03

        NATURAL FREQUENCIES (C.P.S.):

        4.021            4.973            10.328

        EIGENVECTORS BY ROWS

   -0.02183  0.05270  0.00000
    0.00498  0.00206  0.00341
    0.05831  0.02415-0.00163
```

(b) Input Data and Output Results

```
PROGRAM 9: MODAL                 DATA FILE:D15.4B

      NATURAL FREQUENCIES (C.P.S.):

   4.021      4.973      10.328

      MODAL SHAPES BY ROW (EIGENVECTORS):

-0.02183  0.05270   0.00000
 0.00498  0.00206   0.00341
 0.05831  0.02415  -0.00163

      INPUT DATA:

NUMBER OF DEGREES OF FREEDOM                              ND= 3
NUMBER OF EXTERNAL FORCES                                 NF= 1
TIME STEP OF INTEGRATION                                  H= .01
GRAVITATIONAL INDEX                                       G= 0
PRINT TIME HISTORY NPRT = 1; ONLY MAX. VALUES NPRT = 0    NPRT= 0
FORCE #,  COORD.# WHERE FORCE IS APPLIED, NUM. OF POINT DEFINING THE FORCE
    1                    1                    2

        EXCITATION FUNCTION FOR FORCE # 1
```

TIME	EXCITATION
0.00	100000.00
0.20	100000.00

MAXIMUM RESPONSE

COORD.	MAX.DISPL.	MAX. VELOC.	MAX.ACC.
1	0.303	7.025	390.10
2	0.335	6.576	230.79
3	0.003	0.188	10.99

15.5   SUMMARY

The dynamic analysis of plane frames by the stiffness method requires the inclusion of the axial effects in the system matrices (stiffness, mass, etc. matrices). It also requires a transformation of coordinates in order to refer all the element matrices to the same coordinate system, so that the appropriate superposition can be applied to assemble the system matrices.

The required matrices for consideration of axial effects as well as the matrix required for the transformation of coordinates are developed in this chapter. A computer program for modeling structures as plane frames is also presented. This program is organized following the pattern of the BEAM program of the preceding chapter.

PROBLEMS

The following problems are intended for hand calculation, though it is recommended that whenever possible solutions should also be obtained using Program 14 to model the structure as a plane frame, Program 8 to determine natural frequencies and modal shapes, and Program 9 to calculate the response using modal superposition method.

15.1 For the plane frame shown in Fig. 15.8 determine the system stiffness and mass matrices. Base the analysis on the four nodal coordinates indicated in the figure. Use consistent mass method.

15.2 Use the results obtained in Problem 15.1 in performing the static condensation to eliminate the rotational degree of freedom at the support to determine the transformation matrix and the reduced stiffness and mass matrices.

15.3 Determine the natural frequencies and corresponding normal modes for the reduced system in Problem 15.2.

15.4 Determine the response of the frame shown in Fig. 15.1 when it is acted upon by a force $F(t) = 1.0$ kip suddenly applied at nodal coordinate 2 for 0.05 sec. Use results of Problem 15.3 to obtain the modal equations. Neglect damping in the system.

15.5 Determine the maximum response of the frame shown in Fig. 15.8 when subjected to the triangular impulsive load (Fig. 15.9) along the nodal coordinate 2.

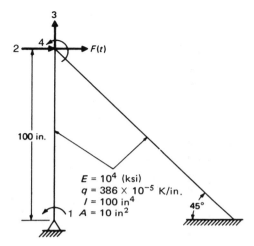

Fig. 15.8

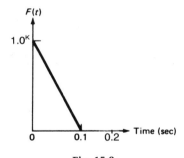

Fig. 15.9

Use results of Problem 15.3 to obtain the modal equations and use the appropriate response spectrum to find maximum modal response (Fig. 4.5). Neglect damping in the system.

15.6 Determine the steady-state response of the frame shown in Fig. 15.8 when subjected to harmonic force $F(t) = 10 \sin 30t$ (kip) along nodal coordinate 2. Neglect damping in the system.

15.7 Repeat Problem 15.6 assuming that the damping is proportional to the stiffness of the system, $[C] = a_0 [K]$, where $a_0 = 0.2$.

15.8 The frame shown in Fig. 15.10 is acted upon by the dynamic forces shown in the figure. Determine the equivalent nodal forces corresponding to each member of the frame.

15.9 Assemble the system equivalent nodal forces $\{F_e\}$ from equivalent member nodal forces which were calculated in Problem 15.8.

15.10 Determine the natural frequencies and corresponding normal modes for the frame shown in Fig. 15.8.

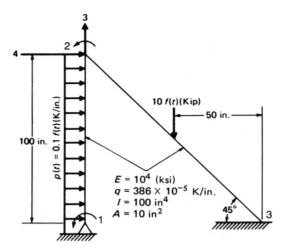

Fig. 15.10

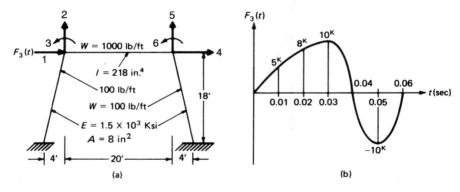

Fig. 15.11

15.11 Determine the response for the frame shown in Fig. 15.11 when subjected to the force $F(t)$ [Fig. 15.11(a)] acting along nodal coordinate 1. Assume 5% damping in all the modes.

15.12 Determine the steady-state response of the frame in Fig. 15.11 acted upon by harmonic force $F_1(t) = 10 \cos 50t$ (kip) as indicated in the figure. Neglect damping in the system.

15.13 Solve Problem 15.11 using step-by-step linear acceleration method (Program 19). Neglect damping.

15.14 Determine the response of the frame shown in Fig. 15.8 when acted upon by the force $F(t)$ (depicted in Fig. 15.12) applied at nodal coordinate 2. Assume 10% damping in all the modes. Use modal superposition method.

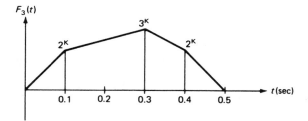

Fig. 15.12

15.15 Find the response in Problem 15.14 using step-by-step linear acceleration method. Assume damping proportional to stiffness by a factor $a_0 = 0.01$.

16

Dynamic Analysis of Grids

In Chapter 15 consideration was given to the dynamic analysis of the plane frame when subjected to forces acting on the plane of the structure. When the planar structural system is subjected to loads applied normally to its plane, the structure is referred to as a *grid*. This structure can also be treated as a special case of the three-dimensional frame to be presented in Chapter 17. The reason for considering the planar frame, whether loaded in its plane or normal to its plane, as a special case, is the immediate reduction of unknown nodal coordinates for a beam element, hence a considerable reduction in the number of unknown displacements for the structural system.

When analyzing the planar frame under action of loads in the plane, the possible components of joint displacements which had to be considered were translations in the X and Y directions and rotation about the Z axis. However, if a plane frame is loaded normal to the plane of the structure, the components of joint displacements required to describe the displacements of a joint are a translation in the Z direction and rotations about the X and Y axes. Thus treating the planar grid structure as a special case, it will be necessary to consider only three components of nodal displacements at each end of a typical grid member.

16.1   LOCAL AND GLOBAL COORDINATE SYSTEMS

For a beam element of a grid, the local orthogonal axes will be established such that the x defines the longitudinal centroidal axis of the member and the x-y plane will coincide with the plane of the structural system, which will be defined by the X-Y plane. In this case, the z axis will define the minor *principal axis* of the cross section while the y axis will define the *major axis* of the cross section. It will be assumed that the shear center of the cross section coincides with the centroid of the cross section. The grid member may have either a variable or constant cross section along its length.

The possible nodal displacements with respect to the local or to the global systems of coordinates are identified in Fig. 16.1. It can be seen that the linear displacements along the z direction for local axes and along the Z direction for the global system are identical since the two axes coincide. However, in general, rotational components at the nodal coordinates differ from these two coordinate systems. Hence, a transformation of coordinates will be required to transform the element matrices from the local to the global coordinates.

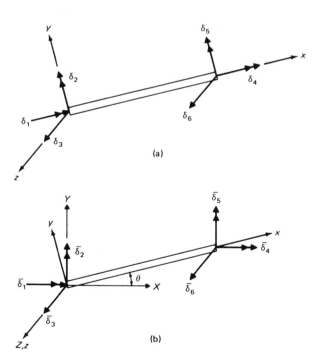

Fig. 16.1 Components of nodal displacements for a grid member. (a) Local coordinate system. (b) Global coordinate system.

16.2   TORSIONAL EFFECTS

The dynamic analysis by the stiffness method for grid structures, that is, for plane frames subjected to normal loads, requires the determination of the torsional stiffness and mass coefficients for a typical member of the grid. The derivation of these coefficients is essentially identical to the derivation of the stiffness and mass coefficients for axial effects on a beam element. Similarity between these two derivations occurs because the differential equations for both problems have the same mathematical form. For the axial problem, the differential equation for the displacement function is given by eq. (15.8) as

$$\frac{du}{dx} = \frac{P}{AE}$$

Likewise, the differential equation for torsional displacement is

$$\frac{d\theta}{dx} = \frac{T}{JG} \tag{16.1}$$

in which θ is the angular displacement, T is the torsional moment, G is the modulus of elasticity in shear, and J is the torsional constant of the cross section (polar moment of inertia for circular sections).

As a consequence of the analogy between eqs. (15.8) and (16.1), we can write the following results already obtained for axial effects. The displacement functions for the torsional effects are the same as the corresponding functions giving the displacements for axial effects; hence by analogy to eqs. (15.10) and (15.11) and in reference to the nodal coordinates of Fig. 16.2, we obtain

$$\theta_1(x) = \left(1 - \frac{x}{L}\right) \tag{16.2}$$

and

$$\theta_2(x) = \frac{x}{L} \tag{16.3}$$

in which the angular displacement function $\theta_1(x)$ corresponds to a unit angular displacement $\delta_1 = 1$ at nodal coordinate 1 and $\theta_2(x)$ corresponds to the displacement function resulting from a unit angular displacement $\delta_2 = 1$ at nodal coordinate 2. Analogous to eq. (15.17), the stiffness influence coefficients for torsional effects may be calculated from

Fig. 16.2 Nodal torsional coordinates for a beam element.

$$k_{ij} = \int_0^L JG\theta_i'(x)\,\theta_j'(x)\,dx \tag{16.4}$$

in which $\theta_1'(x)$ and $\theta_2'(x)$ are the derivatives with respect to x of the displacement functions $\theta_1(x)$ and $\theta_2(x)$. Also analogous to eq. (15.23), the consistent mass matrix coefficients for torsional effects are given by

$$m_{ij} = \int_0^L I_m\theta_i(x)\,\theta_j(x)\,dx \tag{16.5}$$

in which $I_{\bar m}$ is the polar mass moment of inertia, per unit length along the beam element. This moment of inertia may conveniently be expressed as the product of the mass $\bar m$ per unit length times the radius of gyration squared, k^2. The radius of gyration may, in turn, be calculated as the ratio I_0/A. Therefore, the mass polar moment of inertia per unit length $I_{\bar m}$ is given by

$$I_{\bar m} = \bar m\,\frac{I_0}{A} \tag{16.6}$$

in which I_0 is the polar moment of inertia of the cross-sectional area and A the cross-sectional area.

The application of eqs. (16.4) and (16.5) for a uniform beam yields the stiffness and mass matrices such that

$$\begin{Bmatrix} T_1 \\ T_2 \end{Bmatrix} = \frac{JG}{L}\begin{bmatrix} 1 & -1 \\ -1 & 1 \end{bmatrix}\begin{Bmatrix} \delta_1 \\ \delta_2 \end{Bmatrix} \tag{16.7}$$

and

$$\begin{Bmatrix} T_1 \\ T_2 \end{Bmatrix} = \frac{I_{\bar m}L}{6}\begin{bmatrix} 2 & 1 \\ 1 & 2 \end{bmatrix}\begin{Bmatrix} \ddot\delta_1 \\ \ddot\delta_2 \end{Bmatrix} \tag{16.8}$$

in which $I_{\bar m}$ is given by eq. (16.6), and T_1, T_2 are torsional moments at the ends of the beam segments as shown in Fig. 16.2.

16.3   STIFFNESS MATRIX FOR A GRID ELEMENT

The torsional stiffness matrix, eq. (16.7), is combined with the flexural stiffness matrix, eq. (14.20), to obtain the stiffness matrix for a typical member of a grid structure. In reference to the local coordinate system indicated in Fig. 16.1(a), the stiffness equation for a uniform member is then

$$\begin{Bmatrix} P_1 \\ P_2 \\ P_3 \\ P_4 \\ P_5 \\ P_6 \end{Bmatrix} = \frac{EI}{L^3}\begin{Bmatrix} JGL^2/EI & & & & \text{Symmetric} & \\ 0 & 4L^2 & & & & \\ 0 & -6L & 12 & & & \\ -JGL^2/EI & 0 & 0 & JGL^2/EI & & \\ 0 & 2L^2 & -6L & 0 & 4L^2 & \\ 0 & 6L & -12 & 0 & 6L & 12 \end{Bmatrix}\begin{Bmatrix} \delta_1 \\ \delta_2 \\ \delta_3 \\ \delta_4 \\ \delta_5 \\ \delta_6 \end{Bmatrix} \tag{16.9}$$

or in condensed form

$$\{P\} = [K]\{\delta\} \tag{16.10}$$

16.4   CONSISTENT MASS MATRIX FOR A GRID ELEMENT

The combination of the consistent mass matrix for flexural effects (14.34) with the consistent mass matrix for torsional effects (16.8) results in the consistent mass matrix for a typical member of a grid, namely

$$
\begin{Bmatrix} P_1 \\ P_2 \\ P_3 \\ P_4 \\ P_5 \\ P_6 \end{Bmatrix}
= \frac{\bar{m}L}{420}
\begin{bmatrix}
140I_0/A & & & \text{Symmetric} & & \\
0 & 4L^2 & & & & \\
0 & 22L & 156 & 140I_0/A & & \\
70I_0/A & 0 & 0 & 0 & & \\
0 & -3L^2 & -13L & 0 & 4L^2 & \\
0 & 13L & 54 & & -22L & 156
\end{bmatrix}
\begin{Bmatrix} \ddot{\delta}_1 \\ \ddot{\delta}_2 \\ \ddot{\delta}_3 \\ \ddot{\delta}_4 \\ \ddot{\delta}_5 \\ \ddot{\delta}_6 \end{Bmatrix}
\tag{16.11}
$$

or in concise notation

$$\{P\} = [M_c]\{\delta\} \tag{16.12}$$

in which $[M_c]$ is the mass matrix for a typical uniform member of a grid structure.

16.5   LUMPED MASS MATRIX FOR A GRID ELEMENT

The lumped mass allocation to the nodal coordinates of a typical grid member is obtained from static considerations. For a uniform member having a distributed mass along its length, the nodal mass is simply one-half of the total rotational mass $I_{\bar{m}}L$. The matrix equation for the lumped mass matrix corresponding to the torsional effects is then

$$
\begin{bmatrix} P_1 \\ P_2 \end{bmatrix} = \frac{I_{\bar{m}}L}{2} \begin{bmatrix} 1 & 0 \\ 0 & 1 \end{bmatrix} \begin{bmatrix} \ddot{\delta}_1 \\ \ddot{\delta}_2 \end{bmatrix}
\tag{16.13}
$$

The combination of the lumped torsional mass matrix from eq. (16.13) with the flexural mass matrix for a typical member of a grid results in the diagonal matrix which is the lumped mass matrix for the grid element. This matrix, relating forces and accelerations at nodal coordinates, is given by the following equation:

$$
\begin{Bmatrix} P_1 \\ P_2 \\ P_3 \\ P_4 \\ P_5 \\ P_6 \end{Bmatrix}
= \frac{\bar{m}L}{2}
\begin{bmatrix}
I_0/A) & & & & & \\
& 0 & & & & \\
& & 1 & & & \\
& & & I_0/A & & \\
& & & & 0 & \\
& & & & & 1
\end{bmatrix}
\begin{Bmatrix} \ddot{\delta}_1 \\ \ddot{\delta}_2 \\ \ddot{\delta}_3 \\ \ddot{\delta}_4 \\ \ddot{\delta}_5 \\ \ddot{\delta}_6 \end{Bmatrix}
\tag{16.14}
$$

or briefly

$$\{P\} = [M_L]\{\delta\} \qquad (16.15)$$

in which $[M_L]$ is, in this case, the diagonal lumped mass matrix for a grid member.

16.6   TRANSFORMATION OF COORDINATES

The stiffness matrix, eq. (16.9), as well as the consistent and the lumped mass matrix in eqs. (16.11) and (16.14), respectively, are in reference to the local system of coordinates. Therefore, it is necessary to transform the reference of these matrices to the global system of coordinates before their assemblage in the corresponding matrices for the structure. As has been indicated, the z axis for the local coordinate system coincides with the Z axis for the global system. Therefore, the only step left to perform is a rotation of the coordinates in the x-y plane. The corresponding matrix for this transformation may be obtained by establishing the relations between components of the moments at the nodes expressed in these two systems of coordinates. In reference to Fig. 16.3, these relations when written for node ① are

$$P_1 = \bar{P}_1 \cos\theta + \bar{P}_2 \sin\theta$$

$$P_2 = -\bar{P}_1 \sin\theta + \bar{P}_2 \cos\theta$$

$$P_3 = \bar{P}_3 \qquad (16.16a)$$

and for node ②

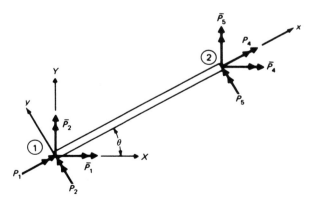

Fig. 16.3 Components of the nodal moments in local and global coordinates.

$$P_4 = \bar{P}_4 \cos \theta + \bar{P}_5 \sin \theta$$

$$P_5 = -\bar{P}_4 \sin \theta + \bar{P}_5 \cos \theta$$

$$P_6 = \bar{P}_6 \tag{16.16b}$$

The identical form of these equations with those derived for the transformation of coordinates for nodal forces of an element of a plane frame, eqs. (15.28) and (15.30), should be noted. Equations (16.16) may be written in matrix notation as

$$
\begin{Bmatrix} P_1 \\ P_2 \\ P_3 \\ P_4 \\ P_5 \\ P_6 \end{Bmatrix} =
\begin{bmatrix}
\cos \theta & \sin \theta & 0 & 0 & 0 & 0 \\
-\sin \theta & \cos \theta & 0 & 0 & 0 & 0 \\
0 & 0 & 1 & 0 & 0 & 0 \\
0 & 0 & 0 & \cos \theta & \sin \theta & 0 \\
0 & 0 & 0 & -\sin \theta & \cos \theta & 0 \\
0 & 0 & 0 & 0 & 0 & 1
\end{bmatrix}
\begin{Bmatrix} \bar{P}_1 \\ \bar{P}_2 \\ \bar{P}_3 \\ \bar{P}_4 \\ \bar{P}_5 \\ \bar{P}_6 \end{Bmatrix} \tag{16.17}
$$

or in short notation

$$\{P\} = [T]\{\bar{P}\} \tag{16.18}$$

in which $\{P\}$ and $\{\bar{P}\}$ are, respectively, the vectors of the nodal forces of a typical grid member in local and global coordinates and $[T]$ the transformation matrix. The same transformation matrix $[T]$ serves also to transform the nodal components of the displacements from a global to a local system of coordinates. In condensed notation, this relation is given by

$$\{\delta\} = [T]\{\bar{\delta}\} \tag{16.19}$$

where $\{\delta\}$ and $\{\bar{\delta}\}$ are, respectively, the components of nodal displacement in local and global coordinates. The substitution of eqs. (16.18) and (16.19) in the stiffness relation eq. (16.10) yields the element stiffness matrix in reference to the global coordinate system, that is,

$$[T]\{\bar{P}\} = [K][T]\{\bar{\delta}\}$$

or, since $[T]$ is an orthogonal matrix, it follows that

$$\{\bar{P}\} = [T]^T [K][T]\{\bar{\delta}\}$$

If we define $[\bar{K}]$ as

$$[\bar{K}] = [T]^T [K][T] \tag{16.20}$$

we obtain

$$\{\bar{P}\} = [\bar{K}]\{\bar{\delta}\} \tag{16.21}$$

Analogously, for the mass matrix, we find

$$\{\bar{P}\} = [\bar{M}] \{\ddot{\bar{\delta}}\}$$ (16.22)

in which

$$[\bar{M}] = [T]^T [M] [T]$$ (16.23)

is the transformed mass matrix.

Example 16.1. Figure 16.4 shows a grid structure in a horizontal plane consisting of two prismatic beam elements with a total of three degrees of freedom as indicated. Determine the natural frequencies and corresponding mode shapes. Use the consistent mass formulation.

The stiffness matrix for elements 1 or 2 of the grid in local coordinates by eq. (16.9) is

$$[K_1] = [K_2] = 10^6 \begin{bmatrix} 40 & 0 & 0 & -40 & 0 & 0 \\ 0 & 200 & -5 & 0 & 100 & 5 \\ 0 & -5 & 0.167 & 0 & -5 & -0.167 \\ -40 & 0 & 0 & 40 & 0 & 0 \\ 0 & 100 & -5 & 0 & 200 & 5 \\ 0 & 5 & -0.167 & 0 & 5 & 0.167 \end{bmatrix}$$

The transformation matrix for element $\triangle$ with $\theta = 0°$ is simply the unit matrix $[T_1] = [I]$. Hence

$$[\bar{K}_1] = [T_1]^T [K_1] [T_1] = [K_1]$$

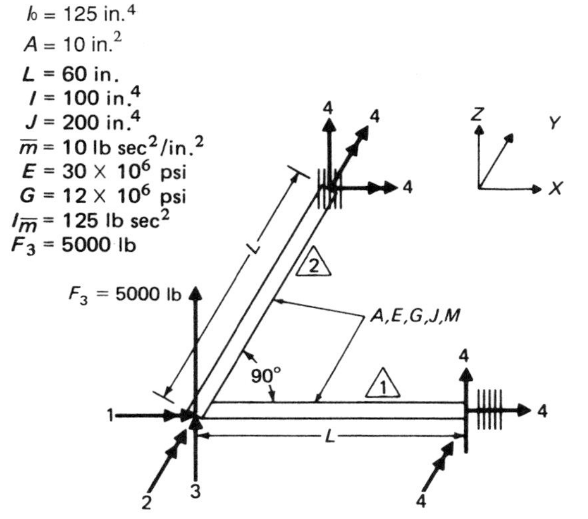

$I_0 = 125$ in.^{4}
$A = 10$ in.^{2}
$L = 60$ in.
$I = 100$ in.^{4}
$J = 200$ in.^{4}
$\bar{m} = 10$ lb sec^2/in.^{2}
$E = 30 \times 10^6$ psi
$G = 12 \times 10^6$ psi
$I_{\bar{m}} = 125$ lb sec^2
$F_3 = 5000$ lb

$F_3 = 5000$ lb

A,E,G,J,M

$90°$

Fig. 16.4

and for element $\triangle$ with $\theta = 90°$

$$[T_2] = \begin{bmatrix} 0 & 1 & 0 & 0 & 0 & 0 \\ -1 & 0 & 0 & 0 & 0 & 0 \\ 0 & 0 & 1 & 0 & 0 & 0 \\ 0 & 0 & 0 & 0 & 1 & 0 \\ 0 & 0 & 0 & -1 & 0 & 0 \\ 0 & 0 & 0 & 0 & 0 & 1 \end{bmatrix}$$

so that

$$[\bar{K}_2] = [T_2]^T [K_2] [T_2]$$

$$= 10^6 \begin{bmatrix} 1 & 2 & 3 & 4 & 4 & 4 \\ 200 & 0 & 5 & 100 & 0 & -5 & & 1 \\ 0 & 40 & 0 & 0 & -40 & 0 & & 2 \\ 5 & 0 & 0.167 & -5 & 0 & -0.167 & & 3 \\ 100 & 0 & 5 & 200 & 0 & -5 & & 4 \\ 0 & -40 & 0 & 0 & 40 & 0 & & 4 \\ -5 & 0 & -0.167 & -5 & 0 & 0.167 & & 4 \end{bmatrix}$$

The system matrix $[K_s]$ assembled from $[\bar{K}_1]$ and $[\bar{K}_2]$ is

$$[K_s] = 10^6 \begin{bmatrix} 240 & 0 & 5 \\ 0 & 240 & -5 \\ 5 & -5 & 0.333 \end{bmatrix} \begin{matrix} \theta_x \\ \theta_Y \\ z \end{matrix}$$
$$ \begin{matrix} \theta_x & \theta_Y & z \end{matrix}$$

Analogously, for the mass, we have from eq. (16.11)

$$[M_1] = [M_2] = \begin{bmatrix} 2500 & 0 & 0 & -1250 & 0 & 0 \\ 0 & 20{,}570 & 1886 & 0 & -15{,}430 & 1114 \\ 0 & 1886 & 223 & 0 & -1114 & 0 \\ -1250 & 0 & 0 & 2500 & 0 & 0 \\ 0 & -15{,}430 & -1114 & 0 & 20{,}570 & -1886 \\ 0 & 1114 & 77 & 0 & -1886 & 223 \end{bmatrix}$$

We then calculate using eq. (16.23)

$$[\bar{M}_1] = [M_1]$$

since

$$[T_1] = [I]$$

and analogously

$$[\bar{M}_2] = \begin{matrix} & 1 & 2 & 3 & 4 & 4 & 4 & \\ \begin{bmatrix} 20{,}570 & 0 & -1886 & 15{,}430 & 0 & 1114 \\ 0 & 2500 & 0 & 0 & 1250 & 0 \\ -1886 & 0 & 223 & 1114 & 0 & 77 \\ 15{,}430 & 0 & 1114 & 20{,}570 & 0 & 1886 \\ 0 & 1250 & 0 & 0 & 2500 & 0 \\ 1114 & 0 & 77 & 1886 & 0 & 223 \end{bmatrix} & \begin{matrix} 1 \\ 2 \\ 3 \\ 4 \\ 4 \\ 4 \end{matrix} \end{matrix}$$

From $[\bar{M}_1]$ and $[\bar{M}_2]$ we assemble the system mass matrix and obtain

$$[M_s] = \begin{bmatrix} 23{,}070 & 0 & 1886 \\ 0 & 23{,}070 & -1886 \\ 1886 & -1886 & 446 \end{bmatrix}$$

The natural frequencies and mode shapes are obtained from the solution of the eigenproblem

$$([K_s] - \omega^2 [M_s]) \{a\} = \{0\}$$

which gives the eigenvalues (squares of the natural frequencies)

$$\omega_1^2 = 396.35, \quad \omega_2^2 = 10{,}402, \quad \text{and} \quad \omega_3^2 = 23{,}866$$

then

$$\omega_1 = 19.91 \text{ rad/sec}, \quad \omega_2 = 101.99 \text{ rad/sec}, \quad \text{and} \quad \omega_3 = 154.49 \text{ rad/sec}$$

and the eigenvectors (modal matrix)

$$[a] = \begin{bmatrix} -1.000 & 1.000 & -1.000 \\ 1.000 & 1.000 & 1.000 \\ 54.285 & 0 & 7.765 \end{bmatrix}$$

The eigenvectors are conveniently normalized by dividing the columns of the modal matrix, respectively, by the factors

$$\sqrt{\{a_1\}^T [M_s] \{a_1\}} = 974.75$$

$$\sqrt{\{a_2\}^T [M_s] \{a_2\}} = 214.81$$

$$\sqrt{\{a_3\}^T [M_s] \{a_3\}} = 120.20$$

The normalized eigenvectors are arranged in columns of the modal matrix, so that

$$[\Phi] = 10^{-2} \begin{bmatrix} -0.1026 & 0.4655 & -0.8320 \\ 0.1026 & 0.4655 & 0.8320 \\ 5.5691 & 0 & 6.4603 \end{bmatrix}$$

Example 16.2. Determine the response of the grid shown in Fig. 16.4 when subjected to a suddenly applied force $F_3 = 5000$ lb as indicated in the figure.

Solution: The natural frequencies and modal shapes for this structure were calculated in Example 16.1. The modal equation is given in general as

$$\ddot{z}_n + \omega_n^2 z_n = P_n$$

where

$$P_n = \sum_i \phi_{in} F_i$$

and F_i the external forces at the nodal coordinates which for this example are $F_1 = F_2 = 0$ and $F_3 = 5000$ lb. Hence, we obtain

$$\ddot{z}_1 + 396.35 z_1 = 278.46$$

$$\ddot{z}_2 + 10,402 z_2 = 0$$

$$\ddot{z}_3 + 23,866 z_3 = 323.01$$

The solution of these equations for zero initial conditions is

$$z_1 = \frac{278.46}{396.35}(1 - \cos 19.91t)$$

$$z_2 = 0$$

$$z_3 = \frac{323.01}{23,866}(1 - \cos 154.49t)$$

The displacements at the nodal coordinates are calculated from

$$\{y\} = [\Phi]\{z\}$$

$$\begin{Bmatrix} y_1 \\ y_2 \\ y_3 \end{Bmatrix} = 10^{-2} \begin{bmatrix} -0.1026 & 0.4655 & -0.8320 \\ 0.1026 & 0.4655 & 0.8320 \\ 5.5691 & 0 & 6.4603 \end{bmatrix} \begin{bmatrix} 0.7026(1 - \cos 19.91t) \\ 0 \\ 0.0135(1 - \cos 154.49t) \end{bmatrix}$$

and finally

$$y_1 = 10^{-3}(-0.8332 + 0.7209 \cos 19.91t + 0.1123 \cos 154.49t) \text{ rad}$$

$$y_2 = 10^{-3}(0.8332 - 0.7209 \cos 19.91t - 0.1123 \cos 154.49t) \text{ rad}$$

$$y_3 = 10^{-3}(40 - 39.13 \cos 19.91t - 0.87 \cos 154.49t) \text{ in}$$

16.7 PROGRAM 15—MODELING STRUCTURES AS GRID FRAMES

Program 15 calculates the stiffness and mass matrices for a grid frame and stores the coefficients of these matrices in a file. Since the stiffness and the mass matrices are symmetric, only the upper triangular portion of these matrices needs to be stored. The program also stores in another file, named by the user, the general information on the grid frame. The information stored in these files is needed by programs which the user may call to perform a dynamic analysis, such as determination of natural frequencies or calculations of the response of the structure subjected to external excitation.

Example 16.3.   For the grid frame shown in Fig. 16.4 and analyzed in the previous examples, (a) use Program 15 to model this structure, (b) use Program 8 to calculate the natural frequencies and mode shapes, and (c) use Program 9 to determine the response to a constant force of 5000 lb suddenly applied for 0.1 sec as indicated in the figure.

Solution:

Problem Data:

Modulus of elasticity: $\qquad E = 30 \times 10^6$ psi

Modulus of rigidity: $\qquad G = 12 \times 10^6$ psi

Distributed mass: $\qquad \bar{m} = 10 \text{ lb} \cdot \sec^2/\text{in}/\text{in}$

Cross-section polar moment of inertia: $\qquad I_0 = 125 \text{ in}^4$

Cross-section torsional constant: $\qquad J = 200 \text{ in}^4$

Cross-section moment of inertia: $\qquad I = 100 \text{ in}^4$

Cross-section area: $\qquad A = 10 \text{ in}^2$

(a) Input Data and Output Results: Modeling the Structure

```
PROGRAM 15: MODELING GRID FRAMES        DATA FILE:D15

     INPUT DATA:

NUMBER OF JOINTS                        NJ = 3
NUMBER OF BEAM ELEMENTS                 NE = 2
NUMBER OF FIXED COORDINATES             NC = 6
MODULUS OF ELASTICITY                    E = 3E+07
MODULUS OF RIGIDITY                     GR = 1.2E+07

     JOINT DATA:

  JOINT #      X-COORDINATE      Y-COORDINATE      CONCENTRATED JOINT MASS
      1           0.00              0.00                  0.00
      2          60.00              0.00                  0.00
      3           0.00             60.00                  0.00
```

BEAM ELEMENT DATA:

BEAM #	FIRST JOINT #	SECOND JOINT #	MASS/ LENGTH	SECTION INERTIA	SECTION AREA	TORSIONAL CONSTANT
1	1	2	10.000	100.00	16.00	200.00
2	1	3	10.000	100.00	16.00	200.00

FIXED COORDINATES DATA:

JOINT #, DIRECTION # (X-ROT = 1 Y-ROT = 2 Z-DISPL = 3)

JOINT #	DIRECTION #
2	1
2	2
2	3
3	1
3	2
3	3

OUTPUT RESULTS:

***SYSTEM STIFFNESS MATRIX**

2.4000E+08	0.0000E+00	5.0000E+06
0.0000E+00	2.4000E+08	−5.0000E+06
5.0000E+06	−5.0000E+06	3.3333E+05

SYSTEM MASS MATRIX

2.3071E+04	0.0000E+00	1.8857E+03
0.0000E+00	2.3071E+04	−1.8857E+03
1.8857E+03	−1.8857E+03	4.4571E+02

(b) Output Results: Natural Frequencies and Modes

PROGRAM 8: NATURAL FREQUENCIES   DATA FILE: SK

EIGENVALUES:

3.963E+02	1.040E+04	2.387E+04

NATURAL FREQUENCIES (C.P.S.):

3.168	16.232	24.587

EIGENVECTORS BY ROWS

−0.00103	0.00103	0.05569
0.00466	0.00466	−0.00000
−0.00832	0.00832	0.06460

(c) Input Data and Output Results: Response by Modal Superposition

PROGRAM 9: PROGRAM 9: MODAL   DATA FILE:D16.3C

INPUT DATA:

NUMBER OF DEGREES OF FREEDOM   D = 3
NUMBER OF EXTERNAL FORCES   NF = 1

```
TIME STEP OF INTEGRATION                                           H = .01
GRAVITATIONAL INDEX (G = 0 for forces applied)                      G = 0
PRINT TIME HISTORY NPRT = 1; ONLY MAX. VALUES NPRT = 0            NPRT = 0

FORCE #,  COORD. # WHERE FORCE IS APPLIED, NUM. OF POINTS DEFINING THE FORCE
    1                       3                           2


               EXCITATION FUNCTION FOR FORCE #1:

         TIME        EXCITATION
         0.00         5000.00
         0.10         5000.00

MODAL DAMPING RATIOS:

     0       0       0

     MAXIMUM RESPONSE:

     COORD.        MAX. DISPL.      MAX. VELOC.      MAX. ACC.
       1            0.00123          0.0309            2.97
       2            0.00123          0.0309            2.97
       3            0.05680          0.8914           36.37
```

The results given by the computer compare very closely to those obtained in Examples 16.1 and 16.2 by hand calculations. For example, the displacement given by the computer (not printed) for nodal coordinate 3, a time $t = 0.1$ sec, is

$$y_3(t = 0.1) = 0.0568 \text{ in}$$

while the hand calculation from Example 16.2 is

$$y_3(t = 0.1) = 0.04 - 0.03913 \cos 1.991 - 0.00087 \cos 15.45$$

$$= 0.0568 \text{ in}$$

16.8   SUMMARY

This chapter has presented the dynamic analysis of plane frames (grids) supporting loads applied normally to its plane. The dynamic analysis of grids requires the inclusion of torsional effects in the element stiffness and mass matrices. It also requires a transformation of coordinates of the element matrices previous to the assembling of the system matrix. The required matrices for torsional effects are developed and a computer program for the dynamic analysis of grids is presented. This program is also organized along the same pattern of the programs in the two preceding chapters for the dynamic analysis of beams and plane frames.

PROBLEMS

The following problems are intended for hand calculation, though it is recommended that whenever possible solutions should also be obtained using Program 15 to model the structure as a grid frame and Programs 8 and 9, respectively, to calculate natural frequencies and to solve for the response.

16.1 For the grid shown in Fig. 16.5 determine the system stiffness and mass matrices. Base the analysis on the three nodal coordinates indicated in the figure. Use consistent mass method.

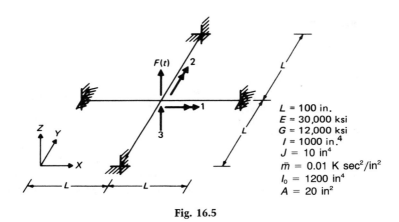

$L = 100$ in.
$E = 30,000$ ksi
$G = 12,000$ ksi
$I = 1000$ in.^{4}
$J = 10$ in^4
$\bar{m} = 0.01$ K sec^2/in^2
$I_0 = 1200$ in^4
$A = 20$ in^2

Fig. 16.5

16.2 Use static condensation to eliminate the rotational degrees of freedom and determine the transformation matrix and the reduced stiffness and mass matrices in Problem 16.1.

16.3 Determine the natural frequency for the reduced system in Problem 16.2.

16.4 Determine the natural frequencies and corresponding normal modes for the grid analyzed in Problem 16.1.

16.5 Determine the response of the grid shown in Fig. 16.5 when acted upon by a force $F(t) = 10$ kip suddenly applied for one second at the nodal coordinate 3 as shown in the figure. Use results of Problem 16.2 to obtain the equation of motion for the condensed system. Assume 10% modal damping.

16.6 Use results from Problem 16.4 to solve Problem 16.5 on the basis of the three nodal coordinates as indicated in Fig. 16.5.

16.7 Determine the steady-state response of the grid shown in Fig. 16.5 when subjected to harmonic force $F(t) = 10 \sin 50t$ (kip) along nodal coordinate 3. Neglect damping in the system.

16.8 Repeat Problem 16.7 assuming that the damping is proportional to the stiffness of the system, $[C] = a_0[K]$, where $a_0 = 0.3$.

16.9 Determine the equivalent nodal forces for a member of a grid loaded with a dynamic force, $P(t) = P_0 f(t)$, uniformly distributed along its length.

16.10 Determine the equivalent nodal forces for a member of a grid supporting a concentrated dynamic force $F(t)$ as shown in Fig. 16.6.

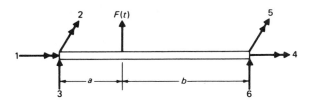

Fig. 16.6

The following problems are intended for computer solution using Program 15, to model the structure and Program 8 to determine natural frequencies and modal shapes and Program 9 to calculate the response.

16.11 Determine the natural frequencies and corresponding normal modes for the grid shown in Fig. 16.5.

16.12 Determine the response of the grid shown in Fig. 16.5 when acted upon by the force depicted in Fig. 16.7 acting along nodal coordinate 3. Neglect damping in the system.

16.13 Repeat Problem 16.12 for 15% damping in all the modes. Use modal superposition method.

16.14 Repeat Problem 16.12. Use step-by-step linear acceleration method, Program 19. Neglect damping.

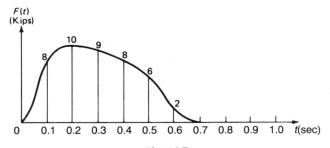

Fig. 16.7

17

Three-dimensional Frames

The stiffness method for dynamic analysis of frames presented in Chapter 15 for plane frames and in Chapter 16 for grids can readily be expanded for the analysis of three-dimensional space frames. Although for the plane frame or for the grid there were only three nodal coordinates at each joint, the three-dimensional frame has a total of six possible nodal displacements at each unconstrained joint: three translations along the x, y, z axes and three rotations about these axes. Consequently, a beam element of a three-dimensional frame or a space frame has for its two joints a total of 12 nodal coordinates; hence the resulting element matrices will be of dimension 12×12.

The dynamic analysis of three-dimensional frames results in a comparatively longer computer program in general, requiring more input data as well as substantially more computational time. However, except for size, the analysis of three-dimensional frames by the stiffness method of dynamic analysis is identical to the analysis of plane frames or plane grids.

17.1 ELEMENT STIFFNESS MATRIX

Figure 17.1 shows a beam segment of a space frame with its 12 nodal coordinates numbered consecutively. The convention adopted is to label first the three translatory displacements of the first joint followed by the three rotational displacements of the same joint, then to continue with the three translatory displacements of the second joint and finally the three rotational displacements of this second joint. The double arrows used in Fig. 17.1 serve to indicate rotational nodal coordinates; hence, these are distinguished from translational nodal coordinates for which single arrows are used.

The stiffness matrix for a three-dimensional uniform beam segment is readily written by the superposition of the axial stiffness matrix from eq. (15.3), the torsional stiffness matrix from eq. (16.6), and the flexural stiffness matrix from eq. (14.20). The flexural stiffness matrix is used twice in forming the stiffness matrix of a three-dimensional beam segment to account for the flexural effects in the two principal planes of the cross section. Proceeding to combine in an appropriate manner these matrices, we obtain in eq. (17.1) the stiffness equation for a uniform beam segment of a three-dimensional frame, namely

$$
\begin{Bmatrix} P_1 \\ P_2 \\ P_3 \\ P_4 \\ P_5 \\ P_6 \\ P_7 \\ P_8 \\ P_9 \\ P_{10} \\ P_{11} \\ P_{12} \end{Bmatrix}
=
\begin{bmatrix}
\frac{EA}{L} & & & & & & & & & & & \\
0 & \frac{12EI_z}{L^3} & & & & \text{Symmetric} & & & & & & \\
0 & 0 & \frac{12EI_y}{L^3} & & & & & & & & & \\
0 & 0 & 0 & \frac{GJ}{L} & & & & & & & & \\
0 & 0 & \frac{-6EI_y}{L^2} & 0 & \frac{4EI_y}{L} & & & & & & & \\
0 & \frac{6EI_z}{L^2} & 0 & 0 & 0 & \frac{4EI_z}{L} & & & & & & \\
\frac{-EA}{L} & 0 & 0 & 0 & 0 & 0 & \frac{EA}{L} & & & & & \\
0 & \frac{-12EI_z}{L^3} & 0 & 0 & 0 & \frac{-6EI_z}{L^2} & 0 & \frac{12EI_z}{L^3} & & & & \\
0 & 0 & \frac{-12EI_y}{L^3} & 0 & \frac{6EI_y}{L^2} & 0 & 0 & 0 & \frac{12EI_y}{L^3} & & & \\
0 & 0 & 0 & \frac{-GJ}{L} & 0 & 0 & 0 & 0 & 0 & \frac{GJ}{L} & & \\
0 & 0 & \frac{-6EI_y}{L^2} & 0 & \frac{2EI_y}{L} & 0 & 0 & 0 & \frac{6EI_y}{L^2} & 0 & \frac{4EI_y}{L} & \\
0 & \frac{6EI_z}{L^2} & 0 & 0 & 0 & \frac{2EI_z}{L} & 0 & \frac{-6EI_z}{L^2} & 0 & 0 & 0 & \frac{4EI_z}{L}
\end{bmatrix}
\begin{Bmatrix} \delta_1 \\ \delta_2 \\ \delta_3 \\ \delta_4 \\ \delta_5 \\ \delta_6 \\ \delta_7 \\ \delta_8 \\ \delta_9 \\ \delta_{10} \\ \delta_{11} \\ \delta_{12} \end{Bmatrix}
$$

(17.1)

[handwritten annotation: Four modes • axial • bending (2) • torsion]

or in condensed notation

$$\{P\} = [K]\{\delta\}$$

(17.2)

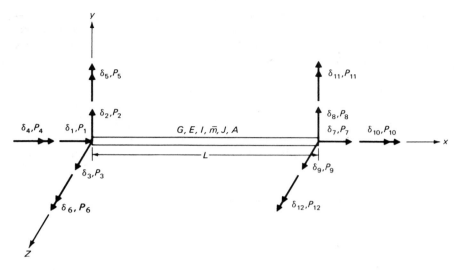

Fig. 17.1 Beam segment of a space frame showing forces and displacements at the nodal coordinates.

in which I_y and I_z are, respectively, the cross-sectional moments of inertia with respect to the principal axes labeled as y and z in Fig. 17.1, and L, A, and J are respectively the length, cross-sectional area, and torsional constant of the beam element.

17.2   ELEMENT MASS MATRIX

The lumped mass matrix for the uniform beam segment of a three-dimensional frame is simply a diagonal matrix in which the coefficients corresponding to translatory and torsional displacements are equal to one-half of the total inertia of the beam segment while the coefficients corresponding to flexural rotations are assumed to be zero. The diagonal lumped mass matrix for the uniform beam of distributed mass $\bar{m}$ and polar mass moment $I_{\bar{m}} = \bar{m} I_0 / A$ of inertia per unit of length may be written conveniently as

$$[M_L] = \frac{\bar{m}L}{2} \lfloor 1 \quad 1 \quad 1 \quad I_0/A \quad 0 \quad 0 \quad 1 \quad 1 \quad 1 \quad I_0/A \quad 0 \quad 0 \rfloor \qquad (17.3)$$

in which I_0 is the polar moment of inertia of the cross-sectional area A.

The consistent mass matrix for a uniform beam segment of a three-dimensional frame is readily obtained combining the consistent mass matrices, eq. (15.26) for axial effects, eq. (16.8) for torsional effects, and eq. (14.34) for flexural effects. The appropriate combination of these matrices results in the consistent mass matrix for the uniform beam segment of a three-dimensional frame, namely,

$$
\begin{Bmatrix} P_1 \\ P_2 \\ P_3 \\ P_4 \\ P_5 \\ P_6 \\ P_7 \\ P_8 \\ P_9 \\ P_{10} \\ P_{11} \\ P_{12} \end{Bmatrix} = \frac{\bar{m}L}{420}
\begin{bmatrix}
140 & & & & & & & & & & & \\
0 & 156 & & & & \text{Symmetric} & & & & & & \\
0 & 0 & 156 & & & & & & & & & \\
0 & 0 & 0 & \dfrac{140I_0}{A} & & & & & & & & \\
0 & 0 & -22L & 0 & 4L^2 & & & & & & & \\
0 & 22L & 0 & 0 & 0 & 4L^2 & & & & & & \\
70 & 0 & 0 & 0 & 0 & 0 & 140 & & & & & \\
0 & 54 & 0 & 0 & 0 & 13L & 0 & 156 & & & & \\
0 & 0 & 54 & 0 & -13L & 0 & 0 & 0 & 156 & & & \\
0 & 0 & 0 & \dfrac{70I_0}{A} & 0 & 0 & 0 & 0 & 0 & \dfrac{140I_0}{A} & & \\
0 & 0 & 13L & 0 & -3L^2 & 0 & 0 & 0 & 22L & 0 & 4L^2 & \\
0 & -13L & 0 & 0 & 0 & -3L^2 & 0 & -22L & 0 & 0 & 0 & 4L^2
\end{bmatrix}
\begin{Bmatrix} \ddot{\delta}_1 \\ \ddot{\delta}_2 \\ \ddot{\delta}_3 \\ \ddot{\delta}_4 \\ \ddot{\delta}_5 \\ \ddot{\delta}_6 \\ \ddot{\delta}_7 \\ \ddot{\delta}_8 \\ \ddot{\delta}_9 \\ \ddot{\delta}_{10} \\ \ddot{\delta}_{11} \\ \ddot{\delta}_{12} \end{Bmatrix}
$$

(17.4)

or in condensed notation

$$\{P\} = [M]\{\ddot{\delta}\} \tag{17.5}$$

17.3   ELEMENT DAMPING MATRIX

The damping matrix for a uniform beam segment of a three-dimensional frame may be obtained in a manner entirely parallel to those of the stiffness, eq. (17.1), and mass, eq. (17.4), matrices. Nevertheless, as was discussed in Section 14.5, in practice, damping is generally expressed in terms of damping ratios for each mode of vibration. Therefore, if the response is sought using the modal superposition method, these damping ratios are directly introduced in the modal equations. When the damping matrix is required explicitly, it may be determined from given values of damping ratios by the methods presented in Chapter 12.

17.4   TRANSFORMATION OF COORDINATES

The stiffness and the mass matrices given by eqs. (17.1) and (17.4), respectively, are referred to local coordinate axes fixed on the beam segment. Inasmuch as the elements of these matrices corresponding to the same nodal coordinates of the structure should be added to obtain the system stiffness and mass matrices, it is necessary first to transform these matrices to the same reference system, the global system of coordinates. Figure 17.2 shows these two reference systems, the x,y,z axes representing the local system of coordinates and the X,Y,Z axes representing the global system of coordinates. Also shown in this figure is a general vector $\vec{A}$ with its components X,Y,Z along the global coordinate. This vector $\vec{A}$ may represent any force or displacement at the nodal coordinates of one of the joints of the structure. To obtain the components of vector $\vec{A}$ along one of the local axes $x,y,$ or z, it is necessary to add the projections along that axis of the components X,Y,Z. For example, the component x of vector $\vec{A}$ along the x coordinate is given by

$$x = X \cos xX + Y \cos xY + Z \cos xZ \qquad (17.6a)$$

in which cos xY is the cosine of the angle between axes x and y and corresponding definitions for other cosines. Similarly, the y and z components of A are

$$y = X \cos yX + Y \cos yY + Z \cos yZ \qquad (17.6b)$$

$$z = X \cos zX + Y \cos zY + Z \cos zZ \qquad (17.6c)$$

These equations are conveniently written in matrix notation as

$$\begin{Bmatrix} x \\ y \\ z \end{Bmatrix} = \begin{bmatrix} \cos xX & \cos xY & \cos xZ \\ \cos yX & \cos yY & \cos yZ \\ \cos zX & \cos zY & \cos zZ \end{bmatrix} \begin{Bmatrix} X \\ Y \\ Z \end{Bmatrix} \qquad (17.7)$$

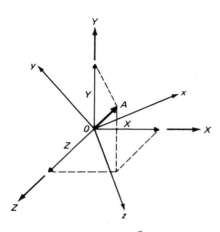

Fig. 17.2 Components of a general vector $\vec{A}$ in local and global coordinates.

in which

$$[T_1] = \begin{bmatrix} \cos xX & \cos xY & \cos xZ \\ \cos yX & \cos yY & \cos yZ \\ \cos zX & \cos zY & \cos zZ \end{bmatrix} \qquad (17.8)$$

is the transformation matrix from global to local coordinates. The direction cosines for the local axis x, $c_1 = \cos xX$, $c_2 = \cos xY$, and $c_3 = \cos xZ$ are given directly by the coordinates of the two end points (X_1, Y_1, Z_1) and (X_2, Y_2, Z_2) of the beam element as

$$c_1 = \frac{X_2 - X_1}{L}$$

$$c_2 = \frac{Y_2 - Y_1}{L}$$

$$c_3 = \frac{Z_2 - Z_1}{L} \qquad (17.9)$$

in which

$$L = \sqrt{(X_2 - X_1)^2 + (Y_2 - Y_1)^2 + (Z_2 - Z_1)^2} \qquad (17.10)$$

For the particular case in which the plane defined by the local axes x-y is vertical, the direction cosines for the other two local axes, y and z, can also be expressed in terms of the coordinates of the two points at the ends of the beam element. Since in this particular case the local axis z is perpendicular to the vertical plane defined by the local axis x, and an axis Y' parallel to the global axis Y as shown in Fig. 17.3, a vector $\vec{z}$ along the local axis z may be found as

$$\vec{z} = \hat{x} * \hat{j} = \begin{vmatrix} \hat{i} & \hat{j} & \hat{k} \\ c_1 & c_2 & c_3 \\ 0 & 1 & 0 \end{vmatrix} = -c_3 \hat{i} + c_1 \hat{k} \qquad (17.11)$$

in which

$$\hat{x} = c_1 \hat{i} + c_2 \hat{j} + c_3 \hat{k} \qquad (17.12)$$

is a unit vector along the local axis x, and $\hat{i}, \hat{j}, \hat{k}$ are respectively the unit vectors along the axes X, Y, Z of the global system.

A unit vector $\hat{z}$ along the local axis z is then calculated from eq. (17.11) as

$$\hat{z} = \frac{\vec{z}}{d} = -\frac{c_3}{d} \hat{i} + \frac{c_1}{d} \hat{k} \qquad (17.13)$$

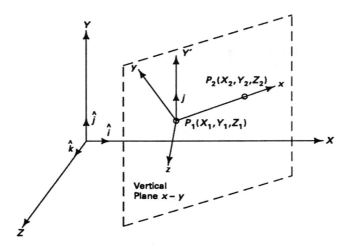

Fig. 17.3 Global system of coordinates (X,Y,Z) and local system of coordinates (x,y,z) with x-y plane vertical.

where

$$d = \sqrt{c_3^2 + c_1^2} \tag{17.14}$$

Finally, a unit vector $\hat{y}$ along the local axis y is given by

$$\hat{y} = \hat{z} * \hat{x} = \begin{vmatrix} i & j & k \\ \dfrac{-c_3}{d} & 0 & \dfrac{c_1}{d} \\ c_1 & c_2 & c_3 \end{vmatrix} \tag{17.15}$$

or

$$\hat{y} = \frac{-c_1c_2}{d}\hat{i} + d\hat{j} - \frac{c_2c_3}{d}\hat{k} \tag{17.16}$$

Therefore, from eqs. (17.12), (17.13), and (17.16), the transformation in eq. (17.7) in the particular case considered is given by

$$\begin{Bmatrix} x \\ y \\ z \end{Bmatrix} = \begin{bmatrix} c_1 & c_2 & c_3 \\ \dfrac{-c_1c_2}{d} & d & \dfrac{-c_2c_3}{d} \\ \dfrac{-c_3}{d} & 0 & \dfrac{c_1}{d} \end{bmatrix} \begin{Bmatrix} X \\ Y \\ Z \end{Bmatrix} \tag{17.17}$$

where d is given by eq. (17.14). Equation (17.17) is the transformation matrix between local coordinates (x,y,z) and global coordinates (X,Y,Z) for the particular case in which the local plane x-y is vertical. If this plane is not vertical, the required transformation may be formulated in the following three steps:

Step 1: Rotate the local coordinate system (x,y,z) around the local axis x until the local y axis and the local axis x form a vertical plane. Let (x',y',z') denote the auxiliary coordinate system resulting from this rotation of the system (x,y,z) about axis x. The transformation matrix between these two local systems, (x',y',z') and (x,y,z), is given by

$$\begin{Bmatrix} x \\ y \\ z \end{Bmatrix} = \begin{bmatrix} 1 & 0 & 0 \\ 0 & \cos\phi & \sin\phi \\ 0 & -\sin\phi & \cos\phi \end{bmatrix} \begin{Bmatrix} x' \\ y' \\ z' \end{Bmatrix} \tag{17.18}$$

in which ϕ is the *angle of rolling* from the axis y to the axis y'. This angle is positive for a clockwise rotation around axis x observing the rotation from the second end point to the first end point of the beam element.

Step 2: Transform from the global system (X,Y,Z) to the auxiliary local system (x',y',z') of coordinates. Since this auxiliary system has its x'-y' plane in a vertical plane, the necessary transformation matrix is given by eq. (17.17) as

$$\begin{Bmatrix} x' \\ y' \\ z' \end{Bmatrix} = \begin{bmatrix} c_1 & c_2 & c_3 \\ \dfrac{-c_1 c_2}{d} & d & \dfrac{-c_2 c_3}{d} \\ \dfrac{-c_3}{d} & 0 & \dfrac{c_1}{d} \end{bmatrix} \begin{Bmatrix} X \\ Y \\ Z \end{Bmatrix} \tag{17.19}$$

Step 3: Obtain the transformation from the global system (X,Y,Z) to the local system (x,y,z), by the combination of the transformations indicated in steps 1 and 2. Hence, the required transformation is obtained by substituting eq. (17.19) into eq. (17.18), namely,

$$\begin{Bmatrix} x \\ y \\ z \end{Bmatrix} = \begin{bmatrix} 1 & 0 & 0 \\ 0 & \cos\phi & \sin\phi \\ 0 & -\sin\phi & \cos\phi \end{bmatrix} \begin{bmatrix} c_1 & c_2 & c_3 \\ \dfrac{-c_1 c_2}{d} & d & \dfrac{-c_2 c_3}{d} \\ \dfrac{-c_3}{d} & 0 & \dfrac{c_1}{d} \end{bmatrix} \begin{Bmatrix} X \\ Y \\ Z \end{Bmatrix} \tag{17.20}$$

Therefore, the general transformation matrix $[T_1]$ from the global system (X,Y,Z) to the local system (x,y,z) is given by eq. (17.21) as the product of the two matrices in eq. (17.20).

$$[T_1] = \begin{bmatrix} c_1 & c_2 & c_3 \\ \dfrac{-c_1 c_2}{d}\cos\phi - \dfrac{c_3}{d}\sin\phi & d\cos\phi & \dfrac{-c_2 c_3}{d}\cos\phi + \dfrac{c_1}{d}\sin\phi \\ \dfrac{c_1 c_2}{d}\sin\phi - \dfrac{c_3}{d}\cos\phi & -d\sin\phi & \dfrac{c_2 c_3}{d}\sin\phi + \dfrac{c_1}{d}\cos\phi \end{bmatrix} \tag{17.21}$$

It should be noted that transformation matrix $[T_1]$ is not defined if the local axis x is parallel to the global axis Y. In this case, eqs. (17.9) and (17.14) result in $c_1 = 0$, $c_3 = 0$, and $d = 0$.

If the centroidal axis of the element is vertical, that is, the local axis x and the global axis Y are parallel, the angle of rolling is defined as the angle of rotation about the local axis x to have the local axis z parallel to the global axis Z. Let us designate by (x',y',z') this auxiliary system of coordinates in which the local axis z' is parallel to the global axis Z as shown in Fig. 17.4.

The transformation of coordinates between this auxiliary system (x',y',z') and the global system (X,Y,Z) obtained from Fig. 17.4 is given by

$$\begin{Bmatrix} x' \\ y' \\ z' \end{Bmatrix} = \begin{bmatrix} 0 & c_2 & 0 \\ -c_2 & 0 & 0 \\ 0 & 0 & 1 \end{bmatrix} \begin{Bmatrix} X \\ Y \\ Z \end{Bmatrix} \tag{17.22}$$

where $c_2 = 1$ when local axis x has the sense of the global axis Y; otherwise, $c_2 = -1$. The simple transformation of coordinates from the (x',y',z') system to the (x,y,z) is

$$\begin{Bmatrix} x \\ y \\ z \end{Bmatrix} = \begin{bmatrix} 1 & 0 & 0 \\ 0 & \cos \phi & \sin \phi \\ 0 & -\sin \phi & \cos \phi \end{bmatrix} \begin{Bmatrix} x' \\ y' \\ z' \end{Bmatrix} \tag{17.23}$$

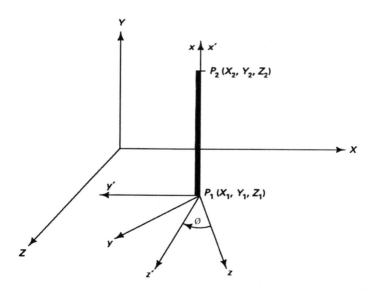

Fig. 17.4 (a) Global axes (X,Y,Z). (b) Local axes (x,y,z) with local axis x parallel to global axis Y. (c) Auxiliary local axis (x',y',z') with local axis z' parallel to global axis Z.

where the angle ϕ is the angle of rolling around the local axis x measured from local axis z to the auxiliary axis z'. This angle is positive for a clockwise rotation observing the x axis from the second end toward the first end of the beam element.

The substitution of eq. (17.22) into eq. (17.23) yields

$$\begin{Bmatrix} x \\ y \\ z \end{Bmatrix} = \begin{bmatrix} 0 & c_2 & 0 \\ -c_2 \cos \phi & 0 & \sin \phi \\ c_2 \sin \phi & 0 & \cos \phi \end{bmatrix} \begin{Bmatrix} X \\ Y \\ Z \end{Bmatrix} \qquad (17.24)$$

Therefore, when the local axis x is parallel to the global axis Y, the transformation matrix $[T_1]$ between the global coordinates (X,Y,Z) and the local coordinates (x,y,z) is given by

$$[T_1] = \begin{bmatrix} 0 & c_2 & 0 \\ -c_2 \cos \phi & 0 & \sin \phi \\ c_2 \sin \phi & 0 & \cos \phi \end{bmatrix} \qquad (17.25)$$

where ϕ is the angle of rolling between the local axis z and the direction of the global axis Z ($c_2 = +1$ when local axis x and global axis Y have the same sense and $c_2 = -1$ for the opposite sense of these axes).

We have, therefore, shown that the knowledge of the coordinates at the two ends of an element and the knowledge of the angle of rolling suffice to calculate the direction cosines of the transformation matrix $[T_1]$ in eq. (17.8). For the beam segment of a three-dimensional frame, the transformation of the nodal displacement vectors involve the transformation of linear and angular displacement vectors at each joint of the segment. Therefore, a beam element of a space frame requires, for the two joints, the transformation of a total of four displacement vectors. This transformation of the 12 nodal displacements $\{\bar{\delta}\}$ in global coordinates to the displacements $\{\delta\}$ in local coordinates may be written in abbreviated form as

$$\{\delta\} = [T]\{\bar{\delta}\} \qquad (17.26)$$

in which

$$[T] = \begin{bmatrix} [T_1] & & & \\ & [T_1] & & \\ & & [T_1] & \\ & & & [T_1] \end{bmatrix} \qquad (17.27)$$

Analogously, the transformation from nodal forces $\{\bar{P}\}$ in global coordinates to nodal forces $\{P\}$ in local coordinates is given by

$$\{P\} = [T]\{\bar{P}\} \qquad (17.28)$$

Finally, to obtain the stiffness matrix $[\bar{K}]$ and the mass matrix $[\bar{M}]$ in reference to the global system of coordinates, we simply substitute, into eq. (17.2), $\{\delta\}$ from eq. (17.26) and $[P]$ from eq. (17.28) to obtain

$$[T]\{\bar{P}\} = [K][T]\{\bar{\delta}\}$$

or

$$\{\bar{P}\} = [T]^T [K][T]\{\bar{\delta}\} \tag{17.29}$$

since $[T]$ is an orthogonal matrix. From eq. (17.29), we may write

$$\{\bar{P}\} = [\bar{K}]\{\bar{\delta}\} \tag{17.30}$$

in which $[\bar{K}]$ is defined as

$$[\bar{K}] = [T]^T [K][T] \tag{17.31}$$

Analogously, the mass matrix in eq. (17.5) is transformed from local to global coordinates by

$$[\bar{M}] = [T]^T [M][T] \tag{17.32}$$

and the damping matrix $[C]$ by

$$[\bar{C}] = [T]^T [C][T] \tag{17.33}$$

17.5 DIFFERENTIAL EQUATION OF MOTION

The direct method which was explained in detail in Chapter 14 may also be used to assemble the stiffness, mass, and damping matrices from the corresponding matrices for a three-dimensional beam segment, eqs. (17.31), (17.32), and (17.33), which are referred to the global system of coordinates. The differential equations of motion which are obtained by establishing the dynamic equilibrium among the inertial, damping, and elastic forces with the external forces may be expressed in matrix notation as

$$[M]\{\ddot{y}\} + [C]\{\dot{y}\} + [K]\{y\} = \{F(t)\} \tag{17.34}$$

in which $[M]$, $[C]$, and $[K]$ are, respectively, the system mass, damping, and stiffness matrices, $\{\ddot{y}\}$, $\{\dot{y}\}$, and $\{y\}$ are the system acceleration, velocity, and displacement vectors, and $\{F(t)\}$ is the force vector which includes the forces applied directly to the joints of the structure and the equivalent nodal forces for the forces not applied at the joints.

17.6 DYNAMIC RESPONSE

The integration of the differential equations of motion, eq. (17.34), may be accomplished by any of the methods presented in previous chapters to obtain the response of structures modeled as beams, plane frames, or grids. The selection of the particular method of solution depends, as discussed previously, on the linearity of the differential equation, that is, whether the stiffness matrix $[K]$ or any other coefficient matrix is constant, and also depends on the complexity of the excitation as a function of time. When the differential equations of motion, eq. (17.34), are linear, the modal superposition method is applicable. This method, as we have seen in the preceding chapters, requires the solution of an eigenproblem to uncouple the differential equations resulting in the modal equations of motion.

If the structure is assumed to follow an elastoplastic behavior or any other form of nonlinearity, it is necessary to resort to some kind of numerical integration in order to solve the differential equations of motion, eq. (17.34). In Chapter 19, the linear acceleration method with a modification known as the Wilson-θ method is presented together with a computer program for analysis of linear structures with an elastic behavior.

17.7   PROGRAM 16—MODELING STRUCTURES AS SPACE FRAMES

Program 16 calculates the stiffness and mass matrices for a three-dimensional frame and stores the coefficients of these matrices in a file. Since the stiffness and mass matrices are symmetric matrices, it is only necessary to store the coefficients in the upper triangle of these matrices. The program also stores in another file the general data of the space frame. The information stored in these files is needed by the programs to undertake the dynamic analysis of the structure, such as the determination of natural frequencies or the response of the structure subjected to external excitation.

Example 17.1.   Determine the stiffness and mass matrices for the three dimensional frame shown in Fig. 17.5.

Solution: As the first step in the analysis, the frame is divided in four elements, as indicated in Fig. 17.5. This division of the structure results in five nodes with a total of 30 nodal coordinates, of which 24 are fixed. The numerical values needed for analysis of this structure are given in Table 17.1.

Input Data and Output Results

```
PROGRAM 16: MODELING SPACE FRAMES     DATA FILE:D16

NUMBER OF JOINTS                                     NJ = 5
NUMBER OF BEAM ELEMENTS                              NE = 4
NUMBER OF FIXED COORDINATES                          NC = 24
MODULUS OF ELASTICITY                                 E = 3E+07
MODULUS OF RIGIDITY                                  GR = 1.2E+07
```

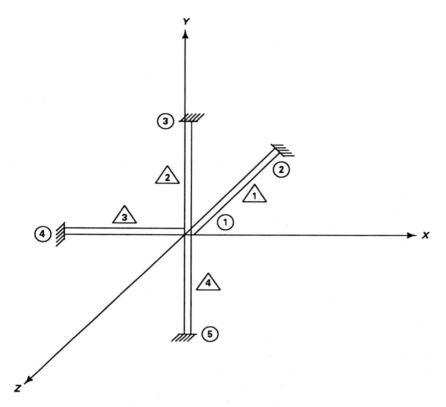

Fig. 17.5 Space Frame of Example 17.1.

JOINT DATA:

JOINT #	X-COORD.	Y-COORD.	Z-COORD.	JOINT MASS
1	0.00	0.00	0.00	0
2	0.00	0.00	−200.00	0
3	0.00	200.00	0.00	0
4	−200.00	0.00	0.00	0
5	0.00	−200.00	0.00	0

BEAM ELEMENT DATA:

MEMBER'S ROLLING ANGLES (DEGREES)

1	0.000
2	0.000
3	0.000
4	0.000

TABLE 17.1   Input Data for Example 17.1.

Quantity	Members 1,3	Members 2,4
Modulus of elasticity (psi)	30×10^6	30×10^6
Modulus of rigidity (psi)	12×10^6	12×10^6
Distributed mass (lb·sec^2/in/in)	.02	.01
Cross-sectional y-moment of inertia (in^4)	200	64
Cross-sectional z-moment of inertia (in^4)	200	64
Torsional constant (in^4)	40.0	12.8
Cross-sectional area (in^2)	50	28

ELEM #	FIRST JOINT #	SECOND JOINT #	MASS/ LENGTH	SECT-Y INERTIA	SECT-Z INERTIA	TORSION CONST	SECTION AREA
1	1	2	0.200	200.00	200.00	40.00	50.00
2	1	3	0.100	64.00	64.00	12.80	28.00
3	4	1	0.200	200.00	200.00	40.00	50.00
4	5	1	0.100	64.00	64.00	12.80	28.00

```
        FIXED COORDINATES DATA:
            DIRECTION # (X=1, Y=2, Z=3, X-ROT=4, Y-ROT=5, Z-ROT=6)
```

NODE	DIR.	NODE	DIR.	NODE	DIR.	NODE	DIR.	NODE	DIR.	NODE	DIR.	NODE	DIR.	NODE	DIR.
2	1	2	2	2	3	2	4	2	5	2	6	3	1	3	2
3	3	3	4	3	5	3	6	4	1	4	2	4	3	4	4
4	5	4	6	5	1	5	2	5	3	5	4	5	5	5	6

```
        OUTPUT RESULTS:
```

SYSTEM STIFFNESS MATRIX

0.7515E+07	0.0000E+00	0.0000E+00	0.0000E+00	−.9000E+06	0.0000E+00
0.0000E+00	0.8418E+07	0.0000E+00	0.9000E+06	0.7813E−02	−.9000E+06
0.0000E+00	0.0000E+00	0.7515E+07	0.0000E+00	0.9000E+06	0.7813E−02
0.0000E+00	0.9000E+06	0.0000E+00	0.1992E+09	0.1000E+01	0.1000E+01
−.9000E+06	0.7813E−02	0.9000E+06	0.1000E+01	0.2415E+09	−.1000E+01
0.0000E+00	−.9000E+06	0.7813E−02	0.1000E+01	−.1000E+01	0.1992E+09

SYSTEM MASS MATRIX

0.4305E+02	0.0000E+00	−.1192E−06	0.0000E+00	−.4190E+03	0.0000E+00
0.0000E+00	0.4305E+02	0.1192E−06	0.4190E+03	0.0000E+00	−.4190E+03
−.1192E−06	0.1192E−06	0.4305E+02	0.0000E+00	0.4190E+03	0.0000E+00
0.0000E+00	0.4190E+03	0.0000E+00	0.3058E+05	0.0000E+00	0.0000E+00
−.4190E+03	0.0000E+00	0.4190E+03	0.0000E+00	0.3054E+05	0.0000E+00
0.0000E+00	−.4190E+03	0.0000E+00	0.0000E+00	0.0000E+00	0.3058E+05

17.8   SUMMARY

This chapter presented the formulation of the stiffness and mass matrices for an element of a space frame, as well as the transformation of coordinates required to refer these matrices to the global system of coordinates. Except for the larger dimensions of the matrices resulting from modeling a space frame, the procedure is identical to the case of a beam, plane frame or grid frame described in the preceding chapters.

PROBLEMS

17.1   Use Program 16 to determine the stiffness and mass matrices of the three dimensional frame modeled as shown in Fig. 17.6. Assume concentrated masses at every joint m = 10 lb · sec²/in.

Problem Data (for all members):

Modulus of elasticity:	$E = 30 \times 10^6$ psi
Modulus of rigidity:	$G = 12 \times 10^6$ psi
Distributed mass:	$\bar{m} = 0.2$ lb · sec²/in/in
Concentrated masses:	$m = 10$ lb · sec²/in
Cross-sectional y-moment of inertia:	$I_y = 300$ in⁴
Cross-sectional z-moment of inertia:	$I_z = 400$ in⁴
Torsional constant:	$J = 500$ in⁴
Cross-sectional area:	$A = 20$ in²

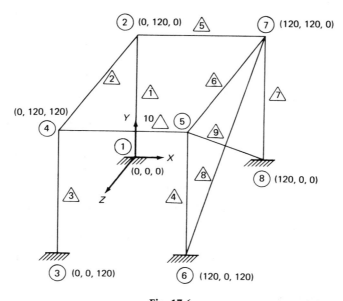

Fig. 17.6

18

Dynamic Analysis of Trusses

The static analysis of trusses whose members are pin-connected reduces to the problem of determining the bar forces due to a set of loads applied at the joints. When the same trusses are subjected to the action of dynamics forces, the simple situation of only axial stresses in the members is no longer present. The inertial forces developed along the members of the truss will, in general, produce flexural bending in addition to axial forces. The bending moments at the ends of the truss members will still remain zero in the absence of external joint moments. The dynamic stiffness method for the analysis of trusses is developed as in the case of framed structures by establishing the basic relations between external forces, elastic forces, damping forces, inertial forces, and the resulting displacements, velocities, and accelerations at the nodal coordinates, that is, by determining the stiffness, damping, and mass matrices for a member of the truss. The assemblage of system stiffness, damping, and mass matrices of the truss as well as the solution for the displacements at the nodal coordinates follows along the standard method presented in the preceding chapters for framed structures.

18.1   STIFFNESS AND MASS MATRICES FOR THE PLANE TRUSS

A member of a plane truss has two nodal coordinates at each joint, that is, a total of four nodal coordinates (Fig. 18.1). For small deflections, it may be assumed that the force–displacement relationship for the nodal coordinates along the axis of the member (coordinates 1 and 3 in Fig. 18.1) are independent of the transverse displacements along nodal coordinates 2 and 4. This assumption is equivalent to stating that a displacement along nodal coordinates 1 or 3 does not produce forces along nodal coordinates 2 or 4 and vice versa.

The stiffness and mass coefficients corresponding to the axial nodal coordinates were derived in Chapter 15 and are given, in general, by eq. (15.17) for the stiffness coefficients and by eq. (15.23) for consistent mass coefficients. Applying these equations to a uniform beam element, we obtain, using the notation of Fig. 18.1, the following coefficients:

$$k_{11} = k_{33} = \frac{AE}{L}, \quad k_{13} = k_{31} = -\frac{AE}{L} \tag{18.1}$$

$$m_{11} = m_{33} = \frac{\bar{m}L}{3}, \quad m_{13} = m_{31} = \frac{\bar{m}L}{6} \tag{18.2}$$

in which $\bar{m}$ is the mass per unit length, A is the cross-sectional area, and L is the length of the element.

The stiffness coefficients, for pin-ended elements, corresponding to the nodal coordinates 2 and 4 are all equal to zero, since a force is not required to produce displacements at these coordinates. Therefore, arranging the coefficients given by eq. (18.1), we obtain the stiffness equation for a uniform member of a truss as

$$\begin{Bmatrix} P_1 \\ P_2 \\ P_3 \\ P_4 \end{Bmatrix} = \frac{AE}{L} \begin{bmatrix} 1 & 0 & -1 & 0 \\ 0 & 0 & 0 & 0 \\ -1 & 0 & 1 & 0 \\ 0 & 0 & 0 & 0 \end{bmatrix} \begin{Bmatrix} \delta_1 \\ \delta_2 \\ \delta_3 \\ \delta_4 \end{Bmatrix} \tag{18.3}$$

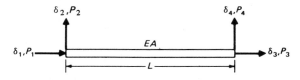

Fig. 18.1 Member of a plane truss showing nodal displacements and forces.

or in condensed notation

$$\{P\} = [K]\{\delta\} \tag{18.4}$$

in which $[K]$ is the element stiffness matrix.

The consistent mass matrix is obtained, as previously demonstrated, using expressions for static displacement functions in the application of the principle of virtual work. The displacement functions corresponding to a unit deflection at nodal coordinates 2 and 4 indicated in Fig. 18.2 are given by

$$u_2 = 1 - \frac{x}{L} \tag{18.5}$$

and

$$u_4 = \frac{x}{L} \tag{18.6}$$

The consistent mass coefficients are given by the general expression, eq. (15.23), which is repeated here for convenience, namely,

$$m_{ij} = \int_0^L \bar{m}(x)\, u_i(x)\, u_j(x)\, dx \tag{18.7}$$

For a uniform member of mass $\bar{m}$ per unit length, the substitution of eqs. (18.5) and (18.6) into eq. (18.7) yields

$$m_{22} = m_{44} = \frac{\bar{m}L}{3}$$

$$m_{24} = m_{42} = \frac{\bar{m}L}{6} \tag{18.8}$$

Finally, the combination of the mass coefficients from eqs. (18.2) and (18.8) forms the consistent mass matrix relating forces to accelerations at the nodal coordinates for a uniform member of a plane truss, namely,

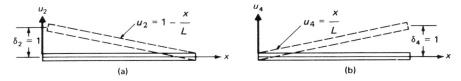

Fig. 18.2 Displacement functions. (a) For a unit displacement $\delta_2 = 1$. (b) For a unit displacement $\delta_4 = 1$.

$$\begin{Bmatrix} P_1 \\ P_2 \\ P_3 \\ P_4 \end{Bmatrix} = \frac{\bar{m}L}{6} \begin{bmatrix} 2 & 0 & 1 & 0 \\ 0 & 2 & 0 & 1 \\ 1 & 0 & 2 & 0 \\ 0 & 1 & 0 & 2 \end{bmatrix} \begin{Bmatrix} \ddot{\delta}_1 \\ \ddot{\delta}_2 \\ \ddot{\delta}_3 \\ \ddot{\delta}_4 \end{Bmatrix} \tag{18.9}$$

or in concise notation

$$\{P\} = [M]\,\{\ddot{\delta}\} \tag{18.10}$$

18.2   TRANSFORMATION OF COORDINATES

The stiffness matrix, eq. (18.3), and the mass matrix, eq. (18.9), were derived in reference to nodal coordinates associated with the local or element system of coordinates. As discussed before in the chapters on framed structures, it is necessary to transform these matrices to a common system of reference, the global coordinate system. The transformation of displacements and forces at the nodal coordinates is accomplished, as was demonstrated in Chapter 15, performing a rotation of coordinates. Deleting the angular coordinates in eq. (15.31) and relabeling the remaining coordinates result in the following transformation for the nodal forces:

$$\begin{Bmatrix} P_1 \\ P_2 \\ P_3 \\ P_4 \end{Bmatrix} = \begin{bmatrix} \cos\theta & \sin\theta & 0 & 0 \\ -\sin\theta & \cos\theta & 0 & 0 \\ 0 & 0 & \cos\theta & \sin\theta \\ 0 & 0 & -\sin\theta & \cos\theta \end{bmatrix} \begin{Bmatrix} \bar{P}_1 \\ \bar{P}_2 \\ \bar{P}_3 \\ \bar{P}_4 \end{Bmatrix} \tag{18.11}$$

or in condensed notation

$$\{P\} = [T]\,\{\bar{P}\} \tag{18.12}$$

in which $\{P\}$ and $\{\bar{P}\}$ are the nodal forces in reference to local and global coordinates, respectively, and $[T]$ the transformation matrix defined in eq. (18.11).

The same transformation matrix $[T]$ also serves to transform the nodal displacement vector $\{\bar{\delta}\}$ in the global coordinate system to the nodal displacement vector $\{\delta\}$ in local coordinates:

$$\{\delta\} = [T]\,\{\bar{\delta}\} \tag{18.13}$$

The substitution of eqs. (18.12) and (18.13) into the stiffness equation (18.4) gives

$$[T]\,[\bar{P}] = [K]\,[T]\,\{\bar{\delta}\}$$

Since $[T]$ is an orthogonal matrix $([T]^{-1} = [T]^T)$, it follows that

$$\{\bar{P}\} = [T]^T\,[K]\,[T]\,\{\bar{\delta}\}$$

or

$${\bar{P}} = [\bar{K}] \{\bar{\delta}\} \tag{18.14}$$

in which

$$[\bar{K}] = [T]^T [K] [T] \tag{18.15}$$

is the element stiffness matrix in the global coordinate system. Analogously, substituting eqs. (18.12) and (18.13) into eq. (18.10) results in

$$\{\bar{P}\} = [T]^T [M] [T] \{\ddot{\bar{\delta}}\} \tag{18.16}$$

or

$$\{\bar{P}\} = [\bar{M}] \{\ddot{\bar{\delta}}\} \tag{18.17}$$

$$[\bar{M}] = [T]^T [M] [T] \tag{18.18}$$

in which $[\bar{M}]$ is the element mass matrix referred to the global system of coordinates. However, there is no need to use eq. (18.18) to calculate matrix $[\bar{M}]$. This matrix is equal to the mass matrix $[M]$ in reference to local axes of coordinates. To verify this fact, we substitute into eq. (18.18) matrices $[M]$ and $[T]$, respectively, from eqs. (18.9) and (18.11) to obtain

$$[\bar{M}] = \frac{\bar{m}L}{6} \begin{bmatrix} c & -s & 0 & 0 \\ s & c & 0 & 0 \\ 0 & 0 & c & -s \\ 0 & 0 & s & c \end{bmatrix} \begin{bmatrix} 2 & 0 & 1 & 0 \\ 0 & 2 & 0 & 1 \\ 1 & 0 & 2 & 0 \\ 0 & 1 & 0 & 2 \end{bmatrix} \begin{bmatrix} c & s & 0 & 0 \\ -s & c & 0 & 0 \\ 0 & 0 & c & s \\ 0 & 0 & -s & c \end{bmatrix}$$

$$[\bar{M}] = \frac{\bar{m}L}{6} \begin{bmatrix} 2 & 0 & 1 & 0 \\ 0 & 2 & 0 & 1 \\ 1 & 0 & 2 & 0 \\ 0 & 1 & 0 & 2 \end{bmatrix} = [M] \tag{18.19}$$

in which we use the notation $c = \cos \theta$, $s = \sin \theta$ and the fact that $\cos^2 \theta + \sin^2 \theta = 1$.

A similar relation is also obtained for the element damping matrix, namely,

$$[\bar{C}] = [T]^T [C] [T] \tag{18.20}$$

in which $[\bar{C}]$ and $[C]$ are the damping matrices referred, respectively, to the global and the local systems of coordinates.

Example 18.1. The plane truss shown in Fig. 18.3 which has only three members is used to illustrate the application of the stiffness method for trusses. For this truss determine the system stiffness and the system consistent mass matrices.

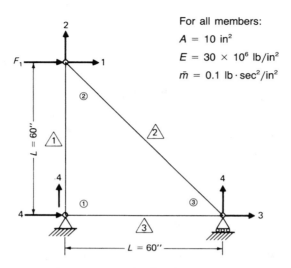

For all members:

$A = 10\ \text{in}^2$

$E = 30 \times 10^6\ \text{lb/in}^2$

$\bar{m} = 0.1\ \text{lb} \cdot \text{sec}^2/\text{in}^2$

Fig. 18.3 Example of a plane truss.

Solution: The stiffness matrix, eq. (18.3), the mass matrix, eq. (18.9), and the transformation matrix, eq. (18.11), are applied to the three members of this truss. For member △, $\theta = 90°$,

$$[K_1] = \frac{AE}{L} \begin{bmatrix} 1 & 0 & -1 & 0 \\ 0 & 0 & 0 & 0 \\ -1 & 0 & 1 & 0 \\ 0 & 0 & 0 & 0 \end{bmatrix}, \quad [\bar{M}_1] = [M_1] = \frac{\bar{m}L}{6} \begin{bmatrix} 2 & 0 & 1 & 0 \\ 0 & 2 & 0 & 1 \\ 1 & 0 & 2 & 0 \\ 0 & 1 & 0 & 2 \end{bmatrix}$$

and

$$[T_1] = \begin{bmatrix} 0 & 1 & 0 & 0 \\ -1 & 0 & 0 & 0 \\ 0 & 0 & 0 & 1 \\ 0 & 0 & -1 & 0 \end{bmatrix}$$

Then from eqs. (18.15)

$$[\bar{K}_1] = [T_1]^T [K_1] [T_1] = \frac{AE}{L} \begin{bmatrix} 0 & 0 & 0 & 0 \\ 0 & 1 & 0 & -1 \\ 0 & 0 & 0 & 0 \\ 0 & -1 & 0 & 1 \end{bmatrix}$$

For member △, $\theta = 135°$,

$$[K_2] = \frac{AE}{\sqrt{2}L} \begin{bmatrix} 1 & 0 & -1 & 0 \\ 0 & 0 & 0 & 0 \\ -1 & 0 & 1 & 0 \\ 0 & 0 & 0 & 0 \end{bmatrix} \quad [\bar{M}_2] = [M_2] = \frac{\bar{m}\sqrt{2}L}{6} \begin{bmatrix} 2 & 0 & 1 & 0 \\ 0 & 2 & 0 & 1 \\ 1 & 0 & 2 & 0 \\ 0 & 1 & 0 & 2 \end{bmatrix}$$

and

$$[T_2] = \frac{1}{\sqrt{2}} \begin{bmatrix} -1 & 1 & 0 & 0 \\ -1 & -1 & 0 & 0 \\ 0 & 0 & -1 & 1 \\ 0 & 0 & -1 & -1 \end{bmatrix}$$

Then from eqs. (18.15)

$$[\bar{K}_2] = [T_2]^T [K_2] [T_2] = \frac{AE}{2\sqrt{2}L} \begin{bmatrix} 1 & -1 & -1 & 1 \\ -1 & 1 & 1 & -1 \\ -1 & 1 & 1 & -1 \\ 1 & -1 & -1 & 1 \end{bmatrix}$$

For member $\triangle$, $\theta = 0°$,

$$[\bar{K}_3] = [K_3] = \frac{AE}{L} \begin{bmatrix} 1 & 0 & -1 & 0 \\ 0 & 0 & 0 & 0 \\ -1 & 0 & 1 & 0 \\ 0 & 0 & 0 & 0 \end{bmatrix}$$

$$[\bar{M}_3] = [M_3] = \frac{\bar{m}L}{6} \begin{bmatrix} 2 & 0 & 1 & 0 \\ 0 & 2 & 0 & 1 \\ 1 & 0 & 2 & 0 \\ 0 & 1 & 0 & 2 \end{bmatrix}$$

substituting the proper numerical values for this example; $L = 60$ in, $A = 10$ in^2, $\bar{m} = 0.1$ lb · sec^2/in^2, $E = 30 \times 10^6$ lb/in^2, and following the rules of the direct method of assembling the system stiffness and mass matrix from the above element matrices, we obtain

$$[K_S] = 10^6 \begin{bmatrix} 1.768 & -1.768 & -1.768 \\ -1.768 & 6.768 & 1.768 \\ -1.768 & 1.768 & 6.768 \end{bmatrix}$$

$$[M_S] = \begin{bmatrix} 4.828 & 0 & 1.414 \\ 0 & 4.828 & 0 \\ 1.414 & 0 & 4.828 \end{bmatrix}$$

where $[K_S]$ and $[M_S]$ are, respectively, the system stiffness and mass matrices for the truss shown in Fig. 18.3.

Example 18.2. Determine the natural frequencies and normal modes for the truss of Example 18.1.

Solution: The differential equations of motion for this system are

$$[M_S] \{\ddot{y}\} + [K_S] \{y\} = 0 \tag{a}$$

Substituting $\{y\} = \{a\}\sin \omega t$, we obtain

$$([K_S] - \omega^2 [M_S])\{a\} = \{0\} \tag{b}$$

For the nontrivial solution, we require

$$\|[K_S] - \omega^2 [M_S]\| = 0 \tag{c}$$

Substituting from Example 18.1 $[K_S]$ and $[M_S]$ and expanding the above determinant give a cubic equation in $\lambda = \omega^2 \bar{m} L^2/6AE$, which has the following roots

$$\lambda_1 = 0.00344 \quad \text{or} \quad \omega_1 = 415 \text{ rad/sec}$$

$$\lambda_2 = 0.0214 \quad \text{or} \quad \omega_2 = 1034 \text{ rad/sec}$$

$$\lambda_3 = 0.0466 \quad \text{or} \quad \omega_3 = 1526 \text{ rad/sec}$$

Substituting in turn ω_1, ω_2, and ω_3 into eq. (b), setting $a_1 = 1$, and solving for a_2 and a_3 give the modal vectors

$$\{a_1\} = \begin{Bmatrix} 1.000 \\ 0.216 \\ 0.274 \end{Bmatrix}, \quad \{a_2\} = \begin{Bmatrix} 1.000 \\ 5.488 \\ -4.000 \end{Bmatrix}, \quad \{a_3\} = \begin{Bmatrix} 1.000 \\ -1.000 \\ -1.524 \end{Bmatrix}$$

which may be normalized using the factors

$$\sqrt{\{a_1\}^T [M_S]\{a_1\}} = 2.489, \quad \sqrt{\{a_2\}^T [M_S]\{a_2\}} = 14.695$$

$$\sqrt{\{a_3\}^T [M_S]\{a_3\}} = 4.066$$

This normalization results in

$$\{\phi_1\} = \begin{Bmatrix} 0.402 \\ 0.087 \\ 0.110 \end{Bmatrix}, \quad \{\phi_2\} = \begin{Bmatrix} 0.068 \\ 0.373 \\ -0.272 \end{Bmatrix}, \quad \{\phi_3\} = \begin{Bmatrix} 0.246 \\ -0.246 \\ -0.375 \end{Bmatrix}$$

These normalized eigenvectors form the modal matrix:

$$[\Phi] = \begin{bmatrix} 0.402 & 0.068 & 0.246 \\ 0.087 & 0.373 & -0.246 \\ 0.110 & -0.272 & -0.375 \end{bmatrix}$$

Example 18.3. Determine the response of the truss in Examples 18.1 and 18.2 when a constant force $F_1 = 5000$ lb is suddenly applied along coordinate 1 as shown in Fig. 18.3.

Solution: The modal equations are given in general [eq. (11.6)] by

$$\ddot{z}_n + \omega_n^2 z_n = P_n$$

in which the modal force

$$P_n = \sum_i \phi_{in} F_i$$

Hence using the results which were calculated in Example 18.2 we obtain

$$\ddot{z}_1 + (415)^2 z_1 = 2010$$

$$\ddot{z}_2 + (1034)^2 z_2 = 340$$

$$\ddot{z}_3 + (1526)^2 z_3 = 1230$$

The solution of the above equations for zero initial conditions $(z_n = 0, \dot{z}_n = 0)$ is given by eqs. (4.5) as

$$z_1 = \frac{2010}{(415)^2} (1 - \cos 415t)$$

$$z_2 = \frac{340}{(1034)^2} (1 - \cos 1034t)$$

$$z_3 = \frac{1230}{(1526)^2} (1 - \cos 1526t)$$

The response at the nodal coordinates is then calculated from

$$\{y\} = [\Phi] \{z\}$$

$$\begin{Bmatrix} y_1 \\ y_2 \\ y_3 \end{Bmatrix} = \begin{bmatrix} 0.402 & 0.068 & 0.246 \\ 0.087 & 0.373 & -0.246 \\ 0.110 & -0.272 & -0.375 \end{bmatrix} \begin{Bmatrix} z_1 \\ z_2 \\ z_3 \end{Bmatrix}$$

or

$$y_1 = 10^{-3} [4.843 - 4.692 \cos 415t - 0.022 \cos 1034t - 0.130 \cos 1526t]$$

$$y_2 = 10^{-3} [1.004 - 1.015 \cos 415t - 0.119 \cos 1034t + 0.130 \cos 1526t]$$

$$y_3 = 10^{-3} [0.999 - 1.284 \cos 415t + 0.087 \cos 1034t + 0.198 \cos 1526t]$$

18.3   PROGRAM 17—MODELING STRUCTURES AS PLANE TRUSSES

Program 17 calculates the stiffness and mass matrices for the plane truss and stores, the coefficients of these matrices in a file. Since the stiffness and the mass matrices are symmetric, only the upper triangular portion of these matrices needs to be stored. The program also stores in another file, which is named by the user, the general information of the plane truss. The information stored in these files is needed by programs, which the user may call, to perform dynamic analysis of the truss, such as determination of natural frequencies or calculation of the response of the structure subjected to external excitation.

Example 18.4. Use Program 17 for the dynamic analysis of the truss shown in Fig. 18.3. Determine the natural frequencies, modal shapes, and the response to a constant force of $F_1 = 5000$ lb suddenly applied at nodal coordinate 1 as shown in the figure.

Solution: In the solution of this problem, it is necessary to model the structure as a plane truss (Program 17), to solve the eigenproblem to determine natural frequencies and modal shapes (Program 8), and finally to calculate the response (Program 9).

Input Data and Output Results

```
PROGRAM 17: MODELING PLANE TRUSSES    DATA FILE:D17

    GENERAL DATA:
```

NUMBER OF JOINTS	NJ= 3
NUMBER OF BEAM ELEMENTS	NE= 3
NUMBER OF FIXED COORDINATES	NC= 3
MODULUS OF ELASTICITY	E= 3E+07

```
    JOINT DATA:
```

JOINT #,	X-COORDINATE,	Y-COORDINATE	JOINT MASS
1	0	0	0
2	0	60	0
3	60	0	0

```
    BEAM ELEMENT DATA:
```

BEAM-ELEMENT #,	FIRST JT. #,	SECOND JT. #,	MASS/LENGTH,	SECT. AREA
1	1	2	0.100	10.00
2	3	2	0.100	10.00
3	1	3	0.100	10.00

```
    FIXED COORDINATES DATA:
```

JOINT #,	DIRECTION # (X=1, Y=2)
1	1
1	2
3	2

OUTPUT RESULTS:

SYSTEM STIFFNESS MATRIX

1.7678E+06	-1.7678E+06	-1.7678E+06
-1.7678E+06	6.7678E+06	1.7678E+06
-1.7678E+06	1.7678E+06	6.7678E+06

SYSTEM MASS MATRIX

4.8284E+00	0.0000E+00	1.4142E+00
0.0000E+00	4.8284E+00	0.0000E+00
1.4142E+00	0.0000E+00	4.8284E+00

PROGRAM 8: NATURAL FREQUENCIES   DATA FILE: SK

INPUT DATA:

OUTPUT RESULTS:

EIGENVALUES:

1.725E+05   1.068E+06   2.329E+06

NATURAL FREQUENCIES (C.P.S.):

66.12   164.52   242.87

EIGENVECTORS BY ROWS:

0.40178 0.08681 0.11035

0.06805 0.37346 -0.27173

-0.24594 0.24516 0.37487

PROGRAM 9: MODAL   DATA FILE:D9.3

INPUT DATA:

NUMBER OF DEGREES OF FREEDOM	ND= 3
NUMBER OF EXTERNAL FORCES	NF= 1
TIME STEP OF INTEGRATION	H= .05
GRAVITATIONAL INDEX	G= 0
PRINT TIME HISTORY NPRT=1; ONLY MAX. VALUES NPRT=0	NPRT= 0

FORCE #, COORD. # WHERE FORCE IS APPLIED, NUM. OF POINTS DEFINING THE FORCE
    1   1   2

EXCITATION FUNCTION FOR FORCE # 1:

TIME	EXCITATION
0.00	5000.00
2.00	5000.00

```
MODAL DAMPING RATIOS:
O          O              O
    OUTPUT RESULTS:
        MAXIMUM RESPONSE:
        COORD.      MAX. DISPL.    MAX. VELOC.    MAX. ACC.
           1          0.00959        2.002         1132.70
           2          0.00211        0.644          517.41
           3          0.00236        0.842          651.70
```

18.4 STIFFNESS AND MASS MATRICES FOR SPACE TRUSSES

The stiffness matrix and the mass matrix for a space truss can be obtained as an extension of the corresponding matrices for the plane truss. Figure 18.4 shows the nodal coordinates in the local system (unbarred) and in the global system (barred) for a member of a space truss. The local x axis is directed along the longitudinal axes of the member while the y and z axes are set to agree with the principal directions of the cross section of the member. The following matrices may then be written for a uniform member of a space truss as an extension of the stiffness, eq. (18.3), and the mass, eq. (18.9), matrices for a member of a plane truss.

Stiffness matrix:

$$[K] = \frac{AE}{L} \begin{bmatrix} 1 & 0 & 0 & -1 & 0 & 0 \\ 0 & 0 & 0 & 0 & 0 & 0 \\ 0 & 0 & 0 & 0 & 0 & 0 \\ -1 & 0 & 0 & 1 & 0 & 0 \\ 0 & 0 & 0 & 0 & 0 & 0 \\ 0 & 0 & 0 & 0 & 0 & 0 \end{bmatrix} \tag{18.21}$$

Consistent mass matrix:

$$[M_c] = \frac{\bar{m}L}{6} \begin{bmatrix} 2 & 0 & 0 & 1 & 0 & 0 \\ 0 & 2 & 0 & 0 & 1 & 0 \\ 0 & 0 & 2 & 0 & 0 & 1 \\ 1 & 0 & 0 & 2 & 0 & 0 \\ 0 & 1 & 0 & 0 & 2 & 0 \\ 0 & 0 & 1 & 0 & 0 & 2 \end{bmatrix} \tag{18.22}$$

Lumped mass matrix:

$$[M_L] = \frac{\bar{m}L}{2} \lceil 1 \quad 1 \quad 1 \quad 1 \quad 1 \quad 1 \rfloor \tag{18.23}$$

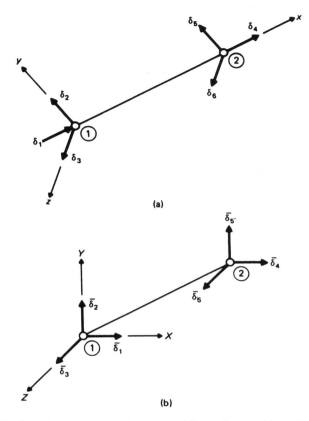

Fig. 18.4 Member of a space truss showing nodal coordinates. (a) In the local system (unbarred). (b) In the global system (barred).

The transformation matrix $[T_1]$ corresponding to three nodal coordinates at a joint is given by eq. (17.8). It is repeated here for convenience.

$$[T_1] = \begin{bmatrix} \cos xX & \cos xY & \cos xZ \\ \cos yX & \cos yY & \cos yZ \\ \cos zX & \cos zY & \cos zZ \end{bmatrix} \qquad (18.24)$$

The direction cosines in the first row of eq. (18.24), $c_1 = \cos xX$, $c_2 = \cos xY$, and $c_3 = \cos xZ$, may be calculated from the coordinates of the two points $P_1\ (X_1, Y_1, Z_1)$ and $P_2\ (X_2, Y_2, Z_2)$ at the ends of the truss element, that is

$$c_1 = \frac{X_1 - X_2}{L}, \; c_2 = \frac{Y_2 - Y_1}{L} \text{ and } c_3 = \frac{Z_2 - Z_1}{L} \qquad (18.25)$$

$$\text{with } L = \sqrt{(X_2 - X_1)^2 + (Y_2 - Y_1)^2 + (Z_2 - Z_1)^2} \qquad (18.26)$$

The transformation matrix for the nodal coordinates at the two ends of a truss member is then given by

$$[T] = \begin{bmatrix} [T_1] & [0] \\ [0] & [T_1] \end{bmatrix}$$

(18.27)

in which $[T_1]$ is given by eq. (18.24). The following transformations are then required to obtain the member stiffness matrix $[\bar{K}]$ and the member mass matrix $[\bar{M}]$ in reference to the global system of coordinates:

$$[\bar{K}] = [T]^T [K] [T]$$

(18.28)

and

$$[\bar{M}] = [T]^T [M] [T]$$

(18.29)

In the case of an element of a space truss, it is only necessary to calculate the direction cosines of the centroidal axis of the element which are given by eq. (18.25). The other direction cosines in eq. (18.24) do not appear in the final expression for the element stiffness matrix $[\bar{K}]$ as may be verified by substituting eqs. (18.24) and (18.27) into eq. (18.28) and proceeding to multiply the matrices indicated in this last equation. The final result of this operation may be written as follows:

$$[\bar{K}] = \frac{AE}{L} \begin{bmatrix} c_1^2 & c_1 c_2 & c_1 c_3 & -c_1^2 & -c_1 c_2 & -c_1 c_3 \\ c_2 c_1 & c_2^2 & c_2 c_3 & -c_2 c_1 & -c_2^2 & -c_2 c_3 \\ c_3 c_1 & c_3 c_2 & c_3^2 & -c_3 c_1 & -c_3 c_2 & -c_3^2 \\ -c_1^2 & -c_1 c_2 & -c_1 c_3 & c_1^2 & c_1 c_2 & c_1 c_3 \\ -c_2 c_1 & c_2^2 & -c_2 c_3 & c_2 c_1 & c_2^2 & c_2 c_3 \\ -c_3 c_1 & -c_3 c_2 & -c_3^2 & c_3 c_1 & c_3 c_2 & c_3^2 \end{bmatrix}$$

(18.30)

Consequently, the determination of the stiffness matrix for an element of a space truss, in reference to the global system of coordinates by eq. (18.30), requires the evaluation by eq. (18.25) of only the direction cosines of the local axis x along the element.

Also, analogously, to eq. (18.19) for an element of a plane truss, the mass matrix $[\bar{M}]$ for an element of a space truss in reference to global coordinates is equal to the mass matrix of the element $[M]$ in local coordinates. Thus, there is no need to perform the operations indicated in eq. (18.29). The substitution into this equation of $[M]$ from eq. (18.22) and $[T]$ from eqs. (18.24) and (18.27) results, after performing the multiplications established by eq. (18.29), in the same matrix $[M]$, that is,

$$[\bar{M}] = [M]$$

(18.31)

18.5   EQUATION OF MOTION FOR SPACE TRUSSES

The dynamic equilibrium conditions at the nodes of the space truss result in the differential equations of motion which in matrix notation may be written as follows:

$$[\bar{M}]\{\ddot{y}\} + [\bar{C}]\{\dot{y}\} + [\bar{K}]\{y\} = \{F(t)\} \tag{18.32}$$

in which $\{y\}$, $\{\dot{y}\}$, and $\{\ddot{y}\}$ are, respectively, the displacement, velocity, and acceleration vectors at the nodal coordinates, $\{F(t)\}$ is the vector of external nodal forces, and $[\bar{M}]$, $[\bar{C}]$, and $[\bar{K}]$ are the system mass, damping, and stiffness matrices.

In the stiffness method of analysis, the system matrices in eq. (18.32) are obtained by appropriate superposition of the corresponding member matrices using the direct method as we have shown previously for the framed structures. As was discussed in the preceding chapters, the practical way of evaluating damping is to prescribe damping ratios relative to the critical damping for each mode. Consequently, when eq. (18.32) is solved using the modal superposition method, the specified modal damping ratios are introduced directly into the modal equations. In this case, there is no need for explicitly obtaining the system damping matrix $[C]$. However, this matrix is required when the solution of eq. (18.32) is sought by other methods of solution, such as the step-by-step integration method. In this case, the system damping matrix $[C]$ can be obtained from the specified modal damping ratios by any of the methods discussed in Chapter 12.

18.6   PROGRAM 18—MODELING STRUCTURES AS SPACE TRUSSES

Program 18 calculates the stiffness and the mass matrices for the space truss and stores the coefficients in a file. Since the stiffness and mass matrices are symmetric, only the upper triangular portion of these matrices needs to be stored. The program also stores in another file, which is named by the user, the general information on the space truss. The information stored in these files is needed by programs, which the user may call, to perform dynamic analysis of the truss, such as determination of natural frequencies or calculation of the response of the structure subjected to external excitation.

Example 18.5.   Determine the stiffness and mass matrices for the space truss shown in Fig. 18.5. The mass per unit length of any member is $\bar{m} = 0.1$ (lb $\cdot$ sec^2/in/in).

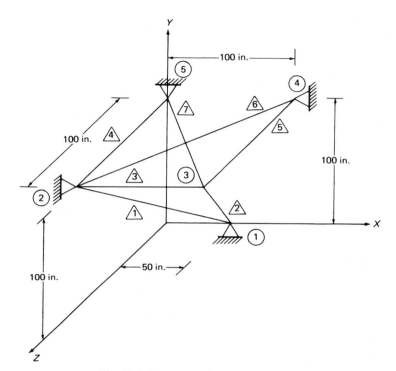

Fig. 18.5 Space truss for Example 18.5.

Solution:

Input Data and Output Results

PROGRAM 18: MODELING SPACE TRUSSES   DATA FILE:D18

NUMBER OF JOINTS   NJ= 5
NUMBER OF BEAM ELEMENTS   NE= 7
NUMBER OF FIXED COORDINATES   NC= 12
MODULUS OF ELASTICITY   E= 3E+07

         JOINT DATA:

JOINT #	X-COORD.	Y-COORD.	Z-COORD.	JOINT MASS
1	50.00	0.00	0.00	0
2	0.00	100.00	100.00	0
3	100.00	100.00	100.00	0
4	100.00	100.00	0.00	0
5	0.00	100.00	0.00	0

INPUT BEAM ELEMENT DATA:

ELEMENT #	FIRST JOINT	SECOND JOINT	MASS/LENGTH	SECT. AREA
1	2	1	0.100	10.00
2	3	1	0.100	10.00
3	2	3	0.100	10.00
4	2	5	0.100	10.00
5	3	4	0.100	10.00
6	2	4	0.100	10.00
7	5	3	0.100	10.00

FIXED COORDINATES DATA:
DIRECTION # (X=1,Y=2,Z=3)

NODE	DIR.	NODE	DIR.	NODE	DIR.	NODE	DIR.	NODE	DIR.	NODE	DIR.	NODE	DIR.	NODE	DIR.
1	1	1	2	1	3	2	1	2	2	2	3	4	1	4	2
4	3	5	1	5	2	5	3								

OUTPUT RESULTS:

SYSTEM STIFFNESS MATRIX

4.2829E+06	4.4444E+05	1.5051E+06
4.4444E+05	8.8889E+05	8.8889E+05
1.5051E+06	8.8889E+05	4.9495E+06

SYSTEM MASS MATRIX

1.6381E+01	0.0000E+00	0.0000E+00
0.0000E+00	1.6381E+01	0.0000E+00
0.0000E+00	0.0000E+00	1.6381E+01

18.7   SUMMARY

The dynamic analysis of trusses by the stiffness matrix method was presented in this chapter. As in the case of framed structures, discussed in the preceding chapters, the stiffness and mass matrices for a member of a truss were developed. The system matrices for a truss are assembled as explained for framed structures by the appropriate superposition of the coefficients in the matrices of the elements.

PROBLEMS

18.1 For the plane truss shown in Fig. 18.6 determine the system stiffness and mass matrices corresponding to the two nodal coordinates indicated in the figure.

18.2 Determine the natural frequencies and corresponding normal modes for the truss shown in Fig. 18.6.

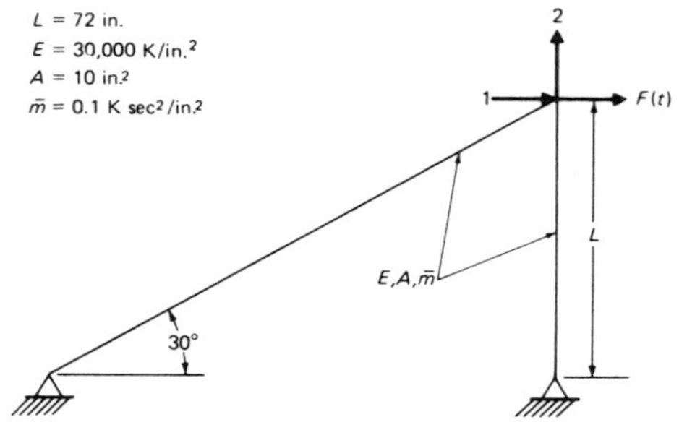

L = 72 in.
E = 30,000 K/in.²
A = 10 in.²
m̄ = 0.1 K sec²/in.²
2
1
F(t)
E,A,m̄
L
30°

Fig. 18.6

18.3 Determine the response of the truss shown in Fig. 18.6 when subjected to a force $F(t) = 10$ kip suddenly applied for 2 sec at nodal coordinate 1. Use the results of Problem 18.2 to obtain the modal equations. Neglect damping in the system.

18.4 Solve Problem 18.3 assuming 10% damping in all the modes.

18.5 Determine the maximum response of the truss shown in Fig. 18.6 when subjected to a rectangular pulse of magnitude $F_0 = 10$ kip and duration $t_d = 0.1$ sec. Use the appropriate response spectrum to determine the maximum modal response (Fig. 4.4). Neglect damping in the system.

18.6 Determine the dynamic response of the frame shown in Fig. 18.6 when subjected to a harmonic force $F(t) = 10 \sin 10t$ (kips) along nodal coordinate 1. Neglect damping in the system.

18.7 Repeat Problem 18.6 assuming that the damping in the system is proportional to the stiffness, $[C] = a_0[K]$ where $a_0 = 0.1$.

18.8 Determine the response of the truss shown in Fig. 18.7 when acted upon by the forces $F_1(t) = 10t$ and $F_2(t) = 5t^2$ during 1 sec. Neglect damping.

18.9 Solve Problem 18.8 assuming 10% modal damping in all the modes.

4
Concentrated mass =
10 K sec²/in.
3
60 in.
F₁ (t)
1
2
Concentrated mass =
10 K sec²/in.
F₂ (t)
E = 10⁴ ksi
m̄ = 0.5 K sec²/in.²
A = 1.0 in.²

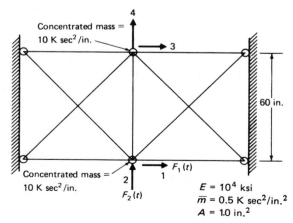

Fig. 18.7

19

Time History Response of Multidegree-of-Freedom Systems

See: Bathe

"Num. Methods in FEMS"

In Chapter 7 the nonlinear analysis of a single degree-of-freedom system using the step-by-step linear acceleration method was presented. The extension of this method, with a modification known as the Wilson-θ method, for the solution of structures modeled as multidegree-of-freedom systems is developed in this chapter. The modification introduced in the method by Wilson et al. 1973 serves to assure the numerical stability of the solution process regardless of the magnitude selected for the time step; for this reason, such a method is said to be *unconditionally stable.* On the other hand, without Wilson's modification, the step-by-step linear acceleration method is only conditionally stable and for numerical stability of the solution may require such an extremely small time step as to make the method impractical if not impossible. The development of the necessary algorithm for the linear and nonlinear multidegree-of-freedom systems by the step-by-step linear acceleration method parallels the presentation for the single degree-of-freedom system in Chapter 7.

Another well-known method for step-by-step numerical integration of the equations of motion of a discrete system is the Newmark beta method. This method which also may be considered an extension of the linear acceleration method is presented later in this chapter after discussing in detail the Wilson-θ method.

19.1   INCREMENTAL EQUATIONS OF MOTION

The basic assumption of the Wilson-θ method is that the acceleration varies linearly over the time interval from t to $t + \theta\Delta t$ where $\theta \geq 1.0$. The value of the factor θ is determined to obtain optimum stability of the numerical process and accuracy of the solution. It has been shown by Wilson that, for $\theta \geq 1.38$, the method becomes unconditionally stable.

The equations expressing the incremental equilibrium conditions for a multidegree-of-freedom system can be derived as the matrix[1] equivalent of the incremental equation of motion for the single degree-of-freedom system, eq. (7.12). Thus taking the difference between dynamic equilibrium conditions defined at time t_i and at $t_i + \tau$, where $\tau = \theta\Delta t$, we obtain the incremental equations

$$M\hat{\Delta}\ddot{y}_i + C(\dot{y})\,\hat{\Delta}\dot{y}_i + K(y)\,\hat{\Delta}y_i = \hat{\Delta}F_i \tag{19.1}$$

in which the circumflex over Δ indicates that the increments are associated with the extended time step $\tau = \theta\Delta t$. Thus

$$\hat{\Delta}y_i = y(t_i + \tau) - y(t_i) \tag{19.2}$$

$$\hat{\Delta}\dot{y}_i = \dot{y}(t_i + \tau) - \dot{y}(t_i) \tag{19.3}$$

$$\hat{\Delta}\ddot{y}_i = \ddot{y}(t_i + \tau) - \ddot{y}(t_i) \tag{19.4}$$

and

$$\hat{\Delta}F_i = F(t_i + \tau) - F(t_i) \tag{19.5}$$

In writing eq. (19.1), we assumed, as explained in Chapter 7 for single degree-of-freedom systems, that the stiffness and damping are obtained for each time step as the initial values of the tangent to the corresponding curves as shown in Fig. 19.1 rather than the slope of the secant line which requires iteration. Hence the stiffness coefficient is defined as

$$k_{ij} = \frac{dF_{si}}{dy_j} \tag{19.6}$$

and the damping coefficient as

$$c_{ij} = \frac{dF_{Di}}{d\dot{y}_j} \tag{19.7}$$

[1]Matrices and vectors are denoted with boldface lettering throughout this chapter.

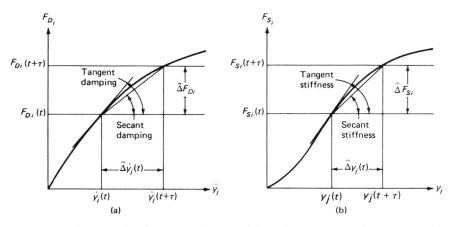

Fig. 19.1 Definition of influence coefficients. (a) Nonlinear viscous damping, c_{ij}. (b) Nonlinear stiffness, k_{ij}.

in which F_{si} and F_{Di} are, respectively, the elastic and damping forces at nodal coordinate i and y_j and $\dot{y}_j$ are, respectively, the displacement and velocity at nodal coordinate j.

19.2   THE WILSON-θ METHOD

The integration of the nonlinear equations of motion by the step-by-step linear acceleration method with the extended step introduced by Wilson is based, as has already been mentioned, on the assumption that the acceleration may be represented by a linear function during the time step $\tau = \theta\Delta t$ as shown in Fig. 19.2. From this figure we can write the linear expression for the acceleration during the extended time step as

$$\ddot{y}(t) = \ddot{y}_i + \frac{\hat{\Delta}\ddot{y}_i}{\tau}(t - t_i) \tag{19.8}$$

in which $\hat{\Delta}\ddot{y}_i$ is given by eq. (19.4). Integrating eq. (19.8) twice between limits t_i and t yields

$$\dot{y}(t) = \dot{y}_i + \ddot{y}_i(t - t_i) + \frac{1}{2}\frac{\hat{\Delta}\ddot{y}_i}{\tau}(t - t_i)^2 \tag{19.9}$$

and

$$y(t) = y_i + \dot{y}_i(t - t_i) + \frac{1}{2}\ddot{y}_i(t - t_i)^2 + \frac{1}{6}\frac{\hat{\Delta}\ddot{y}_i}{\tau}(t - t_i)^3 \tag{19.10}$$

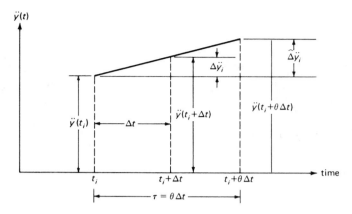

Fig. 19.2 Linear acceleration assumption in the extended time interval.

Evaluation of eqs. (19.9) and (19.10) at the end of the extended interval $t = t_i + \tau$ gives

$$\hat{\Delta}\dot{y}_i = \ddot{y}_i\tau + \tfrac{1}{2}\hat{\Delta}\ddot{y}_i\tau \tag{19.11}$$

and

$$\hat{\Delta}y_i = \dot{y}_i\tau + \tfrac{1}{2}\ddot{y}_i\tau^2 + \tfrac{1}{6}\hat{\Delta}\ddot{y}_i\tau^2 \tag{19.12}$$

in which $\hat{\Delta}y_i$ and $\hat{\Delta}\dot{y}_i$ are defined by eqs. (19.2) and (19.3), respectively. Now eq. (19.12) is solved for the incremental acceleration $\hat{\Delta}\ddot{y}_i$ and substituted in eq. (19.11). We obtain

$$\hat{\Delta}\ddot{y}_i = \frac{6}{\tau^2}\hat{\Delta}y_i - \frac{6}{\tau}\dot{y}_i - 3\ddot{y}_i \tag{19.13}$$

and

$$\hat{\Delta}\dot{y}_i = \frac{3}{\tau}\hat{\Delta}y_i - 3\dot{y}_i - \frac{\tau}{2}\ddot{y}_i \tag{19.14}$$

Finally, substituting eqs. (19.13) and (19.14) into the incremental equation of motion, eq. (19.1), results in an equation for the incremental displacement $\hat{\Delta}y_i$, which may be conveniently written as

$$\bar{K}_i\hat{\Delta}y_i = \overline{\Delta F_i} \tag{19.15}$$

where

$$\bar{K}_i = K_i + \frac{6}{\tau^2}M + \frac{3}{\tau}C_i \tag{19.16}$$

and

$$\overline{\Delta F}_i = \hat{\Delta F}_i + M\left(\frac{6}{\tau}\dot{y}_i + 3\ddot{y}_i\right) + C_i\left(3\dot{y}_i + \frac{\tau}{2}\ddot{y}_i\right) \tag{19.17}$$

Equation (19.15) has the same form as the static incremental equilibrium equation and may be solved for the incremental displacements $\hat{\Delta}y_i$ by simply solving a system of linear equations.

To obtain the incremental accelerations $\hat{\Delta}\ddot{y}_i$ for the extended time interval, the value of $\hat{\Delta}y_i$ obtained from the solution of eq. (19.15) is substituted into eq. (19.13). The incremental acceleration $\Delta\ddot{y}_i$ for the normal time interval Δt is then obtained by a simple linear interpolation. Hence

$$\Delta\ddot{y} = \frac{\hat{\Delta}\ddot{y}}{\theta} \tag{19.18}$$

To calculate the incremental velocity $\Delta\dot{y}_i$ and incremental displacement Δy_i corresponding to the normal interval Δt, use is made of eqs. (19.11) and (19.12) with the extended time interval parameter τ substituted for Δt, that is,

$$\Delta\dot{y}_i = \ddot{y}_i\Delta t + \tfrac{1}{2}\Delta\ddot{y}_i\Delta t \tag{19.19}$$

and

$$\Delta y_i = \dot{y}_i\Delta t + \tfrac{1}{2}\ddot{y}_i\Delta t^2 + \tfrac{1}{6}\Delta\ddot{y}_i\Delta t^2 \tag{19.20}$$

Finally, the displacement y_{i+1} and velocity $\dot{y}_{i+1}$ at the end of the normal time interval are calculated by

$$y_{i+1} = y_i + \Delta y_i \tag{19.21}$$

and

$$\dot{y}_{i+1} = \dot{y}_i + \Delta\dot{y}_i \tag{19.22}$$

As mentioned in Chapter 7 for the single degree-of-freedom system, the initial acceleration for the next step should be calculated from the condition of dynamic equilibrium at the time $t + \Delta t$; thus

$$\ddot{y}_{i+1} = M^{-1}[F_{i+1} - F_D(\dot{y}_{i+1}) - F_S(y_{i+1})] \tag{19.23}$$

in which $F_D(\dot{y}_{i+1})$ and $F_S(y_{i+1})$ represent, respectively, the damping force and stiffness force vectors evaluated at the end of the time step $t_{i+1} = t_i + \Delta t$. Once the displacement, velocity, and acceleration vectors have been determined at time $t_{i+1} = t_i + \Delta t$, the outlined procedure is repeated to calculate these quantities at the next time step $t_{i+2} = t_{i+1} + \Delta t$ and the process is continued to any desired final time.

The step-by-step linear acceleration, as indicated in the discussion for the single degree-of-freedom system, involves two basic approximations: (1) the acceleration is assumed to vary linearly during the time step, and (2) the damping and stiffness characteristics of the structure are evaluated at the initiation of the time step and are assumed to remain constant during this time interval. The algorithm for the integration process of a linear system by the Wilson-θ method is outlined in the next section. The application of this method to linear structures is then developed in the following section.

19.3   ALGORITHM FOR STEP-BY-STEP SOLUTION OF A LINEAR SYSTEM USING THE WILSON-θ METHOD

19.3.1   Initialization

(1) Assemble the system stiffness matrix K, mass matrix M, and damping matrix C.

(2) Set initial values for displacement y_0, velocity $\dot{y}_0$, and forces F_0.

(3) Calculate initial acceleration $\ddot{y}_0$ from

$$M\ddot{y}_0 = F_0 - C\dot{y}_0 - Ky_0$$

(4) Select a time step Δt, the factor θ (usually taken as 1.4), and calculate the constants τ, a_1, a_2, a_3, and a_4 from the relations

$$\tau = \theta\Delta t, \quad a_1 = \frac{3}{\tau}, \quad a_2 = \frac{6}{\tau}, \quad a_3 = \frac{\tau}{2}, \quad a_4 = \frac{6}{\tau^2}$$

(5) Form the effective stiffness matrix $\bar{K}$ [eq. (19.16)], namely,

$$\bar{K} = K + a_4M + a_1C$$

19.3.2   For Each Time Step

(1) Calculate by linear interpolation the incremental load $\hat{\Delta}F_i$ for the time interval t_i to $t_i + \tau$, from the relation

$$\hat{\Delta}F_i = F_{i+1} + (F_{i+2} - F_{i+1})(\theta - 1) - F_i$$

(2) Calculate the effective incremental load $\overline{\hat{\Delta}F_i}$ for the time interval t_i to $t_i + \tau$, from eq. (19.17) as

$$\overline{\hat{\Delta}F_i} = \hat{\Delta}F_i + (a_2M + 3C)\dot{y}_i + (3M + a_3C)\ddot{y}_i$$

(3) Solve for the incremental displacement $\hat{\Delta}y_i$ from eq. (19.15) as

$$\bar{K}\hat{\Delta}y_i = \overline{\hat{\Delta}F_i}$$

(4) Calculate the incremental acceleration for the extended time interval τ, from the relation eq. (19.13) as

$$\hat{\Delta}\ddot{y}_i = a_4\hat{\Delta}y_i - a_2\dot{y}_i - 3\ddot{y}_i$$

(5) Calculate the incremental acceleration for the normal interval from eq. (19.18) as

$$\Delta\ddot{y} = \frac{\hat{\Delta}\ddot{y}}{\theta}$$

(6) Calculate the incremental velocity $\Delta\dot{y}_i$ and the incremental displacement Δy_i from time t_i to $t_i + \Delta t$ from eqs. (19.19) and (19.20) as

$$\Delta\dot{y}_i = \ddot{y}_i\Delta t + \tfrac{1}{2}\Delta\ddot{y}_i\Delta t$$
$$\Delta y_i = \dot{y}_i\Delta t + \tfrac{1}{2}\ddot{y}_i\Delta t^2 + \tfrac{1}{6}\Delta\ddot{y}_i\Delta t^2$$

(7) Calculate the displacement and velocity at time $t_{i+1} = t_i + \Delta t$ using

$$y_{i+1} = y_i + \Delta y_i$$
$$\dot{y}_{i+1} = \dot{y}_i + \Delta\dot{y}_i$$

(8) Calculate the acceleration $\ddot{y}_{i+1}$ at time $t_{i+1} = t_i + \Delta t$ directly from the equilibrium equation of motion, namely,

$$M\ddot{y}_{i+1} = F_{i+1} - C\dot{y}_{i+1} - Ky_{i+1}$$

Example 19.1. Calculate the displacement response for a two-story shear building of Fig. 19.3 subjected to a suddenly applied force of 10 kip at the level of the second floor. Neglect damping and assume elastic behavior.

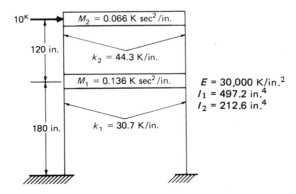

10^K

120 in.

$M_2 = 0.066$ K sec^2/in.

$k_2 = 44.3$ K/in.

$M_1 = 0.136$ K sec^2/in.

180 in.

$k_1 = 30.7$ K/in.

$E = 30,000$ K/in.^{2}
$I_1 = 497.2$ in.^{4}
$I_2 = 212.6$ in.^{4}

Fig. 19.3 Two-story shear building for Examples 19.1 and 19.2.

Solution: The equations of motion for this structure are

$$\begin{bmatrix} 0.136 & 0 \\ 0 & 0.066 \end{bmatrix} \begin{Bmatrix} \ddot{y}_1 \\ \ddot{y}_2 \end{Bmatrix} + \begin{bmatrix} 75.0 & -44.3 \\ -44.3 & 44.3 \end{bmatrix} \begin{Bmatrix} y_1 \\ y_2 \end{Bmatrix} = \begin{Bmatrix} 0 \\ 10 \end{Bmatrix}$$

which, for free vibration, become

$$\begin{bmatrix} 0.136 & 0 \\ 0 & 0.066 \end{bmatrix} \begin{Bmatrix} \ddot{y}_1 \\ \ddot{y}_2 \end{Bmatrix} + \begin{bmatrix} 75.0 & -44.3 \\ -44.3 & 44.3 \end{bmatrix} \begin{Bmatrix} y_1 \\ y_2 \end{Bmatrix} = \begin{Bmatrix} 0 \\ 0 \end{Bmatrix}$$

Substitution of $y_i = a_i \sin \omega t$ results in the eigenproblem

$$\begin{bmatrix} 75.0 - 0.136\omega^2 & -44.3 \\ -44.3 & 44.3 - 0.066\omega^2 \end{bmatrix} \begin{Bmatrix} a_1 \\ a_2 \end{Bmatrix} = \begin{Bmatrix} 0 \\ 0 \end{Bmatrix}$$

which requires for a nontrivial solution

$$\begin{vmatrix} 75.0 - 0.136\omega^2 & -44.3 \\ -44.3 & 44.3 - 0.066\omega^2 \end{vmatrix} = 0$$

Expansion of this determinant yields

$$\omega^4 - 1222.68\omega^2 + 151516 = 0$$

which has the roots

$$\omega_1^2 = 139.94 \quad \text{and} \quad \omega_2^2 = 1082.0$$

Hence, the natural frequencies are

$$\omega_1 = 11.83 \text{ rad/sec}, \quad \omega_2 = 32.90 \text{ rad/sec}$$

or

$$f_1 = 1.883 \text{ cps}, \quad f_2 = 5.237 \text{ cps}$$

and the natural periods are

$$T_1 = 0.531 \text{ sec}, \quad T_2 = 0.191 \text{ sec}.$$

The initial acceleration is calculated from

$$\begin{bmatrix} 0.136 & 0 \\ 0 & 0.066 \end{bmatrix} \begin{Bmatrix} \ddot{y}_{10} \\ \ddot{y}_{20} \end{Bmatrix} + \begin{bmatrix} 75.0 & -44.3 \\ -44.3 & 44.3 \end{bmatrix} \begin{Bmatrix} 0 \\ 0 \end{Bmatrix} = \begin{Bmatrix} 0 \\ 10 \end{Bmatrix}$$

giving

$$\ddot{y}_{10} = 0$$

$$\ddot{y}_{20} = 151.51 \text{ in/sec}^2$$

If we select $\Delta t = 0.02$ and $\theta = 1.4$, $\tau = \theta \Delta t = 0.028$, and calculate the constants, we obtain

$$a_1 = \frac{3}{\tau} = 107.14, \quad a_3 = \frac{\tau}{2} = 0.014$$

$$a_2 = \frac{6}{\tau} = 214.28, \quad a_4 = \frac{6}{\tau^2} = 7653$$

The effective stiffness is then

$$\bar{K} = K + a_4 M + a_1 C \quad (C = 0 \text{ undamped system})$$

$$\bar{K} = \begin{bmatrix} 75.0 & -44.3 \\ -44.3 & 44.3 \end{bmatrix} + 7653 \begin{bmatrix} 0.136 & 0 \\ 0 & 0.066 \end{bmatrix}$$

$$\bar{K} = \begin{bmatrix} 1115.8 & -44.3 \\ -44.3 & 549.4 \end{bmatrix}$$

and the effective force

$$\overline{\hat{\Delta} F} = \hat{\Delta} F + (a_2 M + 3C)\dot{y} + (3M + a_3 C)\ddot{y}$$

$$\overline{\hat{\Delta} F} = \begin{Bmatrix} 0 \\ 0 \end{Bmatrix} + 214.28 \begin{bmatrix} 0.136 & 0 \\ 0 & 0.066 \end{bmatrix} \begin{Bmatrix} 0 \\ 0 \end{Bmatrix} + 3 \begin{bmatrix} 0.136 & 0 \\ 0 & 0.066 \end{bmatrix} \begin{Bmatrix} 0 \\ 151.51 \end{Bmatrix}$$

$$\overline{\hat{\Delta} F} = \begin{bmatrix} 0 \\ 30 \end{bmatrix}$$

Solving for $\hat{\Delta} y$ from $\bar{K} \hat{\Delta} y = \overline{\hat{\Delta} F}$ yields

$$\begin{bmatrix} 1115.8 & -44.3 \\ -44.3 & 549.4 \end{bmatrix} \begin{Bmatrix} \hat{\Delta} y_1 \\ \hat{\Delta} y_2 \end{Bmatrix} = \begin{Bmatrix} 0 \\ 30 \end{Bmatrix}, \quad \hat{\Delta} y = \begin{Bmatrix} 0.002175 \\ 0.054780 \end{Bmatrix}$$

Solving for $\hat{\Delta} \ddot{y}$ from eq. (19.13), we obtain

$$\hat{\Delta} \ddot{y} = 7653 \begin{Bmatrix} 0.002175 \\ 0.054780 \end{Bmatrix} - 214.28 \begin{Bmatrix} 0 \\ 0 \end{Bmatrix} - 3 \begin{Bmatrix} 0 \\ 151.51 \end{Bmatrix}, \quad \hat{\Delta} \ddot{y} = \begin{Bmatrix} 16.645 \\ -35.299 \end{Bmatrix}$$

Then

$$\Delta \ddot{y} = \frac{\hat{\Delta} \ddot{y}}{\theta} = \frac{1}{1.4} \begin{Bmatrix} 16.647 \\ -35.299 \end{Bmatrix} = \begin{Bmatrix} 11.891 \\ -25.21 \end{Bmatrix}$$

From eq. (19.19), it follows that

$$\Delta \dot{y} = \begin{Bmatrix} 0 \\ 151.51 \end{Bmatrix} (0.02) + \frac{0.02}{2} \begin{Bmatrix} 11.891 \\ -25.21 \end{Bmatrix} = \begin{Bmatrix} 0.1189 \\ 2.7781 \end{Bmatrix}$$

From eq. (19.20),

$$\Delta y = \begin{Bmatrix} 0 \\ 0 \end{Bmatrix} (0.02) + \frac{(0.02)^2}{2} \begin{Bmatrix} 0 \\ 151.51 \end{Bmatrix} + \frac{(0.02)^2}{6} \begin{Bmatrix} 11.891 \\ -25.21 \end{Bmatrix} = \begin{Bmatrix} 0.0008 \\ 0.0286 \end{Bmatrix}$$

From eqs. (19.21) and (19.22),

$$\{y\} = \begin{Bmatrix} 0 \\ 0 \end{Bmatrix} + \begin{Bmatrix} 0.0008 \\ 0.0286 \end{Bmatrix} = \begin{Bmatrix} 0.0008 \\ 0.0286 \end{Bmatrix} \tag{a}$$

and

$$\{\dot{y}\} = \begin{Bmatrix} 0 \\ 0 \end{Bmatrix} + \begin{Bmatrix} 0.1189 \\ 2.7781 \end{Bmatrix} = \begin{Bmatrix} 0.1189 \\ 2.7781 \end{Bmatrix} \tag{b}$$

From eq. (19.23),

$$\begin{Bmatrix} 0.136 & 0 \\ 0 & 0.066 \end{Bmatrix} \{\ddot{y}\} = \begin{Bmatrix} 0 \\ 10 \end{Bmatrix} - \begin{Bmatrix} 75.0 & -44.3 \\ -44.3 & 44.3 \end{Bmatrix} \begin{Bmatrix} 0.0008 \\ 0.0286 \end{Bmatrix}$$

which gives

$$\{\ddot{y}\} = \begin{Bmatrix} 8.875 \\ 132.85 \end{Bmatrix} \tag{c}$$

The results given in eqs. (a), (b), and (c) for the displacement, velocity, and acceleration, respectively, at time $t_1 = t_0 + \Delta t$ complete a first cycle of the integration process. The continuation in determining the response for this structure is given in Example 19.2 with the use of the computer program described in the next section.

19.4 PROGRAM 19—RESPONSE BY STEP INTEGRATION

Program 19 performs the step-by-step integration of the equations of motion for a linear system using the linear acceleration method with the Wilson-θ modification. The program requires previous modeling of the structure to determine the stiffness matrix and the mass matrix of the system. Input data include the time step Δt, the maximum time computation, TMAX, and a table of the time–force values for each load applied at the nodal coordinates of the structure.

The program performs a linear interpolation between the load data points, which result in a table giving the magnitude of the applied forces at each nodal coordinate calculated at increments of time equal to the time step Δt. The output consists of a table giving the response for each nodal coordinate in terms of displacement, velocity, and acceleration at increments of time Δt up to the maximum time specified by the duration of the forces including, if desired, an extension with forces set to zero.

Example 19.2.   Use Program 19 to determine the response of the two-story shear building shown in Fig. 19.3. The first cycle of the integration process for this structure has been hand calculated in Example 19.1.

Solution:

Program Input Data and Output Results

```
***PROGRAM 19 RESPONSE BY STEP INTEGRATION***    DATA FILE=D19

   ***RESPONSE BY LINEAR ACCELERATION METHOD***

   INPUT DATA:

NUMBER OF DEGREES OF FREEDOM                      ND= 2
NUMBER OF POINTS DEFINING THE EXCITATION          NEX= 2
COORD. AT EXCITATION (FOR SUPPORT MOTION=0)       NCX= 2
TIME STEP OF INTEGRATION                          H= .02

         SYSTEM STIFFNESS MATRIX

0.7500E+02  -.4430E+02
-.4430E+02   0.4430E+02

         SYSTEM MASS MATRIX

0.1360E+00  0.0000E+00
0.0000E+00  0.6600E-01

         SYSTEM DAMPING MATRIX

0.0000E+00  0.0000E+00
0.0000E+00  0.0000E+00

         EXCITATION FUNCTION

TIME     EXCITATION
0.00       10.000
0.40       10.000

   PRINT TIME HISTORY TABLE (Y/N) ? N
   OUTPUT RESULTS:

         *MAXIMUM RESPONSE*
         COORD.    MAX. DISPL.   MAX. VELOC.   MAX. ACC.
           1         0.692          5.875        89.763
           2         1.087          7.273       147.611
```

19.5   THE NEWMARK BETA METHOD

The Newmark beta method may be considered a generalization of the linear acceleration method. It uses a numerical parameter designated as β. The method, as originally proposed by Newmark (1959), contained in addition to β, a second parameter γ. These parameters replace the numerical coefficients $\frac{1}{2}$ and $\frac{1}{6}$ of the terms containing the incremental acceleration $\Delta\ddot{y}_i$ in eqs. (19.19) and (19.20), respectively. Thus, replacing by γ the coefficient $\frac{1}{2}$ of $\Delta\ddot{y}_i$ in eq. (19.19) and by β the coefficient $\frac{1}{6}$ also of $\Delta\ddot{y}_i$ in eq. (19.20), we have

$$\Delta\dot{y}_i = \ddot{y}_i\Delta t + \gamma\Delta\ddot{y}_i\Delta t^2 \qquad (19.24)$$

and

$$\Delta y_i = \dot{y}_i\Delta t + \frac{1}{2}\ddot{y}_i\Delta t^2 + \beta\Delta\ddot{y}_i\Delta t^2 \qquad (19.25)$$

It has been found that for values of γ different than $\frac{1}{2}$, the method introduces a superfluous damping in the system. For this reason this parameter is generally set as $\gamma = \frac{1}{2}$. The solution of eq. (19.25) for $\Delta\ddot{y}_i$ and its substitution into eq. (19.24) after setting $\gamma = \frac{1}{2}$ yield

$$\Delta\ddot{y}_i = \frac{1}{\beta\Delta t^2}\Delta y_i - \frac{1}{\beta\Delta t}\dot{y}_i - \frac{1}{2\beta}\ddot{y}_i \qquad (19.26)$$

$$\Delta\dot{y}_i = \frac{1}{2\beta\Delta t}\Delta y_i - \frac{1}{2\beta}\dot{y}_i + \left(1 - \frac{1}{4\beta}\right)\Delta t\ddot{y}_i \qquad (19.27)$$

Then the substitution of eqs. (19.26) and (19.27) into the incremental equation of motion

$$M\Delta\ddot{y}_i + C_i\Delta\dot{y}_i + K_i\Delta y_i = \Delta F_i \qquad (19.28)$$

results in an equation to calculate the incremental displacement Δy_i, namely,

$$\bar{K}_i\Delta y_i = \Delta\bar{F}_i \qquad (19.29)$$

where the effective stiffness matrix $\bar{K}_i$ and the effective incremental force vector $\Delta\bar{F}_i$ are given respectively by

$$\bar{K}_i = K_i + \frac{M}{\beta\Delta t^2} + \frac{C_i}{2\beta\Delta t} \qquad (19.30)$$

and

$$\Delta\bar{F}_i = \Delta F_i + \frac{M}{\beta\Delta t}\dot{y}_i + \frac{C_i}{2\beta}\dot{y}_i + \frac{M}{2\beta}y_i - C_i\Delta t\left(1 - \frac{1}{4\beta}\right)\ddot{y}_i \qquad (19.31)$$

In these equations C_i and K_i are respectively the damping and stiffness matrices evaluated at the initial time t_i of the time step $\Delta t = t_{i+1} - t_i$.

In the implementation of the Newmark beta method, the process begins by selecting a numerical value for the parameter β. Newmark suggested a value in the range $\frac{1}{6} \leq \beta \leq \frac{1}{2}$. For $\beta = \frac{1}{6}$ the method is exactly equal to the linear acceleration method and is only conditionally stable.

For $\beta = \frac{1}{4}$ the method is equivalent to assuming that the velocity varied linearly during the time step, which would require that the mean acceleration is maintained for the interval. In this last case, that is, for $\beta = \frac{1}{4}$, the Newmark beta method is unconditionally stable and it provides the satisfactory accuracy.

19.6   ELASTOPLASTIC BEHAVIOR OF FRAMED STRUCTURES

The dynamic analysis of beams and frames having linear elastic behavior was presented in the preceding chapters. To extend this analysis to structures whose members may be strained beyond the yield point of the material, it is necessary to develop the member stiffness matrix for the assumed elastoplastic behavior. The analysis is then carried out by a step-by-step numerical integration of the differential equations of motion. Within each short time interval Δt, the structure is assumed to behave in a linear elastic manner, but the elastic properties of the structure are changed from one interval to another as dictated by the response. Consequently, the nonlinear response is obtained as a sequence of linear responses of different elastic systems. For each successive interval, the stiffness of the structure is evaluated based on the moments in the members at the beginning of the time increment.

The changes in displacements of the linear system are computed by integration of the differential equations of motion over the finite interval and the total displacements by addition of the incremental displacement to the displacements calculated in the previous time step. The incremental displacements are also used to calculate the increment in member end forces and moments from the member stiffness equation. The magnitude of these end moments relative to the yield conditions (plastic moments) determine the characteristics of the stiffness and mass matrices to be used in the next time step.

19.7   MEMBER STIFFNESS MATRIX

If only bending deformations are considered, the force-displacement relationship for a uniform beam segment (Fig. 19.4) with elastic behavior (no hinges) is given by eq. (14.20). This equation may be written in incremental quantities as follows:

$$
\begin{Bmatrix} \Delta P_1 \\ \Delta P_2 \\ \Delta P_3 \\ \Delta P_4 \end{Bmatrix} = \frac{EI}{L^3} \begin{bmatrix} 12 & 6L & -12 & 6L \\ 6L & 4L^2 & -6L & 2L^2 \\ -12 & -6L & 12 & -6L \\ 6L & 2L^2 & -6L & 4L^2 \end{bmatrix} \begin{Bmatrix} \Delta \delta_1 \\ \Delta \delta_2 \\ \Delta \delta_3 \\ \Delta \delta_4 \end{Bmatrix} \tag{19.32}
$$

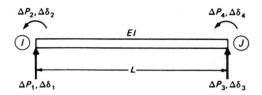

Fig. 19.4 Beam segment indicating incremental end forces and corresponding incremental displacements.

in which ΔP_i and $\Delta \delta_i$ are, respectively, the incremental forces and the incremental displacements at the nodal coordinates of the beam segment. When the moment at one end of the beam reaches the value of the plastic moment M_p, a hinge is formed at that end. Under the assumption of an elastoplastic relation between the bending moment and the angular displacement as depicted in Fig. 19.5, the section that has been transformed into a hinge cannot support a moment higher than the plastic moment M_p but it may continue to deform plastically at a constant moment M_p. The relationship reverses to an elastic behavior when the angular displacement begins to decrease as shown in Fig. 19.5. We note the complete similarity for the behavior between an elastoplastic spring (Fig. 7.4) in a single degree-of-freedom system and an elastoplastic section of a beam (Fig. 19.5).

The stiffness matrix for a beam segment with a hinge at one end (Fig. 19.6) may be obtained by application of eq. (14.16) which is repeated here for convenience, namely

$$k_{ij} = \int_0^L EI \, \psi_i''(x) \, \psi_j''(x) \, dx \qquad (19.33)$$

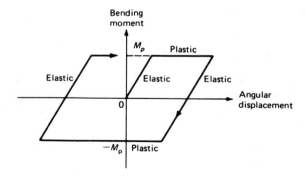

Fig. 19.5 Elastoplastic relation between bending moment and angular displacement at a section of a beam.

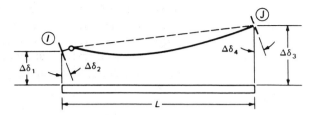

Fig. 19.6 Beam geometry with a plastic hinge at joint $\textcircled{i}$.

where $\psi_i(x)$ and $\psi_j(x)$ are displacement functions. For a uniform beam in which the formation of the plastic hinge takes place at end $\textcircled{i}$ as shown in Fig. 19.6, the deflection functions corresponding to unit displacement at one of the nodal coordinates δ_1, δ_2, δ_3, or δ_4 are given respectively by

$$\psi_1(x) = 1 - \frac{3x}{2L} + \frac{x^3}{2L^3} \tag{19.34a}$$

$$\psi_2(x) = 0 \tag{19.34b}$$

$$\psi_3(x) = \frac{3x}{2L} - \frac{x^3}{2L^3} \tag{19.34c}$$

$$\psi_4(x) = -\frac{x}{2} + \frac{x^3}{2L^2} \tag{19.34d}$$

For example, to calculate k_{11}, we substitute the second derivative $\psi_1''(x)$ from eq. (19.34a) into eq. (19.33) and obtain

$$k_{11} = EI \int_0^L \left(\frac{3x}{L^3}\right)^2 dx = \frac{3EI}{L^3} \tag{19.35}$$

Similarly, all the other stiffness coefficients for the case in which the formation of the plastic hinge takes place at end $\textcircled{i}$ of a beam segment are determined using eq. (19.33) and the deflection functions given by eqs. (19.34). The resulting stiffness equation in incremental form is

$$\begin{Bmatrix} \Delta P_1 \\ \Delta P_2 \\ \Delta P_3 \\ \Delta P_4 \end{Bmatrix} = \frac{EI}{L^3} \begin{bmatrix} 3 & 0 & -3 & 3L \\ 0 & 0 & 0 & 0 \\ -3 & 0 & 3 & -3L \\ 3L & 0 & -3L & 3L^2 \end{bmatrix} \begin{Bmatrix} \Delta\delta_1 \\ \Delta\delta_2 \\ \Delta\delta_3 \\ \Delta\delta_4 \end{Bmatrix} \tag{19.36}$$

It should be pointed out that $\Delta\delta_2$ is the incremental rotation of joint $\textcircled{i}$ at the frame and not the increase in rotation at end $\textcircled{i}$ of the beam under consideration. The incremental rotation of the plastic hinge is given by the difference between $\Delta\delta_2$ and the increase in rotation of the end ① of the member. Hinge rotation may be calculated for the various cases with formulas developed in the next section. Analogous to eq. (19.36), the following equation gives the relationship between incremental forces and incremental displacements for a uniform beam with a hinge at end ⑨ (Fig. 19.7), namely,

$$
\begin{Bmatrix} \Delta P_1 \\ \Delta P_2 \\ \Delta P_3 \\ \Delta P_4 \end{Bmatrix} = \frac{EI}{L^3} \begin{bmatrix} 3 & 3L & -3 & 0 \\ 3L & 3L^2 & -3L & 0 \\ -3 & -3L & 3 & 0 \\ 0 & 0 & 0 & 0 \end{bmatrix} \begin{Bmatrix} \Delta\delta_1 \\ \Delta\delta_2 \\ \Delta\delta_3 \\ \Delta\delta_4 \end{Bmatrix}
\tag{19.37}
$$

Finally, if hinges are formed at both ends of the beam, the stiffness matrix becomes null. Hence in this case the stiffness equation is

$$
\begin{Bmatrix} \Delta P_1 \\ \Delta P_2 \\ \Delta P_3 \\ \Delta P_4 \end{Bmatrix} = \begin{bmatrix} 0 & 0 & 0 & 0 \\ 0 & 0 & 0 & 0 \\ 0 & 0 & 0 & 0 \\ 0 & 0 & 0 & 0 \end{bmatrix} \begin{Bmatrix} \Delta\delta_1 \\ \Delta\delta_2 \\ \Delta\delta_3 \\ \Delta\delta_4 \end{Bmatrix}
\tag{19.38}
$$

19.8   MEMBER MASS MATRIX

The relationship between forces and accelerations at the nodal coordinates of an elastic uniform member considering flexural deformation is given by eq. (14.34). This equation written in incremental quantities is

$$
\begin{Bmatrix} \Delta P_1 \\ \Delta P_2 \\ \Delta P_3 \\ \Delta P_4 \end{Bmatrix} = \frac{\bar{m}L}{420} \begin{bmatrix} 156 & 22L & 54 & -13L \\ 22L & 4L^2 & 13L & -3L^2 \\ 54 & 13L & 156 & -22L \\ -13L & -3L^2 & -22L & 4L^2 \end{bmatrix} \begin{Bmatrix} \Delta\ddot{\delta}_1 \\ \Delta\ddot{\delta}_2 \\ \Delta\ddot{\delta}_3 \\ \Delta\ddot{\delta}_4 \end{Bmatrix}
\tag{19.39}
$$

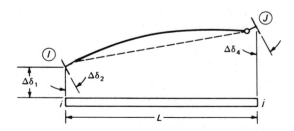

Fig. 19.7 Beam geometry with a plastic hinge at joint ⑨.

where ΔP_i and $\Delta \ddot{\delta}_i$ are, respectively, the incremental forces and the incremental accelerations at the nodal coordinates, L is the length of the member, and $\bar{m}$ is its mass per unit length. Assuming elastoplastic behavior, when the moment at an end of the beam segment reaches the magnitude of the plastic moment M_p and a hinge is formed, the consistent mass coefficients are determined from eq. (14.33) using the appropriate deflection curves. For a uniform beam in which the formation of the plastic hinge develops at end I as shown in Fig. 19.6, the deflection functions corresponding to a unit displacement of the nodal coordinates are given by eqs. (19.34). Analogously, the deflection functions of a beam segment with a plastic hinge at end J as shown in Fig. 19.7 for unit displacement at nodal coordinates δ_1, δ_2, δ_3, or δ_4 are respectively given by

$$\psi_1(x) = 1 + \frac{x^3}{2L^3} - \frac{3x^2}{2L^2}$$

$$\psi_2(x) = \frac{x^3}{2L^2} - \frac{3x^2}{2L} + x$$

$$\psi_3(x) = -\frac{x^3}{2L^3} + \frac{3x^2}{2L^2}$$

$$\psi_4(x) = 0 \qquad\qquad (19.40)$$

The mass coefficients for a beam segment with a hinge at one end are then obtained by application of eq. (14.33) which is repeated here, namely,

$$m_{ij} = \int_0^L \bar{m}\, \psi_i(x)\, \psi_j(x)\, dx \qquad\qquad (19.41)$$

where $\psi_i(x)$ and $\psi_j(x)$ are the corresponding displacement functions from eqs. (19.34) or (19.40).

Application of eq. (19.41) and the use of displacement functions (19.34) results in the mass matrix for a beam segment with a hinge at the $①$ end. The resulting mass matrix relates incremental forces and accelerations at the nodal coordinates, namely,

$$\begin{Bmatrix} \Delta P_1 \\ \Delta P_2 \\ \Delta P_3 \\ \Delta P_4 \end{Bmatrix} = \frac{\bar{m}L}{420} \begin{bmatrix} 204 & 0 & 58.5 & -16.3L \\ 0 & 0 & 0 & 0 \\ 58.5 & 0 & 99 & -36L \\ -16.5L & 0 & -36L & 8L^2 \end{bmatrix} \begin{Bmatrix} \Delta \ddot{\delta}_1 \\ \Delta \ddot{\delta}_2 \\ \Delta \ddot{\delta}_3 \\ \Delta \ddot{\delta}_4 \end{Bmatrix} \qquad (19.42)$$

Analogously to eq. (19.42), the following equation gives the relationship between incremental forces and incremental accelerations for a uniform beam segment with a hinge at the $②$ end:

$$\begin{Bmatrix} \Delta P_1 \\ \Delta P_2 \\ \Delta P_3 \\ \Delta P_4 \end{Bmatrix} = \frac{\bar{m}L}{420} \begin{bmatrix} 99 & 36L & 58.5 & 0 \\ 36L & 8L^2 & 16.5L & 0 \\ 58.5 & 16.5L & 204 & 0 \\ 0 & 0 & 0 & 0 \end{bmatrix} \begin{Bmatrix} \Delta\ddot{\delta}_1 \\ \Delta\ddot{\delta}_2 \\ \Delta\ddot{\delta}_3 \\ \Delta\ddot{\delta}_4 \end{Bmatrix} \qquad (19.43)$$

Finally, if hinges are formed at both ends of the beam segment, the deflection curves are given by

$$\psi_1(x) = -\frac{x}{L} + 1$$

$$\psi_2(x) = 0$$

$$\psi_3(x) = \frac{x}{L}$$

$$\psi_4(x) = 0 \qquad (19.44)$$

and the corresponding relationship in incremental form by

$$\begin{Bmatrix} \Delta P_1 \\ \Delta P_2 \\ \Delta P_3 \\ \Delta P_4 \end{Bmatrix} = \frac{\bar{m}L}{6} \begin{bmatrix} 2 & 0 & 1 & 0 \\ 0 & 0 & 0 & 0 \\ 1 & 0 & 2 & 0 \\ 0 & 0 & 0 & 0 \end{bmatrix} \begin{Bmatrix} \Delta\ddot{\delta}_1 \\ \Delta\ddot{\delta}_2 \\ \Delta\ddot{\delta}_3 \\ \Delta\ddot{\delta}_4 \end{Bmatrix} \qquad (19.45)$$

19.9   ROTATION OF PLASTIC HINGES

In the solution process, at the end of each step interval it is necessary to calculate the end moments of every beam segment to check whether or not a plastic hinge has been formed. The calculation is done using the element incremental moment–displacement relationship. It is also necessary to check if the plastic deformation associated with a hinge is compatible with the sign of the moment. The plastic hinge is free to rotate in one direction only, and in the other direction the section returns to an elastic behavior. The assumed moment rotation characteristics of the member are of the type illustrated in Fig. 19.5. The conditions implied by this model are: (1) the moment cannot exceed the plastic moment; (2) if the moment is less than the plastic moment, the hinge cannot rotate; (3) if the moment is equal to the plastic moment, then the hinge may rotate in the direction consistent with the sign of the moment; and (4) if the hinge starts to rotate in a direction inconsistent with the sign of the moment, the hinge is removed.

The incremental rotation of a plastic hinge is given by the difference between the incremental joint rotation of the frame and the increase in rotation of the end of the member at that joint. For example, with a hinge at end

J only (Fig. 19.6), the incremental joint rotation is $\Delta\delta_2$ and the increase in rotation of this end due to rotation $\Delta\delta_4$ is $-\Delta\delta_4/2$ and that due to the displacements $\Delta\delta_1$ and $\Delta\delta_3$ is $1.5(\Delta\delta_3 - \Delta\delta_1)/L$. Hence the increment in rotation $\Delta\rho_i$ of a hinge formed at end I is given by

$$\Delta\rho_i = \Delta\delta_2 + \frac{1}{2}\Delta\delta_4 - 1.5\frac{\Delta\delta_3 - \Delta\delta_1}{L} \tag{19.46}$$

Similarly, with a hinge formed at end J only (Fig. 19.7), the increment in rotation of this hinge is given by

$$\Delta\rho_j = \Delta\delta_4 + \frac{1}{2}\Delta\delta_2 - 1.5\frac{\Delta\delta_3 - \Delta\delta_1}{L} \tag{19.47}$$

Finally, with hinges formed at both ends of a beam segment (Fig. 19.8), the rotations of the hinges are given by

$$\Delta\rho_i = \Delta\delta_1 - \frac{\Delta\delta_3 - \Delta\delta_1}{L} \tag{19.48}$$

$$\Delta\rho_j = \Delta\delta_2 - \frac{\Delta\delta_3 - \Delta\delta_1}{L} \tag{19.49}$$

19.10 CALCULATION OF MEMBER DUCTILITY RATIO

Nonlinear beam deformations are expressed in terms of the member ductility ratio, which is defined as the ratio of the maximum total end rotation of the member to the end rotation at the elastic limit. The elastic limit rotation is the angle developed when the member is subjected to antisymmetric yield moments M_y as shown in Fig. 19.9. In this case the relation between the end rotation and end moment is given by

$$\phi_y = \frac{M_y L}{6EI} \tag{19.50}$$

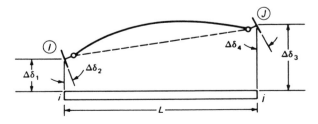

Fig. 19.8 Beam geometry with plastic hinges at both ends.

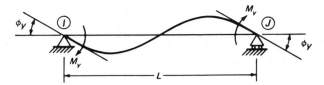

Fig. 19.9 Definition of yield rotation for beam segment.

The member ductility ratio μ is then defined as

$$\mu = \frac{\phi_y + \rho_{max}}{\phi_y} \qquad (19.51)$$

which from eq. (19.50) becomes

$$\mu = 1 + \frac{6EI}{M_yL}\rho_{max} \qquad (19.52)$$

where ρ_{max} is the maximum rotation of the plastic hinge.

19.11   SUMMARY

The determination of the nonlinear response of multidegree-of-freedom structures requires the numerical integration of the governing equations of motion. There are many methods available for the solution of these equations. The step-by-step linear acceleration method with a modification known as the Wilson-θ method was presented in this chapter. This method is unconditionally stable, that is, numerical errors do not tend to accumulate during the integration process regardless of the magnitude selected for the time step. The basic assumption of the Wilson-θ method is that the acceleration varies linearly over the extended interval $\tau = \theta\Delta t$ in which $\theta \geq 1.38$ for unconditional stability.

In the final sections of this chapter, stiffness and mass matrices for elastoplastic behavior of framed structures are presented. Formulas to determine the plastic rotation of hinges and to calculate the corresponding ductility ratios are also presented in this chapter.

PROBLEMS

19.1   The stiffness and the mass matrices for a certain structure modeled as a two-degree-of-freedom system are

$$[K] = \begin{bmatrix} 100 & -50 \\ -50 & 50 \end{bmatrix} \text{(kip/in)}, \quad [M] = \begin{bmatrix} 2 & 0 \\ 0 & 1 \end{bmatrix} \text{(kip} \cdot \text{sec}^2\text{/in)}$$

Use Program 19 to determine the response when the structure is acted upon by the forces

$$\begin{Bmatrix} F_1(t) \\ F_2(t) \end{Bmatrix} = \begin{Bmatrix} 772 \\ 386 \end{Bmatrix} f(t) \text{ (kip)}$$

where $f(t)$ is given graphically in Fig. 19.10. Neglect damping in the system.

19.2 Solve Problem 19.1 considering that the damping present in the system results in the following damping matrix:

$$[C] = \begin{bmatrix} 10 & -5 \\ -5 & 5 \end{bmatrix} \text{ (kip} \cdot \text{sec/in)}$$

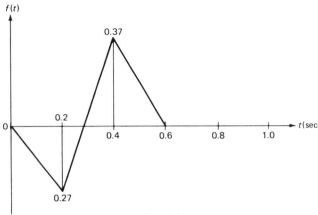

Fig. 19.10

19.3 Use Program 19 to determine the response of the three-story shear building subjected to the force $F_3(t)$ as depicted in Fig. 19.11 applied at the level of the third floor. Neglect damping in the system.

b) damping: $\xi_1 = \text{ } .01$

$\xi_2 = .02$

$\xi_3 = .03$

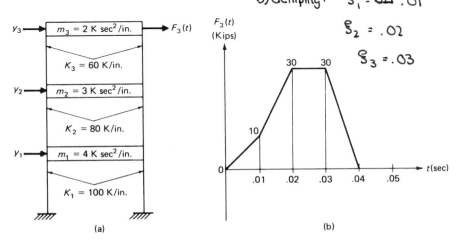

Fig. 19.11

19.4 Solve Problem 19.3 considering damping in the system of 10% in all the modes.

19.5 Use Program 19 to obtain the response in the elastic range for the structure of Problem 19.1 subjected to an acceleration at its foundation given by the function $f(t)$ shown in Fig. 19.10.

19.6 Solve Problem 19.5 considering damping in the system as indicated in Problem 19.2.

19.7 Use Program 19 to obtain the response in the elastic range of the shear building shown in Fig. 19.11 when subjected to an acceleration of its foundation given by the function $f(t)$ depicted in Fig. 19.10. Neglect damping in the system.

PART IV
Structures Modeled with Distributed Properties

20

Dynamic Analysis of Systems with Distributed Properties

The dynamic analysis of structures, modeled as lumped parameter systems with discrete coordinates, was presented in Part I for single degree-of-freedom systems and in Parts II and III for multidegree-of-freedom systems. Modeling structures with discrete coordinates provides a practical approach for the analysis of structures subjected to dynamic loads. However, the results obtained from these discrete models can only give approximate solutions to the actual behavior of dynamic systems which have continuous distributed properties and, consequently, an infinite number of degrees of freedom.

The present chapter considers the dynamic theory of beams and rods having distributed mass and elasticity for which the governing equations of motion are partial differential equations. The integration of these equations is in general more complicated than the solution of ordinary differential equations governing discrete dynamic systems. Due to this mathematical complexity, the dynamic analysis of structures as continuous systems has limited use in practice. Nevertheless, the analysis, as continuous systems, of some simple structures provides, without much effort, results which are of great importance in assessing approximate methods based on discrete models.

20.1   FLEXURAL VIBRATION OF UNIFORM BEAMS

The treatment of beam flexure developed in this section is based on the simple bending theory as it is commonly used for engineering purposes. The method of analysis is known as the Bernoulli–Euler theory which assumes that a plane cross section of a beam remains a plane during flexure.

Consider in Fig. 20.1 the free body diagram of a short segment of a beam. It is of length dx and is bounded by plane faces which are perpendicular to its axis. The forces and moments which act on the element are also shown in the figure: they are the shear forces V and $V + (\partial V/\partial x)\, dx$; the bending moments M and $M + (\partial M/\partial x)\, dx$; the lateral load pdx; and the inertia force $(\bar{m}dx)\partial^2 y/dt^2$. In this notation $\bar{m}$ is the mass per unit length and $p = p(x, t)$ is the load per unit length. Partial derivatives are used to express acceleration and variations of shear and moment because these quantities are functions of two variables, position x along the beam and time t. If the deflection of the beam is small, as the theory presupposes, the inclination of the beam segment from the unloaded position is also small. Under these conditions, the equation of motion perpendicular to the x axis of the deflected beam obtained by equating to zero the sum of the forces in the free body diagram of Fig. 20.1(b) is

$$V - \left(V + \frac{\partial V}{\partial x}\, dx \right) + p(x, t)\, dx - \bar{m}dx \frac{\partial^2 y}{\partial t^2} = 0$$

which, upon simplification, becomes

$$\frac{\partial V}{\partial x} + \bar{m}\, \frac{\partial^2 y}{\partial t^2} = p(x, t) \tag{20.1}$$

From simple bending theory, we have the relations

$$M = EI \frac{\partial^2 y}{\partial x^2} \tag{20.2}$$

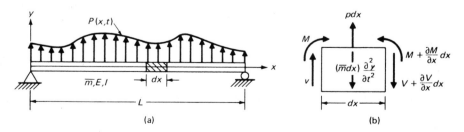

(a)   (b)

Fig. 20.1 Simple beam with distributed mass and load.

and

$$V = \frac{\partial M}{\partial x} \tag{20.3}$$

where E is Young's modulus of elasticity and I is the moment of inertia of the cross-sectional area with respect to the neutral axis through the centroid. For a uniform beam, the combination of eqs. (20.1), (20.2), and (20.3) results in

$$V = EI \frac{\partial^3 y}{\partial x^3} \tag{20.4}$$

and

$$EI \frac{\partial^4 y}{\partial x^4} + \bar{m} \frac{\partial^2 y}{\partial t^2} = p(x, t) \tag{20.5}$$

It is seen that eq. (20.5) is a partial differential equation of fourth order. It is an approximate equation. Only lateral flexural deflections were considered while the deflections due to shear forces and the inertial forces caused by the rotation of the cross section (rotary inertia) were neglected. The inclusion of shear deformations and rotary inertia in the differential equation of motion considerably increases its complexity. The equation taking into consideration shear deformation and rotary inertia is known as Timoshenko's equation. The differential equation (20.5) also does not include the flexural effects due to the presence of forces which may be applied axially to the beam. The axial effects will be discussed in Chapter 21.

20.2 SOLUTION OF THE EQUATION OF MOTION IN FREE VIBRATION

For free vibration $[p(x, t) = 0]$, eq. (20.5) reduces to the homogeneous differential equation

$$EI \frac{\partial^4 y}{\partial x^4} + \bar{m} \frac{\partial^2 y}{\partial t^2} = 0 \tag{20.6}$$

The solution of eq. (20.6) can be found by the method of separation of variables. In this method, it is assumed that the solution may be expressed as the product of a function of position $\Phi(x)$ and a function of time $f(t)$, that is,

$$y(x, t) = \Phi(x) f(t) \tag{20.7}$$

The substitution of eq. (20.7) in the differential equation (20.6) leads to

$$EIf(t) \frac{d^4 \Phi(x)}{dx^4} + \bar{m} \Phi(x) \frac{d^2 f(t)}{dt^2} = 0 \tag{20.8}$$

This last equation may be written as

$$\frac{EI}{\bar{m}} \frac{\Phi^{IV}(x)}{\Phi(x)} = -\frac{\ddot{f}(t)}{f(t)} \qquad (20.9)$$

In this notation Roman indices indicate derivatives with respect to x and overdots indicate derivatives with respect to time. Since the left-hand side of eq. (20.9) is a function only of x while the right-hand side is a function only of t, each side of the equation must equal the same constant value; otherwise, the identity of eq. (20.9) cannot exist. We designate the constant by ω^2 which equated separately to each side of eq. (20.9) results in the two following differential equations:

$$\Phi^{IV}(x) - a^4\Phi(x) = 0 \qquad (20.10)$$

and

$$\ddot{f}(t) + \omega^2 f(t) = 0 \qquad (20.11)$$

where

$$a^4 = \frac{\bar{m}\omega^2}{EI} \qquad (20.12)$$

It is particularly convenient to solve eq. (20.12) for ω and to use the following notation, namely,

$$\omega = C \sqrt{\frac{EI}{\bar{m}L^4}} \qquad (20.13)$$

in which $C = (aL)^2$.

Equation (20.11) is the familiar free-vibration equation for the undamped single degree-of-freedom system and its solution from eq. (1.17) is

$$f(t) = A \cos \omega t + B \sin \omega t \qquad (20.14)$$

where A and B are constants of integration. Equation (20.10) can be solved by letting

$$\Phi(x) = Ce^{sx} \qquad (20.15)$$

The substitution of eq. (20.15) into eq. (20.10) results in

$$(s^4 - a^4)Ce^{sx} = 0$$

which, for a nontrivial solution, requires that

$$s^4 - a^4 = 0 \qquad (20.16)$$

The roots of eq. (20.16) are

$$s_1 = a, \quad s_3 = ai$$

$$s_2 = -a, \quad s_4 = -ai \qquad (20.17)$$

The substitution of each of these roots into eq. (20.15) provides a solution of eq. (20.10). The general solution is then given by the superposition of these four possible solutions, namely,

$$\Phi(x) = C_1 e^{ax} + C_2 e^{-ax} + C_3 e^{iax} + C_4 e^{-iax} \qquad (20.18)$$

where C_1, C_2, C_3, and C_4 are constants of integration. The exponential functions in eq. (20.18) may be expressed in terms of trigonometric and hyperbolic functions by means of the relations

$$e^{\pm ax} = \cosh ax \pm \sinh ax$$
$$e^{\pm iax} = \cos ax \pm i \sin ax \qquad (20.19)$$

Substitution of these relations into eq. (20.18) yields

$$\Phi(x) = A \sin ax + B \cos ax + C \sinh ax + D \cosh ax \qquad (20.20)$$

where A, B, C, and D are new constants of integration. These four constants of integration define the shape and the amplitude of the beam in free vibration; they are evaluated by considering the boundary conditions at the ends of the beam as illustrated in the examples presented in the following section.

20.3   NATURAL FREQUENCIES AND MODE SHAPES FOR UNIFORM BEAMS

20.3.1   Both Ends Simply Supported

In this case the displacements and bending moments must be zero at both ends of the beam; hence the boundary conditions for the simply supported beams are

$$y(0, t) = 0, \quad M(0, t) = 0$$
$$y(L, t) = 0, \quad M(L, t) = 0$$

In view of eqs. (20.2) and (20.7), these boundary conditions imply the following conditions on the shape function $\Phi(x)$.

At $x = 0$,

$$\Phi(0) = 0, \quad \Phi''(0) = 0 \qquad (20.21)$$

At $x = L$,

$$\Phi(L) = 0, \quad \Phi''(L) = 0 \qquad (20.22)$$

The substitution of the first two of these boundary conditions into eq. (20.20) yields

$$\Phi(0) = A0 + B1 + C0 + D1 = 0$$
$$\Phi''(0) = a^2(-A0 - B1 + C0 + D1) = 0$$

which reduce to

$$B + D = 0$$

$$-B + D = 0$$

Hence

$$B = D = 0$$

Similarly, substituting the last two boundary conditions into eq. (20.20) and setting $B = D = 0$ leads to

$$\Phi(L) = A \sin aL + C \sinh aL = 0$$

$$\Phi''(L) = a^2(-A \sin aL + C \sinh aL) = 0 \qquad (20.23)$$

which, when added, give

$$2C \sinh aL = 0$$

From this last relation, $C = 0$ since the hyperbolic sine function cannot vanish except for a zero argument. Thus eqs. (20.23) reduce to

$$A \sin aL = 0 \qquad (20.24)$$

Excluding the trivial solution $(A = 0)$, we obtain the frequency equation

$$\sin aL = 0 \qquad (20.25)$$

which will be satisfied for

$$a_n L = n\pi, \quad n = 0, 1, 2 \ldots \qquad (20.26)$$

Substitution of the roots, eq. (20.26), into eq. (20.13) yields

$$\omega_n = n^2\pi^2 \sqrt{\frac{EI}{\bar{m}L^4}} \qquad (20.27)$$

where the subscript n serves to indicate the order of the natural frequencies.
Since $B = C = D = 0$, it follows that eq. (20.20) reduces to

$$\Phi_n(x) = A \sin \frac{n\pi x}{L}$$

or simply

$$\Phi_n(x) = \sin \frac{n\pi x}{L} \qquad (20.28)$$

We note that in eq. (20.28) the constant A is absorbed by the other constants in the modal response given below by eq. (20.29).
From eq. (20.7) a modal shape or normal mode of vibration is given by

$$y_n(x, t) = \Phi_n(x) f_n(t)$$

or from eqs. (20.14) and (20.28) by

$$y_n(x, t) = \sin \frac{n\pi x}{L} [A_n \cos \omega_n t + B_n \sin \omega_n t] \qquad (20.29)$$

The general solution of the equation of motion in free vibration which satisfies the boundary conditions, eqs. (20.21) and (20.22), is the sum of all the normal modes of vibration, eq. (20.29), namely,

$$y(x, t) = \sum_{n=1}^{\infty} \sin \frac{n\pi x}{L} [A_n \cos \omega_n t + B_n \sin \omega_n t] \qquad (20.30)$$

The constants A_n and B_n are determined, as usual, from the initial conditions. If at $t = 0$, the shape of the beam is given by

$$y(x, 0) = \rho(x)$$

and the velocity by

$$\frac{\partial y(x, 0)}{\partial t} = \psi(x)$$

for $0 \leq x \leq L$, it follows from eq. (20.30) that

$$\sum_{n=1}^{\infty} A_n \sin \frac{n\pi x}{L} = \rho(x)$$

and

$$\sum_{n=1}^{\infty} B_n \omega_n \sin \frac{n\pi x}{L} = \psi(x)$$

Therefore, as shown in Chapter 5, Fourier coefficients are expressed as

$$A_n = \frac{2}{L} \int_0^L \rho(x) \sin \frac{n\pi x}{L} \, dx$$

$$B_n = \frac{2}{\omega_n L} \int_0^L \psi(x) \sin \frac{n\pi x}{L} \, dx \qquad (20.31)$$

The first five values for the natural frequencies and normal modes for the simply supported beam are presented in Table 20.1.

20.3.2   Both Ends Free (Free Beam)

The boundary conditions for a beam with both ends free are as follows.
  At $x = 0$,

TABLE 20.1   Natural Frequencies and Normal Modes for Simply Supported Beams

Natural Frequencies $\omega_n = C_n \sqrt{\dfrac{EI}{\bar{m}L^4}}$			Normal Modes $\Phi_n = \sin \dfrac{n\pi x}{L}$
n	C_n	I_n^*	Shapes
1	π^2	$4/\pi$	(mode shape, span L)
2	$4\pi^2$	0	(mode shape, 0.500L)
3	$9\pi^2$	$4/3\pi$	(mode shape, 0.333L, 0.666L)
4	$16\pi^2$	0	(mode shape, 0.250L, 0.500L, 0.750L)
5	$25\pi^2$	$4/5\pi$	(mode shape, 0.200L, 0.400L, 0.600L, 0.800L)

$^*I_n = \int_0^L \Phi_n(x)\,dx / \int_0^L \Phi^2(x)\,dx$

$$M(0, t) = 0 \quad \text{or} \quad \Phi''(0) = 0$$
$$V(0, t) = 0 \quad \text{or} \quad \Phi'''(0) = 0 \tag{20.32}$$

At $x = L$,

$$M(L, t) = 0 \quad \text{or} \quad \Phi''(L) = 0$$
$$V(L, t) = 0 \quad \text{or} \quad \Phi'''(L) = 0 \tag{20.33}$$

The substitutions of these conditions in eq. (20.20) yield

$$\Phi''(0) = a^2(-B + D) = 0$$
$$\Phi'''(0) = a^3(-A + C) = 0$$

and

$$\Phi''(L) = a^2(-A \sin aL - B \cos aL + C \sinh aL + D \cosh aL) = 0$$
$$\Phi'''(L) = a^3(-A \cos aL + B \sin aL + C \cosh aL + D \sinh aL) = 0$$

From the first two equations we obtain

$$D = B, \quad C = A \tag{20.34}$$

which, substituted into the last two equations, result in

$$(\sinh aL - \sin aL)A + (\cosh aL - \cos aL)B = 0$$

$$(\cosh aL - \cos aL)A + (\sinh aL + \sin aL)B = 0 \qquad (20.35)$$

For nontrivial solution of eqs. (20.35), it is required that the determinant of the unknown coefficients A and B be equal to zero; hence

$$\begin{vmatrix} \sinh aL - \sin aL & \cosh aL - \cos aL \\ \cosh aL - \cos aL & \sinh aL + \sin aL \end{vmatrix} = 0 \qquad (20.36)$$

The expansion of this determinant provides the frequency equation for the free beam, namely

$$\cos aL \cdot \cosh aL - 1 = 0 \qquad (20.37)$$

The first five natural frequencies which are obtained by substituting the roots of eq. (20.37) into eq. (20.13) are presented in Table 20.2. The corresponding normal modes are obtained by letting $A = 1$ (normal modes are determined only to a relative magnitude), substituting in eqs. (20.35) the roots

TABLE 20.2   Natural Frequencies and Normal Modes for Free Beams

Natural Frequencies				Normal Modes
				$\Phi_n(x) = \cosh a_n x + \cos a_n x - \sigma_n(\sinh a_n x + \sin a_n x)$
$\omega_n = C_n \sqrt{\dfrac{EI}{\bar{m}L^4}}$		$\sigma_n = \dfrac{\cosh a_n L - \cos a_n L}{\sinh a_n L - \sin a_n L}$		
n	$C_n = (a_n L)^2$	σ_n	$I_n^{\star}$	Shapes
1	22.3733	0.982502	0.8308	0.224L ... 0.776L
2	61.6728	1.000777	0	0.132L ... 0.868L / 0.500L
3	120.9034	0.999967	0.3640	0.094L 0.644L / 0.356L 0.906L
4	199.8594	1.000001	0	0.073L 0.500L 0.927L / 0.277L 0.723L
5	298.5555	1.000000	0.2323	0.060L 0.409L 0.774L / 0.226L 0.591L 0.940L

$^{\star}I_n = \int_0^L \Phi_n(x)\, dx / \int_0^L \Phi_n^2(x)\, dx$

of a_n of eq. (20.37), solving one of these equations for B, and finally introducing into eq. (20.20) the constants C, D from eq. (20.34) together with B. Performing these operations, we obtain

$$\Phi_n(x) = \cosh a_n x + \cos a_n x - \sigma_n(\sinh a_n x + \sin a_n x) \qquad (20.38)$$

where

$$\sigma_n = \frac{\cosh a_n L - \cos a_n L}{\sinh a_n L - \sin a_n L} \qquad (20.39)$$

20.3.3   Both Ends Fixed

The boundary conditions for a beam with both ends fixed are as follows.
    At $x = 0$,

$$y(0, t) = 0 \quad \text{or} \quad \Phi(0) = 0$$

$$y'(0, t) = 0 \quad \text{or} \quad \Phi'(0) = 0 \qquad (20.40)$$

At $x = L$,

$$y(L, t) = 0 \quad \text{or} \quad \Phi(L) = 0$$

$$y'(L, t) = 0 \quad \text{or} \quad \Phi'(L) = 0 \qquad (20.41)$$

The use of the boundary conditions, eqs. (20.40), into eq. (20.20) gives

$$B + D = 0 \quad \text{and} \quad A + C = 0$$

while conditions, eqs. (20.41), yield the homogeneous system

$$(\cos aL - \cosh aL)B + (\sin aL - \sinh aL)A = 0$$

$$-(\sin aL + \sinh aL)B + (\cos aL - \cosh aL)A = 0 \qquad (20.42)$$

Equating to zero the determinant of the coefficients of this system results in the frequency equation

$$\cos a_n L \cosh a_n L - 1 = 0 \qquad (20.43)$$

From the first of eqs. (20.42), it follows that

$$A = -\frac{\cos aL - \cosh aL}{\sin aL - \sinh aL} B \qquad (20.44)$$

where B is arbitrary. To each value of the natural frequency

$$\omega_n = (a_n L)^2 \sqrt{\frac{EI}{\bar{m}L^4}} \qquad (20.45)$$

obtained by the substitution of the roots of eq. (20.43) into eq. (20.13), there corresponds a normal mode

$$\Phi_n(x) = \cosh a_n x - \cos a_n x - \sigma_n(\sinh a_n x - \sin a_n x) \qquad (20.46)$$

$$\sigma_n = \frac{\cos a_n L - \cosh a_n L}{\sin a_n L - \sinh a_n L} \qquad (20.47)$$

The first five natural frequencies calculated from eqs. (20.43) and (20.45) and the corresponding normal modes obtained from eq. (20.46) are presented in Table 20.3.

20.3.4 One End Fixed and the Other End Free (Cantilever Beam)

At the fixed end $(x = 0)$ of the cantilever beam, the deflection and the slope must be zero, and at the free end $(x = L)$ the bending moment and the shear

TABLE 20.3   Natural Frequencies and Normal Modes for Fixed Beams

Natural Frequencies			Normal Modes	
			$\Phi_n(x) = \cosh a_n x - \cos a_n x - \sigma_n(\sinh a_n x - \sin a_n x)$	
$\omega_n = C_n \sqrt{\dfrac{EI}{\overline{m}L^4}}$		$\sigma_n = \dfrac{\cos a_n L - \cosh a_n L}{\sin a_n L - \sinh a_n L}$		
n	$C_n = (a_n L)^2$	σ_n	I_n^*	Shape
1	22.3733	0.982502	0.8308	
2	61.6728	1.000777	0	
3	120.9034	0.999967	0.3640	
4	199.8594	1.000001	0	
5	298.5555	1.000000	0.2323	

$^*I_n = \int_0^L \Phi_n(x)\, dx / \int_0^L \Phi_n^2(x)\, dx$

force must be zero. Hence the boundary conditions for this beam are as follows.

At $x = 0$,

$$y(0, t) = 0 \quad \text{or} \quad \Phi(0) = 0$$

$$y'(0, t) = 0 \quad \text{or} \quad \Phi'(0) = 0 \tag{20.48}$$

At $x = L$,

$$M(L, t) = 0 \quad \text{or} \quad \Phi''(L) = 0$$

$$V(L, t) = 0 \quad \text{or} \quad \Phi'''(L) = 0 \tag{20.49}$$

These boundary conditions when substituted into the shape equation (20.20) lead to the frequency equation

$$\cos a_n L \cdot \cosh a_n L + 1 = 0 \tag{20.50}$$

To each root of eq. (20.50) corresponds a natural frequency

$$\omega_n = (a_n L)^2 \sqrt{\frac{EI}{\overline{m}L^4}} \tag{20.51}$$

and a normal shape

$$\Phi_n(x) = (\cosh a_n x - \cos a_n x) - \sigma_n(\sinh a_n x - \sin a_n x) \tag{20.52}$$

where

$$\sigma_n = \frac{\cos a_n L + \cosh a_n L}{\sin a_n L + \sinh a_n L} \tag{20.53}$$

The first five natural frequencies and the corresponding mode shapes for cantilever beams are presented in Table 20.4.

20.3.5   One End Fixed and the Other Simply Supported

The boundary conditions for a beam with one end fixed and the other simply supported are as follows.

At $x = 0$,

$$y(0, t) = 0 \quad \text{or} \quad \Phi(0) = 0 \tag{20.54}$$

$$y'(0, t) = 0 \quad \text{or} \quad \Phi'(0) = 0$$

At $x = L$,

$$y(L, t) = 0 \quad \text{or} \quad \Phi(L) = 0$$

$$M(L, t) = 0 \quad \text{or} \quad \Phi''(L) = 0 \tag{20.55}$$

TABLE 20.4   Natural Frequencies and Normal Modes for Cantilever Beams

Natural Frequencies

Normal Modes

$$\Phi_n = (\cosh a_n x - \cos a_n x) - \sigma_n(\sinh a_n x - \sin a_n x)$$

$$\omega_n = C_n \sqrt{\frac{EI}{\bar{m}L^4}}$$

$$\sigma = \frac{\cos a_n L + \cosh a_n L}{\sin a_n L + \sinh a_n L}$$

n	$C_n = (a_n L)^2$	σ_n	I_n^*	Shape
1	3.5160	0.734096	0.7830	
2	22.0345	1.018466	0.4340	
3	61.6972	0.999225	0.2589	
4	120.0902	1.000033	0.0017	
5	199.8600	1.000000	0.0707	

$^*I_n = \int_0^L \Phi_n(x)\,dx / \int_0^L \Phi_n^2(x)\,dx$

The substitution of these boundary conditions into the shape equation (20.20) results in the frequency equation

$$\tan a_n L - \tanh a_n L = 0 \tag{20.56}$$

To each root of this last equation corresponds a natural frequency

$$\omega_n = (a_n L)^2 \sqrt{\frac{EI}{\bar{m}L^4}} \tag{20.57}$$

and a normal mode

$$\Phi_n(x) = (\cosh a_n x - \cos a_n x) + \sigma_n(\sinh a_n x - \sin a_n x) \tag{20.58}$$

where

$$\sigma_n = \frac{\cos a_n L - \cosh a_n L}{\sin a_n L - \sinh a_n L} \tag{20.59}$$

The first five natural frequencies for the fixed simply supported beam and corresponding mode shapes are presented in Table 20.5.

TABLE 20.5   Natural Frequencies and Normal Modes for Fixed Simply Supported Beams

	Natural Frequencies $$\omega_n = C_n \sqrt{\dfrac{EI}{\bar{m}L^4}}$$	Normal Modes $$\Phi(x) = \cosh a_n x - \cos a_n x + \sigma_n(\sinh a_n x - \sin a_n x)$$ $$\sigma_n = \dfrac{\cos a_n L - \cosh a_n L}{\sin a_n L - \sinh a_n L}$$		
n	$C_n = (a_n L)^2$	σ_n	I_n^*	Shape
1	15.4118	1.000777	0.8600	
2	49.9648	1.000001	0.0826	
3	104.2477	1.000000	0.3345	
4	178.2697	1.000000	0.0434	
5	272.0309	1.000000	0.2076	

$^*I_n = \int_o^L \Phi_n(x)\, dx / \int_o^L \Phi_n^2(x)\, dx$

20.4   ORTHOGONALITY CONDITION BETWEEN NORMAL MODES

The most important property of the normal modes is that of orthogonality. It is this property which makes possible the uncoupling of the equations of motion as it has previously been shown for discrete systems. The orthogonality property for continuous systems can be demonstrated in essentially the same way as for discrete parameter systems.

Consider in Fig. 20.2 a beam subjected to the inertial forces resulting from the vibrations of two different modes, $\Phi_m(x)$ and $\Phi_n(x)$. The deflection curves for these two modes and the corresponding inertial forces are depicted in the same figure. Betti's law is applied to these two deflection patterns. Accordingly, the work done by the inertial force, f_{In}, acting on the displacements of mode m is equal to the work of the inertial forces, f_{Im}, acting on the displacements of mode n, that is,

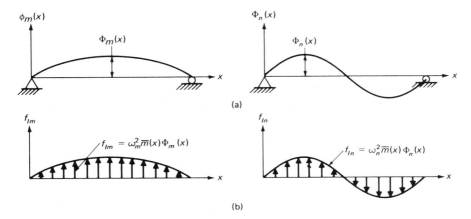

Fig. 20.2 Beam showing two modes of vibration and inertial forces. (a) Displacements. (b) Inertial forces.

$$\int_0^L \Phi_m(x)\, f_{In}(x)\, dx = \int_0^L \Phi_n(x)\, f_{Im}(x)\, dx \qquad (20.60)$$

The inertial force f_{In} per unit length along the beam is equal to the mass per unit length times the acceleration. Inasmuch as the vibratory motion in a normal mode is harmonic, the amplitude of the acceleration is given by $\omega_n^2 \Phi_n(x)$. Hence the inertial force per unit length along the beam for the nth mode is

$$f_{In} = \omega_n^2 \bar{m}(x)\, \phi_n(x)$$

and for the mth mode

$$f_{Im} = \omega_m^2 \bar{m}(x)\, \Phi_m(x) \qquad (20.61)$$

Substituting these expressions in eq. (20.60), we obtain

$$\omega_n^2 \int_0^L \Phi_m(x)\, \bar{m}(x)\, \Phi_n(x)\, dx = \omega_m^2 \int_0^L \Phi_n(x)\, \bar{m}(x)\, \Phi_m(x)\, dx$$

which may be written as

$$(\omega_n^2 - \omega_m^2) \int_0^L \Phi_m(x)\, \Phi_n(x)\, \bar{m}(x)\, dx = 0 \qquad (20.62)$$

It follows that, for two different frequencies $\omega_n \neq \omega_m$, the normal modes must satisfy the relation

$$\int_0^L \Phi_m(x)\, \Phi_n(x)\, \bar{m}(x)\, dx = 0 \qquad (20.63)$$

which is equivalent to the orthogonal condition between normal modes for discrete parameter systems, eq. (10.27).

20.5 FORCED VIBRATION OF BEAMS

For a uniform beam acted on by lateral forces $p(x, t)$, the equation of motion, eq. (20.5), may be written as

$$EI \frac{\partial^4 y}{\partial x^4} = p(x, t) - \bar{m} \frac{\partial^2 y}{\partial t^2} \tag{20.64}$$

in which $p(x, t)$ is the external load per unit length along the beam. We assume that the general solution of this equation may be expressed by the summation of the products of the normal modes $\Phi_n(x)$ multiplied by factors $z_n(t)$ which are to be determined. Hence

$$y(x, t) = \sum_{n=1}^{\infty} \Phi_n(x) z_n(t) \tag{20.65}$$

The normal modes $\Phi_n(x)$ satisfy the differential equation (20.10), which by eq. (20.12) may be written as

$$EI \, \Phi_n^{IV}(x) = \bar{m}\omega_n^2 \, \Phi_n(x), \quad n = 1, 2, 3, \ldots \tag{20.66}$$

The normal modes should also satisfy the specific force boundary conditions at the ends of the beam. Substitution of eq. (20.65) in eq. (20.64) gives

$$EI \sum_n \Phi_n^{IV}(x) z_n(t) = p(x, t) - \bar{m} \sum_n \Phi_n(x) \ddot{z}_n(t) \tag{20.67}$$

In view of eq. (20.66), we can write eq. (20.67) as

$$\sum_n \bar{m}\omega_n^2 \, \Phi_n(x) z_n(t) = p(x, t) - \bar{m} \sum_n \Phi_n(x) \ddot{z}_n(t) \tag{20.68}$$

Multiplying both sides of eq. (20.68) by $\Phi_m(x) \, dx$ and integrating between 0 and L result in

$$\omega_m^2 z_m(t) \int_0^L \bar{m}\Phi_m^2(x) \, dx = \int_0^L \Phi_m(x) \, p(x, t) \, dx - \ddot{z}_m(t) \int_0^L \bar{m}\Phi_m^2(x) \, dx \tag{20.69}$$

We note that all the terms which contain products of different indices ($n \neq m$) vanish from the summations in eq. (20.68) in view of the orthogonality conditions, eq. (20.63), between normal modes. Equation (20.69) may conveniently be written as

$$M_n \ddot{z}_n(t) + \omega_n^2 M_n z_n(t) = F_n(t), \quad n = 1, 2, 3, \ldots, m, \ldots \tag{20.70}$$

where

$$M_n = \int_0^L \bar{m}\Phi_n^2(x) \, dx \tag{20.71}$$

is the modal mass, and

$$F_n(t) = \int_0^L \Phi_n(x) \, p(x, t) \, dx \tag{20.72}$$

is the modal force.

The equation of motion for the nth normal mode, eq. (20.70), is completely analogous to the modal equation, eq. (12.9), for discrete systems. Modal damping could certainly be introduced by simply adding the damping term in eq. (20.70); hence we would obtain

$$M_n\ddot{z}_n(t) + C_n\dot{z}_n(t) + K_n z_n(t) = F_n(t) \tag{20.73}$$

which, upon dividing by M_n, gives

$$\ddot{z}_n(t) + 2\xi_n\omega_n\dot{z}_n(t) + \omega_n^2 z_n(t) = \frac{F_n(t)}{M_n} \tag{20.74}$$

where $\xi_n = c_n/c_{n,\text{cr}}$ is the modal damping ratio and $K_n = M_n\omega_n^2$ is the modal stiffness. The total response is then obtained from eq. (20.65) as the superposition of the solution of the modal equation (20.74) for as many modes as desired. Though the summation in eq. (20.65) is over an infinite number of terms, in most structural problems only the first few modes have any significant contribution to the total response and in some cases the response is given essentially by the contribution of the first mode alone.

The modal equation (20.74) is completely general and applies to beams with any type of load distribution. If the loads are concentrated rather than distributed, the integral in eq. (20.72) merely becomes a summation having one term for each load. The computation of the integral in eqs. (20.71) and (20.72) becomes tedious except for the simply supported beam because the normal shapes are rather complicated functions. Values of the ratios of these integrals needed for problems with uniform distributed load are presented in the last columns of Tables 20.1 through 20.5 for some common types of beams.

Example 20.1.   Consider in Fig. 20.3 a simply supported uniform beam subjected to a concentrated constant force suddenly applied at a section x_1 units from the left support. Determine the response using modal analysis. Neglect damping.

Solution: The modal shapes of a simply supported beam by eq. (20.28) are

$$\Phi_n = \sin\frac{n\pi x}{L}, \quad n = 1, 2, 3, \ldots \tag{a}$$

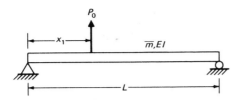

Fig. 20.3 Simply supported beam subjected to a suddenly applied force.

and the modal force by eq. (20.72) is

$$F_n(t) = \int_0^L \Phi_n(x) \, p(x, t) \, dx$$

In this problem $p(x, t) = P_0$ at $x = x_1$; otherwise, $p(x, t) = 0$. Hence

$$F_n(t) = P_0 \Phi_n(x_1)$$

or using eq. (a), we obtain

$$F_n(t) = P_0 \sin \frac{n\pi x_1}{L} \tag{b}$$

The modal mass by eq. (20.71) is

$$M_n = \int_0^L \bar{m}\Phi^2(x) \, dx$$

$$= \int_0^L \bar{m} \sin^2 \frac{n\pi x}{L} \, dx = \frac{\bar{m}L}{2} \tag{c}$$

Substituting the modal force, eq. (b), and the modal mass, eq. (c), into the modal equation (20.70) results in

$$\ddot{z}_n(t) + \omega_n^2 z_n(t) = \frac{P_0 \sin (n\pi x_1/L)}{\bar{m}L/2} \tag{d}$$

For initial conditions of zero displacement and zero velocity, the solution of eq. (d) from eqs. (4.5) is

$$z_n = (z_{st})_n(1 - \cos \omega_n t) \tag{e}$$

in which

$$(z_{st})_n = \frac{2 P_0 \sin (n\pi x_1/L)}{\omega_n^2 \bar{m}L} \tag{f}$$

so that

$$z_n = \frac{2P_0 \sin (n\pi x_1/L)}{\omega_n^2 \bar{m} L} (1 - \cos \omega_n t) \tag{g}$$

The modal deflection at any section of the beam is

$$y_n(x, t) = \Phi_n(x) z_n(t) \tag{h}$$

which, upon substitution of eqs. (a) and (g), becomes

$$y_n(x, t) = \frac{2P_0 \sin (n\pi x_1/L)}{\omega_n^2 \bar{m} L} (1 - \cos \omega_n t) \sin \frac{n\pi x}{L} \tag{i}$$

By eq. (20.65), the total deflection is then

$$y(x, t) = \frac{2P_0}{\bar{m} L} \sum_n \left[\frac{1}{\omega_n^2} \sin \frac{n\pi x_1}{L} (1 - \cos \omega_n t) \sin \frac{n\pi x}{L} \right] \tag{j}$$

As a special case, let us consider the force applied at midspan, i.e., $x_1 = L/2$. Hence eq. (j) becomes in this case

$$y(x, t) = \frac{2P_0}{\bar{m} L} \sum_n \left[\frac{1}{\omega_n^2} \sin \frac{n\pi}{2} (1 - \cos \omega_n t) \sin \frac{n\pi x}{L} \right] \tag{20.75}$$

From the latter (due to the presence of the factor $\sin n\pi/2$) it is apparent that all the even modes do not contribute to the deflection at any point. This is true because such modes are antisymmetrical (shapes in Table 20.1) and are not excited by a symmetrical load.

It is also of interest to compare the contribution of the various modes to the deflection at midspan. This comparison will be done on the basis of maximum modal displacement disregarding the manner in which these displacements combine. The amplitudes will indicate the relative importance of the modes. The dynamic load factor $(1 - \cos \omega_n t)$ in eq. (20.75) has a maximum value of 2 for all the modes. Furthermore, since all sines are unity for odd modes and zero for even modes, the modal contributions are simply in proportion to $1/\omega_n^2$. Hence, from Table 20.1 the maximum modal deflections are in proportion to 1, 1/81, and 1/625 for the first, third, and fifth modes, respectively. It is apparent, in this example, that the higher modes contribute very little to the midspan deflection.

Example 20.2. Determine the maximum deflection at the midpoint of the fixed beam shown in Fig. 20.4 subjected to a harmonic load $p(x, t) = p_0 \sin 300t$ lb/in uniformly distributed along the span. Consider in the analysis the first three modes contributing to the response.

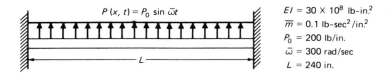

Fig. 20.4 Fixed beam with uniform harmonic load.

Solution: The natural frequencies for uniform beams are given by eq. (20.13) as

$$\omega_n = C_n \sqrt{\frac{EI}{\bar{m}L^4}}$$

or, substituting numerical values for this example, we get

$$\omega_n = C_n \sqrt{\frac{30 \times 10^8}{0.1(240)^4}} \qquad \text{(a)}$$

where the values of C_n are given for the first five modes in Table 20.3. The deflection of the beam is given by eq. (20.65) as

$$y(x, t) = \sum_{n=1}^{\infty} \Phi_n(x) z_n(t) \qquad \text{(b)}$$

in which $\Phi_n(x)$ is the modal shape defined for a fixed beam by eq. (20.46) and $z_n(t)$ is the modal response.

The modal equation by eq. (20.70) (neglecting damping) may be written as

$$\ddot{z}_n(t) + \omega_n^2 z(t) = \frac{\displaystyle\int_0^L p(x, t)\phi_n(x)\, dx}{\displaystyle\int_0^L \bar{m}\phi_n^2(x)\, dx}$$

Then, substituting numerical values to this example, we obtain

$$\ddot{z}_n(t) + \omega_n^2 z_n(t) = \frac{200 \displaystyle\int_0^L \phi_n(x)\, dx}{0.1 \displaystyle\int_0^L \phi_n^2(x)\, dx} \sin 300t$$

or

$$\ddot{z}_n(t) + \omega_n^2 z_n(t) = 2000 I_n \sin 300t \qquad \text{(c)}$$

in which

$$I_n = \frac{\int_0^L \phi_n(x)\, dx}{\int_0^L \phi_n^2(x)\, dx}$$

is given for the first five modes in Table 20.3. The modal steady-state response is

$$z_n(t) = \frac{2000 I_n}{\omega_n^2 - (300)^2} \sin 300t \tag{d}$$

The numerical calculations are conveniently presented in Table 20.6.

The deflections at midspan of the beam are then calculated from eq. (b) and values in Table 20.6 as

$$y\left(\frac{L}{2}, t\right) = [(1.588)(-0.0194) + (-1.410)(0.0173) + (1.414)(0.00065)] \sin 300t$$

$$y\left(\frac{L}{2}, t\right) = -0.0541 \sin 300t \text{ (in)}$$

20.6 DYNAMIC STRESSES IN BEAMS

To determine stresses in beams, we apply the following well-known relationships for bending moment M and shear force V, namely

$$M = EI \frac{\partial^2 y}{\partial x^2}$$

$$V = \frac{\partial M}{\partial x} = EI \frac{\partial^3 y}{\partial x^3}$$

TABLE 20.6   Modal Response at Midspan for the Beam in Fig. 20.4

Mode	$\omega_n \left(\dfrac{rad}{sec}\right)$	$a_n L$	I_n	$z_n = \dfrac{2000\, I_n}{\omega_n^2 - \bar{\omega}^2}$ (in)	$\Phi_n\left(x = \dfrac{L}{2}\right)$
1	67.28	4.730	0.8380	-0.0194	1.588
2	185.45	7.853	0	0	0
3	363.56	10.996	0.3640	0.0173	-1.410
4	600.98	14.137	0	0	0
5	897.76	17.279	0.2323	0.00065	1.414

Therefore, the calculation of the bending moment or the shear force requires only differentiation of the deflection function $y = y(x, t)$ with respect to x. For example, in the case of the simple supported beam with a concentrated load suddenly applied at its center, differentiation of the deflection function, eq. (20.75), gives

$$M = -\frac{2\pi^2 P_0 EI}{\bar{m}L^3} \sum_n \left[\frac{n^2}{\omega_n^2} \sin \frac{n\pi}{2} (1 - \cos \omega_n t) \sin \frac{n\pi x}{L} \right] \qquad (20.76)$$

$$V = -\frac{2\pi^3 P_0 EI}{\bar{m}L^4} \sum_n \left[\frac{n^3}{\omega_n^2} \sin \frac{n\pi}{2} (1 - \cos \omega_n t) \cos \frac{n\pi x}{L} \right] \qquad (20.77)$$

We note that the higher modes are increasingly more important for moments than for deflections and even more so for shear force, as indicated by the factors 1, n^2, and n^3, respectively, in eqs. (20.75), (20.76), and (20.77).

To illustrate, we compare the amplitudes for the first and third modes at their maximum values. Noting that ω_n^2 is proportional to n^4 [eq. (20.27)], we obtain from eqs. (20.75), (20.76), and (20.77) the following ratios:

$$\frac{y_1}{y_3} = 3^4 = 81$$

$$\frac{M_1}{M_3} = 3^2 = 9$$

$$\frac{V_1}{V_3} = 3$$

This tendency in which higher modes have increasing importance in moment and shear calculation is generally true of beam response.

In those cases in which the first mode dominates the response, it is possible to obtain approximate deflections and stresses from static values of these quantities amplified by the dynamic load factor. For example, the maximum deflection of a simple supported beam with a concentrated force at midspan may be closely approximated by

$$y\left(x = \frac{L}{2}\right) = \frac{P_0 L^3}{48EI} (1 - \cos \omega_1 t)$$

If we consider only the first mode, the corresponding value given by eq. (20.75) is

$$y\left(x = \frac{L}{2}\right) = \frac{2P_0}{\bar{m}L\omega_1^2} (1 - \cos \omega_1 t)$$

Since, by eq. (20.12), $\omega_1^2 = \pi^4 EI/\bar{m}L^4$, it follows that

$$y\left(x = \frac{L}{2}\right) = \frac{2P_0L^3}{\pi^4 EI}\left(1 - \cos \omega_1 t\right)$$

$$= \frac{P_0L^3}{48.7EI}\left(1 - \cos \omega_1 t\right)$$

The close agreement between these two computations is due to the fact that static deflections can also be expressed in terms of modal components, and for a beam supporting a concentrated load at midspan the first mode dominates both static and dynamic response.

20.7   SUMMARY

The dynamic analysis of single-span beams with distributed properties (mass and elasticity) and subjected to flexural loading was presented in this chapter. The extension of this analysis to multispan or continuous beams and other structures, though possible, becomes increasingly complex and impractical. The results obtained from these single-span beams are particularly important in evaluating approximate methods based on discrete models, as those presented in preceding chapters. From such evaluation, it has been found that the stiffness method of dynamic analysis in conjunction with the consistent mass formulation provides in general satisfactory results even with a rather coarse discretization of the structure.

The natural frequencies and corresponding normal modes of single-span beams with different supports are determined by solving the differential equation of motion and imposing the corresponding boundary conditions. The normal modes satisfy the orthogonality condition between any two modes m and n, namely,

$$\int_0^L \phi_m(x)\,\phi_n(x)\,\bar{m}\,dx = 0 \quad (m \neq n)$$

The response of a continuous system may be determined as the superposition of modal contributions, that is,

$$y(x, t) = \sum_n \phi_n(x)\,z_n(t)$$

where $z_n(t)$ is the solution of n modal equation

$$\ddot{z}(t) + 2\xi_n\omega_n\dot{z}(t) + \omega_n^2 z(t) = F_n(t)/M_n$$

in which

$$F_n(t) = \int_0^L \phi_n(x)\,p(x, t)\,dx$$

and

$$M_n = \int_0^L \bar{m}\phi_n^2(x)\, dx$$

The bending moment M and the shear for V at any section of a beam are calculated from the well-known relations

$$M = EI \frac{\partial^2 y}{\partial x^2}$$

$$V = EI \frac{\partial^3 y}{\partial x^3}$$

PROBLEMS

20.1 Determine the first three natural frequencies and corresponding modal shapes of a simply supported reinforced concrete beam having a cross section 10 in wide by 24 in deep with a span of 36 ft. Assume the flexural stiffness of the beam, $EI = 3.5 \times 10^9$ lb·in^2 and weight per unit volume $W = 150$ lb/ft^3. (Neglect shear distortion and rotary inertia.)

20.2 Solve Problem 20.1 for the beam with its two ends fixed.

20.3 Solve Problem 20.1 for the beam with one end fixed and the other simply supported.

20.4 Determine the maximum deflection at the center of the simply supported beam of Problem 20.1 when a constant force of 2000 lb is suddenly applied at 9 ft from the left support.

20.5 A simply supported beam is prismatic and has the following properties: $\bar{m} = 0.3$ lb·sec^2/in per inch of span, $EI = 10^6$ lb·in^2, and $L = 150$ in. The beam is subjected to a uniform distributed static load p_0 which is suddenly removed. Write the series expression for the resulting free vibration and determine the amplitude of the first mode in terms of p_0.

20.6 The beam of Problem 20.5 is acted upon by a concentrated force given by $P(t) = 1000 \sin 500t$ lb applied at its midspan. Determine the amplitude of the steady-state motion at a quarter point from the left support in each of the first two modes. Neglect damping.

20.7 Solve Problem 20.6 assuming 10% of critical damping in each mode. Also determine the steady-state motion at the quarter point considering the first two modes.

20.8 The cantilever beam shown in Fig. 20.5 is prismatic and has the following properties: $\bar{m} = 0.5$ lb·sec^2/in per inch of span, $E = 30 \times 10^6$ psi, $L = 100$ in, and $I = 120$ in^4. Considering only the first mode, compute the maximum deflection and the maximum dynamic bending moment in the beam due to the load time function of Fig. 20.5. (Chart in Fig. 4.5 may be used.)

20.9 A prismatic simply supported beam of the following properties: $L = 120$ in, $EI = 10^7$ lb·in^2, and $\bar{m} = 0.5$ lb·sec^2/in per inch of span is loaded as shown in Fig. 20.6. Write the series expression for the deflection at the midsection of the beam.

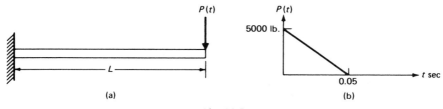

Fig. 20.5

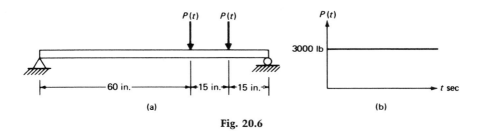

Fig. 20.6

20.10 Assuming that the forces on the beam of Problem 20.9 are applied for only a time duration $t_d = 0.1$ sec, and considering only the first mode, determine the maximum deflection at each of the load points of the beam. (Chart in Fig. 4.4 may be used.)

20.11 A prismatic beam with its two ends fixed has the following properties: $L = 180$ in, $EI = 30 \times 10^8$ lb·in², $\bar{m} = 1$ lb·sec²/in per inch of span. The beam is acted upon by a uniformly distributed impulsive force $p(x, t) = 2000 \sin 400t$ lb during a time interval equal to half of the period of the sinusoidal load function ($t_d = \pi/400$ sec). Determine the maximum deflection at the midsection. Considering only the first mode, determine the maximum deflection at the midsection. (Chart in Fig. 8.3 may be used.)

20.12 Solve Problem 20.11 considering the first two modes.

21

Discretization of Continuous Systems

The modal superposition method of analysis was applied in the preceding chapter to some simple structures having distributed properties. The determination of the response by this method requires the evaluation of several natural frequencies and corresponding mode shapes. The calculation of these dynamic properties is rather laborious, as we have seen, even for simple structures such as one-span uniform beams. The problem becomes increasingly more complicated and unmanageable as this method of solution is applied to more complex structures. However, the analysis of such structures becomes relatively simple if for each segment or element of the structure the properties are expressed in terms of dynamic coefficients much in the same manner as done previously when static deflection functions were used as an approximation to dynamic deflections in determining stiffness, mass, and other coefficients.

In this chapter the dynamic coefficients relating harmonic forces and displacements at the nodal coordinates of a beam segment are obtained from dynamic deflection functions. These coefficients can then be used to assemble the dynamic matrix for the whole structure by the direct method as shown in the preceding chapters for assembling the system stiffness and mass matrices. Also, in the present chapter, the mathematical relationship between the dynamic coefficients based on dynamic displacement functions and the coefficients of the stiffness and consistent mass matrices derived from static displacement functions is established.

21.1   DYNAMIC MATRIX FOR FLEXURAL EFFECTS

As in the case of static influence coefficients (stiffness coefficients, for example), the dynamic influence coefficients also relate forces and displacements at the nodal coordinates of a beam element. The difference between the dynamic and static coefficients is that the dynamic coefficients refer to nodal forces and displacements that vary harmonically while the static coefficients relate static forces and displacements at the nodal coordinates. The dynamic influence coefficient S_{ij} is then defined as the harmonic force of frequency $\bar{\omega}$ at nodal coordinates i, due to a harmonic displacement of a unit amplitude and of the same frequency at nodal coordinate j.

To determine the expressions for the various dynamic coefficients for a uniform beam segment as shown in Fig. 21.1, we refer to the differential equation of motion, eq. (20.5), which in the absence of external loads in the span, that is, $p(x,t) = 0$, is

$$EI \frac{\partial^4 y}{\partial x^4} + \bar{m} \frac{\partial^2 y}{\partial t^2} = 0 \tag{21.1}$$

For harmonic boundary displacements of frequency $\bar{\omega}$, we introduce in eq. (21.1) the trial solution

$$y(x,t) = \Phi(x) \sin \bar{\omega} t \tag{21.2}$$

Substitution of eq. (21.2) into eq. (21.1) yields

$$\Phi^{IV}(x) - \bar{a}^4 \, \Phi(x) = 0 \tag{21.3}$$

where

$$\bar{a}^4 = \frac{\bar{m} \bar{\omega}^2}{EI} \tag{21.4}$$

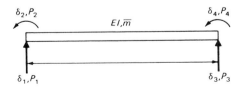

Fig. 21.1 Nodal coordinates of a flexural beam segment.

We note that eq. (21.3) is equivalent to eq. (20.10), which is the differential equation for the shape function of a beam segment in free vibration. The difference between these two equations is that eq. (21.3) is a function of the parameter $\bar{a}$ which, in turn, is a function of the forcing frequency $\bar{\omega}$, while "a" in (20.10) depends on the natural frequency ω. The solution of eq. (21.3) is of the same form as the solution of eq. (20.10). Thus by analogy with eq. (20.20), we can write

$$\Phi(x) = C_1 \sin \bar{a}x + C_2 \cos \bar{a}x + C_3 \sinh \bar{a}x + C_4 \cosh \bar{a}x \qquad (21.5)$$

Now, to obtain the dynamic coefficient for the beam segment, boundary conditions indicated by eqs. (21.6) and (21.7) are imposed:

$$\Phi(0) = \delta_1, \quad \Phi(L) = \delta_3$$

$$\Phi'(0) = \delta_2, \quad \Phi'(L) = \delta_4 \qquad (21.6)$$

Also

$$\Phi'''(0) = \frac{P_1}{EI}, \quad \Phi'''(L) = -\frac{P_3}{EI}$$

$$\Phi''(0) = -\frac{P_2}{EI}, \quad \Phi''(L) = \frac{P_4}{EI} \qquad (21.7)$$

In eqs. (21.6), δ_1, δ_2, δ_3, and δ_4 are amplitudes of linear and angular harmonic displacements at the nodal coordinates while in eqs. (21.7) P_1, P_2, P_3, and P_4 are the corresponding harmonic forces and moments as shown in Fig. 21.1. The substitution of the boundary conditions, eqs. (21.6) and (21.7), into eq. (21.5) results in

$$\begin{bmatrix} \delta_1 \\ \delta_2 \\ \delta_3 \\ \delta_4 \end{bmatrix} = \begin{bmatrix} 0 & 1 & 0 & 1 \\ \bar{a} & 0 & \bar{a} & 0 \\ s & c & S & C \\ \bar{a}c & -\bar{a}s & \bar{a}C & \bar{a}S \end{bmatrix} \begin{bmatrix} C_1 \\ C_2 \\ C_3 \\ C_4 \end{bmatrix} \qquad (21.8)$$

and

$$\begin{bmatrix} P_1 \\ P_2 \\ P_3 \\ P_4 \end{bmatrix} = EI \begin{bmatrix} -\bar{a}^3 & 0 & \bar{a}^3 & 0 \\ 0 & \bar{a}^2 & 0 & -\bar{a}^2 \\ \bar{a}^3c & -\bar{a}^3s & -\bar{a}^3C & -\bar{a}^3S \\ -\bar{a}^2s & -\bar{a}^2c & \bar{a}^2S & \bar{a}^2C \end{bmatrix} \begin{bmatrix} C_1 \\ C_2 \\ C_3 \\ C_4 \end{bmatrix} \qquad (21.9)$$

in which

$$s = \sin \bar{a}L, \quad S = \sinh \bar{a}L$$

$$c = \cos \bar{a}L, \quad C = \cosh \bar{a}L \qquad (21.10)$$

Next, eq. (21.8) is solved for the constants of integration C_1, C_2, C_3, and C_4, which are subsequently substituted into eq. (21.9). We thus obtain the dynamic matrix relating harmonic displacements and harmonic forces at the nodal coordinate of the beam segment, namely

$$\begin{bmatrix} P_1 \\ P_2 \\ P_3 \\ P_4 \end{bmatrix} = B \begin{bmatrix} \bar{a}^2(cS + sC) & & \text{Symmetric} & \\ \bar{a}sS & sC - cS & & \\ -\bar{a}^2(s + S) & \bar{a}(c - C) & \bar{a}^2(cS + sC) & \\ \bar{a}(C - c) & S - s & -\bar{a}sS & sC - cS \end{bmatrix} \begin{bmatrix} \delta_1 \\ \delta_2 \\ \delta_3 \\ \delta_4 \end{bmatrix} \qquad (21.11)$$

where

$$B = \frac{\bar{a}EI}{1 - cC} \qquad (21.12)$$

We require the denominator to be different from zero, that is,

$$1 - \cos \bar{a}L \cosh \bar{a}L \neq 0 \qquad (21.13)$$

The element dynamic matrix in eq. (21.11) can then be used to assemble the system dynamic matrix for a continuous beam or a plane frame in a manner entirely analogous to the assemblage of the system stiffness matrix from element stiffness matrices.

21.2   DYNAMIC MATRIX FOR AXIAL EFFECTS

The governing equation for axial vibration of a beam element is obtained by establishing the dynamic equilibrium of a differential element dx of the beam, as shown in Fig. 21.2. Thus

$$\left(P + \frac{\partial P}{\partial x} dx \right) - P - (\bar{m}dx) \frac{\partial^2 u}{\partial t^2} = 0$$

$$(21.14)$$

$$\frac{\partial P}{\partial x} = \bar{m} \frac{\partial^2 u}{\partial t^2}$$

where u is the displacement at x. The displacement at $x + dx$ will then be $u + (\partial u/\partial x)\, dx$. It is evident that the element dx in the new position has changed length by an amount $(\partial u/\partial x)dx$, and thus the strain is $\partial u/\partial x$. Since from Hooke's law the ratio of stress to strain is equal to the modulus of elasticity E, we can write

$$\frac{\partial u}{\partial x} = \frac{P}{AE} \qquad (21.15)$$

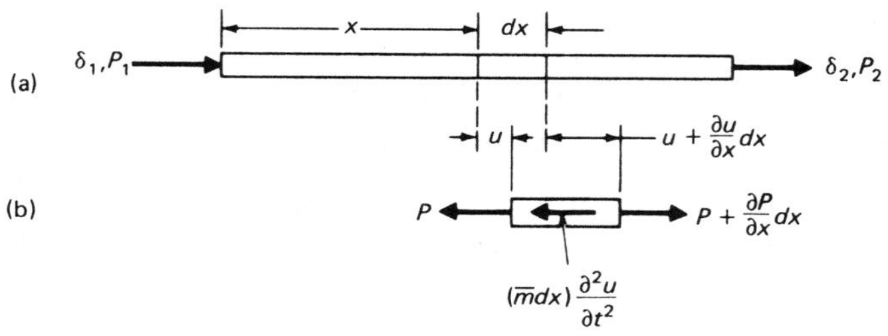

Fig. 21.2 Axial effects on a beam. (a) Nodal axial coordinates. (b) Forces acting on a differential element.

where A is the cross-sectional area of the beam. Differentiating with respect to x results in

$$AE \frac{\partial^2 u}{\partial x^2} = \frac{\partial P}{\partial x} \tag{21.16}$$

and combining eqs. (21.14) and (21.16) yields the differential equation for axial vibration of a beam segment, namely,

$$\frac{\partial^2 u}{\partial x^2} - \frac{\bar{m}}{AE} \frac{\partial^2 u}{\partial t^2} = 0 \tag{21.17}$$

A solution of eq. (21.17) of the form

$$u(x,t) = U(x) \sin \bar{\omega} t \tag{21.18}$$

will result in a harmonic motion of amplitude

$$U(x) = C_1 \sin bx + C_2 \cos bx \tag{21.19}$$

where

$$b = \sqrt{\frac{\bar{m} \bar{\omega}^2}{AE}} \tag{21.20}$$

and C_1 and C_2 are constants of integration.

To obtain the dynamic matrix for the axially vibrating beam segment, boundary conditions indicated by eqs. (21.21) and (21.22) are imposed, namely,

$$U(0) = \delta_1, \qquad U(L) = \delta_2 \tag{21.21}$$

$$U'(0) = -\frac{P_1}{AE}, \qquad U'(L) = \frac{P_2}{AE} \tag{21.22}$$

where δ_1 and δ_2 are the displacements and P_1 and P_2 are the forces at the nodal coordinates of the beam segment as shown in Fig. 21.2.

Substitution of the boundary conditions, eqs. (21.21) and (21.22), into eq. (21.19) results in

$$\begin{bmatrix} \delta_1 \\ \delta_2 \end{bmatrix} = \begin{bmatrix} 0 & 1 \\ \sin bL & \cos bL \end{bmatrix} \begin{bmatrix} C_1 \\ C_2 \end{bmatrix} \qquad (21.23)$$

and

$$\begin{bmatrix} P_1 \\ P_2 \end{bmatrix} = AEb \begin{bmatrix} -1 & 0 \\ \cos bL & -\sin bL \end{bmatrix} \begin{bmatrix} C_1 \\ C_2 \end{bmatrix} \qquad (21.24)$$

Then, solving eq. (21.23) for the constants of integration, we obtain

$$\begin{bmatrix} C_1 \\ C_2 \end{bmatrix} = \begin{bmatrix} -\cot bL & \operatorname{cosec} bL \\ 1 & 0 \end{bmatrix} \begin{bmatrix} \delta_1 \\ \delta_2 \end{bmatrix} \qquad (21.25)$$

subject to the condition

$$\sin bL \neq 0 \qquad (21.26)$$

Finally, the substitution of eq. (21.25) into eq. (21.24) results in eq. (21.27) relating harmonic forces and displacement at the nodal coordinates through the dynamic matrix for an axially vibrating beam segment. Thus we have

$$\begin{bmatrix} P_1 \\ P_2 \end{bmatrix} = EAb \begin{bmatrix} \cot bL & -\operatorname{cosec} bL \\ -\operatorname{cosec} bL & \cot bL \end{bmatrix} \begin{bmatrix} \delta_1 \\ \delta_2 \end{bmatrix} \qquad (21.27)$$

21.3   DYNAMIC MATRIX FOR TORSIONAL EFFECTS

The equation of motion of a beam segment in torsional vibration is similar to that of the axial vibration of beams discussed in the preceding section. Let x (Fig. 21.3) be measured along the length of the beam. Then the angle of twist for any element of length dx of the beam due to a torque T is

$$d\theta = \frac{T dx}{J_T G} \qquad (21.28)$$

where $J_T G$ is the torsional stiffness given by the product of the torsional constant J_T (J_T is the polar moment of inertia for circular sections) and the shear modulus of elasticity G. The torque applied on the faces of the element are T and $T + (\partial T/\partial x)dx$ as shown in Fig. 21.3. From eq. (21.28), the net torque is then

$$\frac{\partial T}{\partial x} dx = J_T G \frac{\partial^2 \theta}{\partial x^2} dx \qquad (21.29)$$

Equating this torque to the product of the mass moment of inertia $I_{\bar{m}} dx$ of the element dx and the angular acceleration $\partial^2 \theta/\partial t^2$, we obtain the differential

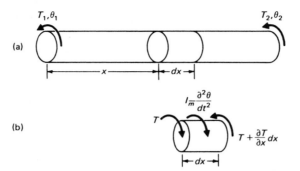

Fig. 21.3 Torsional effects on a beam. (a) Nodal torsional coordinates. (b) Moments acting on a differential element.

equation of motion

$$J_T G \frac{\partial^2 \theta}{\partial x^2} \, dx = I_{\bar{m}} \frac{\partial^2 \theta}{\partial t^2} \, dx$$

or

$$\frac{\partial^2 \theta}{\partial x^2} - \frac{I_{\bar{m}}}{J_T G} \frac{\partial^2 \theta}{\partial t^2} = 0 \qquad (21.30)$$

where $I_{\bar{m}}$ is the mass moment of inertia per unit length about the longitudinal axis x given by

$$I_{\bar{m}} = \bar{m} \frac{I_0}{A} \qquad (21.31)$$

in which I_0 is the polar moment of inertia of the cross-sectional area A.

We seek a solution of eq. (21.30) in the form

$$\theta(x, t) = \theta(x) \sin \bar{\omega} t$$

which, upon substitution into eq. (21.30), results in a harmonic torsional motion of amplitude

$$\theta(x) = C_1 \sin cx + C_2 \cos cx \qquad (21.32)$$

in which

$$c = \sqrt{\frac{I_{\bar{m}} \bar{\omega}^2}{J_T G}} \qquad (21.33)$$

For a circular section, the torsional constant J_T is equal to the polar moment of inertia I_0. Thus eq. (21.33) reduces to

$$c = \sqrt{\frac{\bar{m}\bar{\omega}^2}{AG}} \qquad (21.34)$$

since $I_{\bar{m}} = I_0 \bar{m}/A$ as indicated by eq. (21.31).

We note that eq. (21.30) for torsional vibration is analogous to eq. (21.17) for axial vibration of beam segments. It follows that by analogy to eq. (21.27) we can write the dynamic relation between torsional moments and rotations in a beam segment. Hence

$$\begin{bmatrix} T_1 \\ T_2 \end{bmatrix} = J_T G c \begin{bmatrix} \cot cL & -\csc cL \\ -\csc cL & \cot cL \end{bmatrix} \begin{bmatrix} \theta_1 \\ \theta_2 \end{bmatrix} \qquad (21.35)$$

21.4   BEAM FLEXURE INCLUDING AXIAL-FORCE EFFECT

When a beam is subjected to a force along its longitudinal axis in addition to lateral loading, the dynamic equilibrium equation for a differential element of the beam is affected by the presence of this force. Consider the beam shown in Fig. 21.4 in which the axial force is assumed to remain constant during flexure with respect to both magnitude and direction. The dynamic equilibrium for a differential element dx of the beam [Fig. 21.4(b)] is established by equating to zero both the sum of the forces and the sum of the moments.

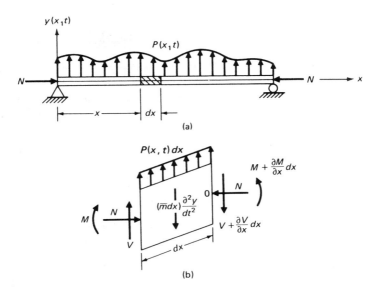

Fig. 21.4 Beam supporting constant axial force and lateral dynamic load. (a) Loaded beam. (b) Forces acting on a differential element.

Summing forces in the y direction, we obtain

$$V + p(x, t)\, dx - \left(V + \frac{\partial V}{\partial x}\, dx\right) - (\bar{m}\, dx)\frac{\partial^2 y}{\partial t^2} = 0 \tag{21.36}$$

which, upon reduction, yields

$$\frac{\partial V}{\partial x} + \bar{m}\,\frac{\partial^2 y}{\partial t^2} = p(x, t) \tag{21.37}$$

The summation of moments about point 0 gives

$$M + V\, dx - \left(M + \frac{\partial M}{\partial x}\, dx\right) + \frac{1}{2}\left(p(x, t) - \bar{m}\,\frac{\partial^2 y}{\partial t^2}\right)dx^2 - N\,\frac{\partial y}{\partial x}\, dx = 0 \tag{21.38}$$

Discarding higher order terms, we obtain for the shear force the expression

$$V = N\,\frac{\partial y}{\partial x} + \frac{\partial M}{\partial x} \tag{21.39}$$

Then using the familiar relation from bending theory,

$$M = EI\,\frac{\partial^2 y}{\partial x^2} \tag{21.40}$$

and combining eqs. (21.37), (21.39), and (21.40), we obtain the equation of motion of a beam segment including the effect of the axial forces, that is,

$$EI\,\frac{\partial^4 y}{\partial x^4} + N\,\frac{\partial^2 y}{\partial x^2} + \bar{m}\,\frac{\partial^2 y}{\partial t^2} = p(x, t) \tag{21.41}$$

A comparison of eqs. (21.41) and (20.5) reveals that the presence of the axial force gives rise to an additional transverse force acting on the beam. As indicated previously in Section 20.1, in the derivation of eq. (21.41) it has been assumed that the deflections are small and that the deflections due to shear forces or rotary inertia are negligible.

In the absence of external loads applied to the span of the beam, eq. (21.41) reduces to

$$EI\,\frac{\partial^4 y}{\partial x^4} + N\,\frac{\partial^2 y}{\partial x^2} + \bar{m}\,\frac{\partial^2 y}{\partial t^2} = 0 \tag{21.42}$$

The solution of eq. (21.42) is found as before by substituting

$$y(x, t) = \Phi(x) \sin \bar{\omega} t \tag{21.43}$$

We thereby obtain the ordinary differential equation

$$\frac{d^4\Phi}{dx^4} + \frac{N}{EI}\frac{d^2\Phi}{dx^2} - \frac{\bar{m}\bar{\omega}^2}{EI}\,\Phi = 0 \tag{21.44}$$

The solution of eq. (21.44) is

$$\Phi(x) = A \sin p_2 x + B \cos p_2 x + C \sinh p_1 x + D \cosh p_1 x \qquad (21.45)$$

where A, B, C, and D are constants of integration and

$$p_1 = \sqrt{\dfrac{-\alpha}{2} + \sqrt{\left(\dfrac{\alpha}{2}\right)^2 + \beta}}$$

$$p_2 = \sqrt{\dfrac{\alpha}{2} + \sqrt{\left(\dfrac{\alpha}{2}\right)^2 + \beta}} \qquad (21.46)$$

$$\alpha = \dfrac{N}{EI} \qquad (21.47)$$

$$\beta = \dfrac{\bar{m}\bar{\omega}^2}{EI} \qquad (21.48)$$

To obtain the dynamic matrix (which in this case includes the effect of axial forces) for the transverse vibration of the beam element, the boundary conditions, eqs. (21.49), are imposed, namely

$$\Phi(0) = \delta_1, \quad \Phi(L) = \delta_3$$

$$\frac{d\Phi(0)}{dx} = \delta_2, \quad \frac{d\Phi(L)}{dx} = \delta_4$$

$$\frac{d^3\Phi(0)}{dx^3} = \frac{P_1}{EI} - \frac{N}{EI}\delta_2, \quad \frac{d^3\Phi(L)}{dx^3} = -\frac{P_3}{EI} - \frac{N}{EI}\delta_4$$

$$\frac{d^2\Phi(0)}{dx^2} = -\frac{P_2}{EI}, \quad \frac{d^2\Phi(L)}{dx^2} = \frac{P_4}{EI} \qquad (21.49)$$

In eqs. (21.49) δ_1, δ_3 and δ_2, δ_4 are, respectively, the transverse and angular displacements at the ends of the beam, while P_1, P_3 and P_2, P_4 are corresponding forces and moments at these nodal coordinates. The substitution into eq. (21.45) of the boundary conditions given by eqs. (21.49) results in a system of eight algebraic equations which upon elimination of the four constants of integration A, B, C, and D yields the dynamic matrix (including the effect of axial forces) relating harmonic forces and displacements at the nodal coordinates of a beam segment. The final result is

$$\begin{bmatrix} P_1 \\ P_2 \\ P_3 \\ P_4 \end{bmatrix} = \begin{bmatrix} S_{11} & & \text{Symmetric} & \\ S_{21} & S_{22} & & \\ S_{31} & S_{32} & S_{33} & \\ S_{41} & S_{42} & S_{43} & S_{44} \end{bmatrix} \begin{bmatrix} \delta_1 \\ \delta_2 \\ \delta_3 \\ \delta_4 \end{bmatrix} \qquad (21.50)$$

where

$$S_{11} = S_{33} = B[(p_1^2 p_2^3 + p_1^4 p_2)cS + (p_1 p_2^4 + p_1^3 p_2^2)sC]$$

$$S_{21} = -S_{43} = B[(p_1 p_2^3 - p_1^3 p_2) + (p_1^3 p_2 - p_1 p_2^3)cC + 2p_1^2 p_2^2 sS]$$

$$S_{22} = S_{44} = B[(p_2^2 p_1 + p_1^3)sC - (p_2^3 + p_1^2 p_2)cS]$$

$$S_{41} = -S_{32} = B[(p_1 p_2^3 + p_1^3 p_2)(C - c)]$$

$$S_{31} = B[(-p_1^2 p_2^3 - p_1^4 p_2)S - (p_1^3 p_2^2 + p_1 p_2^4)s]$$

$$S_{42} = B[(p_1^2 p_2 + p_2^3)S - (p_1 p_2^2 + p_1^3)s] \tag{21.51}$$

In the above, the letters s, c, S, and C denote

$$s = \sin p_2 L, \quad S = \sinh p_1 L$$

$$c = \cos p_2 L, \quad C = \cosh p_1 L$$

and the letter B denotes

$$B = \frac{EI}{2p_1 p_2 - 2p_1 p_2 cC + (p_1^2 - p_2^2)sS} \tag{21.52}$$

Furthermore, eq. (21.50) is subject to the condition

$$2p_1 p_2 - 2p_1 p_2 cC + (p_1^2 - p_2^2)sS \neq 0 \tag{21.53}$$

21.5   POWER SERIES EXPANSION OF THE DYNAMIC MATRIX FOR FLEXURAL EFFECTS

It is of interest to demonstrate that the influence coefficients of the stiffness matrix, eq. (14.20), and of the consistent mass matrix, eq. (14.34), may be obtained by expanding the influence coefficients of the dynamic matrix in a Taylor's series (Paz 1973). For the sake of the discussion, we consider the dynamic coefficient from the second row and first column of the dynamic matrix, eq. (21.11),

$$S_{21} = \frac{\bar{a}^2 EI \sin \bar{a}L \sinh \bar{a}L}{1 - \cos \bar{a}L \cosh \bar{a}L} \tag{21.54}$$

In the following derivation, operations with power series, including addition, subtraction, multiplication, and division, are employed. The validity of these operations and convergence of the resulting series is proved in Knopp.[1] In general, convergent power series may be added, subtracted, or multiplied and the resulting series will converge at least in the common interval of con-

[1]Knopp, K., *Theory and Application of Infinite Series*, Blackie, London, 1963.

vergence of the two original series. The operation of division of two power series may be carried out formally; however, the determination of the radius of convergence of the resulting series is more complicated. It requires the use of theorems in the field of complex variables and it is related to analytical continuation. Very briefly, it can be said that the power series obtained by division of two convergent power series about a complex point Z_0 will be convergent in a circle with center Z_0 and of radius given by the closest singularity to Z_0 of the functions represented by the series in the numerator and denominator.

The known expansions in power series about the origin of trigonometric and hyperbolic functions are used in the intermediate steps in expanding the function in eq. (21.54), namely,

$$\cos x \cosh x = 1 - \frac{x^4}{6} + \frac{x^8}{2520} - \frac{x^{12}}{7,484,400} + \cdots$$

$$(1 - \cos x \cosh x)^{-1} = \frac{6}{x^4} + \frac{1}{70} + \frac{85x^4}{2,910,600} + \cdots$$

$$\sin x \sinh x = x^2 - \frac{x^6}{90} + \frac{x^{10}}{113,400} - \cdots$$

where $x = \bar{a}L$. Substitution of these series equations in the dynamic coefficient, eq. (21.54), yields

$$S_{21} = \frac{\bar{a}^2 EI \sin \bar{a}L \sinh \bar{a}L}{1 - \cos \bar{a}L \cosh \bar{a}L} = \frac{6EI}{L^2} - \frac{11\bar{m}L^2\bar{\omega}^2}{210} - \frac{223\bar{m}^2 L^6 \bar{\omega}^4}{2,910,600EI} - \cdots \qquad (21.55)$$

The first term on the right-hand side of eq. (21.55) is the stiffness coefficient k_{21} in the stiffness matrix, eq. (14.20), and the second term, the consistent mass coefficient m_{21} in the mass matrix, eq. (14.34). The series expansion, eq. (21.55), is convergent in the positive real field for

$$0 < \bar{a}L < 4.73 \qquad (21.56)$$

or from eq. (21.4)

$$0 < \bar{\omega} < (4.73)^2 \sqrt{\frac{EI}{\bar{m}L^4}} \qquad (21.57)$$

In eq. (21.56) the numerical value 4.73 is an approximation of the closest singularity to the origin of the functions in the quotient expanded in eq. (21.54).

The series expansions for all the coefficients in the dynamic matrix, eq. (21.11), are obtained by the method explained in obtaining the expansion of the coefficient S_{21}. These series expansions are:

$$S_{33} = S_{11} = \frac{12EI}{L^3} - \frac{13L\bar{m}\bar{\omega}^2}{35} - \frac{59L^5\bar{m}^2\bar{\omega}^4}{161,700EI} - \cdots$$

$$S_{21} = -S_{43} = \frac{6EI}{L^2} - \frac{11L^2\bar{m}\bar{\omega}^2}{210} - \frac{223L^6\bar{m}^2\bar{\omega}^4}{2,910,600EI} - \cdots$$

$$S_{41} = -S_{32} = \frac{6EI}{L^2} + \frac{13L^2\bar{m}\bar{\omega}^2}{420} + \frac{1681L^6\bar{m}^2\bar{\omega}^4}{23,284,800EI} - \cdots$$

$$S_{22} = S_{44} = \frac{4EI}{L} - \frac{L^3\bar{m}\bar{\omega}^2}{105} - \frac{71L^7\bar{m}^2\bar{\omega}^4}{4,365,900EI} - \cdots$$

$$S_{31} = -\frac{12EI}{L^3} - \frac{9L\bar{m}\bar{\omega}^2}{70} - \frac{1279L^5\bar{m}^2\bar{\omega}^4}{3,880,800EI} - \cdots$$

$$S_{42} = \frac{2EI}{L} + \frac{L^3\bar{m}\bar{\omega}^2}{140} + \frac{1097L^7\bar{m}^2\bar{\omega}^4}{69,854,400EI} - \cdots \tag{21.58}$$

21.6 POWER SERIES EXPANSION OF THE DYNAMIC MATRIX FOR AXIAL AND FOR TORSIONAL EFFECTS

Proceeding in a manner entirely analogous to expansion of the dynamic coefficients for flexural effects, we can also expand the dynamic coefficients for axial and for torsional effects. The Taylor's series expansions, up to three terms, of the coefficients of the dynamic matrix in eq. (21.27) (axial effects) are

$$AEb \cot bL = \frac{AE}{L} - \frac{\bar{m}\bar{\omega}^2L}{3} - \frac{L^3\bar{m}^2\bar{\omega}^4}{45AE} - \cdots$$

$$-AEb \operatorname{cosec} bL = -\frac{AE}{L} - \frac{\bar{m}\bar{\omega}^2L}{6} - \frac{7L^3\bar{m}^2\bar{\omega}^4}{300AE} - \cdots \tag{21.59}$$

It may be seen that the first term in each series of eq. (21.59) is equal to the corresponding stiffness coefficient of the matrix in eq. (15.3), and the second term to the consistent mass coefficient of the matrix in (15.26). Similarly, the Taylor's series expansions of the coefficients of the dynamic matrix for torsional effects, eq. (21.35), are

$$J_T Gc \cot cL = \frac{J_T G}{L} - \frac{LI_{\bar{m}}\bar{\omega}^2}{3} - \frac{L^3 I_{\bar{m}}^2\bar{\omega}^4}{45GJ_T} - \cdots$$

$$-J_T Gc \operatorname{cosec} cL = -\frac{J_T G}{L} - \frac{LI_{\bar{m}}\bar{\omega}^2}{6} - \frac{7L^3 I_{\bar{m}}^2\bar{\omega}^4}{300GJ_T} - \cdots \tag{21.60}$$

Comparing the first two terms of the above series with the stiffness and mass influence coefficients of the matrices in eqs. (16.7) and (16.8), we find that for torsional effects the first term is also equal to the stiffness coefficient, and the second term to the consistent mass coefficient.

21.7   POWER SERIES EXPANSION OF THE DYNAMIC MATRIX INCLUDING THE EFFECT OF AXIAL FORCES

The series expansions of the coefficients of the dynamic matrix, eq. (21.50) (with axial effects), are obtained by the method described in the last two sections. Detailed derivation of these expansions are given by Paz and Dung (1975). The series expansion of the dynamic matrix, eq. (21.50), is

$$[S] = [K] - [G_0]N - [M_0]\,\bar{\omega}^{\,2} - [A_1]\,N\bar{\omega}^{\,2} - [G_1]\,N^2 - [M_1]\,\bar{\omega}^{\,4} - \cdots \quad (21.61)$$

where the first three matrices in this expansion $[K]$, $[G_0]$, and $[M_0]$ are, respectively, the stiffness, geometric, and mass matrices which were obtained in previous chapters on the basis of static displacement functions. These matrices are given, respectively, by eqs. (14.20), (14.45), and (14.34). The other matrices in eq. (21.61) corresponding to higher order terms are represented as follows. The second-order mass-geometrical matrix:

$$[A_1] = \frac{\bar{m}L^3}{EI} \begin{bmatrix} \dfrac{1}{3150} & & \text{Symmetric} & & \\[2mm] \dfrac{L}{1260} & \dfrac{L^2}{3150} & & & \\[2mm] -\dfrac{1}{3150} & \dfrac{L}{1680} & \dfrac{1}{3150} & & \\[2mm] -\dfrac{L}{1680} & \dfrac{L^2}{3600} & -\dfrac{L}{1260} & \dfrac{L^2}{3150} \end{bmatrix}$$

The second-order geometrical matrix:

$$[G_1] = \frac{1}{EI} \begin{bmatrix} \dfrac{L}{700} & & \text{Symmetric} & & \\[2mm] \dfrac{L^2}{1400} & \dfrac{11}{6300}L^3 & & & \\[2mm] -\dfrac{L}{700} & -\dfrac{1}{1400}L^2 & \dfrac{L}{700} & & \\[2mm] \dfrac{L^2}{1400} & -\dfrac{13L^3}{12600} & -\dfrac{1}{1400}L^2 & \dfrac{11L^3}{6300} \end{bmatrix}$$

The second-order mass matrix:

$$[M_1] = \frac{\bar{m}^2 L^3}{1000EI} \begin{bmatrix} \dfrac{59}{161.7} & & \text{Symmetric} & & \\[2mm] \dfrac{223L}{2910.6} & \dfrac{71L^2}{4365.9} & & & \\[2mm] \dfrac{1279}{3880.8} & \dfrac{1681L}{23284.8} & \dfrac{59}{161.7} & & \\[2mm] \dfrac{1681L}{23284.8} & -\dfrac{1097L^2}{69854.4} & -\dfrac{223L}{2910.6} & \dfrac{71L^2}{4365.9} \end{bmatrix}$$

21.8 SUMMARY

The dynamic coefficients relating harmonic forces and displacements at the nodal coordinates of a beam segment were obtained from dynamic deflection equations. These coefficients can then be used in assembling the dynamic matrix for the entire structure by the same procedure (direct method) employed in assembling the stiffness and mass matrices for discrete systems.

In this chapter it has been demonstrated that the stiffness, consistent mass, and other influence coefficients may be obtained by expanding the dynamic influence coefficients in Taylor's series. This mathematical approach also provides higher order influence coefficients and the determination of the radius of convergence of the series expansion.

PART V

Random Vibration

22

Random Vibration

The previous chapters of this book have dealt with the dynamic analysis of structures subjected to excitations which were known as a function of time. Such an analysis is said to be *deterministic*. When an excitation function applied to a structure has an irregular shape which is described indirectly by statistical means, we speak of a *random vibration*. Such a function is usually described as a continuous or discrete function of the exciting frequencies in a manner similar to the description of a function by Fourier series. In structural dynamics, the random excitations most often encountered are either motion transmitted through the foundation or acoustic pressure. Both of these types of loading are usually generated by explosions occurring in the vicinity of the structure. Common sources of these explosions are construction work and mining. Other types of loading, such as earthquake excitation, may also be considered a random function of time. In these cases the structural response is obtained in probabilistic terms using random vibration theory.

A record of random vibration is a time function such as shown in Fig. 22.1. The main characteristic of such a random function is that its instantaneous value cannot be predicted in a deterministic sense. The description and anal-

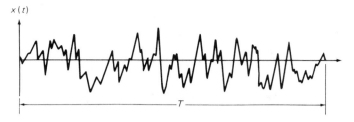

Fig. 22.1 Record of a random function of time.

ysis or random processes are established in a probabilistic sense for which it is necessary to use tools provided by the theory of statistics.

22.1   STATISTICAL DESCRIPTION OF RANDOM FUNCTIONS

In any statistical method a large number of responses is needed to describe a random function. For example, to establish the statistics of the foundation excitation due to explosions in the vicinity of a structure, many records of the type shown in Fig. 22.2 may be needed. Each record is called a *sample*, and the total collection of samples the *ensemble*. To describe an ensemble statistically, we can compute at any time t_i the average value of the instan-

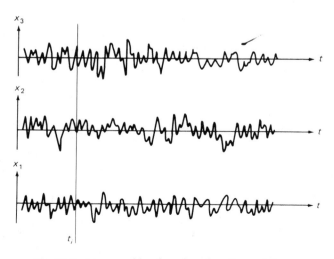

Fig. 22.2 An ensemble of random functions of time.

taneous displacements x_i. If such averages do not differ as we select different values of t_i, then the random process is said to be *stationary*. In addition, if the average obtained with respect to time for any member of the ensemble is equal to the average across the ensemble at an arbitrary time t_i, the random process is called *ergodic*. Thus in a stationary, ergodic process, a single record may be used to obtain the statistical description of a random function. We shall assume that all random processes considered are stationary and ergodic. The random function of time shown in Fig. 22.1 has been recorded during an interval of time T. Several averages are useful in describing such a random function. The most common are the *mean value* $\bar{x}$ which is defined as

$$\bar{x} = \frac{1}{T} \int_0^T x(t)\, dt \tag{22.1}$$

and the *mean-square* value $\overline{x^2}$ defined as

$$\overline{x^2} = \frac{1}{T} \int_0^T x^2(t)\, dt \tag{22.2}$$

Both the mean and the mean-square values provide measurements for the average value of the random function $x(t)$. The measure of how widely the function $x(t)$ differs from the average is given by its *variance*, namely,

$$\sigma_x^2 = \frac{1}{T} \int_0^T [x(t) - \bar{x}]^2\, dt \tag{22.3}$$

When the expression under the integral is expanded and then integrated, we find that

$$\sigma_x^2 = \overline{x^2} - (\bar{x})^2 \tag{22.4}$$

which means that the variance can be calculated as the mean-square minus the square of the mean. Quite often the mean value is zero, in which case variance is equal to the mean square value. The *root mean-square* RMS_x of the random function $x(t)$ is defined as

$$RMS_x = \sqrt{\overline{x^2}} \tag{22.5}$$

The *standard deviation* σ_x of $x(t)$ is the square root of the variance; hence from eq. (22.4)

$$\sigma_x = \sqrt{\overline{x^2} - (\bar{x})^2} \tag{22.6}$$

Example 22.1.   Determine the mean value F, the mean-square value $\overline{F^2}$, the variance σ_F^2, and the root mean square values RMS_F of the forcing function $F(t)$ shown in Fig. 22.3.

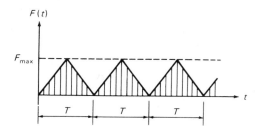

Fig. 22.3 Forcing function for Example 22.1.

Solution: Since the force $F(t)$ is periodic with period T, we can take the duration of the force equal to T; hence by eq. (22.1) we have

$$\bar{F} = \frac{1}{T} \int_0^T F(t)\, dt = \frac{F_{max}}{2}$$

and noting that

$$F(t) = \frac{2F_{max}}{T} t \quad \text{for} \quad 0 < t < \frac{T}{2}$$

we obtain by eq. (22.2)

$$\overline{F^2} = \frac{2}{T} \int_0^{T/2} \left(\frac{2F_{max}}{T}\right)^2 t^2\, dt = \frac{F_{max}^2}{3}$$

The variance may now be calculated from eq. (22.4) as

$$\sigma_F^2 = \overline{F^2} - (\bar{F})^2$$

$$\sigma_F^2 = \frac{F_{max}^2}{3} - \frac{F_{max}^2}{4} = \frac{F_{max}^2}{12}$$

Finally the root mean square value of $F(t)$ is

$$\text{RMS}_F = \sqrt{\overline{F^2}} = \frac{F_{max}}{\sqrt{3}}$$

22.2   THE NORMAL DISTRIBUTION

Figure 22.4 shows a portion of a record of a random function $x(t)$. If we wish to determine the probability of x having a value in the range (x_1, x_2), we may draw horizontal lines through the values x_1 and x_2, and then measure the corresponding time intervals Δt_i. The ratio indicated by

$$P(x_1 \leq x \leq x_2) = \frac{\Delta t_1 + \Delta t_2 + \cdots + \Delta t_n}{T} \tag{22.7}$$

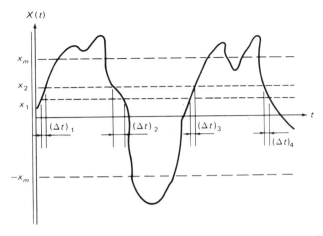

Fig. 22.4 Portion of random record showing determination of probabilities.

and calculated for the entire record length T, is the probability of x having a value between x_1 and x_2 at any selected time t_i during the random process. A somewhat similar question involving the probability that the magnitude of x, that is $|x|$, will be larger than some given value x_m may be answered by drawing lines at levels x_m and $-x_m$ and measuring the corresponding time intervals for which $|x| > x_m$ (see Fig. 22.4). The sum of these time intervals divided by the total time of record T is the probability which is designated by $P(|x| > x_m)$.

In general, the probability that a random variable $x(t)$ has a value between x and $x + dx$ is given by $p(x)\ dx$, where $p(x)$ is the *probability density function*. Having prescribed $p(x)$, for example as the function plotted in Fig. 22.5, the probability of x being in the range (x_1, x_2) at any selected time is given by

$$P(x_1 \leq x \leq x_2) = \int_{x_1}^{x_2} p(x)\ dx \qquad (22.8)$$

and is equal to the shaded area shown between x_1 and x_2 in Fig. 22.5. Similarly, the probability of x being greater than x_m, that is, $P(|x| > x_m)$ can be represented as the two shaded "tail" areas in Fig. 22.5. Since every real x lies in the interval $(-\infty, \infty)$, the area under the entire probability density function is equal to 1, that is,

$$\int_{-\infty}^{\infty} p(x)\ dx = 1 \qquad (22.9)$$

Thus as x tends to infinity in either direction, $p(x)$ must asymptotically diminish to zero.

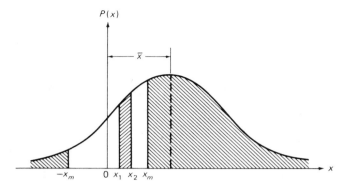

Fig. 22.5 Normal probability density function.

The most commonly used probability density function is the *normal distribution*, also referred to as the *Gaussian distribution*, expressed by

$$p(x) = \frac{1}{\sqrt{2\pi}\sigma} e^{-1/2(x-\bar{x})^2/\sigma^2} \tag{22.10}$$

Figure 22.5 shows the shape of this function. It may be observed that the normal distribution function is symmetric about the mean value $\bar{x}$. In Fig. 22.6 the standard normal distribution is plotted nondimensionally in terms of $(x - \bar{x})/\sigma$. Values of $P(-\infty, x_2)$ [$x_1 = -\infty$ in eq. (22.8)] are tabulated in many sources including mathematical handbooks.[1] The probability of x being between $\bar{x} - \lambda\sigma$ and $\bar{x} + \lambda\sigma$, where λ is any positive number, is given by the equation

$$P[\bar{x} - \lambda\sigma < x < \bar{x} + \lambda\sigma] = \frac{1}{\sqrt{2\pi}\sigma} \int_{\bar{x}-\lambda\sigma}^{\bar{x}+\lambda\sigma} e^{-1/2(x-\bar{x})^2/\sigma^2} \, dx \tag{22.11}$$

[1]*Standard Mathematical Tables*, The Chemical Rubber Co. (CRC) 20th Ed., 1972, pp. 566–575.

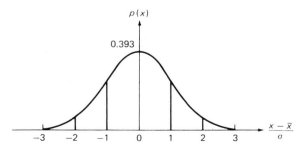

Fig. 22.6 Standard normal probability density function.

Equation (22.11) represents the probability that x lies within λ standard deviations from $\bar{x}$. The probability of x lying more than λ standard deviations from $\bar{x}$ is the probability of $|x - \bar{x}|$ exceeding $\lambda\sigma$, which is 1.0 minus the value given by eq. (22.11). The following table presents numerical values for the normal distribution associated with $\lambda = 1$, 2, and 3:

λ	$P[\bar{x} - \lambda\sigma < x < \bar{x} + \lambda\sigma]$	$P[|x - \bar{x}| > \lambda\sigma]$
1	68.3%	31.7%
2	95.4%	4.6%
3	99.7%	0.3%

22.3   THE RAYLEIGH DISTRIBUTION

Random variables which are positive, such as the absolute value A of the peaks of vibration, often tend to follow the Rayleigh distribution, which is defined by the equation

$$p(A) = \frac{A}{\alpha^2} e^{-A^2/2\alpha^2}, \quad A > 0 \tag{22.12}$$

The probability density $p(A)$ is zero for $A < 0$ and has the shape shown in Fig. 22.7 for positive values A.

The mean and mean square values for the Rayleigh distribution function can be found as

$$\bar{A} = \int_0^\infty A p(A)\, dA = \int_0^\infty \frac{A^2}{\alpha^2} e^{-A^2/2\alpha^2}\, dA = \sqrt{\frac{\pi}{2}}\, \alpha$$

$$\overline{A^2} = \int_0^\infty A^2 p(A)\, dA = \int_0^\infty \frac{A^3}{\alpha^2} e^{-A^2/2\alpha^2}\, dA = 2\alpha^2$$

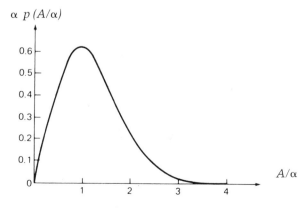

Fig. 22.7 Rayleigh probability density function.

The variance associated with the Rayleigh distribution function is, by eq. (22.4),

$$\sigma_A^2 = \overline{A^2} - (\bar{A})^2 = \frac{4 - \pi}{2}\alpha^2 = 0.429\alpha^2 \qquad (22.13)$$

Also, the probability of A exceeding a specified value $\lambda\sigma_A$ is

$$P[A > \lambda\sigma_A] = \int_{\lambda\sigma_A}^{\infty} \frac{A}{\alpha^2} e^{-A^2/2\alpha^2} \, dA \qquad (22.14)$$

which has the following numerical values:

λ	$P[A > \lambda\sigma_A]$
0	100%
1	80.7%
2	42.4%
3	14.5%
4	3.2%

22.4   CORRELATION

Correlation is a measure of the dependence between two random processes. Consider the two records shown in Fig. 22.8. The *correlation* between them is calculated by multiplying the coordinates of these two records at each time t_i and computing the average over all values of t. It is evident that the correlation so found will be larger when the two records are similar. For dissimilar records with mean zero, some products will be positive and others negative. Hence their average product will approach zero.

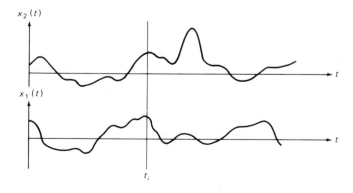

Fig. 22.8 Correlation between $x_1(t)$ and $x_2(t)$.

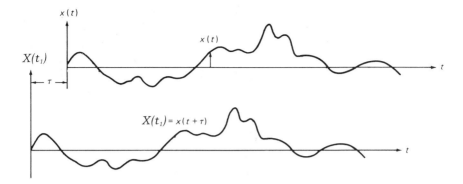

Fig. 22.9 Autocorrelation between $x(t)$ and $x(t + \tau)$.

We consider now the two records shown in Fig. 22.9 in which $x_1(t)$ is identical to $x(t)$ but shifted to the left in the amount τ, that is, $x_1(t) = x(t + \tau)$. The correlation between $x(t)$ and $x_1(t) = x(t + \tau)$ is known as the *autocorrelation* $R(\tau)$ and is given by

$$R(\tau) = \lim_{T \to \infty} \frac{1}{T} \int_0^T x(t)x(t + \tau)\, dt \tag{22.15}$$

When $\tau = 0$, the autocorrelation reduces to the mean square value, that is,

$$R(0) = \lim_{T \to \infty} \frac{1}{T} \int_0^T [x(t)]^2\, dt = \overline{x^2} \tag{22.16}$$

Since the second record of Fig. 22.9 can be considered to be delayed with respect to the first record, or the first record advanced with respect to the second record, it is evident that $R(\tau) = R(-\tau)$ is symmetric about the R axis and that $R(\tau)$ is always less than $R(0)$.

Highly random functions such as the one shown in Fig. 22.10(a) lose their similarity within a short time shift. The autocorrelation of such a function,

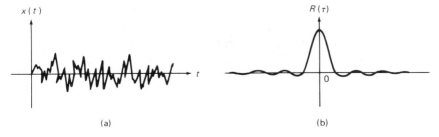

(a)   (b)

Fig. 22.10 Wide-band random process $x(t)$ and its autocorrelation function $R(\tau)$.

therefore, is a sharp spike at $\tau = 0$ that drops off rapidly as τ moves away from zero, as shown in Fig. 22.10(b). For the narrow-band record containing a dominant frequency as shown in Fig. 22.11(a), the autocorrelation has the characteristics indicated in Fig. 22.11(b) in that it is a symmetric function with a maximum at $\tau = 0$ and frequency ω_0 corresponding to the dominant frequency of $x(t)$.

22.5   THE FOURIER TRANSFORM

In Chapter 5, we used Fourier series to obtain the frequency components of periodic functions of time. In general, random vibrations are not periodic, and the frequency analysis requires the extension of Fourier series to the *Fourier integral* for nonperiodic functions. *Fourier transforms*, which result from Fourier integrals, enable a more extensive treatment of the random vibration problem.

We begin by showing that the Fourier integral can be viewed as a limiting case of the Fourier series as the period goes to infinity. Toward this objective we consider the Fourier series in exponential form and substitute the coefficient C_n given by eq. (5.20) into eq. (5.19):

$$F(t) = \sum_{n=-\infty}^{\infty} \frac{1}{T} \int_{-T/2}^{T/2} F(\tau) \, e^{-in\bar{\omega}\tau} \, e^{in\bar{\omega}t} \, d\tau \qquad (22.17)$$

In eq. (22.17) we have selected the integration period from $-T/2$ to $T/2$ and substituted the symbol τ for t as the dummy variable of integration. The frequency $\omega = n\bar{\omega}$ is specified here at discrete, equally spaced values separated by the increment

$$\Delta\omega = (n + 1)\bar{\omega} - n\bar{\omega} = \bar{\omega} = \frac{2\pi}{T}$$

We substitute ω for $n\bar{\omega}$ and $\Delta\omega/2\pi$ for $1/T$ in eq. (22.17) and notice that as T

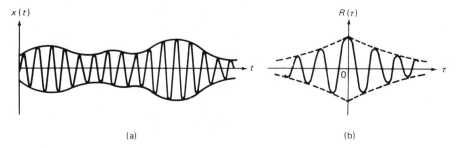

Fig. 22.11 Narrow-band random process $x(t)$ and its autocorrelation function $R(\tau)$.

$\to \infty$, $\Delta\omega \to d\omega$. Thus, in the limit, eq. (22.17) becomes

$$F(t) = \frac{1}{2\pi} \int_{-\infty}^{\infty} \left\{ \int_{-\infty}^{\infty} F(\tau) e^{-i\omega\tau} \, d\tau \right\} e^{i\omega t} \, d\omega \qquad (22.18)$$

which is the Fourier integral of $F(t)$.

Since the function within the inner braces is a function of only ω, we can write this equation in two parts as

$$C(\omega) = \frac{1}{2\pi} \int_{-\infty}^{\infty} F(t) \, e^{-i\omega t} \, dt \qquad (22.19)$$

and

$$F(t) = \int_{-\infty}^{\infty} C(\omega) \, e^{i\omega t} \, d\omega \qquad (22.20)$$

The validity of these relations, according to classical Fourier transform theory, is subject to the condition that

$$\int_{-\infty}^{\infty} |F(t)| \, dt < \infty \qquad (22.21)$$

The function $C(\omega)$ in eq. (22.19) is the *Fourier transform* of $F(t)$ and the function $F(t)$ in eq. (22.20) is the *inverse Fourier transform* of $C(\omega)$. The pair of functions $F(t)$ and $C(\omega)$ is referred to as a *Fourier transform pair*. Equation (22.19) resolves the function $F(t)$ into harmonic components $C(\omega)$, whereas eq. (22.20) synthesizes $F(t)$ from these harmonic components. In practice it is more convenient to use frequency f in cps rather than the angular frequency ω in rad/sec. Mathematically, since $\omega = 2\pi f$ and $d\omega = 2\pi df$, this also has the advantage of reducing the Fourier transform pair into a more symmetric form, namely,

$$F(t) = \int_{-\infty}^{\infty} C(f) \, e^{i2\pi ft} \, df \qquad (22.22)$$

and

$$C(f) = \int_{-\infty}^{\infty} F(t) \, e^{-i2\pi ft} \, dt \qquad (22.23)$$

22.6 SPECTRAL ANALYSIS

We have seen in Chapter 5 that the application of Fourier analysis to a periodic function yields the frequency components of the function given by either trigonometric terms [eq. (5.2)] or exponential terms [eq. (5.19)]. When the periodic function is known at N discrete, equally spaced times, the frequency components are then given by the terms in eq. (5.28). Our purpose in this section is to relate the Fourier analysis for a given function $x(t)$ to its mean square value $\overline{x^2}$.

The contributions of the frequency components of $x(t)$ to the value $\overline{x^2}$ are referred to as the *spectral function* of $x(t)$. Hence spectral analysis consists in expressing $\overline{x^2}$ in terms of either the coefficients of the Fourier series (i.e., a_n and b_n or equivalently C_n) when $x(t)$ is periodic, or in terms of the Fourier transform [i.e., $C(\omega)$] when $x(t)$ is not a periodic function.

We begin by performing the spectral analysis of a periodic function $x(t)$ expressed in Fourier series, eq. (5.2), which is

$$x(t) = a_0 + \sum_{n=1}^{\infty} \{a_n \cos n\bar{\omega}t + b_n \sin n\bar{\omega}t\} \tag{22.24}$$

where the coefficients are given by eq. (5.3) as

$$a_0 = \frac{1}{T} \int_0^T x(t)\, dt$$

$$a_n = \frac{2}{T} \int_0^T x(t) \cos n\bar{\omega}\, dt$$

$$b_n = \frac{2}{T} \int_0^T x(t) \sin n\bar{\omega}\, dt \tag{22.25}$$

In eq. (22.25), T is the period of the function and $\bar{\omega} = 2\pi/T$ its frequency. The substitution of $x(t)$ from eq. (22.24) for one of the factors in the definition of mean square value gives

$$\overline{x^2} = \frac{1}{T} \int_0^T x^2(t)\, dt$$

$$= \frac{1}{T} \int_0^T x(t) \left\{ a_0 + \sum_{n=1}^{\infty} (a_n \cos n\bar{\omega}t + b_n \sin n\bar{\omega}t) \right\} dt$$

$$= \frac{a_0}{T} \int_0^T x(t)dt + \sum_{n=1}^{\infty} \left[\frac{a_n}{T} \int_0^T x(t) \cos n\bar{\omega}t\, dt + \frac{b_n}{T} \int_0^T x(t) \sin n\bar{\omega}t\, dt \right]$$

Finally substituting the integral expressions from eqs. (22.25) we get the desired formula as

$$\overline{x^2} = a_0^2 + \sum_{n=1}^{\infty} \left[\frac{a_n^2}{2} + \frac{b_n^2}{2} \right] \tag{22.26}$$

The *spectrum* of the function $x(t)$ is then given by the terms of the series in eq. (22.26). Each term of this series is the contribution of the corresponding frequency to the mean square value of $x(t)$.

We now consider a discrete time function $F(t_j)$ expressed as a discrete Fourier transform [eq. (5.28)], that is, as

$$F(t_j) = \sum_{n=0}^{N-1} C_n \, e^{2\pi i(nj/N)} \tag{22.27}$$

where C_n is given by eq. (5.27) as

$$C_n = \frac{1}{N} \sum_{j=0}^{N-1} F(t_j) \, e^{-2\pi i(nj/N)} \tag{22.28}$$

As indicated in Chapter 5 [eq. (5.29) or (5.30)], harmonic components of the function $F(t_j)$ higher than the Nyquist frequency, $\omega_{N/2} = \pi/\Delta t$ rad/sec or $f_{N/2} = 1/2\Delta t$ cps, are not included in the discrete Fourier transform [eq. (22.27)]. Also as noted in Chapter 5, if there are harmonic components in $F(t_j)$ higher than this limiting value, these higher frequencies introduce distorting contributions to the lower harmonic frequencies. Hence it is imperative that the value of N be selected sufficiently large to include the frequencies that contribute significantly to the original function. To be certain that this condition is satisfied, one may filter the signal of the function electronically to remove all frequencies higher than the Nyquist frequency.

The mean square value of a discrete function $F(t_j)$ $(j = 0, 1, 2, \ldots, N-1)$ is obtained from eq. (22.2) as

$$\overline{F^2} = \frac{1}{T} \sum_{j=0}^{N-1} F^2(t_j) \, \Delta t \tag{22.29}$$

Substituting $\Delta t/T$ for $1/N$ and using eqs. (22.27) and (22.28) for one factor $F(t_j)$ in eq. (22.29), we obtain

$$\overline{F^2} = \frac{1}{N} \sum_{j=0}^{N-1} F(t_j) \sum_{n=0}^{N-1} C_n \, e^{2\pi i(nj/N)}$$

$$= \sum_{n=0}^{N-1} C_n \left[\frac{1}{N} \sum_{j=0}^{N-1} F(t_j) \, e^{2\pi i(nj/N)} \right]$$

$$= \sum_{n=0}^{N-1} C_n C_n^*$$

where C_n^* is the complex conjugate of C_n.[1] Hence

$$\overline{F^2} = \sum_{n=0}^{N-1} |C_n|^2 = |C_0|^2 + |C_1|^2 + |C_2|^2 + \cdots + |C_{N-1}|^2 \tag{22.30}$$

---

[1]Since $F(t_j)$ is a real function, its conjugate $F(t_j)^* = F(t_j)$ and also $[e^{-2\pi i(nj/N)}]^* = e^{2\pi i(nj/N)}$, it follows that $1/N \sum\limits_{j=0}^{N-1} F(t_j)e^{2\pi i(nj/N)} = C_n^*$.

The terms of the summation in eq. (22.30) are the required spectrum of the discrete function $F(t_j)$; that is, these terms are the frequency contributions to the mean square value $\overline{F^2}$. As we can see in eq. (22.30), the contribution of each frequency is equal to the square of the modulus of the corresponding complex coefficient C_n which is given by eq. (22.28).

Example 22.2.   Determine the spectrum of the function $F(t)$ shown in Fig. 22.12. Assume that the function is defined at eight equally spaced time intervals. Use the spectrum to estimate the mean square value $\overline{F^2}$ and compare this result with the mean square value of $F(t)$ calculated directly from the definition of $\overline{F^2}$.

Solution: We use computer Program 4 with $N = 8$ to determine the discrete Fourier coefficients C_n. The values thus obtained are shown in Table 22.1 together with the calculation needed to obtain the spectrum using eq. (22.30). The summation of the spectral values for $F(t)$ shown in the last column of Table 22.1 is $\overline{F^2} = 0.5400$ E 10. We check this value by calculating the mean square value of $F(t)$ directly from the definition eq. (22.2), namely,

$$\overline{F^2} = \frac{1}{T} \int_0^T F^2(t)\, dt$$

in which we substitute

$$F(t) = 120000t/0.16$$

$$\overline{F^2} = \frac{4}{0.64} \int_0^{0.16} \left[\frac{120000}{0.16} \cdot t \right]^2 dt = 0.4800 \text{ E } 10$$

TABLE 22.1   Spectral Analysis for the Function $F(t)$ in Fig. 22.12

n	Re (C_n)	Im (C_n)	$|C_n|$	$|C_n|^2$
		Fourier Coefficients C_n		Spectrum of $F(t)$
0	0	0		0
1	0	−51210.	51210.	0.2623 E 10
2	0	0		0
3	0	8787.	8787.	0.7721 E 8
4	0	0		0
5	0	8787.	8787.	0.7721 E 8
6	0	0		0
7	0	51210.	51210.	0.2623 E 10
				$\overline{F^2} = 0.5400$ E 10

Considering that we have used in this example a relatively small number of intervals ($N = 8$), the value of $\overline{F^2} = 0.5400$ E 10 obtained from the spectrum of $F(t)$ in Table 22.1 compares fairly well with the exact mean square value $\overline{F^2} = 0.4800$ E 10. As discussed before, errors are introduced in the calculations when the number of time intervals N is not large enough to include the higher frequency components of $F(t)$. In order to improve the calculation of the spectrum, it is necessary to use more time intervals in the discrete Fourier Series. Table 22.2 shows results obtained using Program 4 with $M = 3, 4, 5, 6, 7$, corresponding to $N = 2^M = 8, 16, 32, 64, 128$ time intervals for the function $F(t)$ shown in Fig. 22.12. It may be observed that values displayed in the last column of the table are converging to the exact value $\overline{F^2} = 0.4800$ E 10 as the number of time intervals, N, is increased.

22.7 SPECTRAL DENSITY FUNCTION

If a random process $x(t)$ is normalized (or adjusted) so that the mean value of the process is zero, then, provided that $x(t)$ has no periodic components, the

TABLE 22.2   Mean Square Value for Function $F(t)$ in Fig. 22.12

Exponent M	Time Intervals $N = 2^M$	Mean Square Value $\overline{F^2}$
3	8	0.5400 E 10
4	16	0.4950 E 10
5	32	0.4838 E 10
6	64	0.4809 E 10
7	128	0.4802 E 10
		Exact $\overline{F^2} = 0.4800$ E 10

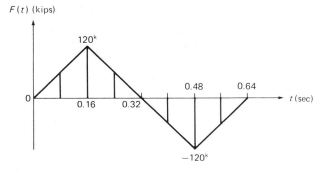

Fig. 22.12 Forcing function for Example 22.2.

autocorrelation function $R_x(\tau)$ approaches zero as τ increases, that is,

$$\lim_{\tau \to \infty} R_x(\tau) = 0 \tag{22.31}$$

We therefore expect that $R_x(\tau)$ should satisfy the condition in eq. (22.21). We can now apply eqs. (22.19) and (22.20) to obtain the Fourier transform of the correlation function $R_x(\tau)$ and its inverse as

$$S_x(\omega) = \frac{1}{2\pi} \int_{-\infty}^{\infty} R_x(\tau)\, e^{-i\omega\tau}\, d\tau \tag{22.32}$$

and

$$R_x(\tau) = \int_{-\infty}^{\infty} S_x(\omega)\, e^{i\omega\tau}\, d\omega \tag{22.33}$$

In eq. (22.32), $S_x(\omega)$ is called the *spectral density function* of $x(t)$. The most important property of $S_x(\omega)$ becomes apparent by letting $\tau = 0$ in eq. (22.33). In this case

$$R_x(0) = \int_{-\infty}^{\infty} S_x(\omega)\, d\omega \tag{22.34}$$

which by eq. (22.16) is equal to mean square value, that is,

$$\overline{x^2} = \int_{-\infty}^{\infty} S_x(\omega)\, d\omega \tag{22.35}$$

The mean square value of a random process is therefore given by the area under the graph of the spectral density function as shown in Fig. 22.13. Consequently, the contribution of an incremental frequency $\Delta\omega$ to the mean square value is

$$\Delta\overline{x^2} = S_x(\omega)\, \Delta\omega \tag{22.36}$$

The spectral density of a given record can be obtained electronically by an instrument called *frequency analyzer* or *spectral density analyzer*. The output of an accelerometer or other vibration transducer is fed into the instrument, which is essentially a variable frequency narrow-band filter with a spectral meter to display the filtered output. With this instrument the experimenter searches for the predominant frequencies present in a vibration signal. The output of the spectral density analyzer is the contribution to the mean square value $\Delta\overline{x^2}$ of the input signal $x(t)$ for a small range $\Delta\omega$ around the set frequency.

When dealing with theory, the natural unit for the frequency is rad/sec. However, in most practical problems the frequency is expressed in cycles per second or Hertz (abbreviated Hz). In the latter case, we rewrite eq. (22.36) as

$$\Delta\overline{x^2} = S_x(f)\Delta f \tag{22.37}$$

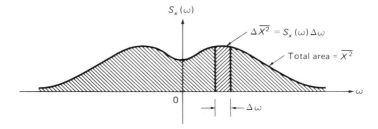

Fig. 22.13 Spectral density function showing area equal to mean-square value.

where f is the frequency in Hertz. Since $\Delta\omega = 2\pi\Delta f$, it follows from eqs. (22.36) and (22.37) that

$$S_x(f) = 2\pi S_x(\omega) \tag{22.38}$$

When the spectral density function for the excitation is known, its mean-square value may be determined from eq. (22.37) as

$$\overline{x^2} = \int_{-\infty}^{\infty} S_x(f)\, df \tag{22.39}$$

The spectral density function $S_x(f)$ is expressed in square units of x per Hertz. Since the autocorrelation function $R_x(\tau)$ is real and even, the use of Euler's relationship

$$e^{i\omega\tau} = \cos \omega\tau + i \sin \omega\tau$$

in eq. (22.32) yields the cosine transform:

$$S_x(\omega) = \frac{1}{2\pi} \int_{-\infty}^{\infty} R_x(\tau) \cos \omega\tau\, d\tau \tag{22.40}$$

It is clear from eq. (22.40) that $S_x(\omega)$ is also an even function of ω; hence eq. (22.33) may be written as

$$R_x(\tau) = \int_{-\infty}^{\infty} S_x(\omega) \cos \omega\tau\, d\omega \tag{22.41}$$

Alternatively, eqs. (22.40) and (22.41) may be written as

$$S_x(\omega) = \frac{1}{\pi} \int_{0}^{\infty} R_x(\tau) \cos \omega\tau\, d\tau \tag{22.42}$$

and

$$R_x(\tau) = 2 \int_{0}^{\infty} S_x(\omega) \cos \omega\tau\, d\omega \tag{22.43}$$

These are the celebrated *Wiener–Kinchin* equations, which describe how the spectral density function can be determined from the autocorrelation function and vice versa.

22.8   NARROW-BAND AND WIDE-BAND RANDOM PROCESSES

A process whose spectral density function has nonzero values only in a narrow frequency range as shown in Fig. 22.14 is called *narrow band* process. In contrast, a *wide-band* process is one whose spectral density function is nonzero over a broad range of frequencies. The time history of such a process is then made up of the superposition of the whole band of frequencies as shown in Fig. 22.15(a). In the limit, when the frequency band extends from $\omega_1 = 0$ to $\omega_2 = \infty$. This spectrum is called *white noise*. From eq. (22.35) the mean square value of a white noise process must be infinite; therefore, the white noise process is only a theoretical concept. In practice a process is called white noise when the bandwidth of its frequencies extends well beyond all the frequencies of interest.

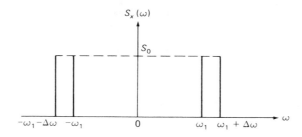

Fig. 22.14 Spectral density function for a narrow-band random process.

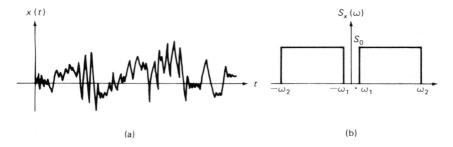

Fig. 22.15 Wide-band process. (a) Time history. (b) Spectral density function.

Example 22.3. Determine the mean square value and the autocorrelation function for the narrow-band random process $x(t)$ whose spectral density function is shown in Fig. 22.14.

Solution: From eq. (22.35)

$$\overline{x^2} = \int_{-\infty}^{\infty} S_x(\omega)\, d\omega = 2S_0\, \Delta\omega$$

and from eq. (22.43)

$$R_x(\tau) = 2 \int_0^{\infty} S_x(\omega) \cos \omega\tau\, d\omega \tag{a}$$

$$= 2 \int_{\omega_1}^{\omega_1 + \Delta\omega} S_0 \cos \omega\tau\, d\omega$$

$$= \frac{2S_0}{\tau} [\sin \omega\tau]_{\omega_1}^{\omega_1 + \Delta\omega}$$

$$= \frac{2S_0}{\tau} [\sin (\omega_1 + \Delta\omega)\tau - \sin \omega_1\tau]$$

$$R_x(\tau) = \frac{4S_0}{\tau} \cos \left(\omega_1 + \frac{\Delta\omega}{2} \right)\tau \cdot \sin \frac{\Delta\omega}{2} \tau \tag{b}$$

The autocorrelation function for a narrow-band random process given by eq. (b) has the form shown in Fig. 22.16, where the predominant frequency of $R_x(\tau)$ is the average value $(\omega_1 + \Delta\omega/2)$. The autocorrelation for such a process has a maximum of $2S_0\, \Delta\omega$ when $\tau = 0$ and decreases like a cosine graph as τ moves from $\tau = 0$.

The autocorrelation function $R_x(\tau)$ for a wide-band random process whose spectral density function extends in the range ω_1 to ω_2 as shown in Fig. 22.15(b)

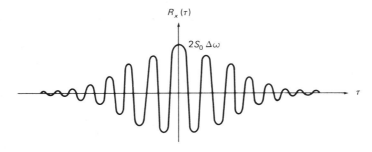

Fig. 22.16 Autocorrelation for a narrow-band random process.

can be obtained from the result of Example 22.3. In this case, letting the lower frequency $\omega_1 = 0$ and the upper frequency $\Delta\omega = \omega_2$, we obtain from eq. (b) of Example 22.3

$$R_x(\tau) = \frac{4S_0}{\tau} \cos\left(\frac{\omega_2\tau}{2}\right) \sin\left(\frac{\omega_2\tau}{2}\right) = \frac{2S_0}{\tau} \sin \omega_2\tau \qquad (22.44)$$

which has the form shown in Fig. 22.17(a).

The autocorrelation function for white noise may be obtained from eq. (22.44) by letting $\omega_2 \to \infty$. In this case adjacent cycles come closer together, resulting in a high peak at $\tau = 0$ and zero value elsewhere as shown in Fig. 22.17(b). This high peak will be of infinite height, zero width, but of a finite area. Such behavior may be described mathematically using *Dirac's delta function* $\delta(\tau)$. The delta function $\delta(\tau)$ is defined as having zero value everywhere except at $\tau = 0$ in such a way that

$$\int_{-\infty}^{\infty} \delta(\tau)f(\tau)\, d\tau = f(0) \qquad (22.45)$$

for any function of time $f(\tau)$ defined at $\tau = 0$.

Using the delta function $\delta(\tau)$ we can express the autocorrelation function for white noise as

$$R_x(\tau) = C\delta(\tau) \qquad (22.46)$$

where C must be determined from eq. (22.32) using the fact that $S_x(\omega)$ must be the constant S_0. Substituting eq. (22.46) into the eq. (22.32) gives

$$S_x(\omega) = \frac{1}{2\pi} \int_{-\infty}^{\infty} C\delta(\tau)\, e^{-j\omega\tau}\, d\tau$$

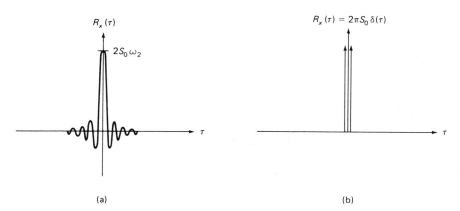

(a)   (b)

Fig. 22.17 Autocorrelation for a wide-band random process becomes a delta function for white noise.

and using eq. (22.45) yields   (22.47)

$$S_x(\omega) = \frac{1}{2\pi} Ce^{-i\omega 0} = \frac{C}{2\pi}$$

Hence

$$S_0 = \frac{C}{2\pi}$$ (22.48)

Finally, solving for C and substituting into eq. (22.46), we obtain the auto-correlation function for white noise as

$$R(\tau) = 2\pi S_0 \delta(\tau)$$ (22.49)

22.9 RESPONSE TO RANDOM EXCITATION

To determine the response of a structural system subjected to a random ex-citation, we need to examine the frequency content of the excitation func-tion. We are mostly interested in estimating the spectral function or the spec-tral density function of the excitation.

Until recently, the procedure for estimating the spectrum of a discrete time series has been to first determine the autocorrelation function [eq. (22.15)] and then apply the Fourier transform to this function to obtain the required spec-trum [eq. (22.32) or eq. (22.42)]. However, the method of calculation has changed since the development of fast Fourier transform (usually abbreviated FFT). As has been indicated in Chapter 5, the FFT is a remarkably efficient method for calculating the Fourier transform of a time series. Rather than estimate the spectrum by first determining the autocorrelation function and then calcu-lating the Fourier transform, it is now more efficient and more accurate to calculate spectra directly from the original time series.

Consider the damped, single degree-of-freedom system shown in Fig. 22.18(a) subjected to a random force $F(t)$, a sample of which is shown in Fig. 22.18(b).

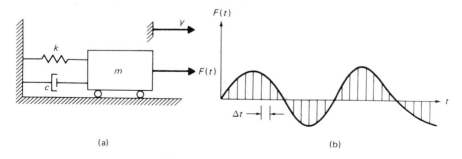

(a)   (b)

Fig. 22.18 Single degree-of-freedom system subject to a random force sampled at regular time intervals.

We will assume that this force is known at N discrete equally spaced times $t_j = j\Delta t$ $(j = 0, 1, 2, ..., N - 1)$. Fourier analysis of $F(t)$ results in the frequency components as given by eqs. (22.27) and (22.28). By superposition the response of the single degree-of-freedom system to the harmonic components of $F(t)$ is given by eq. (5.35) as

$$y(t_j) = \sum_{n=0}^{N-1} \frac{C_n \, e^{2\pi i n j / N}}{k(1 - r_n^2 + 2i r_n \, \xi_n)} \tag{22.50}$$

in which, as discussed in Section 5.6,

$$\omega_n = n\bar{\omega} \qquad \text{for} \quad n \leq N/2$$

$$\omega_n = -(N - n)\bar{\omega} \quad \text{for} \quad n > N/2 \tag{22.51}$$

$$r_n = \frac{\omega_n}{\omega}, \quad \bar{\omega} = \frac{2\pi}{T}, \quad \omega = \sqrt{\frac{k}{m}} \tag{22.52}$$

In these formulas, T is the time duration of the excitation, N the number of equal intervals of the excitation, and ξ_n the damping ratio corresponding to the frequency ω_n.

Equation (22.50) may conveniently be written as

$$y(t_j) = \frac{1}{k} \sum_{n=0}^{N-1} H_n \, C_n \, e^{2\pi i n j / N} \tag{22.53}$$

where H_n is the dimensionless quantity

$$H_n = \frac{1}{1 - r_n^2 + 2i r_n \, \xi_n} \tag{22.54}$$

The mean square value $\overline{y^2}$ of the response can be obtained from eq. (22.30) as

$$\overline{y^2} = \frac{1}{k^2} \sum_{n=0}^{N-1} |H_n|^2 |C_n|^2 \tag{22.55}$$

Alternatively, eq. (22.55) may be expressed in terms of the spectral density function. The frequency contributions $\Delta \overline{F^2}$ to the mean square value $\overline{F^2}$ are given by eq. (22.30) as

$$\Delta \overline{F^2} = |C_n|^2 \tag{22.56}$$

which, by eq. (22.36), may be expressed as

$$\Delta \overline{F^2} = S_F(\omega_n) \, \Delta \omega \tag{22.57}$$

Now, using eqs. (22.56) and (22.57), we may write eq. (22.55) as

$$\overline{y^2} = \frac{1}{k^2} \sum_{n=0}^{N-1} |H_n|^2 S_F(\omega_n) \, \Delta \omega \tag{22.58}$$

or as

$$\overline{y^2} = \sum_{n=0}^{N-1} S_y(\omega_n)\, \Delta\omega \qquad (22.59)$$

where

$$S_y(\omega_n) = \frac{1}{k^2} |H_n|^2 S_F(\omega_n) \qquad (22.60)$$

When the frequency is expressed in cps, we may write eq. (22.59) as

$$\overline{y^2} = \sum_{n=0}^{N-1} S_y(f_n)\, \Delta f \qquad (22.61)$$

where

$$S_y(f_n) = \frac{1}{k^2} |H_n|^2 S_F(f_n) \qquad (22.62)$$

Equation (22.62) states the important result that, when the frequency response H_n/k is known, the spectral density $S_y(f_n)$ for the response can be calculated from the spectral density $S_F(f_n)$ of the excitation.

Example 22.4. Determine the mean square value of the response for the single degree-of-freedom system in which $k = 100000$ lb/in, $m = 100$ lb $\cdot$ sec^2/ in, $c = 632$ lb $\cdot$ sec/in subjected to the $F(t)$ shown in Fig. 22.12. Choose $N = 8$ for the number of intervals.

Solution: The mean square value $\overline{y^2}$ of the response is given by eq. (22.55) as

$$\overline{y^2} = \frac{1}{k^2} \sum_{n=0}^{N-1} |H_n|^2 |C_n|^2$$

From eq. (22.54)

$$|H_n|^2 = \frac{1}{(1 - r_n^2)^2 + (2r_n\, \xi_n)^2}$$

where the frequency ratio is $r_n = \omega_n/\omega$ and the damping ratio is $\xi_n = c_n/c_{cr}$. The natural frequency is $\omega = \sqrt{k/m} = 31.62$ rad/sec and the frequency components by eq. (22.51) are given as

$$\omega_n = n\omega_1, \qquad n \le N/2$$

$$\omega_n = -(N - n)\omega_1, \quad n > N/2$$

in which $\omega_1 = 2\pi/T = 2\pi/0.64 = 9.8175$ rad/sec since the duration of the applied force is $T = 0.64$ sec.

Values of $|C_n|$, for the function $F(t)$ shown in Fig. 22.12, have been deter-mined in Example 22.2 and are shown in Table 22.1. The necessary compu-tations to determine $\overline{y^2}$ are conveniently shown in Table 22.3. From this table the mean square value of the response is $\overline{y^2} = 0.9351$.

Analogous to eq. (22.58), when the excitation is a random acceleration ap-plied to the support of the structure, the mean-square acceleration response $\overline{a_p^2}$ at a point P of the structure is given by

$$\overline{a_p^2} = \frac{1}{k^2} \sum_{n=0}^{N-1} S(\omega_n)|H_n|^2 \, \Delta\omega \qquad (22.63)$$

where H_n/k is now the frequency response in terms of the acceleration at point P resulting from a unit harmonic acceleration at the support of the structure.

The response of any structure subjected to a single point random excitation can be determined by a simple numerical calculation provided that the spec-tral function or the spectral density function of the excitation and the fre-quency response of the structure are known. The frequency response H_n may be obtained experimentally by applying a sinusoidal excitation of varying fre-quency at the foundation and measuring the response at the desired point in the structure. The necessary calculations are explained in the following nu-merical example.

Example 22.5.   Determine the response at a point P of the structure shown schematically in Fig. 22.19(a) when subjected to a random acceleration at its foundation. The spectral density function of the excitation is known and shown in Fig. 22.19(b). The frequency response a_p/a_0 of the structure at point P, ob-tained experimentally, when the foundation is excited by a sinusoidal accel-eration of amplitude a_0 and varying frequency f_n, is shown in Fig. 22.19(c).

TABLE 22.3   Calculation of $\overline{y^2}$ for Example 22.4

n	ω_n (rad/sec)	$r = \omega_n/\omega$	$|H_n|$	$|C_n|$ (lb)	$\Delta\overline{y^2} = |H_n|^2|C_n|^2/k^2$ (in^2)
0	0	0	1.000	0	0
1	9.8175	0.3105	1.104	0.5121 E 5	0.3197
2	19.6350	0.6209	1.595	0	0
3	29.4524	0.9314	4.376	0.8787 E 4	0.1479
4	39.2699	1.2418	1.677	0	0
5	−29.4524	−0.9314	4.376	0.8787 E 4	0.1479
6	−19.6350	−0.6209	1.595	0	0
7	−9.8175	−0.3105	1.104	0.5121 E 5	0.3197

$$\overline{y^2} = 0.9351$$

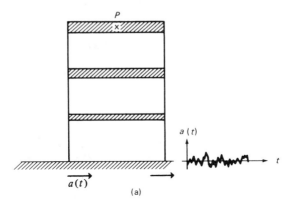

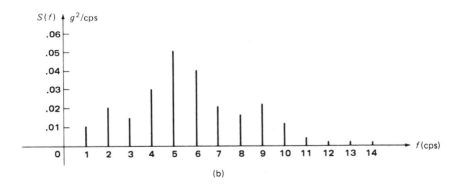

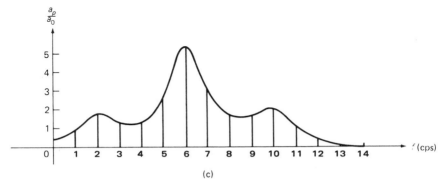

Fig. 22.19 (a) Structure subjected to random acceleration at the base. (b) Spectral density function of the excitation. (c) Relative frequency response at point P.

Solution: The mean square value $\overline{a_p^2}$ of the response at the point P is cal-culated from eq. (22.63) as

$$\overline{a_p^2} = \frac{1}{k^2} \sum_{n=0}^{N-1} S(f_n)|H_n|^2 \, \Delta f \tag{a}$$

where

$$|H_n|/k = a_p/a_0$$

Table 22.4 summarizes the computational procedure. By eq. (a) we obtain from the sum in the last column of this table the mean square value of the response $\overline{a_p^2} = 1.8100g^2$ and by eq. (22.3) (assuming mean value $\bar{a} = 0$)

$$\sigma = \sqrt{1.8100}g = 1.345 \; g$$

The probability of exceeding specified accelerations can now be found using the normal distribution for $|a_p| > \sigma = 1.345 \; g$ and for $|a_p| > 3\sigma = 4.041 \; g$, respectively, as

$$P[|a_p| > 1.345 \; g] = 31.7\%$$

and

$$P[|a_p| > 4.041 \; g] = 0.3\%$$

TABLE 22.4   Calculations of the Response for Example 22.5

f_n (cps)	Δf (cps)	$S(f_n)$ (g^2/cps)	$|H_n|/k$ (in/lb)	$(1/k^2)|H_n|^2 S(f_n)\Delta f$ (g^2 units)
0	1.0	0	0.5	0
1.0	1.0	0.010	1.0	0.0100
2.0	1.0	0.020	1.8	0.0648
3.0	1.0	0.015	1.3	0.0253
4.0	1.0	0.030	1.4	0.0588
5.0	1.0	0.050	2.2	0.2420
6.0	1.0	0.040	5.2	1.0816
7.0	1.0	0.020	3.0	0.1800
8.0	1.0	0.015	1.8	0.0486
9.0	1.0	0.020	1.7	0.0578
10.0	1.0	0.010	1.9	0.0361
11.0	1.0	0.005	1.0	0.0050
12.0	1.0	0	0.4	0
13.0	1.0	0	0	0
14.0	1.0	0	0	0
				sum = 1.8100

Similarly, the probability that the peak acceleration A_p will exceed a specified value is found using the Rayleigh distribution as

$$P[A_p > 1.345 \text{ g}] = 80.7\%$$

$$P[A_p > 4.041 \text{ g}] = 3.2\%$$

22.10 SUMMARY

The objective of this chapter was to introduce the fundamentals of the theory of random vibrations for application in structural dynamics. In structural dynamics the most common source of random vibration is due to explosions occurring in the vicinity of the structure. The response of a structure to earthquakes may also be predicted using random vibration theory.

A random process is described by a function of time whose value at any time is known only as a set of sample records known as an ensemble. Such a function can only be described in probabilistic terms using the tools of statistics. The most important statistics of a random process $x(t)$ are its mean value $\bar{x}$, its mean square value $\overline{x^2}$, and its variance σ_x^2 given, respectively, by eqs. (22.1), (22.2), and (22.3). The most commonly used probability distribution for a random process is the normal distribution. However, when the random variable can only assume positive values (e.g., the absolute values of the peaks of vibration), the process tends to follow the Rayleigh distribution.

The autocorrelation $R_x(\tau)$ of a random variable $x(t)$ is defined by eq. (22.15). The spectral density function $S_x(\omega)$ is defined as the Fourier transform of the autocorrelation function $R(\tau)$ [eq. (22.32)]. Although the spectrum of $x(t)$ can be obtained from $R_x(\tau)$, today it is more efficient to determine the spectrum of a random function from its discrete Fourier series [eq. (22.30)] using the fast Fourier transform (FFT).

If the spectrum of the excitation function and the frequency response of a dynamic system are known, it is a simple matter to calculate the mean-square value of the response using eq. (22.55). Knowing the mean-square value of the response and using standard probability functions (such as the normal or the Rayleigh distributions), we can predict the response in probabilistic terms.

In this introductory chapter on random vibration, the presentation has been restricted to single degree-of-freedom systems. However, the extension to linear multidegree-of-freedom or continuous systems could readily be accomplished using the modal superposition method. This method, as we have seen in previous chapters, transforms a system of differential equations into a set of independent or uncoupled differential equations. Each equation of this set is equivalent to the differential equation for a single degree-of-freedom system and consequently can be solved for random vibration excitation by the methods presented in this chapter.

PROBLEMS

22.1–22.5   Determine the mean and mean square values for the functions shown in Figs. 22.20–22.24.

22.6–22.10   Determine the Fourier series expansions for the periodic functions shown in Figs. 22.20–22.24.

22.11–22.15 Determine and plot the spectral functions for the functions shown in Figs. 22.20–22.24.

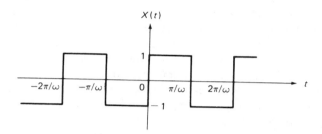

Fig. 22.20

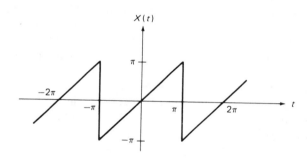

Fig. 22.21

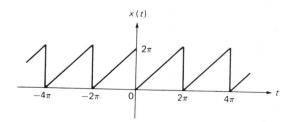

Fig. 22.22

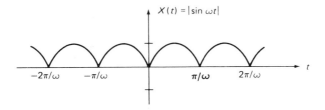

Fig. 22.23

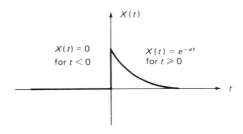

Fig. 22.24

22.16   A sine wave with a steady-state component is given by

$$x(t) = A_0 + A_1 \sin \overline{\omega} t$$

Determine the mean value $\bar{x}$ and the mean-square value $\overline{x^2}$.

22.17   Determine the Fourier coefficients C_n and the spectral function for the periodic function shown in Fig. 22.25.

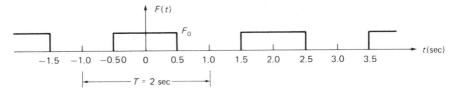

Fig. 22.25

22.18   A random force has a mean value $F = 2$ kips and spectral density function shown in Fig. 22.26. Determine its standard deviation σ_F and its root mean square RMS_F.

Fig. 22.26

22.19   Calculate the autocorrelation function for an ergodic random process $x(t)$. Each sample function is a square wave of amplitude a and period T.

22.20   Determine Fourier transform and Fourier integral representations for the function shown in Fig. 22.27.

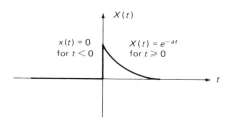

Fig. 22.27

22.21   A single degree-of-freedom system with mass 1.0 lb sec²/in, stiffness 100 lb/in, and damping $\xi = 0.20$ is excited by the force

$$F(t) = 1000 \cos 5t + 1000 \cos 10t + 1000 \cos 15t(\text{ lb})$$

Determine the spectral function and the mean square value of the response. Also plot the spectra for the input force and the output response displacement.

PART VI

Earthquake Engineering

23

Equivalent Static Lateral Force Method: Uniform Building Code—1985

The engineering methods for earthquake resistant design presented in this chapter and in the next chapter are based, respectively, on the requirements of the Uniform Building Codes of 1985 and 1988 [UBC-85 and UBC-88 (International Conference of Building Officials 1985; 1988)]. The Uniform Building Code is the building code most extensively used in the United States, particularly in the western part of the country. The UBC-88 introduced major changes in the previous edition of 1985. However, UBC-85 is also presented because other major codes in use have similar requirements.

In addition to the UBC, there are three major building codes in use in the United States: (1) *The BOCA*, or *Basic Building Code* (Building Officials and Code Administrators International 1987) published by the Building Officials and Code Administrators International, (2) *The National Building Code*, issued by the American Insurance Association (1976), and (3) *The Standard Building Code*, of the Southern Building Code Congress International (1988). In addition to these codes there is *The American National Standard Building Code Requirement for Minimum Design Loads in Buildings and Other Structures*, which for brevity is referred to as ANSI (American National Standards Institute 1982). Also, several organizations involved in earthquake resistant

511

design publish recommendations which form the basis for requirements in the official codes. Four important organizations of this type are: (1) the Structural Engineers Association of California [SEAOC (1988)], (2) the Applied Technology Council [ATC3-06 (1978)], (3) the Building Seismic Safety Council [BSSC (1988)], and (4) The Federal Emergency Management Agency (FEMA). All these organizations periodically issue recommendations and requirements for earthquake resistant design of structures, based on a combination of theory, experiment, and practical observation.

Building codes are intended to provide guidelines, and formulas which constitute minimum legal requirements for design and construction within a particular region. These requirements are intended to achieve satisfactory performance of the structure when subjected to seismic excitation, although they are not optimal: The safety of the structure is not assured in the event of a major earthquake. The objective of the codes is that a minor or moderate earthquake will not damage the structure and that a major earthquake will not produce collapse of the structure. To understand how structures are affected by earthquakes, it is necessary to understand the ground motion produced by earthquakes.

23.1   EARTHQUAKE GROUND MOTION

Most earthquakes are caused by energy release at a dislocation or rupture in crustal plates generated at a point in the interior of the earth known as the *focus* or *hypocenter*. The point on the earth's surface directly above the focus is the *epicenter*. The magnitude of an earthquake is commonly measured by the *Richter magnitude* (M), which is defined as the reading registered by an instrument called a *Wood–Anderson seismograph* at a specified distance of 100 km from the epicenter of the earthquake. Specifically, the Richter magnitude (M) for an earthquake is calculated as

$$M = \log_{10} \frac{A}{A_0} \tag{23.1}$$

where A is the maximum amplitude registered by a Wood–Anderson instrument located at 100 km from the epicenter and A_0 is the reference amplitude of one thousandth of a millimeter (for A expressed in millimeters, $A_0 = 0.001$). When there is no instrument located at this distance, an estimation of the Richter magnitude is calculated based on readings registered at seismographs in the region affected by the earthquake. Earthquakes of magnitude 5.0 or greater generate ground motion sufficiently severe to be potentially damaging to structures. The energy E in Joules released by an earthquake can be estimated by the formula

$$E = 10^{4.8 + 1.5M} \text{ (Joules)} \tag{23.2}$$

where M is the magnitude of the earthquake on the Richter scale. Equation (23.2) reveals that the energy increases by a factor of about 32 for an increase of 1 unit in the magnitude M and by a factor of 1000 for an increase of 2 units in the magnitude of M. Although the Richter magnitude provides a measure of the total energy released by an earthquake, it does not describe the damaging effects caused by the earthquake at a particular location. Such a description is provided by the *modified Mercalli intensity* scale (MMI), with a total of 12 intensity values usually expressed in Roman numerals. Table 23.1 shows the MMI scale. The MMI value assigned to an earthquake at a particular site is based on the observation of resultant damage and is useful in lieu of instrument records of ground motion. Intensities up to VI usually do not produce damage while intensities from VI to XII result in progressively greater damage to buildings and other structures.

Graphical records or time histories of earthquakes are obtained with instruments called *strong-motion accelerographs.* These instruments are installed on the ground. They are commonly designed to register the three orthogonal components of the ground acceleration. Figure 23.1 shows the N–S component of the accelerogram of the El Centro earthquake, which occurred in 1940. The figure also shows, for this earthquake, the velocity and displacement obtained by integration of the accelerogram. The earthquake accelerogram also can be analyzed to obtain direct estimates of peak ground motion, duration of the strong portion of ground shaking, and the frequency content of the earthquake.

It was shown in Chapter 8, that the earthquake response spectrum provides meaningful information for use in structural design. The spectrum, as explained in that chapter, provides values for the maximum absolute acceleration (spectral acceleration), the maximum relative pseudovelocity (spectral velocity), and the maximum relative displacement (spectral displacement) of the single degree-of-freedom system for various damping values. Response spectral diagrams are usually prepared for a specific earthquake, as is the one shown in Fig. 8.8, or are constructed to be used in design on the basis of several past earthquakes, as are the spectral plots shown in Fig. 8.9.

23.2   EQUIVALENT SEISMIC LATERAL FORCE

The primary source for the provisions of the Uniform Building Code is the *Recommended Lateral Force Requirements and Commentaries* written by the Structural Engineering Association of California (SEAOC 1988). The Uniform Building Code of 1985 (UBC-85) (International Conference of Building Officials 1985) embodies the SEAOC recommendations and prescribes that every building shall be designed and constructed to resist a minimum lateral seismic force V, applied statically and independently in the direction of each of the two main axes of the structure. The *lateral seismic force*, or, as it is

TABLE 23.1   Modified Mercalli Intensity Scale (1956 Version)*

Intensity Value	Description
I.	Not felt. Marginal and long-period effects of large earthquakes.
II.	Felt by persons at rest, on upper floors, or favorably placed.
III.	Felt indoors. Hanging objects swing. Vibration like passing of light trucks. Duration estimated. May not be recognized as an earthquake.
IV.	Hanging objects swing. Vibration like passing of heavy trucks; or sensation of a jolt like a heavy ball striking the walls. Standing cars rock. Windows, dishes, doors rattle. Glasses clink. Crockery clashes. In the upper range of IV, wooden walls and frame creak.
V.	Felt outdoors; direction estimated. Sleepers wakened. Liquids disturbed, some spilled. Small unstable objects displaced or upset. Doors swing, close, open. Shutters, pictures move. Pendulum clocks stop, start, change rate.
VI.	Felt by all. Many frightened and run outdoors. Persons walk unsteadily. Windows, dishes, glassware broken. Knickknacks, books, etc. off shelves. Pictures off walls. Furniture moved or overturned. Weak plaster and masonry D** cracked. Small bells ring (church, school). Trees, bushes shaken visibly, or heard to rustle.
VII.	Difficult to stand. Noticed by drivers. Hanging objects quiver. Furniture broken. Damage to masonry D**, including cracks. Weak chimneys broken at roof line. Fall of plaster, loose bricks, stones, tiles, cornices, also unbraced parapets and architectural ornaments. Some cracks in masonry C**. Waves on ponds, water turbid with mud. Small slides and caving in along sand or gravel banks. Large bells ring. Concrete irrigation ditches damaged.
VIII.	Steering of cars affected. Damage to masonry C; partial collapse. Some damage to masonry B; none to masonry A. Fall of stucco and some masonry walls. Twisting, fall of chimneys, factory stacks, monuments, towers, elevated tanks. Frame houses moved on foundations if not bolted down; loose panel walls thrown out. Decayed piling broken off. Branches broken from trees. Changes in flow or temperature of springs and wells. Cracks in wet ground and on steep slopes.
IX.	General panic. Masonry D** destroyed; masonry C** heavily damaged, sometimes with complete collapse; masonry B seriously damaged. General damage to foundations. Frame structures, if not bolted, shifted off foundations. Frames racked. Serious damage to reservoirs. Underground pipes broken. Conspicuous cracks in ground. In alluviated areas sand and mud ejected, earthquake fountains, sand craters.
X.	Most masonry and frame structures destroyed with their foundations. Some well-built wooden structures and bridges

TABLE 23.1   (Continued)

Intensity Value	Description
	destroyed. Serious damage to dams, dikes, embankments. Large landslides. Water thrown on banks of canals, rivers, lakes, etc. Sand and mud shifted horizontally on beaches and flat land. Rails bent slightly.
XI.	Rails bent greatly. Underground pipelines completely out of service.
XII.	Damage nearly total. Large rock masses displaced. Lines of sight and level distorted. Objects thrown into the air.

*Original 1931 version in Wood, H. O., and Neumann, F., 1931, Modified Mercalli intensity scale of 1931: *Seismological Society of America Bulletin*, v. 53, no. 5, pp. 979–987. 1956 version prepared by Charles F. Richter, in *Elementary Seismology*, 1958, pp. 137–138, W. H. Freeman and Company.

**Masonry A, B, C, D. To avoid ambiguity of language, the quality of masonry, brick or otherwise, is specified by the following lettering.

Masonry A. Good workmanship, mortar, and design; reinforced, especially laterally, and bound together by using steel, concrete, etc.; designed to resist lateral forces.

Masonry B. Good workmanship and mortar; reinforced, but not designed in detail to resist lateral forces.

Masonry C. Ordinary workmanship and mortar; no extreme weaknesses like failing to tie in at corners, but neither reinforced nor designed against horizontal forces.

Masonry D. Weak materials, such as adobe; poor mortar; low standards of workmanship; weak horizontally.

also designated, the *base shear force*, of the structure is determined by

$$V = ZIKCSW \qquad (23.3)$$

where W is essentially the dead weight of the structure, Z the *seismic zone factor*, I the *occupancy importance coefficient*, S the *site–structure resonance coefficient*, K the *structural coefficient*, and C the *dynamic factor*.

The *seismic weight W* of the structure includes some permanent live loads, such as storage in warehouses; e.g., 25% of the design storage load is included as dead load. For office buildings, 20 psf is added to the dead load. Snow load is excluded when the design snow load is 30 psf or less. When the design snow load exceeds 30 psf, between 25% and 100% of this load is included as dead load on the base of the snow load duration. The rationale for including a fraction of the snow load in heavy snow areas is that in such areas a significant amount of ice can accumulate on the roofs.

The *seismic zone factor Z* is dependent on the site location, as shown in the seismic map, Fig. 23.2. For zones 0, 1, 2, 3, and 4 the Z values are 0, $3/16$, $3/8$, $3/4$, and 1, respectively. No value is specified for Zone 0, because

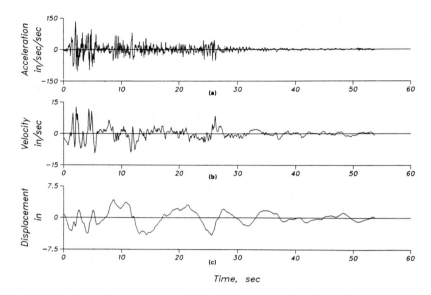

Fig. 23.1 North–south component of the El Centro earthquake, California, 1940. (a) Acceleration. (b) Velocity. (c) Displacement.

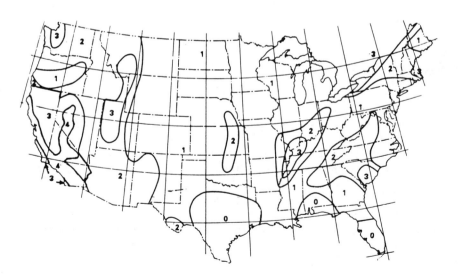

Fig. 23.2 Seismic zone map of the United States (UBC-85). (Reproduced from the 1985 edition of the *Uniform Building Code*, copyright 1985, with permission of the publishers, the International Conference of Building Officials.)

only in exceptional cases would building officials require seismic design in Zone 0.

The *occupancy importance factor I* is (1) 1.5 for essential facilities such as fire stations, hospitals, and police stations, (2) 1.25 for buildings used primarily for assembly of more than 300 persons in one room, and (3) 1.0 for all other buildings. It is obvious that the intention of this factor is to ensure the continuing function of vital facilities to assist and shelter people after a disastrous earthquake.

The *structural factor K* depends on the capacity of the structure to resist lateral forces. The values for K as given by the UBC-85 are shown in Table 23.2 for various structural systems. The values of K for common types of buildings vary from 0.67 for a ductile moment-resisting frame[1] to 1.33 for a box system. The first classification of structural systems in Table 23.2, for which $K = 1.0$, includes shear walls, braced frames, and ordinary moment frames.

Ductile moment-resisting frames are assigned a K value of 0.67, thus reducing the design static lateral forces by 33%. These frames are designed to be more ductile than ordinary frames, braced frames, or shear walls. However, as we will see later, this reduction in forces is eliminated in the story-drift requirements, which limit the relative lateral displacements of consecutive levels of the building.

Ductile moment-resisting frames used in combination with shear walls or braced frames are assigned a K value of 0.8. However, the shear walls or braced frames must be designed to resist the total lateral forces independently of the moment frames. In addition, the moment frames must be able to resist independently 25% of the total lateral force.

The UBC-85 includes an additional requirement related to ductility but not affecting the K factor: All members in braced frames for buildings in Zones 3 and 4, and for buildings in Zone 2 having an importance factor I greater than 1, shall be designed for lateral forces 25% greater than those determined from the code provisions. Also, connections either must be designed to develop the full capacity of the members or must be designed for code forces without the 1/3 increase in stresses as normally permitted when earthquake forces are included in the design.

It is important to recognize that the K factor "rewards" highly ductile structures and "penalizes" nonductile structures. Many studies of the damage produced by earthquakes have repeatedly demonstrated that ductile structures perform very well in large-magnitude earthquakes and can resist drifts far beyond those corresponding to the first yielding of the structure. On the other

---

[1]A ductile moment-resisting frame is a space frame in which the members and joints are capable of resisting forces primarily by flexure and in compliance with provision of Section 2312(j) of the UBC-85.

TABLE 23.2   Structural Factor K*

Type or Arrangement of Resisting Element	K**
1. All building framing systems except as classified next	1.00
2. Buildings with a box system as specified in Section 2312(b) Exception: Buildings not more than three stories in height with stud wall framing and using plywood horizontal diaphragms and plywood vertical shear panels for the lateral force system may use $K = 1.0$.	1.33
3. Buildings with a dual bracing system consisting of a ductile moment resisting space frame and shear walls or braced frames using the following design criteria: a. The frames and shear walls or braced frames resist the total lateral force in accordance with their relative rigidities considering interaction of shear walls and frames. b. The shear walls or braced frames acting independently of the ductile moment-resisting portions of the space frame resist the total required lateral forces. c. The ductile moment-resisting space frames have the capacity to resist not less than 25 percent of the required lateral force.	0.80
4. Buildings with ductile-resisting space frames designed in accordance with the following criteria: The ductile moment-resisting space frames have the capacity to resist the total required lateral force.	0.67
5. Elevated tanks plus full contents, on four or more crossbraced legs and not supported by a building.	2.5
6. Structures other than buildings and other than those set forth in Table No. 23-J of the UBC-1985.	2.0

(Reproduced from the 1985 edition of the Uniform Building Code, ©1985, with permission of the publishers, the International Conference of Building Officials)

*Where wind load as specified in Section 2311 would produce higher stresses, this load shall be used in lieu of the loads resulting from earthquake forces.

**The minimum value of KC shall be 0.12 and the maximum value of KC need not exceed 0.25.

hand, nonductile structures have not fared nearly so well; many failures have occurred when these structures are forced to undergo drifts beyond the first yielding. Furthermore, the code implicitly recognizes the well-accepted fact that the level of forces prescribed by the code represent only a fraction of the actual maximum forces that the structure may experience during a major earthquake. If seismic forces exceed the design strength of the structure, they must be resisted through yielding or ductility of the structure. Thus, the K

factor reflects the potential capability of the structure to resist inelastic deformations.

The *dynamic factor C* is specified as

$$C = \frac{1}{15\sqrt{T}} \leq 0.12 \tag{23.4}$$

where T is the fundamental period of the structure in the direction under consideration. The dynamic factor C decreases the base shear force V [eq. (23.3)] for buildings of longer periods.

To determine the fundamental period, the UBC-85 provides Rayleigh's formula, in which the period is given from eq. (6.66) as

$$T = 2\pi \sqrt{\sum_{i=1}^{N} W_i \delta_i^2 \bigg/ g \sum_{i=1}^{N} f_i \delta_i} \tag{23.5}$$

In this equation δ_i $(i = 1, 2, ..., N)$ are the lateral displacements at the floor levels of the building produced by statically applied lateral forces f_i such as those calculated by eqs. (23.9) and (23.10), and W_i are the seismic weights at the various levels of the building.

The displacements δ_i in eq. (23.5) should be the actual displacements due to the lateral forces f_i and not the displacements amplified by $1/K$ when $K < 1.0$, as indicated by the code. Therefore, the value for T calculated in eq. (23.5) should be multiplied by $\sqrt{K}$, when values for δ_i used in that equation have been amplified by $1/K$.

Since the lateral displacements δ_i are not yet known, eq. (23.5) cannot be used initially. However, the code also provides empirical formulas for estimating the fundamental period:

$$T = 0.05 \, h_N/\sqrt{D} \tag{23.6}$$

in which h_N is the height of the building in feet, and D the plan dimension in feet (in the direction for which the period is being estimated). For ductile moment-resisting frames, the period may be estimated by the empirical formula

$$T = 0.1N \tag{23.7}$$

where N is the number of stories above the base of the building.

Other methods of analysis, "properly substantiated," also are acceptable in determining the period T. The period determined by eqs. (23.6) or (23.7) may differ substantially from the period determined by the analysis. For example, for a steel building classified as a ductile moment-resisting frame, the period predicted by eq. (23.7) is generally shorter than the value obtained using the stiffness method, thus increasing the factor C and the base shear force V in eq. (23.3). On the other hand, an excessively long period will produce a base shear force substantially less than the value calculated using the period determined from eq. (23.6) or (23.7). The UBC-85 does not explicitly limit the value of the period that can be used to determine the base shear force. In practice, however, since the appearance of the first draft of UBC-88 (International Conference of Building Officials 1988), it is common to limit the base shear force to 80% of the value obtained using the empirical formulas, eq. (23.6) or eq. (23.7).

The *site-structure resonance coefficient* S takes into account the site characteristics and the relationship between the soil period and the period of the structure. This coefficient accounts for the likelihood that the building and the site could have natural periods sufficiently close to develop resonance during an earthquake. This condition actually happened in the Mexico City earthquake of 1985 where resonance contributed significantly to severe destruction of the city center.

If the site period T_s has been determined by a geotechnical investigation of the site, the coefficient S can be calculated by the following formulas which are functions of the ratio between the period of the structure, T, and the period of the soil at the site, T_s:

$$S = 1.0 + \frac{T}{T_s} - 0.5\left(\frac{T}{T_s}\right)^2 \qquad \text{for} \quad \frac{T}{T_s} \leq 1.0$$

or
(23.8)

$$S = 1.2 + 0.6\frac{T}{T_s} - 0.3\left(\frac{T}{T_s}\right)^2 \qquad \text{for} \quad \frac{T}{T_s} > 1$$

In using eq. (23.8), the UBC-85 stipulates that T shall be no less than 0.3 sec and that the fundamental site period T_s shall be in the range $0.5 \leq T_s \leq 2.5$. The code also stipulates a minimum value of $S = 1.0$ and a maximum of 1.5. Figure 23.3 shows a plot of S and the ratio T/T_s, where the code limitations for maximum and minimum values are indicated.

In order to use eq. (23.8), the building period T must be determined by analysis and not by the empirical formulas, eq. (23.6) or eq. (23.7). The soil coefficient S calculated from eq. (23.8) reaches a maximum of 1.5 when the building period T is equal to the site period T_s and reaches a minimum of 1.0 when the building period is more than $2\frac{1}{4}$ times longer than the site period.

There is considerable uncertainty in determining the periods T and T_s. Consequently, the calculation of the soil profile coefficient by eq. (23.8) is imprecise; this method of determining S has been eliminated from subsequent codes.

If the site period has not been determined, the UBC-85 provides an alternative method in which the coefficient S is assigned a value equal to 1.0, 1.2, or 1.5 according to the classification of the soil profile at the site, as described in Table 23.3. The code states that the product CS in eq. (23.3) need not exceed 0.14.

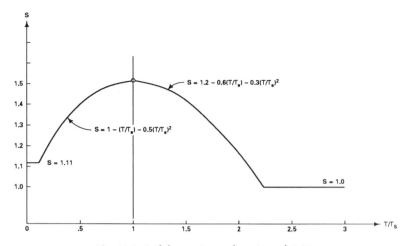

Fig. 23.3 Soil factor S as a function of T/T_s.

23.3   DISTRIBUTION OF THE LATERAL SEISMIC FORCE

For structures having regular shapes or framing systems, the total base shear force V calculated from eq. (23.3) is distributed over the height of the structure as a force at each level, F_x plus an additional force F_t (at the top of the building), as shown in Fig. 23.4:

$$F_x = \frac{(V - F_t)W_x h_x}{\displaystyle\sum_{i=1}^{N} W_i h_i} \tag{23.9}$$

$$F_t = 0.07TV \le 0.25V \quad \text{for} \quad T > 0.7 \text{ sec}$$

$$F_t = 0 \qquad\qquad\qquad \text{for} \quad T \le 0.7 \text{ sec} \tag{23.10}$$

TABLE 23.3   Site-Structure Resonance Coefficient S^1

Soil Profile Type	Description	S
1	Rock or stiff soil less than 200 ft. Shear wave velocity greater than 2500 ft/sec.	1.0
2	Deep cohesionless or stiff clay soil	1.2
3	Soft to medium-stiff clay with or without intervening layers of sand.	1.5

(Reproduced from the 1985 Uniform Building Code, ©1985, with permission of the publishers, the International Conference of Building Officials)

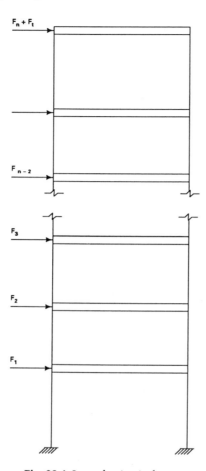

Fig. 23.4 Lateral seismic forces.

$$V = F_t + \sum_{i=1}^{N} F_i \qquad (23.11)$$

In this equation, N is the number of stories above the base, h_x or h_i the height of level x or level i, and W_x or W_i the weight of level x or level i.

At each level, the force F_x is applied over the area of the building according to the mass distribution on that level. The lateral forces given by eqs. (23.9) and (23.10) are applicable only to *regular* structures. For structures having irregular shapes, large differences in lateral resistance (or stiffness) between adjacent stories, or other unusual structural features, lateral forces must be determined considering the dynamic characteristics of the structure. The UBC-

85 gives no guidelines on how to implement this determination. The dynamic analysis may be implemented by an elastic or inelastic time-history analysis of the structure. The ground-motion input can be either a design response spectrum or an earthquake motion, real or artificial. However, it is common practice to determine the dynamic properties of the building, natural frequencies, modal shapes, and damping, and then perform a linear elastic analysis with a design spectrum for the ground motion. This approach is presented in Chapter 25 on the UBC-88: Dynamic Method.

23.4   HORIZONTAL TORSIONAL MOMENT

The Uniform Building Code of 1985 requires that horizontal torsion at the various levels of the structure be taken into account. The inertial force induced by the earthquake at a specific story acts through the center of mass above that story whereas the resultant resisting force is at the center of stiffness of the story. If the structure is not dynamically symmetric, the center of stiffness at a story and the center of the mass above that story may not coincide. Figure 23.5 shows the plan of a building floor in which S is the stiffness center and G is the projection of the center of the mass above that floor. The figure also shows the components of the eccentricity of the center of the mass above that story, measured normal to the direction under con-

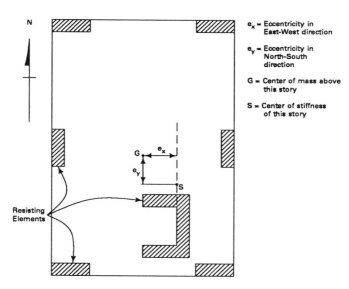

Fig. 23.5 Plan view of a story of a building showing the eccentricity between the mass center and the center of stiffness.

sideration. Furthermore, the code requires design for an assumed accidental torsional moment in each story of no less than 5% of the product of the story shear force and the maximum dimension of the building at that level. This 5% accidental torsional moment is commonly applied as a minimum for the total design torsion and not an addition to the actual calculated torsion.

23.5   OVERTURNING MOMENTS AND STORY SHEAR

The seismic lateral forces produce overturning moments which result in additional axial forces in the columns, especially in the exterior columns of the building. These overturning forces increase the gravity load effect on the exterior columns on one side at a time while the building is subjected to earthquake forces. The overturning moment at any level of the building is determined by statics as the moment produced at that level by the horizontal forces above. Therefore, the overturning moment M_x at level x of the building is given by

$$M_x = F_t(h_N - h_x) + \sum_{i=x}^{N} F_i(h_i - h_x) \qquad (23.12)$$

where the lateral forces F_i and F_t are given by eqs. (23.9) and (23.10) and h_i or h_x is the height of the ith or xth level of the building.

The story shear V_x at level x, is the sum of all the lateral forces at and above that level:

$$V_x = F_t + \sum_{i=x}^{N} F_i \qquad (23.13)$$

23.6   STORY DRIFT

The UBC-85 limits the relative horizontal displacements of consecutive levels of the building (story drift) to 0.5% of the story height, unless an analysis can demonstrate that a greater drift can be tolerated. Also, an amplification factor $1/K$ must be applied to the calculated displacements when the structural factor K is less than 1. This amplification factor recognizes the fact that regardless of the reduced loads permitted for ductile structures, the actual deformation will be produced by forces larger than the design forces, that is, by forces not reduced by the K factor.

23.7   DIAPHRAGM FORCES

The code stipulates that diaphragm forces F_{px} be determined with the following formula:

$$F_{px} = \frac{F_t + \displaystyle\sum_{i=x}^{N} F_i}{\displaystyle\sum_{i=x}^{N} W_i} W_{px} \tag{23.14}$$

The code also stipulates that F_{px} need not exceed $0.30\ ZIW_{px}$ but shall not be less than $0.14\ ZIW_{px}$ (W_{px} is the weight of the diaphragm and attached weights at level x).

Example 23.1. A four-story reinforced concrete framed building has the dimensions in the plan shown in Fig. 23.6. The sizes of the exterior columns (nine each on lines A and C) are 12 in $\times$ 20 in, and the interior columns (nine on line B) are 12 in $\times$ 24 in for the bottom two stories and, respectively, 12 in $\times$ 16 in and 12 in $\times$ 20 in for the top and second highest stories. The height between floors is 12 ft. The dead load per unit area of floor, which consists of floor slab, beam, half the weight of columns above and below the floor, partition walls, etc., is estimated to be 140 psf. The normal live load is assumed as 125 psf. The soil below the foundation is assumed to be hard rock. The building site is located in seismic Zone 3. The building is intended to be used as a warehouse.

Perform the seismic analysis for this structure (in the direction normal to lines A, B, C) in accordance with the Uniform Building Code of 1985.

Solution:

1. Effective Weight at Various Floors: For a warehouse, the design load should include 25% of the live load. No live load needs to be considered in the roof. Hence, the effective weight at all floors, except at the roof, will be $140 + 0.25 \times 125 = 171.25$ psf, and the effective weight for the roof will be 140 psf. The plan area is 48 ft $\times$ 96 ft $= 4608$ ft^2. Hence, the weights of various levels are:

$$W_1 = W_2 = W_3 = 4608 \times 0.17125 = 789.1 \text{ kips}$$
$$W_4 = 4608 \times 0.140 = 645.1 \text{ kips}$$

   The total effective weight of the building is then

$$W = 789.1 \times 3 + 645.1 = 3012.4 \text{ kips}$$

2. Fundamental Period:

$$T = 0.1N \qquad\qquad \text{by eq. (23.7)}$$

$$= 0.40 \text{ sec}$$

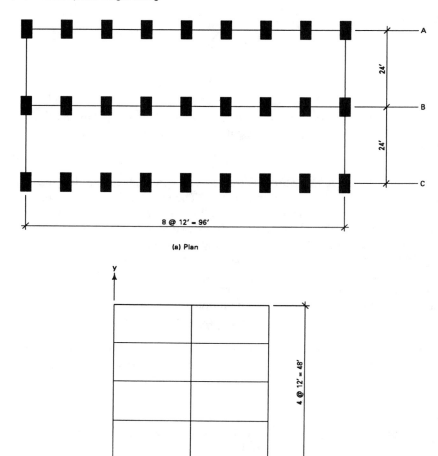

(a) Plan

(b) Elevation

Fig. 23.6 Plan an elevation for a four-story building of Example 23.1.

3.   Base Shear:

$$V = ZIKCSW \qquad \text{by eq. (23.3)}$$

$Z = 0.75$ (for site in Zone 3)

$I = 1.0$ (warehouse)

$K = 0.67$ ductile moment-resisting

space frame (Table 23.2)

$$C = \frac{1}{15\sqrt{T}} \le 0.12 \qquad \text{by eq. (23.4)}$$

$$C = \frac{1}{15\sqrt{0.40}} = 0.1054 < 0.12$$

$S = 1.0 \text{ (rock)} \quad \text{(Table 23.3)}$

$CS = 0.1054 \times 1.0 = 0.1054 < 0.14$

$V = 0.75 \times 1.0 \times 0.67 \times 0.1054 \times 1.0 \times 3012.4$

$\quad = 159.5 \text{ kips}$

4. Lateral Seismic Forces:

$$F_x = \frac{(V - F_t)W_x h_x}{\displaystyle\sum_{i=1}^{N} W_i h_i} \qquad \text{by eq. (23.9)}$$

$$F_t = 0 \quad \text{for} \quad T = 0.40 < 0.7 \text{ sec} \qquad \text{by eq. (23.10)}$$

The necessary calculations to determine the lateral seismic forces are shown in Table 23.4.
5. Shear Force V_x at Story x:

$$V_x = F_t + \sum_{i=x}^{N} F_i \qquad \text{by eq. (23.13)}$$

Calculated values are shown in the last columns of Table 23.4.
6. Overturning Moments M_x:

$$M_x = F_t(h_N - h_x) + \sum_{i=x}^{N} F_i(h_i - h_x) \qquad \text{by eq. (23.12)}$$

Calculated values are shown in Table 23.5.

TABLE 23.4   Lateral Forces

Level	W_i (kip)	h_x (ft)	$W_x h_x$ (k-ft)	F_x (kip)	V_x (kip)
4	645.1	48	30965	56.3	56.3
3	789.1	36	28408	51.6	107.9
2	789.1	24	18938	34.4	142.3
1	789.1	12	9469	17.2	159.5
			$\Sigma = 87780$		

TABLE 23.5   Calculation of Overturning and Torsional Moments

Level	F_x (kip)	M_x (kip-ft)	V_x (kip)	M_{tx} (kip-ft)
4	56.3	—	56.3	270.2
3	51.6	675.6	107.9	517.9
2	34.4	1970.4	142.3	683.0
1	17.2	3678.0	159.5	765.6
Base	—	5593.0	—	—

7.   Minimum Torsional Moment M_{tx}:

$$M_{tx} = 0.05 D_x V_x = (0.05)(96) V_x$$

$$V_x = \text{story shear force}$$

$$D = 96 \text{ ft} = \text{maximum building dimension}$$

Calculated values are shown in Table 23.5.

8.   Lateral Displacement δ_x and Story Drift Δ_x: The lateral displacements
of the building may be determined using a computer program which
implements the static stiffness method of analysis for plane frames.[2]
However, to simplify the hand calculations for this example, it is assumed that the building is a shear building. In this case, the stiffness
for a column between two consecutive floors is given by the formula

$$k = \frac{12EI}{L^3}$$

where

$L = 12$ ft (distance between two floors)

$E = 3 \times 10^3$ ksi (modulus elasticity concrete)

$I = \frac{1}{12} 12 \times 20^3 = 8000$ in^4 (moment of inertia

for the concrete section for columns 12 in $\times$ 20 in)

Therefore, for these columns,

$$k = \frac{12 \times 3 \times 10^3 \times 8000}{144^3} = 96.450 \text{ kip/in}$$

[2] A computer program for the analysis of plane orthogonal frames is included as Program
29 in the set of programs for use with this book.

Similarly, for columns 12 in × 24 in,

$$I = 13824 \text{ in}^4, \quad k = 166.667 \text{ k/in}$$

The total stiffness for the first and second stories is then

$$K_1 = K_2 = 18 \times 96.45 + 9 \times 166.67 = 3236 \text{ kip/in}$$

Similarly, for the third or fourth stories,

$$I = \tfrac{1}{12} 12 \times 16^3 = 4096 \text{ in}^4, \quad k = \frac{12 \times 3 \times 10^3 \times 4096}{144^3} = 49.4 \text{ kip/in}$$

$$I = \tfrac{1}{12} 12 \times 20^3 = 800 \text{ in}^4, \quad k = \frac{12 \times 3 \times 10^3 \times 8000}{144^6} = 96.5 \text{ kip/in}$$

Hence, total stiffness for the third or fourth stories is

$$K_3 = K_4 = 18 \times 49.4 + 9 \times 96.5 = 1757.7 \text{ kip/in}$$

The story drift Δ_x is given by $\Delta_x = V_x/k_x$ divided by the structural factor $K = 0.67$, as the code stipulates when $K < 1.0$. The results of the necessary calculations are shown in Table 23.6. The story drift permitted by the code is equal to 0.5% of the story height. Thus, for this example, the permissible drift is $0.005 \times 144 = 0.72$ (in), which is much larger than the drifts shown in Table 23.6 for this four-story building.

9. Recalculate Fundamental Period Using Rayleigh's Formula: The necessary calculations to determine the fundamental period using eq. (23.5) are shown in Table 23.7. Hence,

$$T = 2\pi \sqrt{\frac{111.64}{386 \times 33.65}} \sqrt{0.67} = 0.48 \text{ sec}$$

where the factor $\sqrt{0.67}$ is required, as explained in relation to eq. (23.5), to correct the displacements δ_i which were amplified by the factor $1/K$. The previously estimated fundamental period, $T = 0.40$ sec, is not sufficiently close to the value obtained by Rayleigh's formula, $T = 0.48$ sec.

TABLE 23.6 Calculation of Story Drift

Level (i)	Story Shear V_x (kip)	Story Stiffness K_x (kip/in)	Story Drift Δ_x (in)	Lateral Displacement δ_x (in)
4	56.3	1757.7	0.048	0.28
3	107.9	1757.7	0.092	0.23
2	142.3	3236.0	0.066	0.14
1	159.5	3236.0	0.074	0.07

TABLE 23.7　Calculations for Rayleigh's Formula

Level	W_i (kip)	δ_i (in)	$W_i\delta_i^2$	F_i (kip)	$F_i\,\delta_i$
4	645.12	0.28	50.58	56.3	15.76
3	789.12	0.23	41.74	51.6	11.87
2	789.12	0.14	15.46	34.4	4.82
1	789.12	0.07	3.86	17.2	1.20
			$\Sigma = 111.64$		$\Sigma = 33.65$

Therefore, the analysis should be repeated using this new value for the fundamental period. Such recalculation is undertaken in the solution of Example 23.3 using the computer program presented in the next section.

10. Diaphragm Forces: Table 23.8 shows the necessary calculations to determine the diaphragm forces from eq. (23.14) at the various levels of the building. The last column of this table contains the minimum values of F_{px} prescribed in the code. These minimum forces are greater than those calculated from eq. (23.14); therefore, they should be used in checking the design of the diaphragms.

23.8　PROGRAM 22—UBC 85

Program 22 follows the provisions of the section on Earthquake Regulation of the Uniform Building Code of 1985. The program has provisions to use eq. (23.6) or eq. (23.7) to determine an approximate value for the fundamental period T of the building. It also has provisions to either accept as input the soil factor S or to calculate this factor using eqs. (23.8), for which it is necessary to enter the value of the natural period of the soil, T_s.

The program calculates at each level, x, of the building, the lateral force F_x, the story shear force V_x, the horizontal torsional moment M_{tx}, the overturning moment M_x, the story drift Δ_x, the lateral displacement δ_x, and the

TABLE 23.8　Calculation for Diaphragm Forces

Level	F_i (kip)	W_i (kip)	$\sum_{i=x}^{N} F_i$	$\sum_{i=x}^{N} W_i$	F_{px} (kip) [eq. (23.14)]	F_{px} (kip), minimum
4	56.3	645.1	56.3	642.1	56.3	67.7
3	51.6	789.1	107.9	1434.2	59.4	82.9
2	34.4	789.1	142.3	2223.3	50.5	82.9
1	17.2	789.1	159.5	3012.4	41.8	82.9

diaphragm design force F_{px}. It also recalculates the fundamental period of the building using Rayleigh's formula, eq. (23.5). When this recalculated value differs significantly from the approximate value obtained from eq. (23.6) or eq. (23.7), it is recommended to repeat the calculations using the fundamental period determined from Rayleigh's formula.

Example 23.2.   Use Program 22 to solve Example 23.1.

Solution: The execution of Program 22 to implement the provisions of UBC-85 requires the previous modeling of the building to determine its stiffness matrix and its mass matrix. These matrices are determined by executing the Main Program and selecting one of the options presented by the MODELING BUILDING MENU. We select option 1, *Modeling as a Shear Building.*

Presented below are the input data and the output results for this example. As expected, the computer provides the same results as those calculated manually in Example 23.1.

Problem Data (from Example 23.1):

Zone number:	$NZ = 3$
Occupancy importance factor:	$I = 1.0$
Structural factor:	$K = 0.67$
Soil factor:	$S = 1.0$
Number of stories:	$N = 4$
Modulus of elasticity:	$E = 3000$ ksi
Maximum building dimension:	$D = 96$ ft
Story height ($i = 1, 2, 3, 4$):	$H_i = 12$ ft
Story weight ($i = 1, 2, 3$):	$W_i = 789.12$ kip
	$W_4 = 645.12$ kip

Moment of Inertia:

Columns 12 in $\times$ 20 in:	$I = 8000$ in^4
Columns 12 in $\times$ 24 in:	$I = 13824$ in^4
Columns 12 in $\times$ 16 in:	$I = 4096$ in^4

Story flexural stiffness:

$$EI_1 = EI_2 = 3000\,[18 \times 8000 + 9 \times 13{,}824] = 8.052 \times 10^8 \text{ kip} \cdot \text{in}^2$$

$$EI_3 = EI_4 = 3000\,[18 \times 4096 + 9 \times 8000] = 4.372 \times 10^8 \text{ kip} \cdot \text{in}^2$$

Input Data and Output Results

PROGRAM 27: SHEAR BUILDING   DATA FILE: D27

GENERAL DATA:

NUMBER OF STORIES (DEGREES OF FREEDOM)   ND = 4

STORY DATA:

STORY #	HEIGHT FT	FLEXURAL STIFFNESS EI (KIP-INCHES**2)	WEIGHT KIP
4	12.00	0.437E+09	645.120
3	12.00	0.437E+09	789.120
2	12.00	0.805E+09	789.120
1	12.00	0.805E+09	789.120

OUTPUT RESULTS:

SYSTEM STIFFNESS MATRIX (KIP/IN)

6.4718E+03	-3.2359E+03	0.0000E+00	0.0000E+00
-3.2359E+03	4.9931E+03	-1.7572E+03	0.0000E+00
0.0000E+00	-1.7572E+03	3.5143E+03	-1.7572E+03
0.0000E+00	0.0000E+00	-1.7572E+03	1.7572E+03

SYSTEM MASS MATRIX (KIP-SEC**2/IN)

2.0444E+00	0.0000E+00	0.0000E+00	0.0000E+00
0.0000E+00	2.0444E+00	0.0000E+00	0.0000E+00
0.0000E+00	0.0000E+00	2.0444E+00	0.0000E+00
0.0000E+00	0.0000E+00	0.0000E+00	1.6713E+00

PROGRAM 22: UBC-1985   DATA FILE: D22

INPUT DATA:

ZONE NUMBER	NZ= 3
OCCUPANCY IMPORTANCE FACTOR	I= 1
STRUCTURAL FACTOR	K= .67
SOIL FACTOR	S= 1

BUILDING DATA:

MAXIMUM BUILDING DIMENSION (FEET)	DM= 96
BUILDING DIMENSION, FORCE DIRECTION (FEET)	D= 48
NUMBER OF STORIES	N= 4

STORY #	HEIGHT FEET	FLEXURAL STIFFNESS (EI) KIP-INCHES**2	WEIGHT KIP
4	12.00	0.437E+09	645.12
3	12.00	0.437E+09	789.12
2	12.00	0.805E+09	789.12
1	12.00	0.805E+09	789.12

ESTIMATE NATURAL PERIOD MENU:

```
1. USE T=0.1*N              (DUCTILE FRAMES)
2. USE T=0.05*H/SQR(D)      (OTHER BUILDINGS)
3. USE ASSUMED VALUE FOR T
       SELECT NUMBER ? 1
```

OUTPUT RESULTS:

```
SEISMIC ZONE FACTOR      Z= .75
FUNDAMENTAL PERIOD       T= .4
SOIL FACTOR              S= 1
DYNAMIC FACTOR           C= .1054093
TOTAL BASE SHEAR         V= 159.5655
```

DISTRIBUTION LATERAL FORCES

LEVEL	LATER FORCE KIP	SHEAR FORCE KIP
4	56.29	56.29
3	51.64	107.93
2	34.43	142.35
1	17.31	159.57

ASSUME ONLY ACCIDENTAL ECCENTRICITY (Y/N) ? Y

TORSIONAL AND OVERTURNING MOMENTS

LEVEL	ECCENTRICITY FT	TORSIONAL MOMENT KIP-FT	OVERTURNING MOMENT KIP-FT
4	—	270.18	675.45
3	—	518.05	1970.57
2	4.80	683.29	3678.80
1	4.80	765.91	5593.59
BASE	4.80	—	—

LATERAL DISPLACEMENTS

LEVEL	DRIFT IN	DISPLACEMENTS IN	MAX. DRIFT PERMITTED IN
4	0.048	0.279	0.720
3	0.092	0.231	0.720

| 2 | 0.066 | 0.139 | 0.720 |
| 1 | 0.074 | 0.074 | 0.720 |

PERIOD BY RAYLEIGH'S FORMULA: T = 0.476 SEC

DIAPHRAGM FORCES

LEVEL	DIAPHRAGM FORCE
	KIP
4	67.74
3	82.86
2	82.86
1	82.86

Example 23.3. Solve Example 23.2 for the fundamental period $T = 0.48$ sec, which was calculated by Rayleigh's formula in that example.

Solution: In the execution of Program 22, we select the option "USE AS-SUMED VALUE FOR T" in the "NATURAL PERIOD MENU." The value $T = 0.48$ is then entered from the keyboard.

Shown below are the input data and output results for the solution of Example 23.3. Comparing these results with those obtained in Example 23.2, in which the assumed period was $T = 0.48$ sec, we observe a decrease in the total base shear force from 159.6 kips to 145.6 kips. This is a 9% decrease in the total lateral force.

Input Data and Output Results

PROGRAM 22: UBC-85   DATA FILE: D22

INPUT DATA:

ZONE NUMBER	NZ= 3
OCCUPANCY IMPORTANCE FACTOR	I = 1
STRUCTURAL FACTOR	K = .67
SOIL FACTOR	S = 1

BUILDING DATA:

MAXIMUM BUILDING DIMENSION (FEET)	DM= 96
NUMBER OF STORIES	N= 4
BUILDING DIMENSION, FORCE DIRECTION (FEET)	D= 48

STORY #	STORY HEIGHT FEET	STORY FLEXURAL STIFFNESS (EI) KIP-INCHES**2	STORY WEIGHT KIP
4	12.00	0.437E+09	645.12
3	12.00	0.437E+09	789.12

| 2 | 12.00 | 0.805E+09 | 789.12 |
| 1 | 12.00 | 0.805E+09 | 789.12 |

ESTIMATE NATURAL PERIOD MENU:

     1. USE T=0.1*N   (DUCTILE FRAMES)
     2. USE T=0.05*H/SQR(D)   (OTHER BUILDINGS)
     3. USE ASSUMED VALUE FOR T

          SELECT NUMBER ? 3

INPUT ASSUMED VALUE FOR FUNDAMENTAL PERIOD: T = ? .48

    OUTPUT RESULTS:

SEISMIC ZONE FACTOR   Z = .75
FUNDAMENTAL PERIOD   T = .48
SOIL FACTOR   S = 1
DYNAMIC FACTOR   C = 0.0962
TOTAL BASE SHEAR   V = 145.66

    DISTRIBUTION LATERAL FORCES

LEVEL	LATERAL FORCE KIP	SHEAR FORCE KIP
4	51.38	51.38
3	47.14	98.52
2	31.43	129.95
1	15.71	145.66

    ASSUME ONLY ACCIDENTAL ECCENTRICITY (Y/N) ? Y

    TORSIONAL AND OVERTURNING MOMENTS

LEVEL	ECCENTRICITY FT	TORSIONAL MOMENT KIP-FT	OVERTURNING MOMENT KIP-FT
4	4.80	246.64	—
3	4.80	472.91	616.60
2	4.80	623.76	1798.88
1	4.80	699.18	3358.27
BASE	—	—	5106.22

    LATERAL DISPLACEMENTS

LEVEL	DRIFT IN	DISPLACEMENTS IN	MAX. DRIFT PERMITTED IN
4	0.044	0.254	0.720
3	0.084	0.211	0.720
2	0.060	0.127	0.720
1	0.067	0.067	0.720

PERIOD BY RAYLEIGH'S FORMULA: T = 0.476 SEC

DIAPHRAGM FORCES

LEVEL	DIAPHRAGM FORCE
	KIP
4	67.74
3	82.86
2	82.86
1	82.86

23.9   DESIGN PROCESS

The earthquake resistant design of the structure is initiated, before considering dynamic forces, by a preliminary estimation of the dead load, live load, snow load, and wind load. This preliminary work allows determination of the effective weight W for the seismic analysis. The type of structural system determines the structural factor K and the use of the building determines the occupancy importance factor I. The geographic location determines the zone factor Z. The fundamental period is then estimated using the empirical formula, eq. (23.6), or the empirical formula, eq. (23.7), provided by the code.

A geotechnical investigation at the site may be conducted to determine the soil period T_s needed to calculate the value of the soil coefficient S from eq. (23.8). Alternatively, a value for this coefficient may be obtained from Table 23.3. This information establishes the coefficient C from eq. (23.4), the total base shear force V from eq. (23.3), and the lateral forces F_x from eqs. (23.9) and (23.10). These lateral forces are applied statically to the various levels of the building. Then the shear force, drift, displacement, overturning moment, and torsional moment are determined at each story of the building. The fundamental period of the building now may be recalculated by Rayleigh's formula, eq. (23.5), and the process repeated if the new calculated period is significantly different from the approximate value used in the preliminary analysis.

Example 23.4.   Solve Example 23.1 after modeling the structure as a plane orthogonal frame (Fig. 23.7). Assume all horizontal members to be 12 in × 24 in in reinforced concrete beams.

Solution: Program 29, which is included in the set of programs for use with this volume, is executed to model the structure as a plane frame. The program uses the stiffness method and static condensation method to determine the reduced stiffness matrix that corresponds to the lateral displacement coordinates at the various levels of the building.

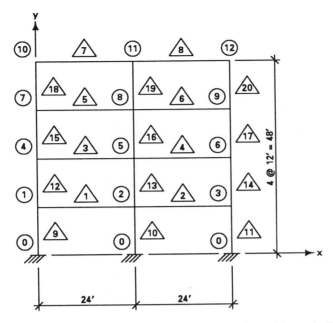

Fig. 23.7 Plane frame modeling the four-story building of Example 23.4.

The implementation of Program 29 to model the building as a plane orthogonal frame requires numbering the joints consecutively excluding the fixed joints (labeled zero) as shown in Fig. 23.7. In this model, each vertical or horizontal member represents all nine elements or members along the buildings. The flexural stiffness EI for the members of the frame are then calculated as follows:

Horizontal members:

$$EI = 9 \times 3000 \times 12 \times 24^3$$
$$= 0.37325 \times 10^9 \; (\text{kip} \cdot \text{in}^2)$$

Exterior columns:

First and second stories: $EI = 0.216 \times 10^9 \; (\text{kip} \cdot \text{in}^2)$

Third and fourth stories: $EI = 0.111 \times 10^9$

Interior columns:

First and second stories: $EI = 0.373 \times 10^9 \; (\text{kip} \cdot \text{in}^2)$

Third and fourth stories: $EI = 0.216 \times 10^9$

Input Data and Output Results

```
PROGRAM 29:  MODELING BLD. AS PLANE FRAME    DATA FILE:  D29

    GENERAL DATA:
```

```
NUMBER OF STORIES                        NST = 4
NUMBER OF JOINTS (EXCLUDING FIXED)       NJ = 12
NUMBER OF HORIZONTAL ELEMENTS            NH = 8
NUMBER OF VERTICAL ELEMENTS              NV = 12
    JOINT COORDINATES
```

JOINT #	X-COORDINATE FT	Y-COORDINATE FT
1	0.00	12.00
2	24.00	12.00
3	48.00	12.00
4	0.00	24.00
5	24.00	24.00
6	48.00	24.00
7	0.00	36.00
8	24.00	36.00
9	48.00	36.00
10	0.00	48.00
11	24.00	48.00
12	48.00	48.00

TOTAL WEIGHT AT EACH LEVEL

LEVEL #	WEIGHT (KIP)
1	789.12
2	789.12
3	789.12
4	645.12

DATA FOR HORIZONTAL ELEMENTS (STARTING AT FIRST LEVEL):

ELEMENT #	LEFT JOINT #	RIGHT JOINT #	FLEX. STIFF. (EI) KIP-IN**2
1	1	2	0.373E+09
2	2	3	0.373E+09
3	4	5	0.373E+09
4	5	6	0.373E+09
5	7	8	0.373E+09
6	8	9	0.373E+09
7	10	11	0.373E+09
8	11	12	0.373E+09

DATA FOR VERTICAL ELEMENTS:

ELEMENT #	STORY #	LOWER JOINT #	UPPER JOINT #	FLEXURAL STIFFNESS EI KIP-IN**2
9	1	0	1	0.216E+09
10	1	0	2	0.373E+09
11	1	0	3	0.216E+09
12	2	1	4	0.216E+09
13	2	2	5	0.373E+09
14	2	3	6	0.216E+09
15	3	4	7	0.111E+09
16	3	5	8	0.216E+09
17	3	6	9	0.111E+09
18	4	7	10	0.111E+09
19	4	8	11	0.216E+09
20	4	9	12	0.111E+09

OUTPUT RESULTS

REDUCED STIFFNESS MATRIX:

5.6114E+03	-2.9162E+03	4.5798E+02	-4.4552E+01
-2.9162E+03	3.8632E+03	-1.8849E+03	2.6869E+02
4.5798E+02	-1.8849E+03	2.8930E+03	-1.3985E+03
-4.4552E+01	2.6869E+02	-1.3985E+03	1.1678E+03

PROGRAM 22: UBC-1985   DATA FILE: D22

The input and output results for this example are identical to those of Example 23.1 except for the values of story drifts and lateral displacements at the various levels of the building. The following are the new values obtained for story drifts and lateral displacements as well as for the period calculated with Rayleigh's formula:

LATERAL DISPLACEMENTS

LEVEL	DRIFT IN	DISPLACEMENT IN	MAX. DRIFT PERMITTED IN
4	0.105	0.624	0.720
3	0.193	0.520	0.720
2	0.190	0.327	0.720
1	0.137	0.137	0.720

PERIOD BY RAYLEIGH'S FORMULA: T = 0.714 SEC

The drift and displacement values obtained for the structure modeled as a plane frame are slightly more than twice the corresponding values shown in

Table 23.6 for the structure modeled as a shear building in Example 23.1. The period now obtained by Rayleigh's formula, $T = 0.714$ sec, is much longer than the value obtained in Example 23.1 ($T = 0.48$ sec) and excessively high compared with the value ($T = 0.40$ sec) given by the empirical formula eq. (23.7). The use of this large value for the period $T = 0.714$ sec would produce a base shear force much lower than the "prudent" limit of 80% of the base shear force obtained when the period is calculated using the empirical formula eq. (23.7).

23.10 SUMMARY

In practice, the design of buildings and other structures to resist the effects of earthquakes is performed to satisfy local building code provisions. This chapter presents the main provisions of the Uniform Building Code of 1985, which is the code most extensively adopted in the different regions of the United States. This code, as well as other seismic codes in use in this country, is based on the response of the structure to lateral forces applied statically. However, the most important dynamic property of the structure, its fundamental period, is considered in the code. The calculations generally required in earthquake resistant design of buildings involve some simple manual calculations as illustrated in the numerical example presented. However, the computer programs available[3] for this volume provide assistance in the process of design since several iterations are needed before arriving at a final, satisfactory design.

The provisions for earthquake resistant design of buildings in the recent edition of the UBC-88 are presented in the next chapter. These new provisions will, in time, supersede those in the 1985 edition. However, it will take some time for local or regional building codes to reflect the new provisions of the UBC-88.

PROBLEMS

23.1 Use Program 22 for seismic resistant design of a 10-story concrete building ($E = 3000$ ksi) having the plan configuration and elevation shown in Fig. 23.8. The size of the exterior columns (nine each on line A and on line C) is 12 in × 20 in and the interior columns (nine on line B) are 12 in × 24 in for all the stories of the building. The height between floors is 12 ft. The building is located in the Los Angeles, California area and is intended to be used as a hospital. The estimated weight at each level, including permanent equipment, is 950 kip, except at the top level (roof), which is 650 kip. A geotechnical investigation of the site gave $T_s = 0.85$ sec for the natural period of the soil. Model the structure as a shear building (Program 27).

[3]See Appendix II.

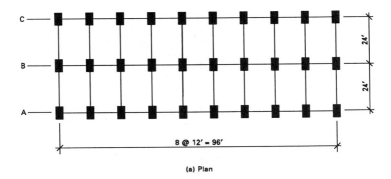

(a) Plan

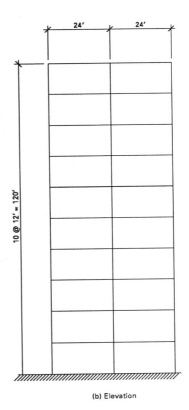

Fig. 23.8

(b) Elevation

23.2 Use Program 22 for the seismic resistant design of the 10-story steel building shown in Fig. 23.9. For each story, the flexural rigidity $EI = 8 \times 10^7$ kip·in^2 and the story weight at the floor level is $W = 500.00$ kip. Model this structure as a shear building (Program 27).

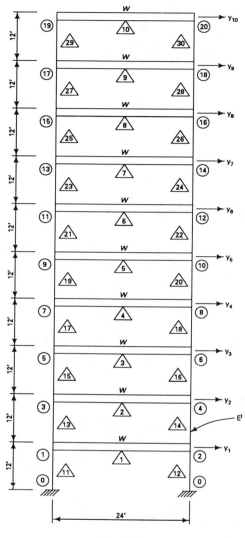

Fig. 23.9

23.3 Repeat Problem 23.2 modeling the structure as a plane orthogonal frame in which the flexural stiffness of the horizontal members is $EI = 6.0 \times 10^7$ kip·in² (Program 29).

23.4 A 12-story moment-resisting steel frame shown in Fig. 23.10 serves to model a building for earthquake analysis. The effective weight on all the levels is 90 psf except on the roof where it is 80 psf. The tributary load for the frames has a 30 ft width. The building is located in Los Angeles, California on a site with a soil profile classified as Type S_3. The intended use of the building is for offices. Use the provisions of the UBC-85 to determine: (a) equivalent lateral forces, (b) story shears, (c) horizontal torsional moments (assume accidental eccentricity), (d) overturning moments, (e) lateral displacements and story drifts, and (f) the design forces for the horizontal diaphragms. The overall length of the building is $D = 6 \times 30 = 180$ ft. Model the building as a plane orthogonal moment-resisting frame. (Use $E = 30,000$ ksi.)

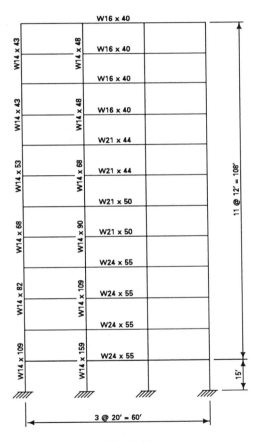

Fig. 23.10

23.5 Solve Problem 23.4 after modeling the structure as a shear building with rigid horizontal diaphragms at all levels of the building.

23.6 A three-story concrete warehouse building with shear walls as shown in Fig.
23.11 is located in San Diego, California on a soil profile of type 2. The story
heights are 20 ft for the first floor and 15 ft for the second and third floors. The
total design load for the first, second, and third levels are respectively 300, 250,
and 220 kip. Using the UBC-85, determine: (a) equivalent lateral seismic forces,
(b) story shears, and (c) horizontal lateral displacements and story drifts. Model
the shear walls as deep cantilever beams (Program 28). Assume modulus of elasticity for concrete $E_c = 3.6 \times 10^3$ ksi.

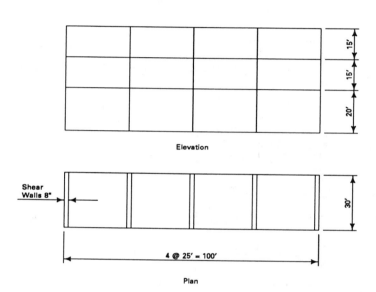

Fig. 23.11

24

Equivalent Static Lateral Force Method: Uniform Building Code—1988

The earthquake resistant design regulations of the 1988 Uniform Building Code [UBC-88 (International Conference of Building Officials 1988)] are based mainly on the publication by the Structural Engineers Association of California (1988) (SEAOC-1988) entitled *Recommended Lateral Force Requirements and Tentative Commentary*, which is based to some extent on the provisions of the Applied Technology Council (1978) recommendations (ATC3-06-1978) and of the Building Seismic Safety Council (1988) guidelines (BSSC-1988). The key UBC-1988 provisions for earthquake resistant design are presented in this chapter.

Compared to the UBC-1985 and previous codes, there is a new philosophy incorporated in the UBC-1988. Seismicity is expressed in terms of zone maps which take into account both the intensity of ground motion and the frequency of earthquake occurrence. The more recent code also imposes additional restrictions on permissible structural systems and on permissible design procedures as compared with the UBC-85.

24.1   EARTHQUAKE-RESISTANT DESIGN METHODS

The seismic zone map of Fig. 24.1 shows the continental United States classified in the following seismic zones: 0, 1, 2A, 2B, 3, and 4. The code provides two methods for earthquake resistant design: (1) The *static lateral force method* and (2) *the dynamic method.* The first method is applicable to all structures in Zone 1 and those in Zone 2 designed for Occupancy Importance Factor IV as described in Table 24.1. It is also applicable, in any seismic zone, to regular structures under 240 ft in height and to irregular structures[1] of not more than five stories nor over 65 ft in height. The dynamic method *may* be used for any structure but *must* be used for structures over 240 ft in height, irregular structures of over five stories or 65 ft in height, and structures located in seismic Zones 3 and 4 with dissimilar structural systems.

24.2   STATIC LATERAL FORCE METHOD

The UBC-88 stipulates that the structure should be designed for a total base shear force given by the following formula:

$$V = \frac{ZIC}{R_W} W \tag{24.1}$$

in which

$$C = \frac{1.25S}{T^{2/3}} \le 2.75 \tag{24.2}$$

The code also stipulates a minimum value for the ratio C/R_W of 0.075 except where the code prescribes scaling of forces by $3R_W/8$. The factors in eqs. (24.1) and (24.2) are defined as follows:

Z is the *seismic zone factor* related to the seismic zones (Fig. 24.1). It is equal to 0.075 for Zone 1, 0.15 for Zone 2A, 0.20 for Zone 2B, 0.30 for Zone 3, and 0.40 for Zone 4. The values of this coefficient can be considered to represent the effective peak ground acceleration (associated with an earthquake that has a 10% probability of being exceeded in 50 years) expressed as a fraction of the acceleration due to gravity.

I is the *occupancy importance factor* related to the anticipated use of the structures as classified in Table 24.1. The importance factor I is equal to 1.25 for essential and hazardous facilities. This value is less than the maximum

---

[1]Irregular structures are those that have significant physical discontinuities in configuration or in their lateral force-resisting systems. Specific features for vertical structural irregularities are described in Table 23-M and for horizontal irregularities in Table 23-N of UBC-88.

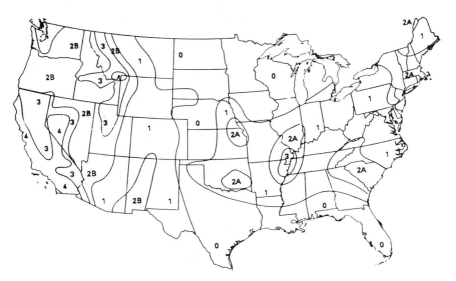

Fig. 24.1 Seismic map of the United States (UBC-88). (Reproduced from the 1988 Uniform Building Code, copyright 1988, with permission of the publishers, the International Conference of Building Officials.)

value of $I = 1.50$ in the UBC-85, but additional requirements are included in UBC-88 which should further increase the margin of safety.

S is the *site coefficient* depending on the characteristics of the soil at the site as described in Table 24.2. This coefficient is no longer specified on the basis of the ratio of the building period to the soil period as in UBC-85.

R_W is the *structural factor* ranging from 4 to 12 as given in Table 24.3. Analogous to the coefficient K in UBC-85, the coefficient R_W is a measure of the capacity of the structural system to absorb energy in the inelastic range through ductility and redundancy. It is based primarily on the performance of similar systems in past earthquakes. The approximate relationship between the two coefficients is $R_W \approx 8/k$.

W is the *seismic weight* which includes the dead weight of the building and 25% of the floor live load for storage or warehouse occupancy. For office buildings 10 psf is added to the dead load. For cases in which the design snow load is greater than 30 psf, the snow load should be included in W, although it could be reduced up to 75% when approved by the building official. The seismic weight W should also include the weight of permanent equipment and partitions as indicated in Section 2312(e) of UBC-88.

T is the *fundamental period* of the building which may be approximated from the following formula (Method A):

$$T = C_t \left[h_N^{3/4} \right] \tag{24.3}$$

TABLE 24.1   Occupancy Importance Factor

Occupancy Categories	Occupancy Type or Functions of Structure	Factor I
I. Essential Facilities	Hospitals and other medical facilities having surgery and emergency treatment areas. Fire and police stations. Tanks or other structures containing, housing, or supporting water or other fire-suppression materials or equipment required for the protection of essential or hazardous facilities, or special occupancy structures. Emergency vehicle shelters and garages. Structures and equipment in emergency-preparedness centers. Standby power-generating equipment for essential facilities. Structures and equipment in government communication centers and other facilities required for emergency response.	1.25
II. Hazardous Facilities	Structures housing, supporting, or containing sufficient quantities of toxic or explosive substances to be dangerous to the safety of the general public if released.	1.25
III. Special Occupancy Structure	Covered structures whose primary occupancy is public assembly—capacity > 300 persons. Buildings for schools through secondary or day-care centers—capacity > 250 students. Buildings for colleges or adult education schools—capacity > 500 students.	1.0

TABLE 24.1   (*Continued*)

Occupancy Categories	Occupancy Type or Functions of Structure	Factor I
	Medical facilities with 50 or more resident incapacitated patients, but not included above.	
	Jails and detention facilities.	
	All structures with occupancy > 5000 persons.	
	Structures and equipment in power-generating stations and other public utility facilities not included above, and required for continued operation.	
IV. Standard Occupancy Structure	All structures having occupancies or functions not listed above.	1.0

(Reprinted from the 1988 Uniform Building Code, © 1988, with permission of the publishers, the International Conference of Building Officials.)

TABLE 24.2   Site Coefficients

Type	Description	Site factor, S
S_1	A rocklike material characterized by a shear-wave velocity greater than 2,500 feet per second or a stiff or dense soil condition where the soil depth is less than 200 feet.	1.0
S_2	A soil profile with dense or stiff soil conditions, where the soil depth exceeds 200 feet.	1.2
S_3	A soil profile 40 feet or more in depth and containing more than 20 feet of soft to medium stiff clay but not more than 40 feet of soft clay.	1.5
S_4	A soil profile containing more than 40 feet of soft clay.	2.0

(Reprinted from the 1988 Uniform Building Code, © 1988, with permission of the publishers, the International Conference of Building Officials.)

where

h_N = total height of the building in feet

C_t = 0.035 for steel moment-resisting frames[2]

C_t = 0.030 for reinforced concrete moment-resisting frames and eccentrically braced frames

C_t = 0.020 for all other buildings[3]

Alternatively, the fundamental period of the structure may be determined from Rayleigh's formula (Method B) as

$$T = 2\pi \sqrt{\sum_{i=1}^{N} W_i \delta_i^2 \Big/ g \sum_{i=1}^{N} f_i \delta_i} \qquad (24.4)$$

where the values of f_i represent any lateral force distribution approximately consistent with results obtained using eqs. (24.5) and (24.6), or any other rational force distribution, and the values of δ_i are the elastic lateral deflections produced by the lateral loads f_i. The value of C in eq. (24.2) resulting from the use of T given by eq. (24.4) shall be not less than 80% of the value obtained by using T from eq. (24.3). This provision of the code is to avoid the possibility of using an excessively long calculated period to justify an unreasonably low base shear.

24.3   DISTRIBUTION OF LATERAL FORCES

The base shear force V calculated from eq. (24.1) is distributed, as in the UBC-85, at the floor levels of the building according to the following formulas:

$$F_x = \frac{(V - F_t)W_x h_x}{\sum_{i=1}^{N} W_i h_i} \qquad (24.5)$$

[2] A moment-resisting frame is a structural frame in which the members and joints are capable of resisting forces primarily by flexure.
[3] Alternatively, the value of C_t for structures with concrete or masonry structural walls (shear walls) may be taken as $0.1/\sqrt{A_c}$. The value of A_c shall be determined from the following formula:

$$A_c = \Sigma A_e[0.2 + (d_e/h_N)^2]$$

where A_e is the minimum cross-sectional area in any horizontal plane in the first story of a structural wall (in square feet), D_e the length, in feet, of a structural wall in the first story in the direction parallel to the applied forces, and h_N the height of the building in feet.

$$F_t = 0.07TV < 0.25V \quad \text{for} \quad T > 0.7 \text{ sec}$$

$$F_t = 0 \qquad\qquad\qquad \text{for} \quad T \le 0.7 \text{ sec} \qquad (24.6)$$

and

$$V = F_t + \sum_{i=1}^{N} F_i \qquad (24.7)$$

where

N = Total number of stories above the base of the building

F_x, F_i, F_N = Lateral force applied at level x, i, or N

F_t = Portion of the base force V at the top of the structure in addition to F_N

h_x, h_i = Height of level x or i above the base

W_x, W_i = Seismic weight of xth or ith level

The distribution of the lateral forces F_x and F_t are shown in Fig. 24.2 for a

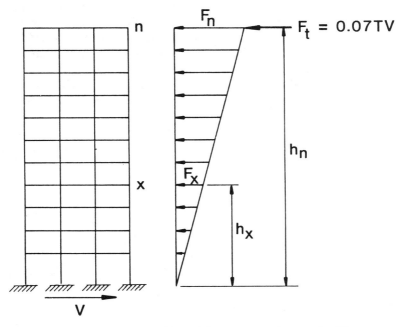

Fig. 24.2 Distribution of lateral forces in a multi-story building.

multistory building. The code stipulates that the force F_x, at level x, be applied over the area of the building according to the mass distribution at that level.

24.4   STORY SHEAR FORCE

The shear force V_x at any story x is given by the sum of the lateral seismic forces at that story and above, that is,

$$V_x = F_t + \sum_{i=x}^{N} F_i \tag{24.8}$$

24.5   HORIZONTAL TORSIONAL MOMENT

The UBC-1988 states that provisions should be made for the increased shear force resulting from horizontal torsion *where diaphragms are not flexible.* Diaphragms are considered flexible when the maximum lateral deformation of the diaphragm is more than twice the average story drift of the associated story.

The torsional moment at a given story results from the eccentricity between the center of the mass above that story and the center of stiffness of the resisting elements of the story. The code also requires that an accidental torsional moment be added to the actual torsional moment. The story accidental torsional moment is now calculated as the summation of the design lateral forces above the story multiplied by 5% of the building dimension perpendicular to the direction under consideration. The assumed accidental torsion is in addition to actual torsion existent in the structure and not a minimum torsion as in UBC-85.

Further provisions in the code account for torsional irregularities in the building by increasing the accidental torsion by an amplification factor A_x, determined as

$$A = \left(\frac{\delta_{max}}{1.2\delta_{avg}}\right)^2 \leq 3.0 \tag{24.9}$$

where

δ_{max} = the maximum displacement at level x
δ_{avg} = the average displacement at the extreme points of the structure at level x

24.6   OVERTURNING MOMENT

The code requires that overturning moments shall be determined at each level of the structure. The overturning moment is determined using the seismic design forces F_x and F_t [eqs. (24.5) and (24.6)] which act on levels above the level under consideration. Hence, the overturning moment M_x at level x of the building is given by

$$M_x = F_t(h_N - h_x) + \sum_{i=x}^{N} F_i(h_i - h_x) \qquad (24.10)$$

24.7 STORY DRIFT LIMITATION

Story drift, the relative displacement between consecutive floor levels produced by the design lateral forces, shall include calculated translational and torsional deflections. The calculated story drift shall not exceed $0.04/R_W$ times the story height, nor 0.005 times the story height for buildings less than 65 ft in height. For buildings greater in height than 65 ft, the calculated drift shall not exceed $0.03/R_W$ times the story height nor 0.004 times the story height. These limitations are more severe than those in UBC-85, but this is partially compensated by the fact that the lateral forces specified by UBC-88 are generally slightly lower than in UBC-85. In addition, the drift limitations are not subject to the 80% limit mentioned in the calculation of the period T, nor to the limitation imposed on the ratio C/R_W cited in relation to use of eq. (24.2).

24.8 P-DELTA EFFECTS (P-Δ)

The so-called P-Δ effect refers to the additional moment produced by the vertical loads and the lateral deflection of columns, or other elements of the building resisting lateral forces. Figure 24.3 shows a column supporting an axial compressive force P, a shear force V, and bending moments M_A and M_B at the two ends. Due to this load, the column undergoes a relative lateral displacement or drift Δ. In this case, the P-Δ effect results in a secondary moment $M_s = P\Delta$, which is resisted by an additional shear force $P\Delta/L$ in the column.

It should be realized that this simple calculation of the P-Δ effect involves an approximation, since the secondary moment $M_s = P\Delta$ will further increase the drift in the column and consequently will produce an increment of the secondary moment and shear force in the column.

An acceptable method to estimate the final drift is to add, for each story, the incremental drifts: Δ_x due to the primary overturning moment M_{xp}; $\Delta_x \cdot \Theta_x$ due to the secondary moment $M_{xs} = M_{xp} \cdot \Theta_x$, $\Delta_x \cdot \Theta_x^2$ due to the next incremental moment $M_{xs} \cdot \Theta_x = (M_{xp} \cdot \Theta_x) \cdot \Theta_x$, etc., where the primary and secondary moments M_{xp} and M_{xs}, as well as their ratio Θ_x, are defined in eq. (24.11).

Hence, the total story drift Δ_{xTOTAL} including the P-Δ effect is given by the geometric series

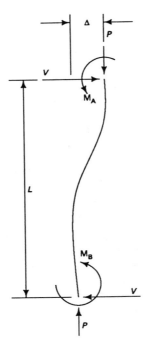

Fig. 24.3 Deflected column showing the P-Δ effect.

$$\Delta_{x\text{TOTAL}} = \Delta_x + \Delta_x\Theta_x + \Delta_x\Theta_x^2 + \cdots$$

which is equal to

$$\Delta_{x\text{TOTAL}} = \Delta_x\left(\frac{1}{1 - \Theta_x}\right)$$

Consequently, the P-Δ effect may be considered by multiplying, for each story, the calculated story drift and the calculated story shear force by the amplification factor $1/(1 - \Theta_x)$ and then proceed to recalculate the overturning moments, and other seismic effects for these amplified story shear forces.

The code specifies that the resulting member forces and moments, as well as the story drift induced by the P-Δ effect, shall be considered in the evaluation of overall structural frame stability.

The P-Δ effect need not be considered when the ratio of secondary moment (resulting from the story drift) to the primary moment (due to the seismic lateral forces), for any story, does not exceed 0.10. This ratio may be evaluated at each story as the product of the total dead and live loads above the story, times the seismic drift divided by the product of the seismic shear force times the height of the story. That is, the ratio Θ_x at level x between the secondary moment M_{xs} resulting from the P-Δ effect and the primary moment M_{xp}, due to the seismic lateral forces, may be calculated from the following formula:

$$\Theta_x = \frac{M_{xs}}{M_{xp}} = \frac{P_x \Delta_x}{V_x H_x} \tag{24.11}$$

where

$P_x =$ total weight at level x and above

$\Delta_x =$ drift of story x

$V_x =$ shear force of story x

$H_x =$ height of story x

24.9 DIAPHRAGM DESIGN FORCE

The code stipulates that floor and roof diaphragms should be designed to resist the forces determined by the following formula:

$$F_{px} = \frac{F_t + \sum\limits_{i=x}^{N} F_i}{\sum\limits_{i=x}^{N} W_i} W_{px} \tag{24.12}$$

in which F_i is the lateral seismic force, W_i is the seismic weight at each level, and W_{px} the weight of the diaphragm and attached parts of the building. The code states that the force F_{px} calculated by eq. (24.12) need not exceed $0.75ZIW_{px}$, but it shall not be less than $0.35ZIW_{px}$.

Example 24.1. Perform the seismic analysis for the four-story building of Example 23.1 using the static lateral force method of the Uniform Building Code–1988.

Solution:

1. Effective Weight at Various Levels: The calculated effective weights W_i in Example 23.1 are:

$$W_1 = W_2 = W_3 = 789.12 \text{ kip}$$

$$W_4 = 645.12 \text{ kip}$$

and total effective weight is

$$W = 3012.4 \text{ kip}$$

2. Fundamental Period:

$$T = C_t h_N^{3/4} \qquad \text{by eq. (24.3)}$$

where

$C_t = 0.030$ (for reinforced concrete moment-resisting frame)

$h_N = 48$ ft   (total height of the building)

$T = 0.030 \times 48^{3/4} = 0.55$ sec

3. Base Shear Force:

$$V = \frac{ZIC}{R_W} W \qquad \text{by eq. (24.1)}$$

$Z = 0.3$ site in Zone 3

$I = 1.0$ warehouse (Table 24.1)

$R_W = 12$ special moment-resisting space frame (Table 24.3)

$S = 1.0$ rock (Table 24.2)

$$C = \frac{1.25S}{T^{2/3}} \leq 2.75$$

$$C = \frac{1.25 \times 1.0}{(0.55)^{2/3}} = 1.862 \qquad \text{by eq. (24.2)}$$

and

$$C/R_W = 0.155 > 0.075$$

Therefore,

$$V = \frac{0.3 \times 1.0 \times 1.862}{12} \, 3012.4 = 140.23 \text{ kip}$$

4. Lateral Forces:

$$F_x = \frac{(V - F)W_x h_x}{\sum\limits_{i=1}^{N} W_i h_i} \qquad \text{by eq. (24.5)}$$

$$F_t = 0 \quad \text{for} \quad T = 0.55 < 0.7 \qquad \text{by eq. (24.6)}$$

Calculated lateral forces are shown in Table 24.4.

TABLE 24.3   Structural Factor

Lateral Load Resisting System	R_W	H_{max}*
A. BEARING WALL SYSTEM		
1. Light-framed walls with shear panels		
a. Plywood walls for structures of three stories or less	8	65
b. All other framed walls	6	65
2. Shear walls		
a. Concrete	6	160
b. Masonry	6	160
3. Light steel-framed bearing walls with tension bracing	4	65
4. Braced frames where bracing carries gravity loads		
a. Steel	6	160
b. Concrete**	4	
c. Heavy timber	4	65
B. BUILDING FRAME SYSTEM		
1. Steel braced frame (EBF)°	10	240
2. Light-framed walls with shear panel		
a. Plywood walls for structures three stories or less	9	65
b. All other framed walls	7	65
3. Shear walls		
a. Concrete	8	240
b. Masonry	8	160
4. Concentrated braced frames		
a. Steel	8	160
b. Concrete**	8	
c. Heavy timber	8	65
C. MOMENT-RESISTING FRAME SYSTEM		
1. Special moment-resisting frames (SMRSF)°		
a. Steel	12	N.L.
b. Concrete	12	N.L.
2. Concrete intermediate moment-resisting frames (IMRSF)°	8	
3. Ordinary moment-resisting frames (OMRSF)		
a. Steel	6	160
b. Concrete***	5	
D. DUAL SYSTEM†		
1. Shear walls		
a. Concrete walls with SMRSF	12	N.L.
b. Concrete with concrete IMRSF	9	160
c. Masonry with SMRSF	8	160
d. Masonry with concrete IMRSF	7	

(continued on next page)

TABLE 24.3   (Continued)

Lateral Load Resisting System	R_W	H_{max}*
2. Steel EBF with steel SMRSF	12	N.L.
3. Concrete braced frames		
a. Steel with steel SMRSF	10	N.L.
b. Concrete with concrete SMRSF**	9	
c. Concrete with concrete IMRSF**	6	

*H = height limit applicable to seismic zones 3 and 4
**Prohibited in seismic zones 3 and 4
***Prohibited in seismic zones 2, 3, and 4
†Structural system defined in the glossary, Appendix 2
NL = No Limit
(Reprinted from the 1988 Uniform Building Code, © 1988, with permission of the publishers, the International Conference of Building Officials.)

5. Shear Force at Story x:

$$V_x = F_t + \sum_{i=x}^{N} F_x \qquad \text{by eq. (24.8)}$$

Calculated shear forces are shown in Table 24.4.
6. Story Lateral Displacement and Drift: The structure is modeled as a shear building. This structure is analyzed also as a plane orthogonal frame, later in the chapter on the application of the computer program for implementation of UBC-88.

The flexural stiffness calculated in Example 23.1 for the first and second stories of the building is

$$K_1 = K_2 = 3236.0 \text{ kip/in}$$

and for the third and fourth stories

$$K_3 = K_4 = 1757.7 \text{ kip/in}$$

TABLE 24.4   Lateral Forces

Level	W_i (kip)	h_x (ft)	$W_x h_x$ (K-ft)	F_x (kip)	V_x (kip)
4	645.12	48	30965	49.5	49.5
3	789.12	36	28408	45.4	94.9
2	789.12	24	18939	30.3	125.2
1	798.12	12	9469	15.1	140.3
			$\Sigma = 87781$		

The drift for the stories is calculated by

$$\Delta_x = V_x/K_i$$

Then, the lateral displacement at any level is given by the sum of the drifts of the above stories. The results of these calculations are shown in Table 24.5.

The code stipulates that the story drift Δ_x should not exceed $(0.04/R_W)$ times the story height nor 0.005 times the story height: Hence, the maximum permissible drift Δ_{max} is given by the smallest of the following results:

$$\Delta_{max} = (0.04/12)144 = 0.48 \text{ in}$$

$$\Delta_{max} = 0.005 \times 144 = 0.72 \text{ in}$$

We observe in Table 24.5 that values of the drift Δ_x for any story are well below the maximum permissible drift.

7. Natural Period Using Rayleigh's Formula:

$$T = 2\pi \sqrt{\frac{\sum\limits_{i=1}^{N} W_i \delta_i^2}{g \sum\limits_{i=1}^{N} f_i \delta_i}} \qquad \text{by eq. (24.4)}$$

$$T = 2\pi \sqrt{\frac{38.72}{386 \times 17.42}} = 0.48 \text{ sec} \qquad \text{by eq. (24.5)}$$

The results of the necessary calculations are shown in Table 24.6.

Because Rayleigh's formula results in a value for the fundamental period ($T = 0.48$ sec) somewhat lower than the approximate value used in the calculations ($T = 0.55$ sec), the seismic analysis could be re-calculated using $T = 0.48$ sec.

TABLE 24.5   Story Drift and Lateral Displacement

Level	Story Shear V_x (kip)	Story Stiffness K_i (kip/in)	Story Drift Δ_x (in)	Lateral Displacement δ_x (in)
4	49.5	1757.7	0.028	0.164
3	94.9	1757.7	0.054	0.136
2	125.2	3236.0	0.039	0.082
1	140.3	3236.0	0.043	0.043

TABLE 24.6   Calculations for Rayleigh's Formula

Level	Weight W_i (kip)	Displacement δ_i (in)	Lateral force F_i (kip)	$W_i\delta_i^2$	$F_i\delta_i$
4	645.12	0.164	49.5	17.35	8.12
3	789.12	0.136	45.4	14.60	6.17
2	789.12	0.082	30.3	5.31	2.48
1	789.12	0.043	15.1	1.46	0.65
				$\Sigma = 38.72$	$\Sigma = 17.42$

8. Overturning Moments: The overturning moment at each level of the building is given by

$$M_x = F_t(h_N - h_x) + \sum_{i=x}^{N} F_i(h_i - h_x) \qquad \text{by eq. (24.10)}$$

Table 24.7 shows the results of the necessary calculations for this example.

9. Accidental Torsional Moments:

$$T_x = 0.05 * D * V_x$$

$$= 0.05 * 96 * V_x$$

Table 24.7 shows the necessary calculations for the torsional moments considering only the accidental torsion at the various levels of the building.

10. The P-Δ Effect: The code stipulates that the P-Δ effect need not be evaluated when the ratio of the secondary moment M_s to the primary moment M_p at each level of the building, is less than 0.10. The results of the necessary calculations to evaluate this ratio [eq. (24.11)] are in Table 24.8. This table shows that the largest moment ratio due to the P-Δ effect is 0.008 which is well below the code stipulation of 0.1. Consequently, there is no need to account for the P-Δ effect.

TABLE 24.7   Overturning Moments and Torsional Moments

Level x	Lateral Force F_x (kip)	Story Height H_x (ft)	Overturning Moment M_x (kip-ft)	Shear Force V_x (kip)	Torsional Moment T_x (kip-ft)
4	49.5	12	—	49.5	238
3	45.4	12	594	94.9	456
2	30.3	12	1724	125.2	601
1	15.1	12	3235	140.3	673
Base	—	—	4919	—	—

TABLE 24.8   Ratio of Secondary Moment to Primary Moment

Level x	Story Weight (kip)	Above Weight P_x (kip)	Story Drift Δ_x (in)	Story Shear V_x (kip)	Story Height H_x (in)	M_s/M_p $= P_x\Delta_x/V_xH_x$
4	645.1	645	0.028	49.5	144	0.002
3	789.1	1434	0.054	94.9	144	0.006
2	789.1	2223	0.039	125.2	144	0.005
1	789.1	3012	0.043	140.3	144	0.008

11. Diaphragm Design Force: The code requires that horizontal diaphragms (floors and roof) be designed to resist the force

$$F_{px} = \frac{F_t + \displaystyle\sum_{i=x}^{N} F_i}{\displaystyle\sum_{i=x}^{N} W_i} W_{px} \qquad \text{by eq. (24.12)}$$

which does not need to be greater than $0.75\ ZIW_{px}$, but shall not be less than $0.35\ ZIW_{px}$. Table 24.9 contains the results of the necessary calculations to determine the diaphragm design forces at the various levels of the building. The minimum values of F_{px}, shown in the last column, in this case, are the design forces for the diaphragms of the building.

24.10   PROGRAM 23 UBC-88: EQUIVALENT STATIC LATERAL FORCE METHOD

Program 23 implements the provisions of the Uniform Building Code of 1988 using the static lateral force method. The program has provisions to estimate the fundamental period T of the building from eq. (24.3) and to calculate the

TABLE 24.9   Diaphragm Forces

Level x	F_i (kip)	W_i (kip)	$\displaystyle\sum_{i=x}^{N} F_i$ (kip)	$\displaystyle\sum_{i=x}^{N} W_i$ (kip)	F_{px} [eq. (24.12)] (kip)	F_{px} Min. (kip)
4	49.5	645.1	49.5	645.1	49.5	67.7
3	45.4	789.1	94.9	1434.2	52.2	82.8
2	30.3	789.1	125.2	2223.3	44.4	82.8
1	15.1	789.1	140.3	3012.4	36.8	82.8

total base shear force V from (24.1) and the lateral forces F_x and F_t from eqs.
(24.5) and (24.6). The program calculates at each level of the building the shear
force V_x, the overturning moment M_x, the torsional moment M_{tx}, the lateral
displacement δ_x, the diaphragm design force F_{px}, and the story drift Δ_x. It also
calculates the secondary moment M_{xs}, the primary moment M_{xp}, and their
ratio Θ_x using eq. (24.11). Finally, the fundamental period T is recalculated
using Rayleigh's formula, eq. (24.4).

Example 24.2. Use Program 23 to solve Example 24.1 after modeling the
building as a plane orthogonal frame. Values for the flexural stiffness for the
horizontal and for the vertical members of the frame are given in Example
23.4.

Solution: As indicated in Example 23.4, Program 29 is used first to model
the structure as a plane orthogonal frame. The values for the flexural stiffness
of the horizontal and of the vertical members of the frame are given in Example 23.4. Next, program 23 is executed to implement UBC-88. The numerical results provided by the computer are the same as the values obtained
by hand calculation in Example 24.1, except for the values of story-drifts and
lateral displacements. For the building modeled as a plane frame, the story-
drifts and lateral displacements are slightly more than twice the values obtained in Example 24.1 for the structure modeled as a shear building.

Input Data and Output Results

```
PROGRAM 23: UBC-88 (STATICS)           DATA FILE: D23

      INPUT DATA:

SEISMIC ZONE                                    NZ$ = 3
OCCUPANCY IMPORTANCE FACTOR                      I = 1
STRUCTURAL SYSTEM FACTOR                        RW = 12
SITE COEFFICIENT                                 S = 1

      BUILDING DATA:

BUILDING DIMENSION NORMAL TO LATERAL FORCES (FT)      D = 96
NUMBER OF STORIES                                    N = 4

STORY          STORY          STORY
  #            HEIGHT         WEIGHT
               FEET            KIP

  4            12.00          645.12
  3            12.00          789.12
  2            12.00          789.12
  1            12.00          789.12
```

ESTIMATE NATURAL PERIOD MENU:

1. USE CT = 0.035 FOR STEEL MOMENT-RESISTING FRAMES

2. USE CT = 0.030 FOR REINFORCED CONCRETE MOMENT-RESISTING
              FRAMES AND ECCENTRIC BRACED FRAMES

3. USE CT = 0.020 FOR ALL OTHER BUILDINGS

4. USE ASSUMED VALUE FOR THE PERIOD T

              SELECT NUMBER ? 2

OUTPUT RESULTS:

SEISMIC ZONE FACTOR	Z =	.3
FUNDAMENTAL PERIOD	T =	.547 SEC
SOIL FACTOR	S =	1
DYNAMIC FACTOR	C =	1.8689
TOTAL BASE SHEAR	V =	140.75 KIP

DISTRIBUTION LATERAL FORCES:

LEVEL	LATERAL FORCE KIP	SHEAR FORCE KIP
4	49.65	49.65
3	45.55	95.20
2	30.37	125.57
1	15.18	140.75

FLEXIBLE DIAPHRAGMS (Y/N) ? N

NOTE:

THE TORSIONAL ECCENTRICITY AT A LEVEL IS EQUAL TO THE DISTANCE FROM
THE CENTER OF THE ABOVE MASS TO THE CENTER OF STIFFNESS AT THAT LEVEL

CONSIDER ONLY ACCIDENTAL ECCENTRICITY (Y/N) ? Y

TORSIONAL AND OVERTURNING MOMENTS:

LEVEL	ECCENTRICITY FEET	OVERTURNING MOMENT KIP-FT	TORSIONAL MOMENT KIP-FT
4	4.80	0.00	238.32
3	4.80	595.80	456.96
2	4.80	1738.21	602.72
1	4.80	3245.01	675.60
0	0.00	4934.01	0.00

DISPLACEMENT AND STORY DRIFT:

LEVEL	DRIFT IN	DISPLACEMENTS IN	MAX. DRIFT PERMITTED IN
4	0.062	0.369	0.480
3	0.114	0.307	0.480
2	0.112	0.193	0.480
1	0.081	0.081	0.480

PERIOD BY RAYLEIGH'S FORMULA: T = 0.714 SEC

DIAPHRAGM FORCES:

LEVEL	DIAPHRAGM FORCE
4	67.74
3	82.86
2	82.86
1	82.86

P-DELTA EFFECT:

LEVEL	LATERAL FORCE DEFLEC. IN	P-DELTA DEFLEC. IN	TOTAL DEFLEC. IN	RATIO OF MOMENTS
4	0.369	0.004	0.373	0.006
3	0.307	0.004	0.311	0.012
2	0.193	0.003	0.196	0.014
1	0.081	0.001	0.082	0.012

Example 24.3. Solve Example 24.2 for the period $T = .714$ calculated by Rayleigh's formula in that example.

Solution: Program 23 is executed after modeling the building as a plane frame using Program 29. Option 4 is selected in the Estimate Natural Frequency Menu and the value $T = 0.714$ is input.

From the results that follow, the base shear force is now equal to 117.84 kip. This value is greater than 80% of the base shear force ($0.80 \times 140.75 = 112.8$) obtained in Example 24.2 using the empirical formula to estimate the period of the building. Therefore, the reduced value for the shear force ($V = 117.84$ kip) is acceptable by UBC-88.

Input Data and Output Results

PROGRAM 23: UBC-88 (STATICS)   DATA FILE: D23

    INPUT DATA:

```
SEISMIC ZONE                                        NZ$ = 3
OCCUPANCY IMPORTANCE FACTOR                          I = 1
STRUCTURAL SYSTEM FACTOR                            RW = 12
SITE COEFFICIENT                                     S = 1

        BUILDING DATA:

BUILDING DIMENSION NORMAL TO LATERAL FORCES (FT)    D = 96
NUMBER OF STORIES                                   N = 4

STORY #        STORY HEIGHT        STORY WEIGHT
                  FEET                 KIP

    4            12.00              645.12
    3            12.00              789.12
    2            12.00              789.12
    1            12.00              789.12
```

ESTIMATE NATURAL PERIOD MENU:

1. USE CT = 0.035 FOR STEEL MOMENT-RESISTING FRAMES

2. USE CT = 0.030 FOR REINFORCED CONCRETE MOMENT-RESISTING
                   FRAMES AND ECCENTRIC BRACED FRAMES

3. USE CT = 0.020 FOR ALL OTHER BUILDINGS

4. USE ASSUMED VALUE FOR THE PERIOD T

                SELECT NUMBER ? 4

INPUT ASSUMED VALUE FOR FUNDAMENTAL PERIOD: T = ? 0.714

        OUTPUT RESULTS:

```
SEISMIC ZONE FACTOR                    Z = 0.3
FUNDAMENTAL PERIOD                     T = 0.714  SEC
SOIL FACTOR                            S = 1
DYNAMIC FACTOR                         C = 1.564
TOTAL BASE SHEAR                       V = 117.84 KIP
```

        DISTRIBUTION LATERAL FORCES:

```
LEVEL         LATERAL FORCE         SHEAR FORCE
                  KIP                   KIP

    4            45.38              45.38
    3            36.23              81.61
    2            24.15              105.76
    1            12.08              117.84
```

        FLEXIBLE DIAPHRAGMS (Y/N) ? N

NOTE:

THE TORSIONAL ECCENTRICITY AT A LEVEL IS EQUAL TO THE DISTANCE FROM
THE CENTER OF THE ABOVE MASS TO THE CENTER OF STIFFNESS AT THAT LEVEL

CONSIDER ONLY ACCIDENTAL ECCENTRICITY (Y/N) ? Y

TORSIONAL AND OVERTURNING MOMENTS:

LEVEL	ECCENTRICITY FEET	OVERTURNING MOMENT KIP-FT	TORSIONAL MOMENT KIP-FT
4	4.80	0.00	217.83
3	4.80	544.57	391.73
2	4.80	1523.89	507.66
1	4.80	2793.06	565.63
0	0.00	4207.14	0.00

DISPLACEMENT AND STORY DRIFT:

LEVEL	DRIFT IN	DISPLACEMENTS IN
4	0.055	0.316
3	0.098	0.260
2	0.095	0.163
1	0.068	0.068

PERIOD BY RAYLEIGH'S FORMULA: T = 0.714 SEC

DIAPHRAGM FORCES:

LEVEL	DIAPHRAGM FORCE
4	67.74
3	82.86
2	82.86
1	82.86

P-DELTA EFFECT:

LEVEL	LATERAL FORCE DEFLEC. IN	P-DELTA DEFLEC. IN	TOTAL DEFLEC. IN	RATIO OF MOMENTS
4	0.316	0.004	0.320	0.005
3	0.260	0.003	0.264	0.012
2	0.163	0.002	0.165	0.014
1	0.068	0.001	0.069	0.012

24.11   SUMMARY

The Uniform Building Code of 1988 provides two methods for earthquake resistant design of buildings, the static lateral force method and the dynamic method. The static lateral force method is applicable to (1) structures in Zone 1 and to those structures in Zone 2 designed for occupancy importance factor IV, (2) regular structures under 240 ft in height, and (3) irregular structures of no more than five stories nor over 65 ft in height. The *dynamic method* may be used for any structure including tall structures over 240 ft with irregular features, as permitted by the code.

The first of these methods is presented in this chapter. The second method, which is based on the modal superposition method and the use of response spectra, is presented in Chapter 25. Computer Program 23 implements the provisions of UBC-88 using the Static Lateral Force Method.

PROBLEMS

24.1 Use Program 23 to solve Problem 23.1 by the equivalent static lateral force method of the UBC-88.

24.2 Use Program 23 to solve Problem 23.2 by the equivalent static lateral force method of the UBC-88.

24.3 Repeat Problem 24.2 modeling the structure as a plane orthogonal frame.

24.4 Solve Problem 23.4 using UBC-88.

24.5 Solve Problem 24.4 using the value of the natural period determined with Rayleigh's formula observing the limitation of 80% in the value of coefficient C in eq. (24.2).

24.6 Solve Problem 23.6 using the UBC-88 by the equivalent static lateral force method.

24.7 Solve Problem 24.6 using the value of the natural period determined with Rayleigh's formula. Observe the limitation of 80% in the value of coefficient C in eq. (24.2).

25

Uniform Building Code–1988: Dynamic Method

In Chapter 8, we introduced the concept of response spectrum as a plot of the maximum response (spectral displacement, spectral velocity, or spectral acceleration) versus the natural frequency or natural period of a single degree-of-freedom system subjected to a specific excitation. In the present chapter, we will use seismic response spectra for earthquake resistant design of buildings modeled as discrete systems with concentrated masses at each level of the building.

It is necessary to perform a transformation of coordinates to obtain the modal equations of motion, then combine their spectral responses to obtain the maximum response of the structure because response spectral charts are prepared for single degree-of-freedom systems. In earthquake resistant design of buildings, the maximum responses include displacements, accelerations, shear forces, overturning moments, and torsional moments.

25.1 MODAL SEISMIC RESPONSE OF BUILDINGS

The equations of motion of a building modeled with lateral displacement coordinates at the N levels and subjected to seismic excitation at the base may be written, neglecting damping, from eq. (11.35) as

$$[M]\{\ddot{u}\} + [K]\{u\} = -[M]\{1\}\,\ddot{y}_s(t) \tag{25.1}$$

In eq. (25.1), $[M]$ and $[K]$ are respectively the mass and the stiffness matrices of the system, $\{u\}$ and $\{\ddot{u}\}$ are respectively the displacement and acceleration vectors (relative to the base), $\ddot{y}_s(t)$ is the function of the seismic acceleration at the base of the building, and $\{1\}$ a vector with all its elements equal to 1.

25.1.1   Modal Equation and Participation Factor

As presented in Chapter 11, the solution of eq. (25.1) may be found by solving the corresponding eigenproblem

$$[[K] - \omega^2[M]]\{\Phi\} = \{0\} \tag{25.2}$$

to determine the natural frequencies ω_1, ω_2, ..., ω_N (or natural periods T_1, T_2, ..., T_N) and the modal matrix $[\Phi]$ with its columns containing the normalized modal shapes. Then the linear transformation

$$\{u\} = [\Phi]\{z\} \tag{25.3}$$

is introduced in eq. (25.1) to yield the modal equations

$$\ddot{z}_m + \omega_m^2 z_m = -\Gamma_m \ddot{y}_s(t) \qquad (m = 1, 2, ..., N) \tag{25.4}$$

in which Γ_m is the participation factor given by eq. (11.39) as

$$\Gamma_m = \frac{\displaystyle\sum_{i=1}^{N} W_i \phi_{im}}{\displaystyle\sum_{i=1}^{N} W_i \phi_{im}^2} \tag{25.5}$$

For normalized eigenvectors, the participation factor reduces to

$$\Gamma_m = \frac{1}{g}\sum_{i=1}^{N} \phi_{im} W_i \tag{25.6}$$

because in this case

$$\sum_{i=1}^{N} \frac{W_i}{g}\phi_{im}^2 = 1$$

where g is the acceleration due to gravity. Damping may be introduced in the modal equation (25.4) by simply adding the damping term to this equation, namely,

$$\ddot{z}_m + 2\xi_m \omega_m \dot{z}_m + \omega_m^2 z_m = -\Gamma_m \ddot{y}_s(t) \tag{25.7}$$

where ξ_m is the modal damping ratio. Equation (25.7) can be written for convenience with omission of the participation factor as

$$\ddot{q}_m + 2\xi_m\omega_m\dot{q}_m + \omega_m^2 q_m = \ddot{y}_s(t) \qquad (25.8)$$

with the substitution

$$z_m = -\Gamma_m q_m \qquad (25.9)$$

25.1.2   Modal Shear Force

The value of the maximum response in eq. (25.8) for the modal spectral acceleration, $S_{am} = (\ddot{q}_m)_{max}$, is found from an appropriate response spectral chart, such as the charts in Chapter 8 or the spectral chart provided by UBC-88 (Fig. 25.1).

From eqs. (25.3) and (25.9), the maximum acceleration a_{xm} of the mth mode at the level x of the building is given by

$$a_{xm} = \Gamma_m\phi_{xm}S_{am} \qquad (25.10)$$

in which S_{am} and a_{xm} are usually expressed in units of the gravitational acceleration g.

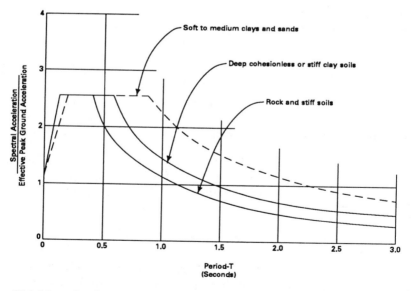

Fig. 25.1 Normalized response spectra shapes. (Reproduced from the *1988 Uniform Building Code*, © 1988, with permission of the publishers, the International Conference of Building Officials.)

As stated in Chapter 8, the modal values of the spectral acceleration S_{am}, the spectral velocity S_{vm}, and the spectral displacement S_{dm} are related by an apparent harmonic relationship:

$$S_{am} = \omega_m S_{vm} = \omega_m^2 S_{dm}$$

or in terms of the modal period $T_m = 2\pi/\omega_m$ by

$$S_{am} = \frac{2\pi}{T_m} S_{vm} = \left(\frac{2\pi}{T_m}\right)^2 S_{dm}$$

On the basis of these relations, the spectral acceleration S_{am} in eq. (25.10) may be replaced by the spectral displacement S_{dm} times ω_m^2 or by the spectral velocity S_{vm} times ω_m.

The modal lateral force F_{xm} at the level x of the building is then given by Newton's Law as

$$F_{xm} = a_{xm} W_x$$

or by eq. (25.10) as

$$F_{xm} = \Gamma_m \phi_{xm} S_{am} W_x \qquad (25.11)$$

in which S_{am} is the modal spectral acceleration in g units and W_x is the weight attributed to the level x of the building.

The modal shear force V_{xm} at the level x of the building is equal to the sum of the seismic forces F_{xm} above that level, namely,

$$V_{xm} = \sum_{i=x}^{N} F_{im} \qquad (25.12)$$

The total modal shear force V_m at the base of the building is then calculated as

$$V_m = \sum_{i=1}^{N} F_{im} \qquad (25.13)$$

or using eq. (25.11)

$$V_m = \sum_{i=1}^{N} \Gamma_m \phi_{im} W_i S_{am} \qquad (25.14)$$

25.1.3   Effective Modal Weight

The effective modal weight W_m is defined by the equation

$$V_m = W_m S_{am} \qquad (25.15)$$

Then, from eq. (25.14), the modal weight is

$$W_m = \Gamma_m \sum_{i=1}^{N} \phi_{im} W_i \qquad (25.16)$$

Combining eqs. (25.5) and (25.16) results in the following important expression for the effective modal weight:

$$W_m = \frac{\left[\sum_{i=1}^{N} \phi_{im} W_i \right]^2}{\sum_{i=1}^{N} \phi_{im}^2 W_i} \qquad (25.17)$$

It can be proven (Clough and Penzien 1975, pp. 559–560) analytically that the sum of the effective modal weights for all the modes of the building is equal to the total design weight of the building, that is,

$$\sum_{m=1}^{N} W_m = \sum_{i=1}^{N} W_i \qquad (25.18)$$

Equation (25.18) is most convenient in assessing the number of significant modes of vibration to consider in the design. Specifically, the UBC-88 requires that, in applying the dynamic method of analysis, all the significant modes of vibration be included. This requirement can be satisfied by including a sufficient number of modes such that their total effective modal weight is at least 90% of the total design weight of the building. Thus, this requirement can be satisfied by simply adding a sufficient number of effective modal weights [eq. (25.17)] until their total weight is 90% or more of the seismic design weight of the building.

25.1.4   Modal Lateral Forces

By combining eq. (25.11) with eqs. (25.15) and (25.16), we may express the modal lateral force F_{xm} as

$$F_{xm} = C_{xm} V_m \qquad (25.19)$$

where the modal seismic coefficient C_{xm} at level x is given by

$$C_{xm} = \frac{\phi_{xm} W_x}{\sum_{i=1}^{N} \phi_{im} W_i} \qquad (25.20)$$

25.1.5 Modal Displacements

The modal displacement δ_{xm} at the level x of the building may be expressed, in view of eqs. (25.3) and (25.9), as

$$\delta_{xm} = \Gamma_m \phi_{xm} S_{dm} \tag{25.21}$$

where Γ_m is the participation factor for the mth mode, ϕ_{xm} is the component of the modal shape at level x of the building, and S_{dm} is the spectral displacement for that mode.

Alternatively, the modal displacement δ_{xm} may be calculated from Newton's Law of Motion in the form

$$F_{xm} = \frac{W_x}{g} \omega_m^2 \delta_{xm} \tag{25.22}$$

because the magnitude of the modal acceleration corresponding to the modal displacement δ_{xm} is $\omega_m^2 \delta_{xm}$. Hence, from eq. (25.22)

$$\delta_{xm} = \frac{g}{\omega_m^2} \cdot \frac{F_{xm}}{W_x} \tag{25.23}$$

or substituting $\omega_m = 2\pi/T_m$

$$\delta_{xm} = \frac{g}{4\pi^2} \cdot \frac{T_m^2 F_{xm}}{W_x} \tag{25.24}$$

where T_m is the mth natural period.

25.1.6 Modal Drift

The modal drift Δ_{xm} for the xth story of the building, defined as the relative displacement of two consecutive levels, is given by

$$\Delta_{xm} = \delta_{xm} - \delta_{(x-1)m} \tag{25.25}$$

with $\delta_{0m} = 0$.

As indicated in Article 24.7, the UBC-88 stipulates that the calculated story drift, which must include translational and torsional deflections, shall not exceed $0.04/R_w^1$ times the story height nor 0.005 times the story height for buildings less than 65 ft in height. For buildings greater in height than 65 ft, the calculated drift shall not exceed $0.03/R_w$ times the story height nor 0.004 times the story height.

[1]R_w is the structural factor given in Table 24.3.

25.1.7   Modal Overturning Moment

The modal overturning moment M_{xm} at the level x of the building which is calculated as the sum of the moments of the seismic forces F_{xm} above that level is given by

$$M_{xm} = \sum_{i=x}^{N} F_{im}(h_i - h_x) \tag{25.26}$$

where h_i is the height of level i.

The modal overturning moment M_m at the base of the building then is given by

$$M_m = \sum_{i=1}^{N} F_{im}h_i \tag{25.27}$$

25.1.8   Modal Torsional Moment

The modal torsional moment M_{txm} at level x, which is due to eccentricity e_x between the center of the above mass and the center of stiffness at that level (measured normal to the direction considered), is calculated as

$$M_{txm} = e_x V_{xm} \tag{25.28}$$

where V_{xm} is the modal shear force at level x.

As mentioned in Chapter 24, the UBC-88 requires that an accidental torsional moment be added to the torsional moment existent at each level. The recommended way to add the accidental torsion is to offset the center of mass at each level by 5% of the dimension of the building normal to the direction under consideration.

25.2   TOTAL DESIGN VALUES

The design values for the base shear, story shear, lateral deflection, story drift, overturning moment, and torsional moment are obtained by combining corresponding modal responses. Such combination has been performed by application of the technique SRSS. This technique, as mentioned in Chapter 11, estimates the maximum modal response by calculating the square root of the sum of the squared values of the modal contributions. However, as discussed in Section 11.6 of Chapter 11, the SRSS technique may result in relatively large errors when some of the natural frequencies are closely spaced. This situation generally occurs in the analysis of space structures. At the present, the more refined technique described in that section, CQC (complete quadratic contribution), is becoming the technique of choice for implementation in computer programs. For preliminary or hand calculation, however, the simpler technique SRSS commonly is used. The following formulas may be used to estimate maximum design values by application of the SRSS technique:

1. Design Base Shear:

$$V = \sqrt{\sum_{m=1}^{N} V_m^2} \qquad (25.29)$$

   where the modal base shear V_m is given by eq. (25.15).

2. Design Lateral Seismic Force at Level x:

$$F_x = \sqrt{\sum_{m=1}^{N} F_{xm}^2} \qquad (25.30)$$

   where the seismic modal force F_{xm} is given by eq. (25.19).

3. Design Shear Force at Story x:

$$V_x = \sqrt{\sum_{m=1}^{N} V_{xm}^2} \qquad (25.31)$$

   where the modal shear force V_{xm} at level x is given by eq. (25.12).

4. Design Lateral Deflection at Level x:

$$\delta_x = \sqrt{\sum_{m=1}^{N} \delta_{xm}^2} \qquad (25.32)$$

   where the modal displacement δ_{xm} at level x is given by eq. (25.23) or eq. (25.24).

5. Design Drift for Story x:

$$\Delta_x = \sqrt{\sum_{m=1}^{N} \Delta_{xm}^2} \qquad (25.33)$$

   where the modal drift Δ_{xm} at story x is given by eq. (25.25).

6. Design Overturning Moment at Level x:

$$M_x = \sqrt{\sum_{m=1}^{N} M_{xm}^2} \qquad (25.34)$$

   where the modal overturning moment M_{xm} at level x is given by eq. (25.26).

7. Design Torsional Moment at Level x:

$$M_{tx} = \sqrt{\sum_{m=1}^{N} M_{txm}^2} \qquad (25.35)$$

   where the modal torsional moment M_{txm} at level x is given by eq. (25.28).

25.3 PROVISIONS OF UBC-88: DYNAMIC METHOD

In Chapter 24, we defined the various factors needed to calculate the base shear force given by eqs. (24.1) and (24.2) of the UBC-88 using the equivalent lateral force method. The following is a recapitulation of these factors which are also pertinent to the Dynamic Method:

1. *Seismic Zone Coefficient Z.* This coefficient has been defined by classifying the United States in five seismic zones on the basis of past historical earthquake activity. These zones are designated as 1, 2A, 2B, 3, and 4 as shown in Fig. 24.1. The numerical values assigned to the coefficient are:

Zone	Coefficient
1	0.075
2A	0.15
2B	0.20
3	0.30
4	0.40

   The seismic coefficient indicates the effective peak ground acceleration of the design earthquake.
2. *The Occupancy Importance Factor I.* This factor is equal to 1.25 for essential or hazardous facilities and 1.0 for special or standard occupancy structures as described in Table 24.1.
3. *The Site Coefficient S.* The site coefficient is defined by four soil profiles S_1, S_2, S_3, and S_4 described in Table 24.2 with numerical values 1.0, 1.2, 1.5, and 2.0, respectively.
4. *The Structural Factor R_w.* Numerical values for the structural factor R_w are given in Table 24.3 for various structural systems. As described in Chapter 24, the R_w factor is related to the ductility of the structural system.
5. *The Seismic Weight W.* The seismic weight W includes the dead weight of the building, permanent equipment, and partitions. In specific cases, it also may include some of the snow load and some of the live load as indicated in Chapter 24.

The Uniform Building Code of 1988 introduced, for the first time, the dynamic method of analysis. The code stipulates that when this method is used it shall be performed using accepted principles of dynamics and it shall conform to the criteria established in the code. These criteria include the following provisions:

1. The ground motion representation may be one of the following:

(i)   The response spectra[1] given in the code and reproduced in Fig. 25.1.

(ii)   An elastic response spectrum developed specifically for the site. The preparation of such spectra is mandatory for soil type S_4.

(iii)   Ground motion time histories developed for the specific site.

(iv)   The vertical component of the ground motion may be defined by scaling the corresponding horizontal acceleration by a factor of two-thirds.

2. The adopted mathematical model should represent the spatial distribution of the mass and stiffness of the structure to an extent which is adequate for the calculation of the significant features of its dynamic response.

3. The code indicates that the dynamic analysis may be performed using:

   (a)   An elastic response spectrum in which all the significant maximum modal contributions are combined in a statistical manner to obtain approximate total structural response. As indicated in Section 25.1.3, this requirement is readily satisfied by including a sufficient number of modes with a total effective modal weight equal to 90% or more of the seismic weight.

   (b)   A time history response in which the dynamic response of the structure to a specific ground motion is obtained at each increment of time.

25.4   SCALING OF RESULTS

When the base shear force calculated by the dynamic method is less than that determined by the equivalent lateral force method, for irregular buildings, the base shear calculated by dynamic analysis shall be scaled up to match 100% of the base shear determined by the static lateral force procedure. For regular buildings, the base shear force shall be scaled up to match 90% of the base shear determined by static lateral force procedure, except that it shall not be less than 80% of the value obtained when eq. (24.3) is used to determine the fundamental period of the structure. All corresponding response parameters, including member forces and moments, shall be adjusted proportionally.

   The code also stipulates that the base shear for a given direction, determined using dynamic analysis, need not exceed the value obtained by the static lateral force method. In this case, all corresponding response parameters are adjusted proportionately.

[1]Values obtained from the UBC-88 Response Spectra should be multiplied by the modification factor Z/R_w, where Z is the *seismic zone coefficient* and R_w the *structural factor*. The UBC-88 indicates explicitly the modification factor Z but not R_w, although such factor was intended in the code. Also, the response spectrum curves provided by UBC-88 have a 5% damping as stated in the Commentary of *Recommended Lateral Force Requirement and Tentative Commentary* (Source: SEAOC-1990).

Example 25.1. Consider again the four-story reinforced concrete building of Examples 23.1 and 24.1. Model this building as a plane orthogonal frame and perform the seismic analysis by the UBC-88: dynamic method. Use the response spectra provided by the code (Fig. 25.1).

Solution:

Problem Data (from Example 24.1)

Seismic weights:

$$W_1 = W_2 = W_3 = 789.1 \text{ kip}, \quad W_4 = 645.1 \text{ kip}$$

Total seismic weight: $W = 3012.4$ kip
Seismic zone factor: $Z = 0.3$ (for Zone 3)
Occupancy importance factor: $I = 1.0$ (warehouse)
Site coefficient: $S = 1.0$ (rock)

1. *Modeling the Structure:* This structure has been modeled as a plane orthogonal frame (using Program 29) in the solution of Example 23.4. The reduced stiffness matrix in reference to the four lateral displacement coordinates of the building, from Example 23.4, is given by

$$[K] = \begin{bmatrix} 5611 & -2916 & 458 & -45 \\ -2916 & 3863 & -1885 & 269 \\ 458 & -1885 & 2893 & -1398 \\ -45 & 259 & -1398 & 1168 \end{bmatrix} \text{(kip/in)}$$

and the mass matrix (W_i/g) by

$$[M] = \begin{bmatrix} 2.036 & 0 & 0 & 0 \\ 0 & 2.036 & 0 & 0 \\ 0 & 0 & 2.036 & 0 \\ 0 & 0 & 0 & 1.671 \end{bmatrix} \text{(kip} \cdot \text{sec}^2/\text{in)}$$

2. *Natural Periods and Modal Shapes:* The natural frequencies and the normalized modal shapes are obtained by solving the eigenproblem

$$[[K] - \omega^2 [M]] \{\phi\} = \{0\} \tag{e}$$

The roots of the corresponding characteristic equation

$$\|[K] - \omega^2 [M]\| = 0 \tag{f}$$

are

$$\omega_1^2 = 77.22, \quad \omega_2^2 = 1939.75$$

$$\omega_3^2 = 678.60, \quad \omega_4^2 = 4052.78 \tag{g}$$

resulting in the natural frequencies $(f = \omega/2\pi)$

$$f_1 = 1.40 \text{ cps}, \quad f_3 = 7.01 \text{ cps}$$

$$f_2 = 4.14 \text{ cps}, \quad f_4 = 10.13 \text{ cps} \qquad \text{(h)}$$

or natural periods $(T = 1/f)$

$$T_1 = 0.715 \text{ sec}, \quad T_3 = 0.143 \text{ sec}$$

$$T_2 = 0.241 \text{ sec}, \quad T_4 = 0.099 \text{ sec} \qquad \text{(i)}$$

and the corresponding modal shapes arranged in the columns of the modal matrix are

$$[\Phi] = \begin{bmatrix} 0.11277 & -0.31074 & -0.35308 & 0.50518 \\ 0.27075 & -0.46790 & -0.11497 & -0.42860 \\ 0.43123 & -0.06907 & 0.50310 & 0.21286 \\ 0.51540 & 0.45452 & -0.34652 & -0.07766 \end{bmatrix} \qquad \text{(j)}$$

3. *Spectral Accelerations:* The spectral accelerations (scaled down by the effective peak ground acceleration of 0.3 g and divided by the structural factor $R_w = 12$) for the natural periods in eq. (i) obtained from the spectral chart for rock or stiff soils in Fig. 25.1 are

$$S_{a1} = 0.034 \text{ g}, \quad S_{a3} = 0.061 \text{ g}$$

$$S_{a2} = 0.063 \text{ g}, \quad S_{a4} = 0.050 \text{ g} \qquad \text{(k)}$$

4. *Effective Modal Weights:* The effective modal weight is given by eq. (25.17) as

$$W_m = \frac{\left[\displaystyle\sum_{i=1}^{N} \phi_{im} W_i \right]^2}{\displaystyle\sum_{i=1}^{N} \phi_{im}^2 W_i}$$

Values obtained for W_m ($m = 1, 2, 3, 4$) are shown in Table 25.1. This table also shows the effective modal weight as a percentage of the total seismic weight of the building.

5. *Modal Base Shear:* The modal base shear is defined by eq. (25.15) as

$$V_m = W_m S_{am}$$

Numerical values of V_m also are given in Table 25.1. The total base shear force given by eq. (25.29) then is calculated from values in the last column of Table 25.1 as

$$V = \sqrt{(84)^2 + (23)^2 + (6)^2 + (4)^2} = 87.4 \text{ (kip)}$$

TABLE 25.1   Modal Effective Weight and Modal Base Shear

Mode m	Modal Effective Weight W_m (kip)	(%)	Modal Base Shear V_m (kip)
1	2465	82	84
2	366	12	23
3	99	3	6
4	82	3	4
Total Weight =	3012 kips		

6. *Scaling Modal Effective Weight and Modal Base Shear:* As required by the UBC-1988, the modal values for the effective weight and for the base shear force in Table 25.1 are scaled up by the ratio r of 90% of the base shear determined by the equivalent static lateral force method and the value for the base shear force calculated by the dynamic method. The base shear determined in Example 24.2 using the equivalent static lateral force method was equal to 140.75 kip, while the value calculated in this example using the dynamic method is equal to 87.4 kip. Therefore, the scaling ratio is

$$r = \frac{0.90 \times 140.75}{87.4} = 1.45$$

Table 25.2 shows the result of scaling (by the factor $r = 1.45$) the values in Table 25.1 for the modal effective weight W_m and the modal base shear V_m.

7. *Modal Seismic Force:* The numerical values for the seismic coefficients C_{xm} and for the seismic lateral forces F_{xm} calculated from eqs. (25.20) and (25.19) are shown respectively in Tables 25.3 and 25.4. The design seismic forces calculated by eq. (25.30) are shown in the last column of Table 25.4.

TABLE 25.2   Scaled Values for Modal Effective Weight and for Modal Base Shear

Mode m	Modal Effective Weight W_m (kip)	(%)	Modal Base Shear V_m (kip)
1	3584	82	122
2	531	12	33
3	144	3	8
4	119	3	5
	Σ = 4378 kips		

TABLE 25.3   Modal Seismic Coefficient C_{xm}

Level	Mode 1	Mode 2	Mode 3	Mode 4
4	0.340	−0.780	1.141	−0.281
3	0.348	0.145	−2.027	0.942
2	0.219	0.982	0.463	−1.897
1	0.091	0.652	1.422	2.235

TABLE 25.4   Modal Seismic Force F_{xm} (kip)

Level	Mode 1	Mode 2	Mode 3	Mode 4	Design F_x Values
4	41	−26	9	−1	50
3	42	5	−16	5	46
2	27	32	4	−9	43
1	11	22	11	11	29

8. *Modal Shear Force:* Values for the modal shear force V_{xm} at level x calculated using eq. (25.31) and results from Table 25.4 are given in Table 25.5. Design values for the story shear forces V_x calculated by eq. (25.31) are given in the last column of Table 25.5. It should be noticed that the values shown in Table 25.5 for the modal shear force at level $x = 1$ are precisely the values for the base shear force calculated by eq. (25.15) and shown in Table 25.2.

9. *Modal Lateral Displacement:* Values for the modal lateral displacement δ_{xm} at level x calculated from eq. (25.24) are shown in Table 25.6. This table also shows in the last column design values for lateral displacements δ_x calculated by eq. (25.32).

10. *Modal Story Drift:* Table 25.7 shows the values for modal story drift Δ_{xm} calculated by eq. (25.25). Design values for story drift obtained from eq. (25.33) are given in the last column of this table. The max-

TABLE 25.5   Modal Shear Force V_{xm} (kip)

Level x	Mode 1	Mode 2	Mode 3	Mode 4	Design V_x Values
4	41	−26	9	−1	49
3	84	−21	−7	3	86
2	111	11	−3	−6	111
1	122	33	8	5	126

TABLE 25.6   Modal Lateral Displacement δ_{xm} (in)

Level	Mode 1	Mode 2	Mode 3	Mode 4	Design δ_x Values
4	0.321	−.023	0.002	0.001	0.320
3	0.268	0.003	−0.005	0.000	0.266
2	0.169	0.023	0.000	−0.002	0.167
1	0.070	0.015	0.002	0.001	0.063

imum story drift permitted by the UBC-1988 should not exceed $(0.04/R_w)H_x = [(0.04/12)\,144 = 0.48$ in] or $0.005\,H_x\,(0.005 \times 144 = 0.72$ in), where H_x is the story height and R_w the structural factor from Table 24.3. The design values Δ_x calculated in Table 25.7 for this example are well below these limits for story drift.

11. *Modal Overturning Moments:* Table 25.8 shows the values for modal overturning moment calculated using eq. (25.26). The last column of this table gives the design values for overturning moments M_x calculated by eq. (25.34).

12. *Modal Torsional Moments:* Table 25.9 shows the values calculated from eq. (25.28) for the modal torsional moments assuming only accidental eccentricity e_x of 5% at each level x of the building (for this example $e_x = 0.05 \times 96$ ft $= 4.8$ ft). Design values for torsional moments M_{tx} calculated from eq. (25.35) are shown in the last column of Table 25.9.

13. *The P-Δ Effect:* The UBC-88 specifies that the P-Δ effect does not need to be considered when the ratio of the secondary moment $M_s = P_x\Delta_x$ to the overturning or primary moment M_p is less than 0.10 calculated for each level x of the building. The results of the necessary calculations to evaluate this ratio [eq. (24.11)] are shown in Table 25.10. The largest moment ratio in Table 25.10 is 0.013, which is well below the code limit of 0.1. Consequently, there is no need to account for the P-Δ effect.

TABLE 25.7   Modal Story Drift Δ_{xm} (in)

Story	Mode 1	Mode 2	Mode 3	Mode 4	Design Δ_x Values
4	0.053	−0.026	0.007	−0.001	0.059
3	0.099	−0.020	−0.005	0.002	0.101
2	0.099	0.008	−0.002	−0.003	0.099
1	0.070	0.015	0.003	0.001	0.071

TABLE 25.8   Modal Overturning Moments M_{xm} (kip · ft)

Level x	Mode 1	Mode 2	Mode 3	Mode 4	Design M_x Values
4	—	—	—	—	—
3	497	−310	109	−17	596
2	1504	−562	24	22	1605
1	2832	−425	−17	−52	2864
Base	4293	−30	79	8	4293

TABLE 25.9   Modal Torsional Moments M_{txm} (kip · ft)

Level x	Mode 1	Mode 2	Mode 3	Mode 4	Design M_{tx} Values
4	199	−124	43	−7	238
3	402	−101	−35	15	416
2	531	54	−17	−30	534
1	584	157	38	23	606

TABLE 25.10   Calculation of Ratio of Secondary to Primary Moments

Level x	Story Weight W_x (kip)	Above Weight P_x (kip)	Story Drift Δ_x (in)	Story Shear V_x (kip)	Story Height H_x (in)	$M_s/M_p = P_x\Delta_x/V_xH_x$
4	645.1	645	0.059	49	144	0.005
3	789.1	1434	0.101	86	144	0.012
2	789.1	2223	0.099	111	144	0.015
1	789.1	3012	0.071	126	144	0.013

25.5   PROGRAM 24—UBC-1988: DYNAMIC LATERAL FORCE METHOD

Program 24 can be used to calculate, using modal superposition and response spectra, the response of buildings subjected to seismic excitation at the base. The use of this program requires the previous determination of the stiffness and mass matrices of the structure as well as the solution of the corresponding eigenproblem to determine the natural frequencies (or natural periods) and the modal shapes. The stiffness and the mass matrices may be obtained by the execution of one of the following programs for modeling buildings: (1)

Program 27 for modeling structures as shear buildings; (2) Program 28 for modeling structures as cantilever buildings; or (3) Program 29 for modeling structures as plane orthogonal frames. The solution of the corresponding eigenproblem may then be obtained by the execution of Program 30 or Program 31 to determine natural frequencies and normal modes, using the Jacobi Method or using the subspace iteration method.

Program 24 uses the following analytical expressions[2] of the response spectra provided by the UBC-88 in Fig. 24.1:

Soil Type S_1:

$$S_a = 1 + 10T \quad \text{for} \quad 0 < T \le 0.15 \text{ sec}$$

$$S_a = 2.5 \quad \quad \text{for} \quad 0.15 < T \le 0.39 \text{ sec}$$

$$S_a = 0.975/T \quad \text{for} \quad T > 0.39 \text{ sec}$$

Soil Type S_2:

$$S_a = 1 + 10T \quad \text{for} \quad 0 < T \le 0.15 \text{ sec}$$

$$S_a = 2.5 \quad \quad \text{for} \quad 0.15 < T \le 0.585 \text{ sec}$$

$$S_a = 1.463/T \quad \text{for} \quad T > 0.585 \text{ sec}$$

Soil Type S_3:

$$S_a = 1 + 75T \quad \text{for} \quad 0 < T \le 0.2 \text{ sec}$$

$$S_a = 2.5 \quad \quad \text{for} \quad 0.2 < T \le 0.915 \text{ sec}$$

$$S_a = 2.288/T \quad \text{for} \quad T > 0.915$$

where S_a is the spectral acceleration normalized to a peak ground acceleration of one g and T is the fundamental period of the building.

The program calculates the effective modal weights W_m from eq. (25.17) and the modal base shear force V_m from eq. (25.15). Then it requests information for scaling these effective values when the base shear calculated by the equivalent lateral force method differs from the value obtained from the dynamic method.

The output from Program 24 includes the modal values for the lowest four modes and the design values for each level of the building as follows:

1. Seismic lateral forces F_x from eqs. (25.19), (25.20), and (25.30).
2. Shear forces V_x from eqs. (25.12) and (25.31).
3. Lateral displacements δ_x from eqs. (25.24) and (25.32).
4. Story drifts Δ_x from eqs. (25.25) and (25.33).

[2]From *Recommended Lateral Force Requirements and Tentative Commentary*, SEAOC-90.

5. Overturning moments M_x from eqs. (25.26) and (25.34).
6. Torsional moments M_{tx} from eqs. (25.28) and (25.35).
7. Diaphragm forces F_{px} from eq. (24.12).
8. P-Δ effect, that is, the secondary moment M_{sx}, the primary moment M_{px}, and their ratio $\Theta_x = M_{sx}/M_{px}$ from eq. (24.11).

Example 25.2. Solve Example 25.1 using: (a) Program 29 to model the structure as a plane orthogonal frame; (b) Program 30 to calculate the natural periods and normal modes; (c) Program 24 for earthquake resistant design by means of the Dynamic Method of the UBC-88.

Solution: (a) *Modeling the Building as a Plane Frame:* The input data and output results for Program 29 to model the structure as a plane orthogonal frame are contained in the solution of example 23.4.

(b) *Natural Periods and Modal Shapes:*

```
PROGRAM 30:  NATURAL PERIODS AND MODAL SHAPES
             DATA FILE: SK

*OUTPUT RESULTS

NATURAL PERIODS (SEC):

0.715  0.241  0.143  0.099

                    EIGENVECTORS BY ROWS:

        0.11277         0.27075         0.43123         0.51540
       -0.31074        -0.46790        -0.06907         0.45452
       -0.35308        -0.11497         0.50310        -0.34652
        0.50518        -0.42860         0.21286        -0.07766
```

(c) *Analysis by UBC-88: Dynamic Method:*

```
      INPUT DATA AND OUTPUT RESULTS FOR EXAMPLE 25.2

PROGRAM 24: UBC-88 (DYNAMICS)        DATA FILE: D24


                RESULTS FROM PREVIOUS PROGRAM:

NUMBER OF STORIES                                    ND = 4

                STORY DATA:

STORY #           STORY HEIGHT          STORY WEIGHT
                      FT                    KIP

   4                12.00                 645.12
```

```
    3                  12.00                  789.12
    2                  12.00                  789.12
    1                  12.00                  789.12
```

THE NATURAL PERIODS (SEC) ARE:

```
0.715            0.241            0.143            0.099
```

INPUT DATA:

```
SEISMIC FACTOR                                       Z = .3
IMPORTANCE FACTOR                                    I =  1
STRUCTURAL FACTOR                                    RW = 12
BUILDING LENGTH NORMAL TO FORCES (FEET)              LBN = 96
```

SOIL PROFILE MENU

```
    1.   ROCK AND STIFF SOILS
    2.   DEEP COHESIONLESS OR STIFF CLAY SOILS
    3.   SOFT MEDIUM CLAYS OR SANDS
         SELECT NUMBER ? 1
```

NUMBER OF MODES IN THE RESPONSE = 4

OUTPUT RESULTS:

SPECTRAL ACCELERATION:

```
0.034            0.063            0.061            0.050
```

MODAL EFFECTIVE WEIGHT Wm AND EFFECTIVE BASE SHEAR Vm:

MODE	MODAL EFFECTIVE WEIGHT (KIP)	%	MODAL BASE SHEAR KIP
1	2464.98	81.83	84.03
2	365.73	12.14	22.86
3	99.41	3.30	6.03
4	82.37	2.73	4.09

DESIGN BASE SHEAR (KIP) = 87.39186

SCALE THE RESPONSE, AS INDICATED BY THE UBC-88 (Y/N)? Y

INPUT THE TOTAL BASE SHEAR FORCE CALCULATED USING THE

STATIC LATERAL FORCE METHOD   ?126.7

SCALE FACTOR FOR BASE SHEAR = 1.45

MODAL EFFECTIVE WEIGHT Wm AND EFFECTIVE BASE SHEAR Vm:

MODE	MODAL EFFECTIVE WEIGHT (KIP)	%	MODAL BASE SHEAR KIP
1	3573.71	81.83	121.83
2	530.23	12.14	33.14
3	144.12	3.30	8.74
4	119.41	2.73	5.93

DESIGN BASE SHEAR (KIP) = 126.7

MODAL SEISMIC COEFFICIENTS C_{xm}:

LEVEL	MODE 1	MODE 2	MODE 3	MODE 4
4	0.3409	−0.7804	1.1412	−0.2810
3	0.3489	0.1451	−2.0267	0.9421
2	0.2190	0.9827	0.4631	−1.8968
1	0.0912	0.6526	1.4223	2.2358

MODAL SEISMIC LATERAL FORCES F_{xm}(KIP):

LEVEL	MODE 1	MODE 2	MODE 3	MODE 4	DESIGN FX VALUES
4	41.53	−25.86	9.98	−1.67	49.96
3	42.50	4.81	−17.72	5.59	46.63
2	26.69	32.57	4.05	−11.25	43.77
1	11.12	21.63	12.44	13.26	30.36

MODAL SHEAR FORCES V_{xm}(KIP):

LEVEL	MODE 1	MODE 2	MODE 3	MODE 4	DESIGN VX VALUES
4	41.53	−25.86	9.98	−1.67	49.96
3	84.03	−21.05	−7.74	3.92	87.06
2	110.72	11.51	−3.69	−7.33	111.62
1	121.83	33.14	8.74	5.93	126.70

MODAL LATERAL DISPLACEMENTS D_{xm}(IN):

LEVEL	MODE 1	MODE 2	MODE 3	MODE 4	DESIGN DX VALUES
4	0.3218	−.0228	0.0031	−.0002	0.3226
3	0.2692	0.0035	−.0045	0.0007	0.2693
2	0.1690	0.0235	0.0010	−.0014	0.1707
1	0.0704	0.0156	0.0031	0.0016	0.0722

MODAL STORY DRIFT (IN):

LEVEL	MODE 1	MODE 2	MODE 3	MODE 4	DESIGN DRX VALUES	MAX. PERMITTED
4	0.0525	−.0263	0.0075	−.0009	0.0592	0.4800
3	0.1002	−.0200	−.0055	0.0020	0.1023	0.4800
2	0.0986	0.0079	−.0021	−.0030	0.0990	0.4800
1	0.0704	0.0156	0.0031	0.0016	0.0722	0.4800

MODAL OVERTURNING MOMENT Mxm(KIP-FEET):

LEVEL	MODE 1	MODE 2	MODE 3	MODE 4	DESIGN MX VALUES
3	498.35	−310.35	119.73	−20.00	599.50
2	1506.72	−563.00	26.83	27.05	1608.93
1	2835.33	−424.87	−17.48	−60.91	2867.68
0	4297.31	−27.20	87.44	10.27	4298.30

FLEXIBLE DIAPHRAGMS (Y/N) ? N

DIAPHRAGMS NOT FLEXIBLE: CALCULATE TORSION

NOTE:

THE TORSIONAL ECCENTRICITY AT A LEVEL IS EQUAL TO THE DISTANCE FROM THE CENTER OF THE ABOVE MASS TO THE CENTER OF STIFFNESS AT THAT LEVEL

DESIGN FOR ONLY ACCIDENTAL ECCENTRICITY (Y/N) ? Y

MODAL TORSIONAL MOMENTS MTxm(KIP-FEET):

LEVEL	MODE 1	MODE 2	MODE 3	MODE 4	DESIGN MTX VALUES
4	199.34	−124.14	47.89	−8.00	239.80
3	403.35	−101.06	−37.16	18.82	417.90
2	531.44	55.25	−17.72	−35.19	535.76
1	584.79	159.07	41.97	28.47	608.16

DIAPHRAGM FORCES:

LEVEL	DIAPHRAGM FORCE Wpx(KIP)
4	67.74
3	82.86
2	82.86
1	82.86

P-DELTA EFFECT:

LEVEL	SECONDARY MOMENT KIP-FT	PRIMARY MOMENT KIP-FT	RATIO OF MOMENTS
4	3.18	599.50	0.0053
3	12.23	1009.43	0.0121
2	18.34	1258.76	0.0146
1	18.12	1430.62	0.0127

25.6 SUMMARY

The provisions of the Uniform Building Code of 1988 for earthquake resistant design by the dynamic method require the modeling of the building as a discrete system with one coordinate (horizontal displacement) at each level of the building. Possible models include: (1) the shear building in which the horizontal diaphragms are assumed absolutely rigid, (2) the cantilever building in which the horizontal diaphragms are considered absolutely flexible, and (3) the plane orthogonal frame in which the stiffness of horizontal diaphragms is considered as part of the effect of horizontal members of the frame. The implementation of the dynamic method also requires the solution of the corresponding eigenproblem for the modeled structure to determine its natural periods and modal shapes.

The maximum response of the modal equations is then obtained from a spectral chart such as the one provided by the code. Modal seismic forces and modal response in terms of shear forces, overturning moments, torsional moments, lateral displacements, and story drifts are determined at each level of the building. The final design values are calculated using the SRSS technique or the CQC technique to combine the modal contributions.

The modal method in which the modal responses are combined is valid while the structure remains in linear elastic behavior as expected when subjected to an earthquake of moderate intensity. However, for a strong earthquake, the provisions of the UBC-88 are not intended to maintain the structure vibrating within its linear elastic behavior. When subjected to a strong earthquake, the structure will deform in the inelastic range producing plastic deformations and structural damage. However, its ductility will provide a mechanism to absorb energy and structural stability will be maintained, although the structure may continue to undergo large deflections which may create further damage.

PROBLEMS

25.1 Use Program 24 for the seismic resistant design of the 10-story building of Problem 23.1.

25.2 Solve Problem 23.2 using the dynamic method of the UBC-88. The total length of the building is 120 ft and its width 24 ft. Consider only the accidental torsional eccentricity.

25.3 Solve Problem 23.3 using the dynamic method of the UBC-88. The length of the building is 96 ft and its width is 36 ft. Estimated eccentricity at each level $e_x = 6$ ft.

25.4 Solve Problem 25.3 modeling the structure as a plane orthogonal frame.

APPENDICES

Appendix I

Answers to Problems in Part I

1.1. $T = 2\pi L \sqrt{\dfrac{W}{g} \cdot \dfrac{L}{3EI + 2kL^3}}$

1.2. $y(t = 1) = -0.89$ in
$\dot{y}(t = 1) = 22.66$ in/sec

1.3. $f = 2.24$ cps.

1.4. (a) $f = 2.87$ cps.
(b) $f = 2.74$ cps.

1.5. $f = \dfrac{4}{\pi} \sqrt{\dfrac{3EIg}{L^3 W}}$

1.6. $y(t = 2) = -0.474$ in
$\dot{y}(t = 2) = -21.05$ in/sec
$\ddot{y}(t = 2) = 4065$ in/sec^2

1.7. $\theta = \theta_0 \cos \sqrt{\dfrac{g}{L}} t + \dfrac{\dot{\theta}_0}{\omega} \sin \sqrt{\dfrac{2}{L}} t$

1.8. $f = \dfrac{1}{2\pi} \sqrt{\dfrac{ka^2 - mgL}{mL^2}}$

1.9. $f = \dfrac{1}{2\pi} \sqrt{\dfrac{3EI}{mL^3} - \dfrac{3g}{2L}}$

1.10. (a) $f = \dfrac{1}{2\pi} \sqrt{\dfrac{3EIkg}{(3EI + kL^3)W}}$

  (b) $f = \dfrac{1}{2\pi} \sqrt{\dfrac{48EIkg}{(48EI + kL^3)W}}$

  (c) $f = \dfrac{1}{2\pi} \sqrt{\dfrac{3EILg}{a^2b^2W}}$

  (d) $f = \dfrac{1}{2\pi} \sqrt{\dfrac{3EILkg}{(3EIL + a^2b^2K)W}}$

1.11. $f = \dfrac{1}{2\pi} \sqrt{k\left(\dfrac{1}{m_1} + \dfrac{1}{m_2}\right)}$

2.1. $y(t = 1) = -0.0373$ in
  $\dot{y}(t = 1) = 0.570$ in/sec

2.2. $y(t = 2) = -4.65 \times 10^{-9}$ in
  $\dot{y}(t = 2) = -4.083 \times 10^{-8}$ in/sec
  $\ddot{y}(t = 2) = 4.18 \times 10^{-5}$ in/sec^2

2.3. $c = 0.73$ lb. sec/in

2.4. $\xi = 1.5\%$

2.5. (a) for $\xi = 1$, $\quad y = [y_0(1 + \omega t) + v_0 t]e^{-\omega t}$

  (b) for $\xi > 1$, $y = e^{-\xi\omega t}\left[y_0 \cosh \omega_D' t + \dfrac{v_0 + y_0\xi\omega}{\omega_D'} \sinh \omega_D' t\right]$

  where $\omega_D' = \omega\sqrt{\xi^2 - 1}$

2.6. (a) $\xi \quad = 0.4167$
  (b) $T_D = 0.2765$ sec
  (c) $\delta \quad = 2.8801$
  (d) $\dfrac{y_1}{y_2} = 17.8161$

2.9. (a) $\xi = 0.076$
(b) $f_D = 8.69$ cps
(c) $\delta = 0.48$
(d) $\dfrac{y_1}{y_2} = 1.61$

2.10. (a) $\xi = 0.018$
(b) $\omega_D = 57.76$ rad/sec
(c) $\delta = 0.113$
(d) $\dfrac{y_1}{y_2} = 1.12$

2.11. $m_1 m_2 \ddot{u} + (m_1 + m_2)c\dot{u} + (m_1 + m_2)ku = 0$

2.12. $\ddot{u} + 2\xi\omega\dot{u} + \omega^2 u = 0$

$$\text{where: } \omega = \sqrt{\frac{k}{M}}, \quad M = \frac{m_1 m_2}{m_1 + m_2}, \quad \omega_D = \omega\sqrt{1 - \xi^2},$$

$$\xi = \frac{c}{c_{cr}}, \quad c_{cr} = 2\sqrt{kM}$$

3.1. $Y = 0.0037$ in
3.2. $A_T = 51.2$ lb
3.3. $Y = 0.823$ in
3.4. $Y = 0.746$ in
3.5. (a) $A_T = 15,803$ lb
(b) $T_R = 3.16$

3.6. $k = 93$ lb/in
3.7. $u = 0.013$ in
3.8. $T_R = 0.34$

3.9. (a) $Y_1 = 0.064$ in for $f_1 = 800$ RPM
$Y_2 = 0.0446$ in for $f_2 = 1000$ RPM
$Y_3 = 0.0302$ in for $f_3 = 1200$ RPM
(b) $Y(r = 1) = 0.076$ in

3.10. $f = f_r \sqrt{1 + \dfrac{m_3}{m}}$

3.11. $\omega_p = \omega\sqrt{1 - 2\xi^2}$

$$Y_p = \frac{y_{st}}{2\xi\sqrt{1 - \xi^2}}$$

3.12. (a) f = 18.58 cps
(b) ξ = 0.0735
(c) F_0 = 4825 lb
(d) F_0 = 4840 lb.

3.13. $\xi = \dfrac{Y_1(1 - r_1^2)}{2r_1\sqrt{Y_r^2 r_1^2 - Y_1^2}}$

$F_r = \dfrac{Y_r Y_1(1 - r_1^2)k}{r_1\sqrt{Y_r^2 r_1^2 - Y_1^2}}$

3.14. (a) $M\ddot{u} + c\dot{u} + ku = \dfrac{m_1 F_0}{m_1 + m_2}\sin \bar{\omega}t$

where $M = \dfrac{m_1 m_2}{m_1 + m_2}$

(b) $u = \dfrac{m_1 F_0 \sin (\bar{\omega}t - \theta)}{k(m_1 + m_2)\sqrt{(1 - r^2)^2 + (2r\xi)^2}}$

4.1. (a) $y(t = 0.5) = -0.407$ in
(b) $y_{max} = 1.35$ in

4.2. (a) $y(t = 0.5) = -0.102$ in
(b) $y_{max} = 1.17$ in

4.3. $\text{DLF} = \dfrac{t}{t_d} - \dfrac{\sin \omega t}{\omega t_d}\; t \le t_d$

$\text{DLF} = 1 + \dfrac{1}{\omega t_d}(\sin \omega t - \sin \omega(t + t_d))\; t \ge t_d$

4.4. V_{max} = 18,093 lb for left column
V_{max} = 1,908 lb for right column

4.5. V_{max} = 15,640 lb for left column
V_{max} = 1,649 lb for right column

4.6. $y(t = 0.5) = -1.903$ in

4.7. $y(t = 0.5) = -0.060$ in

4.8. $u(t = 1) = -2.809$ in

4.9. $u(t = 1) = -2.397$ in

4.10. u_{max} = 6.03 in (undamped system)
u_{max} = 4.59 in (with 20% damping)

4.11. $y_{max} = 1.51$ in

4.12. $y_{max} = 1.42$ in

4.13. $y_{max} = 0.58$ in

4.14. $y_{max} = 1.00$ in

4.15. $y_{max} = 0.76$ in

4.16. $y_{max} = 0.66$ in

4.17. $y_{max} = 0.34$ in

4.18. $\tau_{1max} = 3842$ psi
$\tau_{2max} = 6831$ psi

4.19. $y_{max} = 0.71$ in

4.20. (a) $\sigma = \pm 6193$ psi
(b) $F_{max} = 11{,}376$ lb

5.1. $F(t) = \dfrac{120}{\pi}\left[\sin 2\pi t + \dfrac{1}{3}\sin 6\pi t + \dfrac{1}{5}\sin 10\pi t \ldots\right]$

5.2. $F(t) = 10^{-6}[357 \sin 2\pi t - 26 \cos 2\pi t$
$+ 36 \sin 6\pi t - 532 \cos 6\pi t$
$- 35 \sin 10\pi t - 7 \cos 10\pi t + \ldots]$

5.3. $u(t = 0.5) = 0.3518$ in

5.4. (a) $a_n = \dfrac{720}{\pi(1 - n^2)}$, $n = 2, 4, 6, \ldots$

$a_n = 0$, $n = 1, 3, 5, \ldots$
$b_n = 0$, $n = 1, 2, 3, \ldots$

5.5. $u(t = 0.05) = -0.2065$ in

5.6. $u(t = 0.05) = 0.2064$ in

5.7. $u(t = 0.05) = 0.1295$ in

5.8. (a) $a_0 = 0.0350$
$a_1 = 0.0069$ $b_1 = -0.0361$
$a_2 = -0.0724$ $b_2 = -0.0402$
(b) $y(t = 0.35) = 0.2570$ in

5.9. $y(t = 0.35) = 0.2327$ in

5.10. $y(t = 0.5) = 0.027$ in

5.11. $y(t = 0.5) = (0.02842 - 0.00011\, i)$ in

5.12. (a) $a_0 = \dfrac{P_0}{\pi}$

$$a_n = 0, \quad n = 1, 3, 5, \ldots, \quad b_1 = \dfrac{P_0}{2}$$

$$a_n = \dfrac{P_0}{\pi} \cdot \dfrac{2}{1 - \pi^2}, \quad n = 2, 4, 6, \ldots, \quad b_n = 0, \quad n > 1$$

5.13. $y(t = 0.5) = 0.0731$ in

5.14. $y(t = 0.5) = 0.0543$ in

6.1. $M^* = 4.23$ lb sec^2/in
$C^* = 225$ lb sec/in
$K^* = 45{,}000$ lb/in
$F^*(t) = 625\, f(t)$ lb

6.2. $M^* = 10m$
$C^* = c$
$K^* = k$
$F^*(t) = \dfrac{M(t)}{L}$

6.3. $M^* = \dfrac{7}{12}\, \bar{m} L^2$
$C^* = cL$
$K^* = kL$
$F^*(t) = \dfrac{P_0 L}{6}\, f(t)$

6.4. $M^* = \dfrac{m}{2\pi}\,(5\pi - 8)$

$K^* = \dfrac{EI\pi^4}{32L^3}$

$F^*(t) = 0.2929 F_0 f(t)$

6.5. $K_G^* = -\dfrac{N\pi^2}{8L}$

6.6. $M^* = 0.1237\, \dfrac{\gamma d}{g}$

$K^* = \dfrac{E_c \pi d^4}{128L^3}$

$F^*(t) = -0.1807\, P_0(t) L d$

6.7. $\omega = \sqrt{\dfrac{48EI}{L^3\left(m + \dfrac{17}{35}m_b\right)}}$ rad/sec

6.8. $\omega = 7.854\sqrt{\dfrac{EI}{m_b L^4}}$ rad/sec

6.9. $f = 3.51$ cps

6.10. $\omega = 2.15\sqrt{\dfrac{gEI}{WL^3}}$ rad/sec

6.11. $\omega = 0.62\sqrt{\dfrac{gAG}{WL}}$ rad/sec

6.12. $f = 0.507$ cps

6.13. $f = 0.696$ cps

7.1. $u_{max} = -10.27$ in

7.2. $y_{max} = 2.56$ in

7.3. $y_{max} = 6.47$ in

7.4. $u_{max} = 1.03$ in

7.5. $u_{max} = 5.19$ in

7.6. $y_{max} = 2.36$ in

7.7. $\mu = 1.7$

7.8. $y(t = 0.5) = 0.4477$ in

7.9. $y(t = 0.5) = 0.3043$ in

7.10. $y(t = 0.5) = 0.1423$ in

7.11. $y(t = 0.5) = 0.1365$ in

7.12. $a_0 = 2.81$

8.1. $y_{max} = 0.374$ in

8.2. $\sigma_{max} = 7.246$ ksi

8.3. $y_{max} = 0.418$ in

8.4. $\sigma_{max} = 12.788$ ksi

8.5. $S_D = 1.9$ in
$S_v = 22.4$ in/sec
$S_a = 0.68$ g

8.6. $S_D = 1.28$ in
$S_v = 15.36$ in/sec
$S_a = 0.48$ g

8.7. $S_D = 11.0$ in
$(F_S)_{max} = 88.0$ kip

8.8. $(F_S)_{max} = 36.0$ kip

8.9. $(F_T)_{max} = 44.0$ kip

8.10. $S_D = 8.0$ in

8.11. $\mu = 1.8$

8.12.   (a) $S_D = 4.0$ in
$S_v = 50.3$ in/sec
$S_a = 1.63$ g
(b) $S_D = 4.8$ in
$S_v = 60.0$ in/sec
$S_a = 1.96$ g

8.13. $S_D = 0.46$ in
$S_v = 5.8$ in/sec
$S_a = 0.19$ g

8.14. $S_D = 0.78$ in   (at $f = 0.5$ cps)
$S_v = 2.46$ in/sec
$S_a = 7.72$ in/sec^2

8.15. $S_D = 8.03$ in   (at $f = 1.00$ cps)
$S_v = 50.45$ in/sec
$S_a = 317.00$ in/sec^2

8.16. $S_D = 3.26$ in   (at $f = 1.00$ cps)
$S_v = 20.38$ in/sec
$S_a = 127.40$ in/sec^2

Appendix II

Computer Programs

The computer programs used throughout this book are written in the BASIC language. These programs are organized in two sets: structural dynamics programs and earthquake engineering programs. Table AII.1 presents a brief description of the computer programs in these two sets. The programs are written for IBM or compatible microcomputers. They are available directly from the author in a compiled or noncompiled version of BASIC (BASICA, QUICK BASIC, or TURBO BASIC). A convenient form for ordering diskettes with the programs of interest is provided in the back of this book.

Table AII. 1   Computer Programs

Program	File	Chapter	Purpose
		Structural Dynamics	
1	MASTER	*	Menus with options and commands
2	PP2	4	Response simple oscillator by direct integration
3	PP3	4	Response simple oscillator to impulses
4	PP4	5	Response simple oscillator in frequency domain
5	PP5	7	Response oscillator for elastoplastic behavior
6	PP6	8	Seismic response spectra
7	PP7	9	Modeling structures as shear buildings
8	PP8	10	Natural frequencies and normal modes
9	PP9	11	Response by modal superposition
10	PP10	11	Response to harmonic excitation
11	PP11	12	Absolute damping from modal damping
12	PP12	13	Reduction of the dynamic problem
13	PP13	14	Modeling structures as beams
14	PP14	15	Modeling structures as plane frames
15	PP15	16	Modeling structures as grid frames
16	PP16	17	Modeling structures as space frames
17	PP17	18	Modeling structures as plane trusses
18	PP18	18	Modeling structures as space trusses
19	PP19	19	Response by step integration
20	PP20	*	Maximum member's end forces and moments
X1	X1	*	File for structures
X2	X2	*	File for eigensolution
		Earthquake Engineering	
21	MAIN	*	Menus with options and commands
22	EQ2	23	Earthquake-resistant design: UBC-85
23	EQ3	24	Earthquake-resistant design: UBC-88 (statics)
24	EQ4	25	Earthquake-resistant design: UBC-88 (dynamics)
25	EQ5	*	Earthquake-resistant design: NEHRP-88** (statics)
26	EQ6	*	Earthquake-resistant design: NEHRP-88 (dynamics)
27	EQ7	9	Modeling structures as shear buildings
28	EQ8	*	Modeling structures as cantilever buildings
29	EQ9	15	Modeling structures as plane frames

Table AII.1   (Continued)

Program	File	Chapter	Purpose
30	EQ10	10	Natural frequencies and modal shapes (Jacobi Method)
31	EQ11	*	Natural frequencies and modal shapes (Subspace Iteration Method)

*Program not described in the book.
**National Earthquake Hazard Reduction Program.

Appendix III

Organizations and Their Acronyms

ACI   American Concrete Institute
P.O. Box 19150
Redford Station
Detroit, MI 48219
Phone (313)532-2600

AISC   American Institute of Steel Construction
400 North Michigan Avenue
Chicago, IL 60611
Phone (312)670-5432

ANSI   American National Standards Institute
1430 Broadway
New York, NY 10018
Phone (212)354-3300

BOCA   Building Officials and Code Administrators
4051 West Flossmoor Road
Country Club Hills, IL 60477
Phone (312)799-2300

BSSC   Building Seismic Safety Council
1201 L Street, N.W.
Suite 400
Washington, DC 20005
Phone (202)289-8000

EERC   Earthquake Engineering Research Center
University of California
1301 South 46th Street
Richmond, CA 94304
Phone (415)525-3668

EERI   Earthquake Engineering Research Institute
2620 Telegraph Avenue
Berkeley, CA 94704
Phone (415)525-3668

FEMA   Federal Emergency Management Agency
Administrators of the National Earthquake Hazard Reduction
  Program (NEHRP)
National Emergency Training Center
Emmitsburg, MD 21727
Phone (301)447-6771

ICBO   International Conference of Building Officials
Publishers of Uniform Building Code (UBC)
5360 South Workman Mill Road
Whittier, CA 90601
Phone (213)385-4424

MITA   Micro-Text Association
P.O. Box 35101
Louisville, KY 40292
Phone (502)895-1369

NBS   National Bureau of Standards
Department of Commerce
Washington, DC 20234
Phone (301)975-2000

PCA   Portland Cement Association
5420 Old Orchard Road
Skokie, IL 60077
Phone (708)966-6200

SBCCI   Southern Building Code Congress International
             Publisher of the Standard Building Code (SBC)
             900 Mont Clair
             Birmingham, AL 35213
             Phone (205)591-1853

SEAOC   Structural Engineering Association of California, with a
                 research group—Applied Technology Council (ATC)
             217 Second Street
             San Francisco, CA 94105
             Phone (415)974-5147

SEAOW   Structural Engineers Association of Washington
             P.O. Box 4250
             Seattle, WA 98104
             Phone (206)682-6026

Glossary

Accelerometer. A seismograph for measuring ground acceleration as a function of time.

Aliasing. The phenomenon in which higher harmonies introduce spurious low-frequency components. This occurs when the number of sampled points of a function is insufficient to describe the function. (See Nyquist Frequency.)

Amplitude. Maximum value of a function as it varies with time. If the variation with time can be described by either a sine or cosine function, it is said to vary harmonically.

Angular Frequency/Circular Frequency. The frequency of periodic function in cycles per second (Hertz) multiplied by 2π, expressed in rad/sec.

Autocorrelation of a Random Function $x(t)$. Correlation between the function $x(t)$ and the out-of-phase function $x(t + \tau)$ as defined by eq. (22.15).

Base Shear Force. The total lateral force on the structure equivalent to the earthquake excitation at the base of the structure.

Basic Design Spectra. Smooth or average plots of maximum response of single degree-of-freedom systems used in seismic design of structures.

Bearing Wall System. A structural system without a complete vertical load–carrying space frame.

Boundary Condition. A constraint applied to the structure independent of time.

Braced Frame. An essentially vertical truss system of the concentric or eccentric type which is provided to resist lateral forces.

Building Frame System. An essentially complete space frame which provides support for gravity loads.

Characteristic Equation. An equation whose roots are the natural frequencies.

Circular Frequency. See Angular Frequency.

Complementary Solution. The solution of a homogeneous differential equation (no excitation).

Complete Quadratic Combination (CQC). A method of combining values of modal contributions which is based on random vibration theory and includes cross correlation terms.

Concentric Braced Frame. A braced frame in which the members are subjected primarily to axial forces.

Consistent Mass. Mass influence coefficients determined by assuming that the dynamic displacement functions are equal to the static displacement functions.

Correlation between Random Variables $x_1(t)$ and $x_2(t)$. The time average of the product of the functions $x_1(t)$ and $x_2(t)$.

Coupled Equations. A system of differential equations in which the equations are not independent from each other.

Critical Damping. Minimum amount of viscous damping for which the system will not vibrate.

D'Alembert Principle. This principle states that a dynamic system may be assumed to be in equilibrium provided that the inertial forces are considered as external forces.

Damped Frequency. The frequency at which a viscously damped system oscillates in free vibration.

Damping. The property of the structure to absorb vibrational energy.

Damping Ratio. The ratio of the viscous damping coefficient to the critical damping.

Degrees of Freedom. The number of independent coordinates required to completely define the position of the system at any time.

Deterministic Vibration. A process which can be predicted by an exact mathematical expression.

Dirac's Delta Function. A generalized function having the properties described in eq. (22.45).

Direct Stiffness Method. The method of assembling the system stiffness matrix by proper summation of the stiffness coefficients of the elements in the system.

Discrete Fourier Transform. A summation of harmonic terms to express the Fourier transform for a function defined by a finite number of points.

Dual System. A combination of a special or intermediate moment-resisting space frame and shear walls or braced frames.

Ductility Ratio. The ratio between the maximum displacement for elastoplastic behavior and the displacement corresponding to yield point.

Dynamic Condensation. A method of reducing the dimension of the eigenproblem by establishing the dynamic relationship between primary and secondary coordinates.

Dynamic Magnification Factor. The ratio of the maximum displacement of a single degree of freedom excited by a harmonic force to the deflection that would result if a force of that magnitude were applied statically.

Earthquake. The vibrations of the Earth caused by the passage of seismic waves radiating from some source of elastic energy.

Eigenproblem. The problem of solving a homogeneous system of equations containing a parameter which should be determined to provide nontrivial solutions.

Elastic Rebound Theory. The theory of earthquake generation proposing that faults remain locked while strain energy slowly accumulates in the surrounding rock, and then suddenly slip, releasing this energy.

Elastoplastic. A system which behaves elastically for a force that does not exceed a maximum value and plastically above this maximum.

Ensemble. A set of samples or records of a random process.

Epicenter. The point on the Earth's surface directly above the focus.

Ergodic Process. A stationary random process for which the time average of any record is equal to the average across the ensemble.

Fast Fourier Transform (FFT). A very efficient algorithm implemented in a computer program for the calculation of the response in the frequency domain.

Flexibility Coefficient. f_{ij} is the displacement at coordinate i due to a unit force (or moment) applied at coordinate j.

Forced Vibration. Vibration in which the response is due to external excitation of the system.

Fourier Analysis. Method of determining the response by superposition of the responses to the harmonic components of the excitation.

Fourier Transform. The Fourier transform $C(\omega)$ of a function $F(t)$ is defined by eq. (22.19).

Fourier Transform Pair. In reference to the function $F(t)$, the Fourier transform pair is given by eqs. (22.19) and (22.20).

Free Body Diagram. A sketch of the system, isolated from all other bodies, in which all the forces external to the body are shown.

Free Vibration. The vibration of a system in absence of external excitation.

Frequency Analyzer or Spectral Density Analyzer. An instrument that measures electronically the spectral density function of a signal.

Frequency Ratio. The ratio between the forcing frequency to the natural frequency for a system excited by a harmonic load.

Fundamental Frequency. The lowest natural frequency of a multidegree-of-freedom vibrating system.

Gauss–Jordan Reduction or Elimination. A computational technique in which elementary row operations are applied systematically to solve a linear system of equations.

Generalized Coordinates. A set of independent quantities which describe the dynamic system at any time. These quantities are generally functions of the geometric coordinates.

Geometric Stiffness Coefficient. k_{Gij} is the force at coordinate i due to a unit displacement at coordinate j and resulting from the axial forces in the structure.

Harmonic. A sinusoidal function having a frequency that is an integral multiple of the fundamental frequency.

Harmonic Force. A force expressed by a sine, cosine (or equivalent exponential) function.

Hypocenter or Focus. The point in the interior of the earth at which rupture is initiated during an earthquake.

Impulsive Load. A load that is applied during a relatively short time interval producing an instantaneous change in velocity.

Initial Conditions. The initial values of specific functions such as displacement, velocity, or acceleration evaluated at time $t = 0$.

Intensity (of Earthquakes). A measure of ground shaking obtained from the damage done to structures built by man, changes in the Earth's surface, and felt reports.

Intermediate Moment-Resisting Space Frame (IMRSF). A concrete space frame designed in conformance with Section 2625 (k) of UBC-88.

Isolation. The reduction of severity of the response, usually attained by proper use of a resilient support.

Lateral Force Method. A method of analysis in which lateral horizontal forces at various levels of the structure are equivalent to seismic excitation at the base of the structure.

Lateral Force–Resisting System. That part of the structural system assigned to resist lateral forces.

Linear Acceleration Method. A step-by-step method for the integration of the differential equations of motion in which the acceleration is assumed to be a linear function during each time step.

Linear System. A system of differential equations in which no term contains products (or exponents) of the dependent variables or their derivatives.

Logarithmic Decrement. The natural logarithm of the ratio of any two successive amplitudes of the same sign obtained in the decay curve in a free vibration test.

Lumped Mass. A method of discretization in which the distributed mass of the elements is lumped at the nodes or joints.

Mathematical Model. The idealization of a system including all the assumptions imposed on the physical problem.

Mean-Square Value. The time average of the square of a random function as defined by eq. (22.2).

Mean Value. The time average of a random function defined by eq. (22.1).

Modal Shapes (also Normal Modes). The relative displacements at the coordinates of a multidegree-of-freedom system vibrating at one of the natural frequencies.

Modal Superposition Method. A method of solution of multidegree-of-freedom systems in which the response is determined from the solution of independent modal (or normal) equations.

Modified Mercalli Intensity (MMI). A measure of the effect of an earthquake at a particular location.

Moment-Resisting Space Frame. A space frame in which the members and joints are capable of resisting forces primarily by flexure.

Narrow-Band Process. A random process whose spectral density function has nonzero values only in a narrow frequency range.

Natural Frequency. The number of cycles per second at which a single degree-of-freedom system vibrates freely or a multidegree-of-freedom system vibrates in one of the normal modes.

Natural Period. The time interval for a vibrating system in free vibration to do one oscillation.

Newmark Beta Method. A numerical method to calculate the response of a structure subjected to external excitation (force or motion).

Node or Joint: A point joining elements of the structure and at which displacements are known or to be determined.

Normal Distribution or Gaussian Distribution. A function whose probability density function is given by eq. (22.10).

Normal Modes. See Modal Shapes.

Nyquist Frequency. The maximum frequency component that can be detected from a function sampled at time spacing Δt $[N_y = 1/2\Delta t$ (Hz)$]$.

Occupancy Factor. A numerical factor in the calculation of the base shear force that depends on the intended use of the structure.

Ordinary Moment-Resisting Space Frame (OMRSF). A moment-resisting space frame not meeting special detailing requirements for ductile behavior.

P-Delta Effect The secondary effect on shears, axial forces, and moments of frame members induced by the vertical loads on the laterally displaced building frame.

Periodic Function. A function that repeats itself at a fixed time interval known as the period of the function.

Power Spectral Density. A term used to describe the intensity of random vibration at a given frequency, measured in g^2/Hz.

Principle of Virtual Work. The work done by all the forces acting on a system in static or dynamic equilibrium, which occurs during a virtual displacement compatible with the constraints of the system that is equal to zero.

Probability Density Function. A function $p(x)$ such that the probability of $x(t)$ of being in the range $(x, x + dx)$ is $p(x)\ dx$.

Pseudo-Velocity. The velocity calculated by analogy with the apparent harmonic motion for a system seismically excited.

Random Function. A function (as opposed to a deterministic function) whose value at any time cannot be determined exactly, but can only be predicted in probabilistic terms by statistical methods.

Random Vibration or Random Process. A process which cannot be predicted in a deterministic sense, but only probabilistically using the theory of statistics.

Rayleigh Distribution. A function whose probability density function is given by eq. (22.12).

Rayleigh's Formula. A formula to estimate the fundamental period of the structure.

Resonance. The condition in which the frequency of the excitation equals the natural frequency of the vibrating system.

Response. The force or motion that results from external excitation on the structure.

Response Spectrum. A plot of maximum response (displacement, velocity, or acceleration) for a single degree-of-freedom system defined by its natural frequency (or period) subjected to a specific excitation.

Richter Magnitude (M). A measure related to total energy released during an earthquake.

Root Mean Square (RMS). The square root of the mean-square value of a random function [eq. (22.5)].

Sample. A record of random process.

Seismic Zone Factor. A numerical factor in the calculation of the base shear force at a given geographic location.

Seismograph. An instrument for recording as a function of time the motions of the Earth's surface that are caused by seismic waves.

Shear Wall. A wall designed to resist lateral forces parallel to the plane of the wall (sometimes referred to as a vertical diaphragm or structural wall).

Shock Spectrum. See Response Spectrum.

Simple Harmonic Motion. The motion of a system which may be expressed by a sine or cosine function of time.

Site-Structure Resonance Coefficient. A numerical factor in the calculation of the base shear force that depends on the condition of the soil.

Space Frame. A three-dimensional structural system, without bearing walls, composed of members interconnected so as to function as a complete self-contained unit with or without the aid of horizontal diaphragms or floor-bracing systems.

Special Moment-Resisting Space Frame (SMRSF). A moment-resisting space frame specially detailed to provide ductile behavior and comply with the requirements given in Chapter 26 or 27 of UBC-88 (International Conference of Building Officials 1988).

Spectral Analysis or Spectrum. A description of contributions of the frequency components to the mean-square value of a random function.

Spectral Density Function. A function that describes the intensity of random vibration in terms of the mean-square value per unit of frequency.

Square Root Sum of Squares (SRSS). A method of combining maximum values of modal contributions by taking the square root of the sum of the squared modal contributions.

Standard Deviation. The square root of the variance. It may be calculated by eq. (22.6).

Static Condensation. A method of reducing the dimensions of the stiffness and of the mass matrices by establishing the static relation between primary and secondary coordinates.

Stationary Process. A random process for which the average across the ensemble has the same value at any selected time.

Steady-State Vibration. The motion of the system that remains after the transient motion existing at the initiation of the motion has vanished.

Stiffness Coefficient. k_{ij} is the force at coordinate i due to a unit displacement at coordinate j.

Story Drift. The relative lateral displacements of consecutive levels of a building.

Spring Constant. The change in load on a linear elastic structure required to produce a unit increment of deflection.

Strong Motion Accelerograph. An instrument to register seismic motions higher than a specified amplitude.

Structure. An assemblage of framing members designed to support gravity loads and resist lateral forces. Structures may be categorized as building structures or nonbuilding structures.

Structural Factor. A numerical factor in the calculation of the base shear force that depends on the type of structural system.

Tectonic Earthquakes. Earthquakes resulting from the sudden release of energy stored by a major deformation of the earth.

Time History Response. The response (motion or force) of the structure evaluated as a function of time.

Transient Vibration. The initial portion of the motion which vanishes due to the presence of damping forces in the system.

Transmissibility. The nondimensional ratio, in the steady-state condition, of the response motion to the input motion. Or the nondimensional ratio of the amplitude of the force transmitted to the foundation to the amplitude of the force exciting the system.

Variance of $x(t)$. The average of the squares of the deviations of $x(t)$ values from the mean value $\bar{x}$ [eq. (22.3)].

Viscous Damping. Dissipation of energy such that the motion is resisted by a force proportional to the velocity but in the opposite direction.

White Noise. A wide-band random process for which the spectral density function is constant over the whole frequency range.

Wide-band Process. A random process whose spectral function has nonzero value over a large range of frequencies.

Wiener–Kinchin Equations. These are equations that relate the autocorrelation function and the spectral density function [eqs. (22.42) and (22.43)].

Wilson's θ Method. A modification of the step-by-step linear acceleration method in which the time step is multiplied by a factor necessary to render the method unconditionally stable.

Wood-Anderson Seismograph. An instrument used to register seismic motions.

Selected Bibliography

EARTHQUAKE ENGINEERING

Blume, J. A., Newmark, N. M., and Corning, L. (1961), *Design of Multistory Reinforced Concrete Buildings for Earthquake Motions*, Portland Cement Association, Chicago.

Hart, Gary C., and Englekirk, Robert E. (1982), *Earthquake Design of Concrete Masonry Buildings: Response Spectra Analysis and General Earthquake Modeling Considerations*, Prentice-Hall, Englewood Cliffs, NJ.

Housner, G. W. (1970), Design spectrum, in *Earthquake Engineering* (R. L. Weigel, Ed.), Prentice-Hall, Englewood Cliffs, NJ.

Housner, G. W., and Jennings, P. C. (1982), *Earthquake Design Criteria*, Earthquake Engineering Institute, Berkeley, CA.

Hudson, D. E. (1970), Dynamic tests of full scale structures, in *Earthquake Engineering* (R. L. Weigel, Ed.), Prentice-Hall, Englewood Cliffs, NJ.

Naeim, Farzad (1989), *The Seismic Design Handbook*, Van Nostrand Reinhold, New York.

Newmark, N. M., and Hall, W. J. (1973), *Procedures and Criteria for Earthquake Resistant Design: Building Practices for Disaster Mitigation*, Building Science Series 46, National Bureau of Standards, Washington, DC, pp. 209–237.

Newmark, N. M., and Hall, W. J. (1982), *Earthquake Spectra and Design*, Earthquake Engineering Research Institute, Berkeley, CA.

Newmark, N. M., and Riddell, R. (1980), in *Inelastic Spectra for Seismic Design*, Proceedings of 7th World Conference on Earthquake Engineering, Istanbul, Turkey, Vol. 4, pp. 129–136.

Newmark, N. M., and Rosenblueth, E. (1971), *Fundamentals of Earthquake Engineering*, Prentice-Hall, Englewood Cliffs, NJ.

Popov, E. P., and Bertero, V. V. (1980), Seismic analysis of some steel building frames, *J. Eng. Mech., ASCE*, **106**, 75–93.

Steinbrugge, Karl V. (1970), Earthquake damage and structural performance in the United States, in *Earthquake Engineering* (R. L. Weigel, Ed.), Prentice-Hall, Englewood Cliffs, NJ.

Wakabayashi, Minoru (1986), *Design of Earthquake-Resistant Buildings*, McGraw-Hill, New York.

STRUCTURAL DYNAMICS

Bathe, K. J. (1982), *Finite Element Procedures in Engineering Analysis*, Prentice-Hall, Englewood Cliffs, NJ.

Berg, Glen V. (1989), *Elements of Structural Dynamics*, Prentice-Hall, Englewood Cliffs, NJ.

Biggs, J. M. (1964), *Introduction to Structural Dynamics*, McGraw-Hill, New York.

Blevins, R. D. (1979), *Formulas for Natural Frequency and Mode Shape*, Van Nostrand Reinhold, New York.

Chopra, A. (1981), *Dynamics of Structures: A Primer*, Earthquake Engineering Research Institute, Berkeley, CA.

Clough, R. W., and Penzien, J. (1975), *Dynamics of Structures*, McGraw-Hill, New York.

Gallagher, R. H. (1975), *Finite Element Analysis*, Prentice-Hall, Englewood Cliffs, NJ, p. 115.

Guyan, R. J. (1965), Reduction of stiffness and mass matrices, *AIAA J.*, **13**, 380.

Harris, Cyril M. (1987), *Shock and Vibration Handbook*, 3rd ed., McGraw-Hill, New York.

Kiureghian, A. D. (1980), A response spectrum method for random vibration, Report No. UCB/EERC-80/15, Earthquake Engineering Research Center, University of California, Berkeley, CA.

Nashif, A. D., Jones, D. I. C., and Henderson, J. P. (1985), *Vibration Damping*, Wiley, New York.

Newmark, N. M. (1959), A method of computation for structural dynamics, *Trans. ASCE*, **127**, 1406–35.

Paz, Mario (1973), Mathematical observations in structural dynamics, *Int. J. Comput. Struct.*, **3**, 385–396.

Paz, Mario (1983), Practical reduction of structural problems, *J. Struct. Eng.*, *ASCE*, **109**(111), 2590–2599.

Paz, Mario (1984a), Dynamic condensation, *AIAA J.*, **22**(5), 724–727.

Paz, Mario (1984b), in *Structural Mechanics Software Series*, The University Press of Virginia, Charlottesville, Vol. V, pp. 271–286.

Paz, Mario (1985), *Micro-Computer Aided Engineering: Structural Dynamics*, Van Nostrand Reinhold, New York.

Paz, Mario (1989), Modified dynamic condensation method, *J. Struct. Eng.*, *ASCE*, **115**(1), 234–238.

Paz, M., and Dung, L. (1975), Power series expansion of the general stiffness matrix for beam elements, *Int. J. Numer. Methods Eng.*, **9**, 449–459.

Wilson, E. L., Farhoomand, I., and Bathe, K. J. (1973). Nonlinear dynamic analysis of complex structures, *Int. J. Earthquake and Structural Dynamics*, Vol. 1, pp. 241–252.

Wilson, E. L., Der Kiureghian, A., and Bayo, E. P. (1981), A replacement for the SRSS method in seismic analysis, *Int. J. Earthquake Eng. Struct. Dyn.*, **9**, 187–194.

BUILDING CODES

American Insurance Association (1976), *The National Building Code*, New York.

American National Standards Institute (1982), *American National Standard Building Code Requirements for Minimum Design Loads in Buildings and Other Structures*, ANSI A58.1982, New York.

Applied Technology Council (1978), *Tentative Provisions for the Development of Seismic Regulations for Buildings*, ATC 3-06, National Bureau of Standards Special Publication 510, U.S. Government Printing Office, Washington, DC.

Berg, Glen V. (1983), *Seismic Design Codes and Procedures*, Earthquake Engineering Research Institute, Berkeley, CA.

Building Officials and Code Administrators International (1987), *BOCA Basic Building Code*, Homewood, IL.

Building Seismic Safety Council (BSSC) (1988), *Recommended Provisions for the Development of Seismic Regulations for New Buildings*, NEHRP (National Earthquake Hazard Reduction Program), Washington, DC.

International Conference of Building Officials (1985), *Uniform Building Code (UBC)*, Whittier, CA.

International Conference of Building Officials (1988), *Uniform Building Code (UBC)*, Whittier, CA.

Southern Building Code Congress International (1976), *Standard Building Code*, Birmingham, AL.

Structural Engineering Association of California [SEAOC (1990)], *Recommended Lateral Force Requirements and Tentative Commentary*, San Francisco, CA.

RANDOM VIBRATIONS AND FAST FOURIER TRANSFORM

Cooley, P. M., and Tukey, J. W. (1965), An algorithm for the machine computation of complex Fourier series, *Math. Comput.*, **19,** pp. 297–301.

Cooley, J. W., Lewis, P. A. W., and Welch, P. D. (1969), The Fast Fourier Transform and its Applications, *IEEE Transactions on Education*, Vol. 12, No. 1, pp. 27–34.

Newland, D. E. (1984), *An Introduction to Random Vibrations and Spectral Analysis*, 2nd ed., Longman, New York.

Ramirez, R. W. (1985), *The FFT: Fundamentals and Concepts*, Prentice-Hall, Englewood Cliffs, NJ.

Yang, C. Y. (1986), *Random Vibration of Structures*, Wiley, New York.

INDEX

DISKETTE ORDERING FORM

Professor Mario Paz
P.O. Box 35101
Louisville, KY 40232

Please send to:

NAME _____

STREET ADDRESS _____

CITY, STATE, AND ZIP CODE _____

COUNTRY _____

STRUCTURAL DYNAMICS, SET OF 20 PROGRAMS:

_____BASICA (Advanced BASIC)	$20
_____QUICK TURBO BASIC (source and compiled programs)	45

EARTHQUAKE ENGINEERING, SET OF 10 PROGRAMS:

_____BASICA (Advanced BASIC)	$25
_____QUICK TURBO BASIC (source and compiled programs)	45
_____Shipping and handling per order	5
_____Overseas shipping add $25 per order	
_____Canada add $10 per order	

Total enclosed (check or money order) . . . . $ . . . .

Disk size (check one)

5 1/4 DISKETTES _____

3 1/2 DISKETTES _____